THE REINFORCEMENT LEARNING WORKSHOP

Learn how to apply cutting-edge reinforcement learning algorithms to a wide range of control problems

Alessandro Palmas, Emanuele Ghelfi, Dr. Alexandra Galina Petre, Mayur Kulkarni, Anand N.S., Quan Nguyen, Aritra Sen, Anthony So, and Saikat Basak

THE REINFORCEMENT LEARNING WORKSHOP

Authors: Alessandro Palmas, Emanuele Ghelfi, Dr. Alexandra Galina Petre, Mayur Kulkarni, Anand N.S., Quan Nguyen, Aritra Sen, Anthony So, and Saikat Basak

Reviewers: Alberto Boschetti, Richard Brooker, Alekhya Dronavalli, Harshil Jain, Sasikanth Kotti, Nimish Sanghi, Shanmuka Sreenivas, and Pritesh Tiwari

Managing Editors: Snehal Tambe, Aditya Shah, and Ashish James

Acquisitions Editors: Manuraj Nair, Kunal Sawant, Sneha Shinde, Anindya Sil, Archie Vankar, Karan Wadekar, and Alicia Wooding

Production Editor: Shantanu Zagade

Editorial Board: Megan Carlisle, Samuel Christa, Mahesh Dhyani, Heather Gopsill, Manasa Kumar, Alex Mazonowicz, Monesh Mirpuri, Bridget Neale, Dominic Pereira, Shiny Poojary, Abhishek Rane, Brendan Rodrigues, Erol Staveley, Ankita Thakur, Nitesh Thakur, and Jonathan Wray

First published: August 2020

Production reference: 1140820

ISBN: 978-1-80020-045-6

Published by Packt Publishing Ltd.

Livery Place, 35 Livery Street

Birmingham B3 2PB, UK

EXPERIENCE THE WORKSHOP ONLINE

Thank you for purchasing the print edition of *The Reinforcement Learning Workshop*. Every physical print copy includes free online access to the premium interactive edition. There are no extra costs or hidden charges.

Figure A: An example of the companion video in the Workshop course player (dark mode)

With the interactive edition you'll unlock:

- **Screencasts**: Supercharge your progress with screencasts of all exercises and activities.

- **Built-In Discussions**: Engage in discussions where you can ask questions, share notes and interact. Tap straight into insight from expert instructors and editorial teams.

- **Skill Verification**: Complete the course online to earn a Packt credential that is easy to share and unique to you. All authenticated on the public Bitcoin blockchain.

- **Download PDF and EPUB**: Download a digital version of the course to read offline. Available as PDF or EPUB, and always DRM-free.

To redeem your free digital copy of *The Reinforcement Learning Workshop* you'll need to follow these simple steps:

1. Visit us at https://courses.packtpub.com/pages/redeem.

2. Login with your Packt account, or register as a new Packt user.

3. Select your course from the list, making a note of the three page numbers for your product. Your unique redemption code needs to match the order of the pages specified.

4. Open up your print copy and find the codes at the bottom of the pages specified. They'll always be in the same place:

EXERCISE 4.02: PERFORMING MISSING VALUE ANALYSIS FOR THE DATAFRAMES

In this section, we will be implementing a missing value analysis on the first DataFrame to find the missing values. This exercise is a continuation of *Exercise 4.01, Importing Data into DataFrames*. Follow these steps to complete this exercise:

1. Import the **missingno** package:

```
# To analyze the missing data
!pip install missingno
import missingno as msno
```

2. Find the missing values in the first DataFrame and visualize the missing values in a plot:

```
# Missing Values in the first DataFrame
msno.bar(dataframes[0],color='red',labels=True,sort="ascending")
```

A B 2 1 C

**Figure B: Example code in the bottom-right corner, to be used
for free digital redemption of a print workshop**

5. Merge the codes together (without spaces), ensuring they are in the correct order.

6. At checkout, click **Have a redemption code?** and enter your unique product string. Click **Apply**, and the price should be free!

Finally, we'd like to thank you for purchasing the print edition of *The Reinforcement Learning Workshop*! We hope that you finish the course feeling capable of tackling challenges in the real world. Remember that we're here to help if you ever feel like you're not making progress.

If you run into issues during redemption (or have any other feedback) you can reach us at workshops@packt.com.

Table of Contents

Chapter 3: Deep Learning in Practice with TensorFlow 2   131

Chapter 5: Dynamic Programming 259

Chapter 6: Monte Carlo Methods

Chapter 8: The Multi-Armed Bandit Problem 423

Chapter 12: Evolutionary Strategies for RL   623

PREFACE

ABOUT THE BOOK

Various intelligent applications such as video games, inventory management software, warehouse robots, and translation tools use **Reinforcement Learning (RL)** to make decisions and perform actions that maximize the probability of the desired outcome. This book will help you to get to grips with the techniques and the algorithms for implementing RL in your machine learning models.

Starting with an introduction to RL, you'll be guided through different RL environments and frameworks. You'll learn how to implement your own custom environments and use OpenAI baselines to run RL algorithms. Once you've explored classic RL techniques such as Dynamic Programming, Monte Carlo, and TD Learning, you'll understand when to apply the different deep learning methods in RL and advance to deep Q-learning. The book will even help you understand the different stages of machine-based problem-solving by using DARQN on a popular video game Breakout. Finally, you'll find out when to use a policy-based method to tackle an RL problem.

By the end of *The Reinforcement Learning Workshop*, you'll be equipped with the knowledge and skills needed to solve challenging machine learning problems using reinforcement learning.

AUDIENCE

If you are a data scientist, machine learning enthusiast, or a Python developer who wants to learn basic to advanced deep reinforcement learning algorithms, this workshop is for you. A basic understanding of the Python language is necessary.

ABOUT THE CHAPTERS

Chapter 1, Introduction to Reinforcement Learning, introduces you to RL, which is one of the most exciting fields in machine learning and artificial intelligence.

Chapter 2, Markov Decision Processes and Bellman Equations, teaches you about Markov chains, Markov reward processes, and Markov decision processes. You will learn about state values and action values, as well as using the Bellman equation to calculate these quantities.

Chapter 3, Deep Learning in Practice with TensorFlow 2, introduces you to TensorFlow and Keras, giving you an overview of their key features and applications and how they work in synergy.

Chapter 4, Getting Started with OpenAI and TensorFlow for Reinforcement Learning, sees you working with two popular OpenAI tools, Gym and Universe. You will learn how to formalize the interfaces of these environments, how to interact with them, and how to create a custom environment for a specific problem.

Chapter 5, Dynamic Programming, teaches you how to use dynamic programming to solve problems in RL. You will learn about the concepts of policy evaluation, policy iteration, and value iteration, and see how to implement them.

Chapter 6, Monte Carlo Methods, teaches you how to implement the various types of Monte Carlo methods, including the "first visit" and "every visit" techniques. You will see how to use these Monte Carlo methods to solve the frozen lake problem.

Chapter 7, Temporal Difference Learning, prepares you to implement TD(0), SARSA, and TD(λ) Q-learning algorithms in both stochastic and deterministic environments.

Chapter 8, The Multi-Armed Bandit Problem, introduces you to the popular multi-armed bandit problem and shows you some of the most commonly used algorithms to solve the problem.

Chapter 9, What Is Deep Q-Learning?, educates you on deep Q-learning and covers some hands-on implementations of advanced variants of deep Q-learning, such as double deep Q-learning, with PyTorch.

Chapter 10, Playing an Atari Game with Deep Recurrent Q-Networks, introduces you to **Deep Recurrent Q-Networks** and its variants. You will get hands-on experience in training RL agents to play an Atari game.

Chapter 11, Policy-Based Methods for Reinforcement Learning, teaches you how to implement different policy-based methods of RL, such as policy gradients, deep deterministic policy gradients, trust region policy optimization, and proximal policy optimization.

Chapter 12, Evolutionary Strategies for RL, combines evolutionary strategies with traditional machine learning methods, specifically in the selection of neural network hyperparameters. You will also identify the limitations of these evolutionary methods.

> **NOTE**
>
> The interactive version of *The Reinforcement Learning Workshop* contains a bonus chapter, *Recent Advancements* and *Next Steps*. This chapter teaches you novel methods of implementing reinforcement learning algorithms with an emphasis on areas of further exploration such as one-shot learning and transferable domain priors. You can find the interactive version here: courses.packtpub.com.

CONVENTIONS

Code words in text, database table names, folder names, filenames, file extensions, pathnames, dummy URLs, user input, and Twitter handles are shown as follows: "Recall that an algorithm class' implementation needs two specific methods to interact with the bandit API, **decide()** and **update()**, the latter of which is simpler and is implemented."

Words that you see onscreen (for example, in menus or dialog boxes) also appear in the text like this: "The **DISTRIBUTIONS** tab provides an overview of how the model parameters are distributed across epochs."

A block of code is set as follows:

```
class Greedy:
    def __init__(self, n_arms=2):
        self.n_arms = n_arms
        self.reward_history = [[] for _ in range(n_arms)]
```

New terms and important words are shown like this: "Its architecture allows users to run it on a wide variety of hardware, from CPUs to **Tensor Processing Units** (**TPUs**), including GPUs as well as mobile and embedded platforms."

CODE PRESENTATION

Lines of code that span multiple lines are split using a backslash (\ ). When the code is executed, Python will ignore the backslash, and treat the code on the next line as a direct continuation of the current line.

For example:

```
history = model.fit(X, y, epochs=100, batch_size=5, verbose=1, \
                     validation_split=0.2, shuffle=False)
```

Comments are added into code to help explain specific bits of logic. Single-line comments are denoted using the # symbol, as follows:

```
# Print the sizes of the dataset
print("Number of Examples in the Dataset = ", X.shape[0])
print("Number of Features for each example = ", X.shape[1])
```

Multi-line comments are enclosed by triple quotes, as shown below:

```
"""
Define a seed for the random number generator to ensure the
result will be reproducible
"""
seed = 1
np.random.seed(seed)
random.set_seed(seed)
```

SETTING UP YOUR ENVIRONMENT

Before we explore the book in detail, we need to set up specific software and tools. In the following section, we shall see how to do that.

INSTALLING ANACONDA FOR JUPYTER NOTEBOOK

Jupyter notebooks are available once you install Anaconda on your system. Anaconda can be installed on Windows systems using the steps available at https://docs.anaconda.com/anaconda/install/windows/.

For other systems, navigate to the respective installation guide from https://docs.anaconda.com/anaconda/install/.

INSTALLING A VIRTUAL ENVIRONMENT

In general, it is good practice to use separate virtual environments when installing Python modules, to be sure that the dependencies of different projects do not conflict with one another. So, it is recommended that you adopt this approach before executing these instructions.

Since we are using Anaconda here, it is highly recommended that you use conda-based environment management. Run the following commands in Anaconda Prompt to create an environment and activate it:

```
conda create --name [insert environment name here]
conda activate [insert environment name here]
```

INSTALLING GYM

To install Gym, please make sure you have Python 3.5+ installed on your system. You can simply install Gym using **pip**. Run the code in Anaconda Prompt, as shown in the following code snippet:

```
pip install gym
```

You can also build the Gym installation from source, by cloning the Gym Git repository directly. This type of installation proves useful when modifying Gym or adding environments if required. Use the following code to install Gym from source:

```
git clone https://github.com/openai/gym
cd gym
pip install -e .
```

Run the following code to perform a full installation of Gym. This installation may need you to install other dependencies, which include **cmake** and a recent version of **pip**:

```
pip install -e .[all]
```

In *Chapter 11, Policy-Based Methods for Reinforcement Learning*, you will be working in the **Box2D** environment available in Gym. You can install the **Box2D** environment by using the following command:

```
pip install gym "gym[box2d]"
```

INSTALLING TENSORFLOW 2

To install TensorFlow 2, run the following command in Anaconda Prompt:

```
pip install tensorflow
```

If you are using a GPU, you can use the following command:

```
pip install tensorflow-gpu
```

INSTALLING PYTORCH

PyTorch can be installed on Windows using the steps available at https://pytorch.org/.

In the case of non-availability of a GPU on your system, you can install the CPU version of PyTorch by running the following code in Anaconda Prompt:

```
conda install pytorch-cpu torchvision-cpu -c pytorch
```

INSTALLING OPENAI BASELINES

OpenAI Baselines can be installed using the instructions at https://github.com/openai/baselines.

Download the OpenAI Baselines repository, check out the TensorFlow 2 branch, and install it as follows:

```
git clone https://github.com/openai/baselines.git
cd baselines
git checkout tf2
pip install -e .
```

We use OpenAI Baselines in *Chapter 1, Introduction to Reinforcement Learning*, and *Chapter 4, Getting Started with OpenAI and TensorFlow* for Reinforcement Learning. As OpenAI Baselines uses a version of Gym that is not the latest version, **0.14**, you might get an error as follows:

```
AttributeError: 'EnvSpec' object has no attribute '_entry_point'
```

The solution to this bug is to change the two **env.entry_point** attributes in **baselines/run.py** back to **env._entry_point**.

The detailed solution is available at https://github.com/openai/baselines/issues/977#issuecomment-518569750.

Alternatively, you can also use the following command to upgrade the Gym installation in that environment:

```
pip install --upgrade gym
```

INSTALLING PILLOW

Use the following command in Anaconda Prompt to install Pillow:

```
conda install -c anaconda pillow
```

Alternatively, you can also run the following command using **pip**:

```
pip install pillow
```

You can read more about Pillow at https://pypi.org/project/Pillow/2.2.1/.

INSTALLING TORCH

Use the following command to install **torch** using **pip**:

```
pip install torch==0.4.1 -f https://download.pytorch.org/whl/torch_
stable.html
```

Note that you will be using version **0.4.1** of **torch** only in *Chapter 11, Policy-Based Methods for Reinforcement Learning*. You can revert to the updated version of PyTorch by using the command under the *Installing PyTorch* section for the other chapters.

INSTALLING OTHER LIBRARIES

pip comes pre-installed with Anaconda. Once Anaconda is installed on your machine, all the required libraries can be installed using **pip**, for example, **pip install numpy**. Alternatively, you can install all the required libraries using **pip install -r requirements.txt**. You can find the **requirements.txt** file at https://packt.live/311jIlu.

The exercises and activities will be executed in Jupyter Notebooks. Jupyter is a Python library and can be installed in the same way as the other Python libraries – that is, with **pip install jupyter**, but fortunately, it comes pre-installed with Anaconda. To open a notebook, simply run the command **jupyter notebook** in the Terminal or Command Prompt.

ACCESSING THE CODE FILES

You can find the complete code files of this book at https://packt.live/2V1MwHi.

We've tried to support interactive versions of all activities and exercises, but we recommend a local installation as well for instances where this support isn't available.

If you have any issues or questions about installation, please email us at `workshops@packt.com`.

1

INTRODUCTION TO REINFORCEMENT LEARNING

OVERVIEW

This chapter introduces the **Reinforcement Learning** (**RL**) framework, which is one of the most exciting fields of machine learning and artificial intelligence. You will learn how to describe the characteristics and advanced applications of RL to show what can be achieved within this framework. You will also learn to differentiate between RL and other learning approaches. You will learn the main concepts of this discipline both from a theoretical point of view and from a practical point of view using Python and other useful libraries.

By the end of the chapter, you will understand what RL is and know how to use the Gym toolkit and Baselines, two popular libraries in this field, to interact with an environment and implement a simple learning loop.

INTRODUCTION

Learning and adapting to new circumstances is a crucial process for humans and, in general, for all animals. Usually, learning is intended as a process of trial and error through which we improve our performance in particular tasks. Our life is a continuous learning process, that is, we start from simple goals (for example, walking), and we end up pursuing difficult and complex tasks (for example, playing a sport). As humans, we are always driven by our reward mechanism, which awards good behaviors and punishes bad ones.

Reinforcement Learning (**RL**), inspired by the human learning process, is a subfield of machine learning and deals with learning from interaction. With the term "interaction," we mean the process of trial and error through which we, as humans, understand the consequences of our actions and build up our own experiences.

RL, in particular, considers sequential decision-making problems. These are problems in which an agent has to take a sequence of decisions, that is, actions, to maximize a certain performance measure.

RL considers tasks to be **Markov Decision Processes** (**MDPs**), which are problems arising in many real-world scenarios. In this setting, the decision-maker, referred to as the agent, has to make decisions accounting for environmental uncertainty and experience. Agents are goal-directed; they need only a notion of a goal, such as a numerical signal, to be maximized. Unlike supervised learning, in RL, there is no need to provide good examples; it is the agent who learns how to map situations to actions. The mapping from situations (states) to actions is called "policy" in literature, and it represents the agent's behavior or strategy. Solving an MDP means finding the agent's policy by maximizing the desired outcome (that is, the total reward). We will study MDPs in more detail in future chapters.

RL has been successfully applied to various kinds of problems and domains, showing exciting results. This chapter is an introduction to RL. It aims to explain some applications and describe concepts both from an intuitive perspective and from a mathematical point of view. Both of these aspects are very important when learning new disciplines. Without intuitive understanding, it is impossible to make sense of formulas and algorithms; without mathematical background, it is tough to implement existing or new algorithms.

In this chapter, we will first compare the three main machine learning paradigms, namely supervised learning, RL, and unsupervised learning. We will discuss their differences and similarities and define some example problems.

Second, we will move on to a section that contains the theory of RL and its notations. We will learn about concepts such as what an agent is, what an environment is, and how to parameterize different policies. This section represents the fundamentals of this discipline.

Third, we will begin using two RL frameworks, namely Gym and Baselines. We will learn that interacting with a Gym environment is extremely simple, as is learning a task using Baselines algorithms.

Finally, we will explore some RL applications to motivate you to study this discipline, showing various techniques that can be used to face real-world problems. RL is not bound to the academic world. However, it is still crucial from an industrial point of view, allowing you to solve problems that are almost impossible to solve using other techniques.

LEARNING PARADIGMS

In this section, we will discuss the similarities and differences between the three main **learning paradigms** under the umbrella of machine learning. We will analyze some representative problems in order to understand the characteristics of these frameworks better.

INTRODUCTION TO LEARNING PARADIGMS

For a learning paradigm, we implement a problem and a solution method. Usually, learning paradigms deal with data and rephrase the problem in a way that can be solved by finding parameters and maximizing an objective function. In these settings, the problem can be faced using mathematical and optimization tools, allowing a formal study. The term "learning" is often used to represent a dynamic process of adapting the algorithm's parameters in such a way as to optimize their performance (that is, to learn) on a given task. Tom Mitchell defined learning in a precise way, as follows:

"A computer program is said to learn from experience E with respect to some class of tasks T and performance measure P, if its performance at tasks in T, as measured by P, improves with experience E."

Let's rephrase the preceding definition more intuitively. To define whether a program is learning, we need to set a task; that is the goal of the program. The task can be everything we want the program to do, that is, play a game of chess, do autonomous driving, or carry out image classification. The problem should be accompanied by a performance measure, that is, a function that returns how well the program is performing on that task. For the chess game, a performance function can simply be represented by the following:

$$P = \begin{cases} +1 \ if \ the \ program \ wins \ against \ the \ opponent \\ -1 \ if \ the \ program \ loses \ against \ the \ opponent \end{cases}$$

Figure 1.1: A performance function for a game of chess

In this context, the experience is the amount of data collected by the program at a specific moment. For chess, the experience is represented by the set of games played by the program.

The same input presented at the beginning of the learning phase or the end of the learning phase can result in different responses (that is, outputs) from the algorithm; the differences are caused by the algorithm's parameters being updated during the process.

In the following table, we can see some examples of the experience, task, and performance tuples to better understand their concrete instantiations:

Experience	Task	Performance
Chess games	Playing chess	Percentage of winning games
Driving sequences	Autonomous driving	Race time
Labeled images	Image classification	Classification accuracy

Figure 1.2: Table for instantiations

It is possible to classify the learning algorithms based on the input they have and on the feedback they receive. In the following section, we will look at the three main learning paradigms in the context of machine learning based on this classification.

SUPERVISED VERSUS UNSUPERVISED VERSUS RL

The three main learning paradigms are supervised learning, unsupervised learning, and RL. The following figure represents the general schema of each of these learning paradigms:

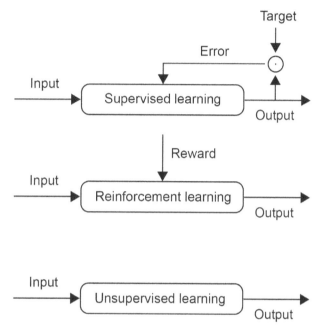

Figure 1.3: Representation of learning paradigms

From the preceding figure, we can derive the following information:

- Supervised learning minimizes the error of the output of the model with respect to a target specified in the training set.

- RL maximizes the reward signal of the actions.

- Unsupervised learning has no target and no reward; it tries to learn a data representation that can be useful.

Let's go more in-depth and elaborate on these concepts further, particularly from a mathematical perspective.

Supervised learning deals with learning a function by mapping an input to an output when the correspondences between the input and output (sample, label) are given by an external teacher (supervisor) and are contained in a training set. The objective of supervised learning is to generalize to unseen samples that are not included in the dataset, resulting in a system (for example, a function) that is able to respond correctly in new situations. Here, the correspondences between the sample and label are usually known (for example, in the training set) and given to the system. Examples of supervised learning tasks include regression and classification problems. In a regression task, the learner has to find a function, f, of the input, $x \in \mathbb{R}^d$, producing a (or n, in general) real output, y. In mathematical notation, we have to find f such that:

$$y = f(x) \in \mathbb{R}$$

Figure 1.4: Regression

Here, y is known for the examples in the training set. In a classification task, the function to be learned is a discrete mapping; y belongs to a finite and discrete set. Formalizing the problem, we search for a discrete-valued function, f, such that:

$$y = f(x) \in \left\{ c_1, c_2, \ldots, c_n \right\}$$

Figure 1.5: Classification

Here, the set, $\left\{ c_1, c_2, \ldots, c_n \right\}$, represents the set of possible classes or categories.

Unsupervised learning deals with learning patterns in the data when the target label is not present or is unknown. The objective of unsupervised learning is to find a new, usually smaller, representation of data. Examples of unsupervised learning algorithms include clustering and **Principal Component Analysis (PCA)**.

In a clustering task, the learner should split the dataset into clusters (a group of elements) according to some similarity measure. At first glance, clustering may seem very similar to classification; however, as an unsupervised learning task, the labels, or classes, are not given to the algorithm inside the training set. Indeed, it is the algorithm itself that should make sense of its inputs, by learning a representation of the input space in such a way that similar samples are close to each other.

For example, in the following figure, we have the original data on the left; on the right, we have the possible output of a clustering algorithm. Different colors denote different clusters:

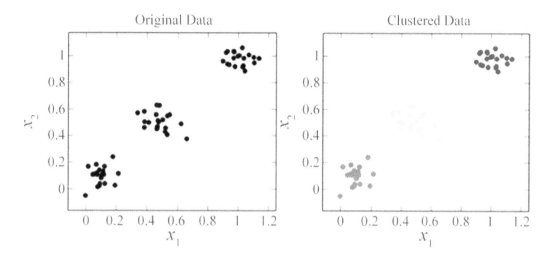

Figure 1.6: An example of a clustering application

In the preceding example, the input space is composed of two dimensions, that is, $x \in R^2$, and the algorithm found three clusters or three groups of similar elements.

PCA is an unsupervised algorithm used for dimensionality reduction and feature extraction. PCA tries to make sense of data by searching for a representation that contains most of the information from the given data.

RL is different from both supervised and unsupervised learning. RL deals with learning control actions in a sequential decision-making problem. The sequential structure of the problem makes RL challenging and different from the two other paradigms. Moreover, in supervised and unsupervised learning, the dataset is fixed. In RL, the dataset is continuously changing, and dataset creation is itself the agent's task. In RL, different from supervised learning, no teacher provides the correct value for a given sample or the right action for a given situation. RL is based on a different form of feedback, which is the environment's feedback evaluating the behavior of the agent. It is precisely the presence of feedback that also makes RL different from unsupervised learning.

We will explore these concepts in more detail in future sections:

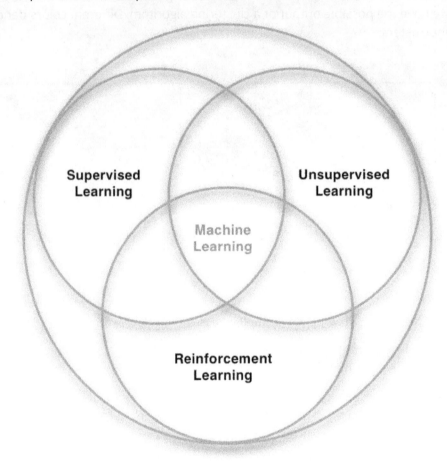

Figure 1.7: Machine learning paradigms and their relationships

RL and supervised learning can also be mixed up. A common technique (also used by AlphaGo Zero) is called **imitation learning** (or behavioral cloning). Instead of learning a task from scratch, we teach the agent in a supervised way how to behave (or which action to take) in a given situation. In this context, we have an expert (or multiple experts) demonstrating to the agent the desired behavior. In this way, the agent can start building its internal representation and its initial knowledge. Its actions won't be random at all when the RL part begins, and its behavior will be more focused on the actions shown by the expert.

Let's now look at a few scenarios that will help us to classify the problems in a better manner.

CLASSIFYING COMMON PROBLEMS INTO LEARNING SCENARIOS

In this section, we will understand how it is possible to frame some common real-world problems into a learning framework by defining the required elements.

PREDICTING WHETHER AN IMAGE CONTAINS A DOG OR A CAT

Predicting the content of an image is a standard classification example; therefore, it lies under the umbrella of supervised learning. Here, we are given a picture, and the algorithm should decide whether the image contains a dog or a cat. The input is the image, and the associated label can be 0 for cats and 1 for dogs.

For a human, this is a straightforward task, as we have an internal representation of dogs and cats (as well as an internal representation of the world), and we are trained extensively in our life to recognize dogs and cats. Despite this, writing an algorithm that is able to identify whether an image contains a dog or a cat is a difficult task without machine learning techniques. For a human, it is elementary to know whether the image is of a dog or cat; it is also easy to create a simple dataset of images of cats and dogs.

Why Not Unsupervised Learning?

Unsupervised learning is not suited to this type of task as we have a defined output we need to obtain from an input. Of course, supervised learning methods build an internal representation of the input data in which similarities are better exploited. This representation is only implicit; it is not the output of the algorithm as is the case in unsupervised learning.

Why Not RL?

RL, by definition, considers sequential decision-making problems. Predicting the content of an image is not a sequential problem, but instead a one-shot task.

DETECTING AND CLASSIFYING ALL DOGS AND CATS IN AN IMAGE

Detection and classification are two examples of supervised learning problems. However, this task is more complicated than the previous one. The detection part can be seen as both a regression and classification problem at the same time. The input is always the image we want to analyze, and the output is the coordinate of the bounding boxes for each dog or cat in the picture. Associated with each bounding box, we have a label to classify the content in the region of interest as a dog or a cat:

Figure 1.8: Cat and dog detection and classification

Why Not Unsupervised Learning?

As in the previous example, here, we have a determined output given an input (an image). We do not want to extract unknown patterns in the data.

Why Not RL?

Detection and classification are not tasks that are suited to the RL framework. We do not have a set of actions the agent should take to solve a problem. Also, in this case, the sequential structure is absent.

PLAYING CHESS

Playing chess can be seen as an RL problem. The program can perceive the current state of the board (for example, the positions and types of pawns), and, based on that, it should decide which action to take. Here, the number of possible actions is vast. Selecting an action means to understand and anticipate the consequences of the move to defeat the opponent:

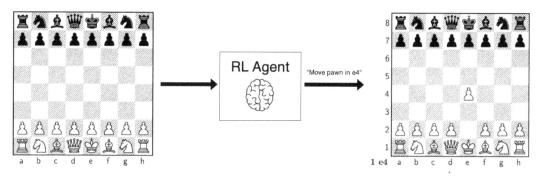

Figure 1.9: Chess as an RL problem

Why Not Supervised?

We can think of playing chess as a supervised learning problem, but we would need to have a dataset, and we should incorporate the sequential structure of the game into the supervised learning problem. In RL, there is no need to have a dataset; it is the algorithm itself that builds up a dataset through interaction and, possibly, self-play.

Why Not Unsupervised?

Unsupervised learning does not fit in this problem as we are not dealing with learning a representation of the data; we have a defined objective, which is winning the game.

In this section, we compared the three main learning paradigms. We saw the kind of data they have at their disposal, the type of interaction each algorithm has with the external world, and we analyzed some particular problems to understand which learning paradigm is best suited.

When facing a real-world problem, we always have to remember the distinction between these techniques, selecting the best one based on our goals, our data, and on the problem structure.

FUNDAMENTALS OF REINFORCEMENT LEARNING

In RL, the main goal is to learn from interaction. We want agents to learn a behavior, a way of selecting actions in given situations, to achieve some goal. The main difference between classical programming or planning is that we do not want to code the planning software explicitly on our own, as this would require a great effort; it can be very inefficient and even impossible. The RL discipline was born precisely for this reason.

RL agents start (usually) with no idea of what to do. They typically do not know the goal, they do not know the game's rules, and they do not know the dynamics of the environment or how their actions influence the state.

There are three main components of RL: perception, actions, and goals.

Agents should be able to perceive the current environment state to deal with a task. This perception, also called observation, might be different from the actual environment state, can be subject to noise, or can be partial.

For example, think of a robot moving in an unknown environment. For robotic applications, usually, the robot perceives the environment using cameras. Such a perception does not represent the environment state completely; it can be subject to occlusions, poor lighting, or adverse conditions. The system should be able to deal with this incomplete representation and learn a way of moving in the environment.

The other main component of an agent is the ability to act; the agent should be able to take actions that affect the environment state or the agent's state.

Agents should also have a goal defined through the environment state. Goals are described using high-level concepts such as winning a game, moving in an environment, or driving correctly.

One of the challenges of RL, a challenge that does not arise in other types of learning, is the exploration-exploitation trade-off. In order to improve, the agent has to exploit its knowledge; it should prefer actions that have demonstrated themselves as useful in the past. There's a problem here: to discover better actions, the agent should continue exploring, trying moves they have never done before. To estimate the effect of an action reliably, an agent has to perform each action many times. The critical thing to notice here is that neither exploration nor exploitation can be performed individually in order to learn a task.

The aforementioned is very similar to the challenges we face as babies when we have to learn how to walk. At first, we try different types of movement, and we start from a simple movement yielding satisfactory results: crawling. Then, we want to improve our behavior to become more efficient. To learn a new behavior, we have to do movements we never did before: we try to walk. At first, we perform different actions yielding unsatisfactory results: we fall many times. Once we discover the correct way of moving our legs and balancing our body, we become more efficient in walking. If we did not explore further and we stopped at the first behavior that yields satisfactory results, we would crawl forever. By exploring, we learn that there can be different behaviors that are more efficient. Once we learn how to walk, we can stop exploring, and we can start exploiting our knowledge.

ELEMENTS OF RL

Let's introduce the main elements of the RL framework intuitively.

AGENT

In RL, the agent is the abstract concept of the entity that moves in the world, takes actions, and achieves goals. An agent can be a piece of autonomous driving software, a chess player, a Go player, an algorithmic trader, or a robot. The agent is everything that can perceive and influence the state of the environment and, therefore, can be used to accomplish goals.

ACTIONS

An agent can perform actions based on the current situation. Actions can assume different forms depending on the specific task.

Actions can be to steer, to push the accelerator pedal, or to push the brake pedal in an autonomous driving context. Other examples of actions include moving the horse to the H5 position or moving the king to the A5 position in a chess context.

Actions can be low-level, such as controlling the voltage of the motors of a vehicle, but they can also be high-level, or planning actions, such as deciding where to go. The decision on the action level is the responsibility of the algorithm's designer. Actions that are too high-level can be challenging to implement at a lower level; they might require extensive planning at lower levels. At the same time, low-level actions make the problem difficult to learn.

ENVIRONMENT

The environment represents the context in which the agent moves and takes decisions. An environment is composed of three main elements: states, dynamics, and rewards. They can be explained as follows:

- **State**: This represents all of the information describing the environment at a particular timestep. The state is available to the agent through observations, which can be a partial or full representation.

- **Dynamics**: The dynamics of an environment describe how actions influence the state of the environment. The environment dynamic is usually very complex or unknown. An RL algorithm using the information of the environment dynamic to learn how to achieve a goal belongs to the category of model-based RL, where the model represents the mathematical description of the environment. Most of the time, the environment dynamic is not available to the agent. In this case, the algorithm belongs to the model-free category. Even if the environment model is not available, too complicated, or too approximated, the agent can learn a model of the environment during training. Also, in this case, the algorithm is said to be model-based.

- **Rewards**: Rewards are scalar values associated with each timestep describing the agent's goal.
  Rewards can also be described as environmental feedback, providing information to an agent about its behavior; it is, therefore, necessary for making learning possible. If the agent receives a high reward, it means that it performed a good move, a move bringing it closer to its goal.

POLICY

A policy describes the behavior of the agent. Agents select actions by following their policies. Mathematically, a policy is a function mapping states to actions. What does this mean? Well, it means that the input of the policy is the current state, and its output is the action to take. A policy can have different forms. It can be a simple set of rules, a lookup table, a neural network, or any function approximator. A policy is the core of the RL framework, and the goal of all RL algorithms (implicit or explicit) is to improve the agent's policy to maximize the agent's performance on a task (or on a set of tasks). A policy can be stochastic, involving a distribution over actions, or it can be deterministic. In the latter case, the selected action is uniquely determined by the environment's state.

AN EXAMPLE OF AN AUTONOMOUS DRIVING ENVIRONMENT

To better understand the environment's role and its characteristics in the RL framework, let's formalize an autonomous driving environment, as shown in the following figure:

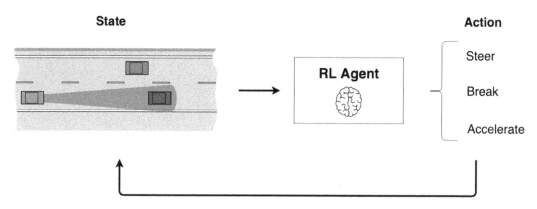

Figure 1.10: An autonomous driving scenario

Considering the preceding figure, let's now look at each of the components of the environment:

- **State**: The state can be represented by the 360-degree image of the street around our car. In this case, the state is an image, that is, a matrix of pixels. It can also be represented by a series of images covering the whole space around the car. Another possibility is to describe the state using features and not images. The state can be the current velocity and acceleration of our vehicle, the distance from other cars, or the distance from the street border. In this case, we are using preprocessed information to represent the state more easily. These features can be extracted from images or other types of sensors (for example, **Light Detection and Ranging – LIDAR**).

- **Dynamics**: The dynamics of the environment in an autonomous car scenario are represented by the equations describing how the system changes when the car accelerates, breaks, or steers. For instance, the vehicle is going at 30 km/h, and the next vehicle is 100 meters away from it. The state is represented by the car's speed and the proximity information concerning the next vehicle. If the car accelerates, the speed changes according to the car's properties (included in the environment dynamics). Also, the proximity information changes since the next vehicle can be closer or further away (according to the speed). In this situation, at the next timestep, the car's speed can be 35 km/h, and the next vehicle can be closer, for example, only 90 meters away.

- **Reward**: The reward can represent how well the agent is driving. It's not easy to formalize a reward function. A natural reward function should award states in which the car is aligned to the street and should avoid states in which the car crashes or goes off the road. The reward function definition is an open problem and researchers are putting efforts into developing algorithms where the reward function is not needed (self-motivation or curiosity-driven agents), where the agent learns from demonstrations (imitation learning), and where the agent recovers the reward function from demonstrations (**Inverse Reinforcement Learning or IRL**).

> **NOTE**
>
> For further reading on curiosity-driven agents, please refer to the following paper: https://pathak22.github.io/large-scale-curiosity/resources/largeScaleCuriosity2018.pdf.

We are now ready to design and implement our first environment class using Python. We will demonstrate how to implement the state, the dynamics, and the reward of a toy problem in the following exercise.

EXERCISE 1.01: IMPLEMENTING A TOY ENVIRONMENT USING PYTHON

In this exercise, we will implement a simple toy environment using Python. The environment is illustrated in *Figure 1.11*. It is composed of three states (1, 2, 3) and two actions (A and B). The initial state is state 1. States are represented by nodes. Edges represent transitions between states. On the edges, we have an action causing the transition and the associated reward.

The representation of the environment in *Figure 1.11* is the standard environment representation in the context of RL. In this exercise, we will become acquainted with the concept of the environment and its implementation:

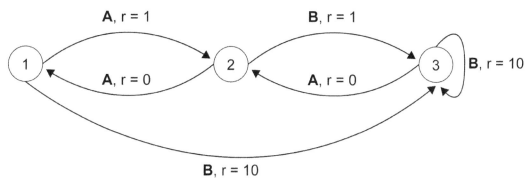

Figure 1.11: A toy environment composed of three states (1, 2, 3) and two actions (A and B)

In the preceding figure, the reward is associated with each state-action pair.

The goal of this exercise is to implement an **Environment** class with a **step()** method that takes as input the agent's actions and returns a state-action pair (next state, reward). In addition to this, we will write a **reset()** method that restarts the environment's state:

1. Create a new Jupyter notebook or a simple Python script to enter the code.

2. Import the **Tuple** type from **typing**:

    ```
    from typing import Tuple
    ```

3. Define the class constructor by initializing its properties:

    ```
    class Environment:
        def __init__(self):
            """
            Constructor of the Environment class.
            """
            self._initial_state = 1
            self._allowed_actions = [0, 1]   # 0: A, 1: B
            self._states = [1, 2, 3]
            self._current_state = self._initial_state
    ```

> **NOTE**
>
> The triple-quotes (**"""**) shown in the code snippet above are used to denote the start and end points of a multi-line code comment. Comments are added into code to help explain specific bits of logic.

We have two allowed actions, the action 0 and the action 1 representing the actions A and B. We have three environment states: 1, 2, and 3. We define the **current_state** variable to be equal to the initial state (state 1).

4. Define the step function, which is responsible for updating the current state based on the previous state and the action taken by the agent:

```python
def step(self, action: int) -> Tuple[int, int]:
    """
    Step function: compute the one-step dynamic from the \
    given action.

    Args:
        action (int): the action taken by the agent.

    Returns:
        The tuple current_state, reward.
    """

    # check if the action is allowed
    if action not in self._allowed_actions:
        raise ValueError("Action is not allowed")

    reward = 0
    if action == 0 and self._current_state == 1:
        self._current_state = 2
        reward = 1
    elif action == 1 and self._current_state == 1:
        self._current_state = 3
        reward = 10
    elif action == 0 and self._current_state == 2:
        self._current_state = 1
        reward = 0
    elif action == 1 and self._current_state == 2:
        self._current_state = 3
        reward = 1
    elif action == 0 and self._current_state == 3:
        self._current_state = 2
        reward = 0
    elif action == 1 and self._current_state == 3:
        self._current_state = 3
```

```
        reward = 10

    return self._current_state, reward
```

> **NOTE**
>
> The **#** symbol in the code snippet above denotes a code comment.
> Comments are added into code to help explain specific bits of logic.

We first check that the action is allowed. Then, we define the new current state and reward based on the action and the previous state by looking at the transition in the previous figure.

5. Now, we need to define the **reset** function, which simply resets the environment state:

```
def reset(self) -> int:
    """
    Reset the environment starting from the initial state.

    Returns:
        The environment state after reset (initial state).
    """
    self._current_state = self._initial_state
    return self._current_state
```

6. We can use our environment class to understand whether our implementation is correct for the specified environment. We can do this with a simple loop, using a predefined set of actions to test the transitions of our environment. A possible action set, in this case, is **[0, 0, 1, 1, 0, 1]**. Using this set, we will test all of the environment's transitions:

```
env = Environment()
state = env.reset()

actions = [0, 0, 1, 1, 0, 1]

print(f"Initial state is {state}")

for action in actions:
    next_state, reward = env.step(action)
```

```
    print(f"From state {state} to state {next_state} \
with action {action}, reward: {reward}")
    state = next_state
```

> **NOTE**
>
> The code snippet shown here uses a backslash (\ ) to split the logic
> across multiple lines. When the code is executed, Python will ignore the
> backslash, and treat the code on the next line as a direct continuation of the
> current line.

The output should be as follows:

```
Initial state is 1
From state 1 to state 2 with action 0, reward: 1
From state 2 to state 1 with action 0, reward: 0
From state 1 to state 3 with action 1, reward: 10
From state 3 to state 3 with action 1, reward: 10
From state 3 to state 2 with action 0, reward: 0
From state 2 to state 3 with action 1, reward: 1
```

To understand this better, compare the output with *Figure 1.11* to discover whether
the transitions and rewards are compatible with the selected actions.

> **NOTE**
>
> To access the source code for this specific section, please refer
> to https://packt.live/2Arr9rO.
>
> You can also run this example online at https://packt.live/2zpMul0.

In this exercise, we implemented a simple RL environment by defining the step
function and the reset function. These functions are at the core of every environment,
representing the interaction between the agent and the environment.

THE AGENT-ENVIRONMENT INTERFACE

RL considers sequential decision-making problems. In this context, we can refer to the agent as the "decision-maker." In sequential decision-making problems, actions taken by the decision-maker do not only influence the immediate reward and the immediate environment's state, but they also affect future rewards and states. MDPs are a natural way of formalizing sequential decision-making problems. In MDPs, an agent interacts with an environment through actions and receives rewards based on the action, on the current state of the environment, and on the environment's dynamics. The goal of the decision-maker is to maximize the cumulative sum of rewards given a horizon (which is possibly infinite). The task the agent has to learn is defined through the rewards it receives, as you can see in the following figure:

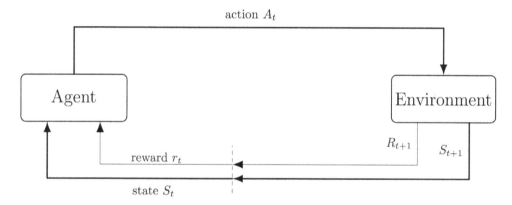

Figure 1.12: The Agent-Environment interface

In RL, an episode is divided into a sequence of discrete timesteps: $0, 1, \ldots, T$. Here, T represents the horizon length, which is possibly infinite. The interaction between the agent and the environment happens at each timestep. At each timestep, the agent receives a representation of the current environment's state, $S_t \in \mathcal{S}$. Based on this state, it selects an action, A_t, belonging to the action space given the current state, $\mathcal{A}(S_t)$, $A_t \in \mathcal{A}(S_t)$. The action affects the environment. As a result, the environment changes its state, transitioning to the next state, S_{t+1}, according to its dynamics. At the same time, the agent receives a scalar reward, $r_t \in R \subset R$ quantifying how good the action taken in that state was.

Let's now try to understand the mathematical notations used in the preceding example:

- Time horizon T: If a task has a finite time horizon, then T is an integer number representing the maximum duration of an episode. In infinite tasks, T can also be $+\infty$.

- Action A_t is the action taken by the agent in the timestep, t. The action belongs to the action space, $\mathcal{A}$, defined by the current state, s_t.

- State s_t is the representation of the environment's state received by the agent at time t. It belongs to the state space, $\mathcal{S}$, defined by the environment. It can be represented by an image, a sequence of images, or a simple vector assuming different shapes. Note that the actual environment state can be different and more complex than the state perceived by the agent.

- Reward r_t is represented by a real number, describing how good the taken action was. A high reward corresponds to a good action. The reward is fundamental for the agent to understand how to achieve a goal.

In episodic RL, the agent-environment interaction is divided into episodes; the agent has to achieve the goal within the episode. The interaction is finalized to learn better behavior. After several episodes, the agent can decide to update its behavior by incorporating its knowledge of past interactions. Based on the effect of the action on the environment and the received rewards, the agent will perform more frequent actions yielding higher rewards.

WHAT'S THE AGENT? WHAT'S IN THE ENVIRONMENT?

An important aspect to take into account when dealing with RL is the difference between the agent and the environment. This difference is not typically defined in terms of a physical distinction. Usually, we model the environment as everything that's not under the control of the agent. The environment can include physical laws, other agents, or an agent's properties or characteristics.

However, this does not imply that the agent does not know the environment. The agent can also be aware of the environment and the effect of its actions on it, but it cannot change the way the environment reacts. Also, the reward computation belongs to the environment, as it must be entirely outside the agent's control. If this is not the case, the agent can learn how to modify the reward function in such a way as to maximize its performance without learning the task. The boundary between the agent and environment is a control boundary, meaning that the agent cannot control the reaction of the environment. It is not a knowledge boundary since the agent can know the environment model perfectly and still find difficulties in learning the task.

ENVIRONMENT TYPES

In this section, we will examine some possible environment dichotomies. The characterization of the environment depends on the state space (finite or continuous), on the type of transitions (deterministic or stochastic), on the information available to the agent (fully or partially observable), and the number of agents involved in the learning problem (single versus multi-agent).

FINITE VERSUS CONTINUOUS

The state space gives the first distinction. The state space can be divided into two main categories: a finite state space and a continuous state space. A finite state space has a finite number of possible states in which the agent can be, and it's the more straightforward case. An environment with a continuous state space has infinite possible states. In these types of environments, the generalization properties of the agent are fundamental to solve a task because the probability of arriving at the same state twice is almost zero. In continuous environments, an agent cannot use the experience due to the previous presence in that state; it has to generalize using some kind of similarity with respect to the previously experienced states. Note that generalization is also essential for finite state spaces with a considerable number of states (for example, when the state space is represented by the set of all possible images).

Consider the following examples:

- Chess is finite. There is a finite number of possible states in which an agent can be. The state, for chess, is represented by the chessboard situation at a given time. We can calculate all the possible states by varying the situation of the chessboard. The number of states is very high but still finite.

- Autonomous driving can be defined as a continuous problem. If we describe the autonomous driving problem as a problem in which the agent has to make driving decisions based on the sensors' input, we obtain a continuous problem. The sensors provide continuous input in a given range. The agent state, in this case, can be represented by the agent's speed, the agent's acceleration, or the rotation of the wheels per minute.

DETERMINISTIC VERSUS STOCHASTIC

A deterministic environment is an environment in which, given a state, an action is performed by the agent; the following state is uniquely determined as well as the following reward. Deterministic environments are simple types of environments, but they are also rarely used due to their limited applicability in the real world.

Almost all real-world environments are stochastic. In stochastic environments, a state and an action performed by the agent determines the probability distribution over the next state and the next reward. The following state is not uniquely determined, but it's uncertain. In these types of environments, the agent should act many times to obtain a reliable estimate of its consequences.

Notice that, in a deterministic environment, the agent could perform each action in each state exactly once, and based on the acquired knowledge, it can solve the task. Also, notice that solving the task does not mean taking actions that yield the highest immediate return, because this action can also bring the agent to an inconvenient part of the environment where future rewards are always low. To solve the task correctly, the agent should take actions with the highest associated future return (called a state-action value). The state-action value does not take into account only the immediate reward but also the future rewards, giving the agent a farsighted view. We will define later what a state-action value is.

Consider the following examples:

- Rubik's Cube is deterministic. To a given action, it corresponds a defined state transition.

- Chess is deterministic but opponent-dependent. The successive state does not depend only on the agent's action but also on the opponent's action.

- Texas Hold'em is stochastic and opponent-dependent. The transition to the next state is stochastic and depends on the deck, which is not known by the agent.

FULLY OBSERVABLE VERSUS PARTIALLY OBSERVABLE

The agent, to plan actions, has to receive a representation of the environment state, S_t (refer to *Figure 1.12, The Agent-Environment interface*). If the state representation received by the agent completely defines the state of the environment, the environment is **Fully Observable**. If some parts of the environment are outside the representation observed by the agent, the environment is **Partially Observable**, also called the **Partially Observable Markov Decision Process** (**POMDP**). Partially observable environments are, for example, multi-agent environments. In the case of partially observable environments, the information perceived by the agents, together with the action taken, is not sufficient for determining the next state of the environment. A technique to improve the perception of the agent, making it more accurate, is to keep the history of taken actions and observations, but this requires some memory techniques (such as a **Recurrent Neural Network**, or **RNN**, or **Long Short-Term Memory**, or **LSTM**) embedded in the agent's policy.

> **NOTE**
>
> For more information on LSTMs, please refer to https://www.bioinf.jku.at/publications/older/2604.pdf.

POMDP VERSUS MDP

Consider the following figure:

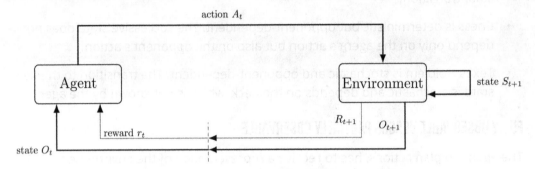

Figure 1.13: A representation of a partially observable environment

In the preceding figure, the agent does not receive the full environment state but only an observation, O_t.

To better understand the differences between these two types of environments, let's look at *Figure 1.13*. In partially observable environments (POMDP), the representation given to the agent is only a part of the actual environment state, and it is not enough to understand the actual environment state without uncertainty.

In fully observable environments (MDPs), the state representation given to the agent is semantically equivalent to the state of the environment. Notice that, in this case, the state given to the agent can assume a different form (for example, an image, a vector, a matrix, or a tensor). However, from this representation, it is always possible to reconstruct the actual state of the environment. The meaning of the state is precisely the same, even if under a different form.

Consider the following examples:

- Chess (and, in general, board games) is fully observable. The agent can perceive the whole environment state. In a chess game, the environment state is represented by the chessboard, and the agent can exactly perceive the position of each pawn.

- Poker is partially observable. A poker agent cannot perceive the whole state of the game, which includes the opponent cards and deck cards.

SINGLE AGENTS VERSUS MULTIPLE AGENTS

Another useful characteristic of environments is the number of agents involved in a task. If there is only one agent, the subject of our study, the environment is a single-agent environment. If the number of agents is more than one, the environment is a multi-agent environment. The presence of multiple agents increases the complexity of the problem since the action that influences the state becomes a joint action, the set of all the agents' actions. Usually, agents only know their individual actions and do not know another agent's actions. For this reason, the multi-agent environment is an instance of POMDP in which the partial visibility is due to the presence of other agents. Notice that each agent has its own observation, which can differ from the other agent's observation, as shown in the following figure:

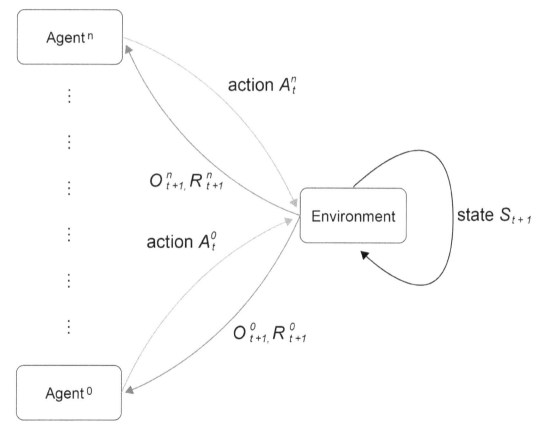

Figure 1.14: A schematic representation of the multi-agent decentralized MDP

Consider the following examples:

- Robot navigation is usually a single-agent task. We may have only one agent moving in a possible unknown environment. The goal of the agent can be to reach a given position in the environment while avoiding crashes as much as possible in the minimum amount of time.

- Poker is a multi-agent task where we have two agents competing against each other. The perceived state is different in this case and the perceived reward is also different.

AN ACTION AND ITS TYPES

The action set of an agent in an environment can be finite or continuous. If the action set is finite, the agent has at its disposal a finite number of actions. Consider the MountainCar-v0 (discrete) example, described in more detail later. This has a discrete action set; the agent only has to select the direction in which to accelerate, and the acceleration is constant.

If the action set is continuous, the agent has at its disposal infinite actions from which it should select the best actions in a given state. Usually, tasks with continuous action sets are more challenging to solve than those with finite actions.

Let's look at the example of MountainCar-v0:

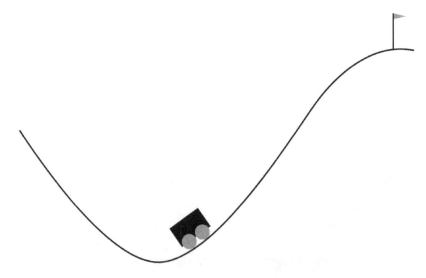

Figure 1.15: A MountainCar-v0 task

As you can see in the preceding figure, a car is positioned in a valley between two mountains. The goal of the car is to arrive at the flag on the mountain to its right.

The MountainCar-v0 example is a standard RL benchmark in which there is a car trying to ramp itself up a mountain. The car's engine doesn't have enough strength to ramp upward. For this reason, the car should use the inertia given from the shape of the valley, that is, it should go to the left to gain speed. The state is composed of the car velocity, acceleration, and *x* position. There are two versions of this task based on the action set we define, as follows:

- **MountainCar-v0 discrete**: We have only two possible actions, (-1, +1) or (0, 1), depending on the parameterization.

- **MountainCar-v0 continuous**: A continuous set of actions from -1 to +1.

POLICY

We define the policy as the behavior of the agent. Formally, a policy is a function that takes as input the history of the current episode and outputs the current action. The concept of policies has huge importance in RL; all RL algorithms focus on learning the best policy for a given task.

An example of a winning policy for the MountainCar-v0 task is a policy that brings the agent up on the left mountain and then uses the cumulated potential to ramp up the mountain on the right. For negative velocities, the optimal action is LEFT, as the agent should go as high as possible on the left mountain. For positive velocities, the agent should take the action RIGHT, as its goal is to ramp up the mountain on its right.

A Markovian policy is simply a policy depending only on the current state and not the whole history.

We denote a stationary Markovian policy with π as follows:

$$\pi : \mathcal{S} \rightarrow \mathcal{A},$$

Figure 1.16: Stationary Markovian policy

The Markovian policy goes from the state space to the action space. If we evaluate the policy in a given state, s_t, we obtain the selected action, a_t, in that state:

$$\pi\left(s_t\right) = a_t$$

Figure 1.17: Stationary Markovian policy in state s_t

A policy can be implemented in different ways. The most straightforward policy is just a rule-based policy, which is essentially a set of rules or heuristics.

Policies that are a subject of interest in RL are usually parametric. Parametric policies are (differentiable) functions depending on a set of parameters. Usually, the policy parameters are identified as $\theta = \left\{\theta_0, \theta_1, \ldots, \theta_{d-1}\right\} \in \mathbb{R}^d$:

$$\pi_\theta : \mathcal{S} \to \mathcal{A}$$

Figure 1.18: Parametric policies

The set of policy parameters can be represented by a vector in a *d*-dimensional space. The selected action is determined by the policy structure (we will explore some possible policy structures later on), by the policy parameters, and, of course, by the current environment state.

STOCHASTIC POLICIES

The policies presented so far are merely deterministic policies because the output is precisely an action. Stochastic policies are policies that output a distribution over the action space. Stochastic policies are usually powerful policies that mix both exploration and exploitation. With stochastic policies, it is possible to obtain complex behaviors.

A stochastic policy assigns a certain probability to each action. The actions will be selected according to the associated probability.

Figure 1.19 explains, graphically, and with an example, the differences between a stochastic policy and a deterministic policy. The policy in the figure has three possible actions.

The stochastic policy (upper part) assigns to actions, respectively, a probability of 0.2, 0.7, and 0.1. The most probable action is the second action, which is associated with the highest probability. However, all of the actions could also be selected.

In the bottom part, we have the same set of actions with a deterministic policy. The policy, in this case, selects only one action (the second in the figure) with a probability of 1. In this case, actions 1 and 3 will not be selected, having an associated probability of 0.

Note that we can obtain a deterministic policy from a stochastic one by taking the action associated with the highest probability:

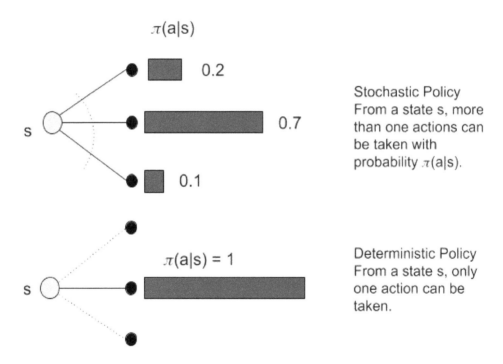

Figure 1.19: Stochastic versus deterministic policies

POLICY PARAMETERIZATIONS

In this section, we will analyze some possible policy parameterizations. Parameterizing a policy means giving a structure to the policy function and considering how parameters affect our output actions. Based on the parameterization, it is possible to obtain simple policies or even complex stochastic policies starting from the same input state.

Linear (Deterministic)

The resulting action is a linear combination of the state features, $\phi(s)$:

$$\pi(s) = \theta^T \phi(s)$$

Figure 1.20: An expression of a linear policy

A linear policy is a very simple policy represented by a matrix multiplication.

Consider the example of MountainCar-v0. The state space is represented by the position, speed, and acceleration: $[x;v;\dot{v}]$. We usually add a constant, 1, that corresponds to the bias term. Therefore, $s = [x;v;\dot{v};1]$. Policy parameters are defined by $[\theta_0;\theta_1;\theta_2;\theta_3]$. We can simply use as state features the identity function, $\phi(s) = s$.

The resulting policy is as follows:

$$\pi(s) = \left[\theta_0;\theta_1;\theta_2;\theta_3;\right]^T \cdot \left[x;v;\dot{v};1\right]$$

$$\pi(s) = \theta_0 \cdot x + \theta_1 \cdot v + \theta_2 \cdot \dot{v} + \theta_3 \cdot 1$$

Figure 1.21: A linear policy for MountainCar-v0

> **NOTE**
>
> Using a comma, $,$, we can denote the column separator, and with a semicolon, $;$, we can denote the row separator.
>
> Therefore, $[a,b,c]$ is a row vector, and $[a;b;c]$ is a column vector that is equivalent to $\begin{bmatrix} a \\ b \\ c \end{bmatrix}$.

If the environment state is [1, 2, 0.1], the cart is in position $x=1$ with velocity $v=2$ and acceleration $\dot{v}=0.1$, and the policy parameters are defined by [4, 5, 1, 1], we obtain an action, $a=4+10+0.1+1=15.1$.

Since the action space of MountainCar-v0 is defined in the interval, [-1, +1], we need to squash the resulting action using a squashing function such as *tanh* (hyperbolic tangent). In our case, *tanh* applied to the output of the multiplication results in approximately +1:

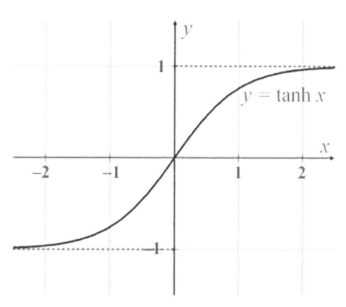

Figure 1.22: A hyperbolic tangent plot; the hyperbolic tangent squashes the real numbers in the interval, [-1, +1]

Even if linear policies are simple, they are usually enough to solve most tasks, given that the state features represent the problem.

Gaussian Policy

In the case of Gaussian parameterization, the resulting action has a Gaussian distribution in which the mean, μ_θ, and the variance, σ_θ^2, depend on state features:

$$\pi(\cdot | s, \theta) \sim \mathcal{N} \left(\mu_\theta(\phi(s)) , \sigma_\theta(\phi(s))^2 \right)$$

Figure 1.23: Expression for a Gaussian policy

Here, with the symbol $|$, we denote the conditional distribution; therefore, with $|$S, we denote the distribution conditioned on state S.

Remember, the functional form of the Gaussian distribution, $\mathcal{N}(\mu, \sigma^2)$, is as follows:

$$p(x) = \frac{1}{\sqrt{2\pi\sigma^2}} e^{-\frac{(x-\mu)^2}{2\sigma^2}}$$

Figure 1.24: A Gaussian distribution

In the case of a Gaussian policy, this becomes the following:

$$\pi(a | s, \theta) = \frac{1}{\sqrt{2\pi\sigma_\theta(\phi(s))^2}} e^{-\frac{\left(x - \mu_\theta(\phi(s))\right)^2}{2\sigma_\theta(\phi(s))^2}}$$

Figure 1.25: A Gaussian policy

Gaussian parameterization is useful for continuous action spaces. Note that we are giving the agent the possibility of also changing the variance of the distribution. This means that it can decide to increase the variance, enabling it to explore scenarios where it's not sure what the best action to take is, or it can reduce the variance by increasing the amount of exploitation when it's very sure about which action to take in a given state. The effect of the variance can be visualized as follows:

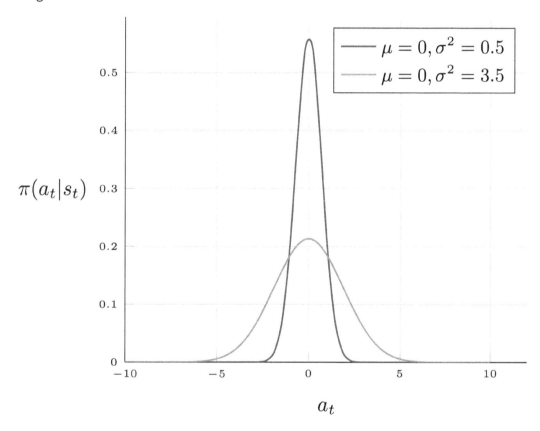

Figure 1.26: The effect of the variance on a Gaussian policy

In the preceding figure, if the variance increases (the lower curve), the policy becomes more exploratory. Additionally, actions that are very far from the mean have nonzero probabilities. When the variance is small (the higher curve), the policy is highly exploitative. This means that only actions that are very close to the mean have nonzero probabilities.

In the preceding diagram, the smaller Gaussian represents a highly explorative policy with respect to the larger policy. Here, we can see the effect of the variance on the policy exploration attitude.

While learning a task, in the first training episodes, the policy needs to have a high variance in order for it to explore different actions. The variance will be reduced once the agent gains some experience and becomes more and more confident about the best actions.

The Boltzmann Policy

Boltzmann parameterization is used for discrete action spaces. The resulting action is a softmax function acting on the weighted state features, as stated in the following expression:

$$\pi\left(a_i\middle|s\right) = \frac{e^{\theta_i^{T}\phi(s)}}{\sum_j e^{\theta_j^{T}\phi(s)}}$$

Figure 1.27: Expression for a Boltzmann policy

Here, θ_i is the set of parameters associated with action a_i.

The Boltzmann policy is a stochastic policy. The motivation behind this is very simple; let's sum the policy over all the actions (the denominator does not depend on the action, i), as follows:

$$\sum_{i=0}\pi\left(a_i\middle|s\right) = \frac{\sum_i e^{\theta_i^{T}\phi(s)}}{\sum_j e^{\theta_j^{T}\phi(s)}} = 1$$

Figure 1.28: A Boltzmann policy over all of the actions

The Boltzmann policy becomes deterministic if we select the action with the highest probability, which is equivalent to selecting the mean action in a Gaussian distribution. What the Boltzmann parameterization represents is simply a normalization of the value, $e^{\theta_i^{T}\phi(s)}$, corresponding to the score of action i. The score is thus normalized by considering the value of all the other actions obtaining a distribution.

In all of these parametrizations, the state features might be non-linear features depending on several parameters, for example, whether it is coming from a neural network, the **radial basis function (RBF)** features, or the tile coding features.

EXERCISE 1.02: IMPLEMENTING A LINEAR POLICY

In this exercise, we will practice with the implementation of a linear policy. The goal is to write the presented parameterizations in the case of a state composed of n components. In the first case, the features can be represented by the identity function; in the second case, the features are represented by a polynomial function of order 2:

1. Open a new Jupyter notebook and import NumPy to implement all of the requested policies:

```
from typing import Callable, List

import matplotlib
from matplotlib import pyplot as plt

import numpy as np
import scipy.stats
```

2. Let's now implement the linear policy. A linear policy can be efficiently represented by a dot product between the policy parameters and state features. The first step is to write the constructor:

```
class LinearPolicy:
    def __init__(
        self, parameters: np.ndarray, \
        features: Callable[[np.ndarray], np.ndarray]):
        """
        Linear Policy Constructor.

        Args:
            parameters (np.ndarray): policy parameters
            as np.ndarray.
            features (Callable[[np.ndarray], np.ndarray]):
            function used to extract features from the
            state representation.
        """

        self._parameters = parameters
```

```
self._features = features
```

The constructor simply sets the attribute's parameters and features. The feature parameter is actually a callable that takes, as input, a NumPy array and returns another NumPy array. The input is the environment state, whereas the output is the state features.

3. Next, we will implement the call method. The **__call__** method takes as input the state, and returns the selected action according to the policy parameters. The call represents a real policy implementation. What we have to do in the linear case is to first apply the feature function and then compute the dot product between the parameters and the features. A possible implementation of the call function is as follows:

```python
def __call__(self, state: np.ndarray) -> np.ndarray:
    """
    Call method of the Policy.

    Args:
        state (np.ndarray): environment state.

    Returns:
        The resulting action.
    """

    # calculate state features
    state_features = self._features(state)


    """
    the parameters shape [0] should be the same as the
    state features as they must be multiplied
    """
    assert state_features.shape[0] == self._parameters.shape[0]


    # dot product between parameters and state features
    return np.dot(self._parameters.T, state_features)
```

4. Let's try the defined policy with a state composed of a 5-dimensional array. Sample a random set of parameters and a random state vector. Create the policy object. The constructor needs the callable features, which, in this case, is the identity function. Call the policy to obtain the resulting action:

```
# sample a random set of parameters
parameters = np.random.rand(5, 1)

# define the state features as identity function
features = lambda x: x

# define the policy
pi: LinearPolicy = LinearPolicy(parameters, features)

# sample a state
state = np.random.rand(5, 1)

# Call the policy obtaining the action
action = pi(state)

print(action)
```

The output will be as follows:

```
[[1.33244481]]
```

This value is the action selected by our agent, given the state and the policy parameters. In this case, the selected action is `[[1.33244481]]`. The meaning of the action depends on the RL task.

Of course, you will obtain different results based on the sampled parameters and sampled state. It is always possible to seed the NumPy random number generator to obtain reproducible results.

> **NOTE**
>
> To access the source code for this specific section, please refer to https://packt.live/2Yvrku7. You can also refer to the Gaussian and Boltzmann policies that are implemented in the same notebook.
>
> You can also run this example online at https://packt.live/3dXc4Nc.

In this exercise, we practiced with different policies and parameterizations. These are simple policies, but they are the building blocks of more complex policies. The trick is just to substitute the state features with a neural network or any other feature extractor.

GOALS AND REWARDS

In RL, the agent's goal is to maximize the total amount of reward it receives during an episode.

This is based on the famous reward hypothesis in *Sutton & Barto 1998*:

"That all of what we mean by goals and purposes can be well thought of as the maximization of the expected value of the cumulative sum of a received scalar signal (called reward)."

The important thing here is that the reward should not describe how to achieve the goal; instead, it should describe the goal of the agent. The reward function is an element of the environment, but it can also be designed for a specific task. In principle, there are infinite reward functions for each task. Usually, reward functions that are characterized by a lot of information help the agent to learn. Sparse reward functions (with no information) makes learning difficult or, sometimes, impossible. Sparse reward functions are functions in which, most of the time, the reward is constant (or zero).

Sutton's hypothesis, which we explained earlier, is the basis of the RL framework. This hypothesis may be wrong; probably, a scalar reward signal (and its maximization) is not enough to define complex goals; however, still, this hypothesis is very flexible, simple, and it can be applied to a wide range of tasks. At the time of writing, the reward function design is more art than engineering; there are no formal practices regarding how to write a reward function, rather there are only best practices based on experience. Usually, a simple reward function works very well. Usually, we associate a positive value with good actions and behavior and negative values with bad actions or actions that are not important at that particular moment.

In a locomotion task (for example, teaching a robot how to move), the reward may be defined as proportional to the robot's forward movement. In chess, the reward may be defined as 0 for each timestep: +1 if the agent wins and -1 if the agent loses. If we want our agent to solve Rubik's Cube, the reward may be defined similarly: 0 every step and +1 if the cube is solved.

Sometimes, as we learned earlier, defining a scalar reward function for a task is not easy, and, nowadays, it is more art than engineering or science.

In each of these tasks, the final objective is to learn a policy, a way of selecting actions, maximizing the total rewards received by the agent. Tasks can be episodic or continuous. Episodic tasks have a finite length, that is, a finite number of timesteps (for example, T is finite). Continuous tasks can last forever or until the agent reaches its goal. In the first case, we can simply define the total reward (**return**) received by an agent as the sum of the individual rewards:

$$G = r_0 + r_1 + \ldots + r_T$$

Figure 1.29: Expression for a total reward

Usually, we are interested in the return from a certain timestep, G_t. In other words, the return, G_t, quantifies the agent's performance in the long term, and it can be calculated as the sum of immediate rewards following time t until the end of the episode (timestep T):

$$G_t = r_{t+1} + r_{t+2} + \cdots + r_T$$

$$G_t = \sum_{k=t+1}^{T} r_k$$

Figure 1.30: Expression for a return from timestep t

It is straightforward to see that, with this formulation, the return for continuing tasks diverges to infinity.

In order to deal with continuing tasks, we need to introduce the notion of a discounted return. This concept formalizes, in mathematical terms, the principle that the immediate reward (sometimes) is more valuable than the same amount of reward after many steps. This principle is widely known in economics. The discount factor, $\gamma \in [0, 1]$, quantifies the present value of future rewards. We are ready to present the unified notation for the return in episodic and continuing tasks.

The discounted return is the cumulative, discounted sum of rewards until the end of the episode. In mathematical terms, it can be formalized as follows:

$$G_t = r_{t+1} + \gamma r_{t+2} + \cdots + \gamma^{T-t-1} r_T$$

$$G_t = \sum_{k=t+1}^{T} \gamma^{k-t-1} r_k$$

Figure 1.31: Expression for the discounted return from timestep t

To understand how the discount affects the return, it is possible to see that the value of receiving reward r after a $k+1$ timestep is $\gamma^k r$, since $\gamma \in [0, 1]$ is less than or equal to r. It is worth introducing the effect of the discount on the return. If $\gamma < 1$, the return, even if composed by an infinite sum, has a bounded value. If $\gamma = 0$, the agent is myopic since it cares only about the immediate reward, and it does not care about future rewards. A myopic agent can cause problems: the only thing it learns is to select the action yielding the highest immediate return. A myopic chess player can, for example, eat the opponent's pawn causing the game's loss. Notice that, for some tasks, this isn't always a problem. This includes tasks in which the current action does not affect the future reward and has no consequences for the agent's future. These tasks can be solved by finding the action that causes a higher immediate reward for each state independently. Most of the time, the current action influences the future of the agent and its rewards. If the discount factor is near to 1, the agent is farsighted; it is possible for them to sacrifice an action yielding to a good immediate reward now for a higher reward in future steps.

It is important to understand the relationship between returns at different timesteps, both from a theoretical point of view but also from an algorithmic point of view, because many RL algorithms are based on this principle:

$$G_t = r_{t+1} + \gamma r_{t+2} + \cdots + \gamma^T r_T$$

$$G_t = r_{t+1} + \gamma \left(r_{t+2} + \cdots + \gamma^{T-1} r_T \right)$$

$$G_t = r_{t+1} + \gamma G_{t+1}.$$

Figure 1.32: Relationship between returns at different timesteps

By following these simple steps, we can see that the return from the timestep equals the immediate reward plus the return at the following step scaled by gamma. This simple relationship will be extensively used in RL algorithms.

WHY DISCOUNT?

The following describes the motivations as to why many RL problems are discounted:

- It is convenient from a mathematical perspective to have a bounded return, and also in the case of continuing tasks.

- If the task is a financial task, immediate rewards may gain more interest than delayed rewards.

- Animal and human behavior show a preference for immediate rewards.

- A discounted reward may also represent uncertainty about the future.

- It is also possible to use an undiscounted return $(\gamma = 1)$ if all of the episodes terminate after a finite number of steps.

This section introduced the main elements of RL, including agents, actions, environments, transition functions, and policies. In the next section, we will practice with these concepts by defining agents, environments, and measuring the performance of agents on some tasks.

REINFORCEMENT LEARNING FRAMEWORKS

In the previous sections, we learned the basic theory behind RL. In principle, an agent or an environment can be implemented in any way or any language. For RL, the primary language used by both academic and industrial people is Python, as it allows you to focus on the algorithms and not on the language details, making it very simple to use. Implementing, from scratch, an algorithm or a complex environment (that is, an autonomous driving environment) might be very difficult and error-prone. For this reason, several well-established and well-tested libraries make RL very easy for newcomers. In this section, we will explore the main Python RL libraries. We will present OpenAI Gym, a set of environments that is ready to use and easy to modify, and OpenAI Baselines, a set of high quality, state-of-the-art algorithms. By the end of this chapter, you will have learned about and practiced with environments and agents.

OPENAI GYM

OpenAI Gym (https://gym.openai.com) is a Python library that provides a set of RL environments ranging from toy environments to Atari environments and more complex environments, such as **MuJoCo** and **Robotics** environments. OpenAI Gym, besides providing this large set of tasks, also provides a unified interface for interacting with RL tasks and a set of interfaces that are useful for describing the environment's characteristics, such as the action space and the state space. An important property of Gym is that its only focus is on environments; it makes no assumption of the type of agent you have or the computational framework you use. We will not cover the installation details in this chapter for ease of presentation. Instead, we will focus on the main concepts and learn how to interact with these libraries.

GETTING STARTED WITH GYM – CARTPOLE

CartPole is a classical control environment provided by Gym and used by researchers as a starting point of algorithms. It consists of a cart that moves along the horizontal axis (1-dimensional) and a pole anchored to the cart on one endpoint:

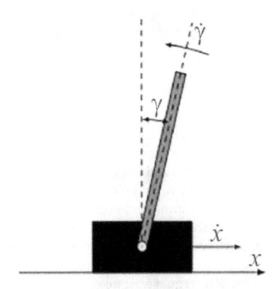

Figure 1.33: CartPole environment representation

The agent has to learn how to move the cart to balance the pole (that is, to stop the pole from falling). The episode ends when the pole angle ((γ)) becomes higher than a certain threshold (($\bar{\gamma}$)). The state space is represented by the position of the cart along the axis, x; the velocity along the axis, $\dot{x}$; the pole angle, γ; and the pole angular velocity, $\dot{\gamma}$. The state space is continuous in this case, but it can also be discretized to make learning simpler.

In the following steps, we will practice with Gym and its environments.

Let's create a CartPole environment using Gym and analyze its properties in a Jupyter notebook. Please refer to the *Preface* for Gym installation instructions:

```
# Import the gym Library
import gym


# Create the environment using gym.make(env_name)
env = gym.make('CartPole-v1')


"""
Analyze the action space of cart pole using the property action_space
"""

print("Action Space:", env.action_space)


"""
Analyze the observation space of cartpole using the property observation_
space
"""

print("Observation Space:", env.observation_space)
```

If you run these lines, you will get the following output:

```
Action Space: Discrete(2)
Observation Space: Box(4,)
```

Discrete(2) means that the action space of CartPole is a discrete action space composed of two actions: Go Left and Go Right. These actions are the only actions available to the agent. The action of Go Left, in this case, is represented by action 0, and the action of Go Right by action 1.

Box(4,) means that the state space (the observation space) of the environment is represented by a 4-dimensional box, a subspace of $\mathbb{R}^n$. Formally, it is a Cartesian product of n intervals. The state space has a lower bound and an upper bound. The bounds may also be infinite, creating an unbounded box.

To inspect the observation space better, we can use the properties of **high** and **low**:

```
# Analyze the bounds of the observation space
print("Lower bound of the Observation Space:", \
    env.observation_space.low)
print("Upper bound of the Observation Space:", \
    env.observation_space.high)
```

This will print the following:

```
Lower bound of the Observation Space: [-4.8000002e+00 -3.4028235e+38
-4.1887903e-01 -3.4028235e+38]
Upper bound of the Observation Space: [4.8000002e+00  3.4028235e+38
4.1887903e-01  3.4028235e+38]
```

Here, we can see that upper and lower bounds are arrays of 4 elements; one element for each state dimension. The following are some observations:

- The lower bound of the cart position (the first state dimension) is -4.8, while the upper bound is 4.8.

- The lower bound of the velocity (the second state dimension) is -3.10^{38}, basically $-\infty$; and the upper bound is $+3.10^{38}$, basically $+\infty$.

- The lower bound of the pole angle (the third state dimension) is -0.4 radians, representing an angle of -24 degrees. The upper bound is 0.4 radians, representing an angle of +24 degrees.

- The lower and upper bounds of the pole angular velocity (the fourth state dimension) are, respectively, $-\infty$ and $+\infty$, similar to the lower and upper bounds for the cart policy's angular velocity.

GYM SPACES

The Gym **Space** class represents the way Gym describes actions and state spaces. The most used spaces are the **Discrete** and **Box** spaces.

A discrete space is composed of a fixed number of elements. It can represent both a state space but also an action space, and it describes the number of elements through the **n** attribute. Its elements range from 0 to **n-1**.

A **Box** space describes its shape through the **shape** attribute. It can have an n-dimensional shape that corresponds to an **n**-dimensional box. A **Box** space can also be unbounded. Each interval has the form of one of $[a,b], (-\infty, b], [a, +\infty), (-\infty, +\infty)$.

It is possible to sample from the action space to gain insight into the elements it is composed of using the **space.sample()** method.

> **NOTE**
>
> For the sampling distribution of box environments, to create a sample of the box, each coordinate is sampled according to the form of the interval in the following distributions:
>
> - $[a, b]$: A uniform distribution
> - $[a, +\infty)$: A shifted exponential distribution
> - $(-\infty, b]$: A shifted negative exponential distribution
> - $(-\infty, +\infty)$: A normal distribution

Let's now demonstrate how to create simple spaces and how to sample from spaces:

```python
# Type hinting
from typing import Tuple

import gym

# Import the spaces module
from gym import spaces

# Create a discrete space composed by N-elements (5)
n: int = 5
discrete_space = spaces.Discrete(n=n)

# Sample from the space using .sample method
print("Discrete Space Sample:", discrete_space.sample())

"""
Create a Box space with a shape of (4, 4)
Upper and lower Bound are 0 and 1
"""
```

```
box_shape: Tuple[int, int] = (4, 4)
box_space = spaces.Box(low=0, high=1, shape=box_shape)

# Sample from the space using .sample method
print("Box Space Sample:", box_space.sample())
```

This will print the samples from our spaces:

```
Discrete Space Sample: 4
Box Space Sample: [[0.09071387 0.4223234  0.09272052 0.15551752]
 [0.8507258  0.28962377 0.98583364 0.55963445]
 [0.4308358  0.8658449  0.6882108  0.9076272 ]
 [0.9877584  0.7523759  0.96407163 0.630859  ]]
```

Of course, the samples will change according to your seeds.

As you can see, we have sampled element 4 from our discrete space composed of 5 elements (from 0 to 4). We sampled a random 4 x 4 matrix with elements between 0 and 1, the lower and the upper bound of our space.

To obtain reproducible results, it is also possible to set the seed of an environment using the **seed** method:

```
# Seed spaces to obtain reproducible samples
discrete_space.seed(0)
box_space.seed(0)

# Sample from the seeded space
print("Discrete Space (seed=0) Sample:", discrete_space.sample())

# Sample from the seeded space
print("Box Space (seed=0) Sample:", box_space.sample())
```

This will print the following:

```
Discrete Space (seed=0) Sample: 0
Box Space (seed=0) Sample: [[0.05436005 0.9653909
0.63269097 0.29001734]
 [0.10248426 0.67307633 0.39257675 0.66984606]
 [0.05983897 0.52698725 0.04029069 0.9779441 ]
 0.46293673 0.6296479  0.9470484  0.6992778 ]]
```

The previous statement will always print the same sample since we set the seed to 0. Seeding an environment is very important in order to guarantee reproducible results.

EXERCISE 1.03: CREATING A SPACE FOR IMAGE OBSERVATIONS

In this exercise, we will create a space to represent an image observation. Image-based observations are essential in RL since they allow the agent to learn from pixels and require minimal feature engineering or need to go through the feature extraction phase. The agent can focus on what is important for its task without being limited by manually decided heuristics. We will create a space representing RGB images with dimensions equal to 256 x 256:

1. Open a new Jupyter notebook and import the desired modules – **gym** and NumPy:

```
import gym
from gym import spaces

import matplotlib.pyplot as plt
%matplotlib inline

import numpy as np # used for the dtype of the space
```

2. We are dealing with 256 x 256 RGB images, so the space has a shape of (256, 256, 3). In addition, the images range from 0 to 255 (if we consider the **uint8** images):

```
"""
since the Space is RGB images with shape 256x256 the final shape is
(256, 256, 3)
"""
shape = (256, 256, 3)

# If we consider uint8 images the bounds are 0-255
low = 0
high = 255

# Space type: unsigned int
dtype = np.uint8
```

3. We are now ready to create the space. An image is a **Box** space since it has defined bounds:

```
# create the space
space = spaces.Box(low=low, high=high, shape=shape, dtype=dtype)

# Print space representation
print("Space", space)
```

This will print the representation of our space:

```
Space Box(256, 256, 3)
```

The first dimension is the image width, the second dimension is the image height, and the third dimension is the number of channels.

4. Here is a sample from the space:

```
# Sample from the space
sample = space.sample()
print("Space Sample", sample)
```

This will return the space sample; in this case, it is a huge tensor of 256 x 256 x 3 unsigned integers (between 0 and 255). The output (fewer lines are presented now) should be similar to the following:

```
Space Sample [[[ 37 254 243]
  [134 179  12]
  [238  32   0]
  ...
  [100  61  73]
  [103 164 131]
  [166  31  68]]

 [[218 109 213]
  [190  22 130]
  [ 56 235 167]
```

5. To visualize the returned sample, use the following code:

```
plt.imshow(sample)
```

The output will be as follows:

Figure 1.34: A sample from a Box space of (256, 256) RGB

The preceding is not very informative because it is a random image.

6. Now, suppose we want to give our agent the opportunity to see the last **n=4** frames. By adding the temporal component, we can obtain a state representation composed of 4 dimensions. The first dimension is the temporal one, the second is the width, the third is the height, and the last one is the number of channels. This is a very useful technique that allows the agent to understand its movement:

```
# we want a space representing the last n=4 frames
n_frames = 4  # number of frames
width = 256  # image width
height = 256  # image height
channels = 3  # number of channels (RGB)
shape_temporal = (n_frames, width, height, channels)
# create a new instance of space
space_temporal = spaces.Box(low=low, high=high, \
                            shape=shape_temporal, dtype=dtype)

print("Space with temporal component", space_temporal)
```

This will print the following:

```
Space with temporal component Box(4, 256, 256, 3)
```

As you can see, we have successfully created a space and, on inspecting the space representation, we notice that we have another dimension: the temporal dimension.

> **NOTE**
>
> To access the source code for this specific section, please refer to https://packt.live/2AwJm7x.
>
> You can also run this example online at https://packt.live/2UzxoAY.

Image-based environments are very important in RL. They allow the agent to learn salient features for solving the task directly from raw pixels, without any preprocessing. In this exercise, we learned how to create a Gym space for image observations and how to deal with image spaces.

RENDERING AN ENVIRONMENT

In the *Getting Started with Gym – CartPole* section, we saw a sample from the CartPole state space. However, visualizing or understanding the CartPole state from a vector representation is not an easy task, at least for a human. Gym also allows you to visualize a given task (if possible) through the **env.render()** function.

> **NOTE**
>
> The **env.render()** function is usually slow. Rendering an environment is done primarily to understand the behavior learned by the agent after the training or on intervals of many training steps. Usually, we train agents without rendering the environment state to improve the training speed.

If we just call the **env.render()** function, we will always see the same scene, that is, the environment state does not change. To see the evolution of the environment in time, we must call the **env.step()** function, which takes as input an action belonging to the action space and applies the action in the environment.

RENDERING CARTPOLE

The following code demonstrates how to render the CartPole environment. The action is a sample from the action space. For RL algorithms, the action will be smartly selected from the policy:

```
# Create the environment using gym.make(env_name)
env = gym.make("CartPole-v1")

# reset the environment (mandatory)
env.reset()

# render the environment for 100 steps
n_steps = 100
for i in range(n_steps):
    action = env.action_space.sample()
    env.step(action)
    env.render()

# close the environment correctly
env.close()
```

If you run this script, you will see that **gym** opens a window and displays the CartPole environment with random actions, as shown in the following figure:

Figure 1.35: A CartPole environment rendered in Gym (the initial state)

A REINFORCEMENT LEARNING LOOP WITH GYM

To understand the consequences of an action, and to come up with a better policy, the agent observes its new state and a reward. Implementing this loop with **gym** is easy. The key element is the **env.step()** function. This function takes an action as input. It applies the action and returns four values, which are described as follows:

- **Observation**: The observation is the next environmental state. This is represented as an element belonging to the observation space of the environment.

- **Reward**: The reward associated with a step is a float value that is related to the action given as input to the function.

- **Done**: This return value assumes the **True** value when the episode is finished, and it's time to call the **env.reset()** function to reset the environment state.

- **Info**: This is a dictionary containing debugging information; usually, it is ignored.

Let's now implement the RL loop within the Gym environment.

EXERCISE 1.04: IMPLEMENTING THE REINFORCEMENT LEARNING LOOP WITH GYM

In this exercise, we will implement a basic RL loop with episodes and timesteps using the CartPole environment. You can change the environment and use other environments as well; nothing changes as the main goal of Gym is to unify the interfaces of all possible environments in order to build agents that are as environment-agnostic as possible. The transparency with respect to the environment is a very peculiar thing in RL: the algorithms are not usually suited to the task but are task-agnostic so that they can be applied successfully to a variety of environments and still solve them.

We need to create the Gym CartPole environment as before using the **gym.make()** function. After that, we can loop for a defined number of episodes; for each episode, we loop for a defined number of steps or until the episode is terminated (by checking the **done** value). For each timestep, we have to call the **env.step()** function by passing an action (we will pass a random action for now), and then we collect the desired information:

1. Open a new Jupyter notebook and define the import, the environment, and the desired number of steps:

```
import gym

import matplotlib.pyplot as plt
```

```
%matplotlib inline

env = gym.make("CartPole-v1")

# each episode is composed by 100 timesteps
# define 10 episodes
n_episodes = 10
n_timesteps = 100
```

2. Loop for each episode:

```
# loop for the episodes
for episode_number in range(n_episodes):
    # here we are inside an episode
```

3. Reset the environment and get the first observation:

```
"""
the reset function resets the environment and returns
the first environment observation
"""
observation = env.reset()
```

4. Loop for each timestep:

```
"""
loop for the given number of timesteps or
until the episode is terminated
"""
for timestep_number in range(n_timesteps):
```

5. Render the environment, select the action (randomly by using the **env. action_space.sample()** method), and then take the action:

```
# render the environment
env.render(mode="rgb-array")

# select the action
action = env.action_space.sample()

# apply the selected action by calling env.step
observation, reward, done, info = env.step(action)
```

6. Check whether the episode has been terminated using the **done** variable:

```
"""if done the episode is terminated, we have to reset
the environment
"""
if done:
    print(f"Episode Number: {episode_number}, \
Timesteps: {timestep_number}")
    # break from the timestep loop
    break
```

7. After the episode loop, close the environment in order to release the associated memory:

```
# close the environment
env.close()
```

If you run the previous code, the output should, approximately, be like this:

```
Episode Number: 0, Timesteps: 34
Episode Number: 1, Timesteps: 10
Episode Number: 2, Timesteps: 12
Episode Number: 3, Timesteps: 21
Episode Number: 4, Timesteps: 16
Episode Number: 5, Timesteps: 17
Episode Number: 6, Timesteps: 12
Episode Number: 7, Timesteps: 15
Episode Number: 8, Timesteps: 16
Episode Number: 9, Timesteps: 16
```

We have the episode number and the number of steps taken in that episode. We can see that the average number of timesteps for an episode is approximately 17. This means that, using the random policy, after 17 episodes on average, the pole falls and the episode finishes.

> **NOTE**
>
> To access the source code for this specific section, please refer to https://packt.live/2MOs5t5.
>
> This section does not currently have an online interactive example, and will need to be run locally.

The goal of this exercise was to understand the bare bones of each RL algorithm. The only different thing here is that the action selection phase should take into account the environment state in order for it to be useful, and it should not be random.

Let's now move toward completing an activity to measure the performance of an agent.

ACTIVITY 1.01: MEASURING THE PERFORMANCE OF A RANDOM AGENT

The measurement of the performance and the design of an agent is an essential phase of every RL experiment. The goal of this activity is to practice with these two concepts by designing an agent that is able to interact with an environment using a random policy and then measure the performance.

You need to design a random agent using a Python class to modularize and keep the agent independent from the main loop. After that, you have to measure the mean and the variance of the discounted return using a batch of 100 episodes. You can use every environment you want, taking into account that the agent's action should be compatible with the environment. You can design two different types of agents for discrete action spaces and continuous action spaces. The following steps will help you to complete the activity:

1. Import the required libraries: **abc**, **numpy**, and **gym**.

2. Define the **Agent** abstract class in a very simple way, defining only the **pi ()** function that represents the policy. The input should be an environment state. The **__init__** method should take as input the action space and build the distribution accordingly.

3. Define a **ContinuousAgent** deriving from the **Agent** abstract class. The agent should check that the action space is coherent with it, and it should be a continuous action space. The agent should also initialize a probability distribution for sampling actions (you can use NumPy to define probability distributions). The continuous agent can change the distribution type according to the distributions defined by the Gym spaces.

4. Define a **DiscreteAgent** deriving from the **Agent** abstract class. The discrete agent should, of course, initialize a uniform distribution.

5. Implement the **pi ()** function for both agents. This function is straightforward and should only sample from the distribution defined in the constructor and return it, ignoring the environment state. Of course, this is a simplification. You can also implement the **pi ()** function in the **Agent** base class.

6. Define the main RL loop in another file by importing the agent.

7. Instantiate the correct agent according to the selected environment. Examples of environments are "CartPole-v1" or "MountainCar-Continuous-v0."

8. Take actions according to the **pi** function of the agent.

9. Measure the performance of the agent collecting (in a list or a NumPy array) the discounted return for each episode. Then, take the average and the standard deviation (you can use NumPy for this). Remember to apply the discount factor (user-defined) to the immediate reward. You have to keep a cumulated discount factor by multiplying the discount factor at each timestep.

The output should be similar to the following:

```
Episode Number: 0, Timesteps: 27, Return: 28.0
Episode Number: 1, Timesteps: 9, Return: 10.0
Episode Number: 2, Timesteps: 13, Return: 14.0
Episode Number: 3, Timesteps: 16, Return: 17.0
Episode Number: 4, Timesteps: 31, Return: 32.0
Episode Number: 5, Timesteps: 10, Return: 11.0
Episode Number: 6, Timesteps: 14, Return: 15.0
Episode Number: 7, Timesteps: 11, Return: 12.0
Episode Number: 8, Timesteps: 10, Return: 11.0
Episode Number: 9, Timesteps: 30, Return: 31.0
Statistics on Return: Average: 18.1, Variance: 68.89000000000001
```

> **NOTE**
>
> The solution to this activity can be found on page 680.

The solution to this activity can be found on page 680.

OPENAI BASELINES

OpenAI Baselines (https://github.com/openai/baselines) is a set of state-of-the-art RL algorithms. The main goal of Baselines is to make it easier to reproduce results on a set of benchmarks, to evaluate new ideas, and to compare them to existing algorithms. In this section, we will learn how to use Baselines to run an existing algorithm on an environment taken from Gym (refer to the previous section) and how to visualize the behavior learned by the agent. As for Gym, we will not cover the installation instructions; these can be found in the *Preface* section. The implementation of the Baselines' algorithm is based on TensorFlow, one of the most popular libraries for machine learning.

GETTING STARTED WITH BASELINES – DQN ON CARTPOLE

Training a **Deep Q Network (DQN)** on CartPole is straightforward with Baselines; we can do it with just one line of Bash.

Just use the terminal and run this command:

```
# Train model and save the results to cartpole_model.pkl
python -m baselines.run -alg=deepq -env=CartPole-v0 -save_path=./
cartpole_model.pkl -num_timesteps=1e5
```

Let's understand the parameters, as follows:

- **--alg=deepq** specifies the algorithm to be used to train our agent. In our case, we selected **deepq**, that is, DQN.

- **--env=CartPole-v0** specifies the environment to be used. We selected CartPole, but we can also select many other environments.

- **--save_path=./cartpole_model.pkl** specifies where to save the trained agent.

- **--num_timesteps=1e5** is the number of training timesteps.

After having trained the agent, it is also possible to visualize the learned behavior using the following:

```
# Load the model saved in cartpole_model.pkl
# and visualize the learned policy
python -m baselines.run --alg=deepq --env=CartPole-v0 --load_path=./
cartpole_model.pkl --num_timesteps=0 --play
```

DQN is a very powerful algorithm; using it for a simple task such as CartPole is almost overkill. We can see that the agent has learned a stable policy, and the pole almost never falls. We will explore DQN in more detail in the following chapters.

In the following steps, we will train a DQN agent on the CartPole environment using Baselines:

1. First, we import **gym** and **baselines**:

```
import gym


# Import the desired algorithm from baselines
from baselines import deepq
```

2. Define a callback to inform **baselines** when to stop training. The callback should return **True** if the reward is satisfying:

```
def callback(locals, globals):
    """
    function called at every step with state of the algorithm.
    If callback returns true training stops.
    stop training if average reward exceeds 199
    time should be greater than 100 and the average of
    last 100 returns should be >= 199
    """
    is_solved = (locals["t"] > 100 and \
                sum(locals["episode_rewards"]\
                    [-101:-1]) / 100 >= 199)
    return is_solved
```

3. Now, let's create the environment and prepare the algorithm's parameters:

```
# create the environment
env = gym.make("CartPole-v0")


"""
Prepare learning parameters: network and learning rate
the policy is a multi-layer perceptron
"""
network = "mlp"
# set learning rate of the algorithm
learning_rate = 1e-3
```

4. We can use the **deep.learn()** method to start the training and solve the task:

```
"""
launch learning on this environment using DQN
ignore the exploration parameter for now
"""
actor = deepq.learn(env, network=network, lr=learning_rate, \
                    total_timesteps=100000, buffer_size=50000, \
                    exploration_fraction=0.1, \
                    exploration_final_eps=0.02, print_freq=10, \
                    callback=callback,)
```

After some time, depending on your hardware (it usually takes a few minutes), the learning phase terminates, and you will have the CartPole agent saved to your current working directory.

We should see the **baselines** logs reporting the agent's performance over time.

Consider the following example:

```
------------------------------------
| % time spent exploring  | 2        |
| episodes                | 770      |
| mean 100 episode reward | 145      |
| steps                   | 6.49e+04 |
```

The following are the observations from the preceding logs:

*   The **episodes** parameter reports the episode number we are referring to.

*   **mean 100 episode reward** is the average return obtained in the last 100 episodes.

*   **steps** is the number of training steps the algorithm has performed.

Now we can save our actor so that we can reuse it without retraining it:

```
print("Saving model to cartpole_model.pkl")
actor.save("cartpole_model.pkl")
```

After the **actor.save** function, the **"cartpole_model.pkl"** file contains the trained model.

Now it is possible to use the model and visualize the agent's behavior.

The actor returned by **deepq.learn** is actually a callable that returns the action given the current observation – it is the agent policy. We can use it by passing the current observation, and it returns the selected action:

```python
# Visualize the policy
n_episodes = 5
n_timesteps = 1000
for episode in range(n_episodes):
    observation = env.reset()
    episode_return = 0
    for timestep in range(n_timesteps):
        # render the environment
        env.render()

        # select the action according to the actor
        action = actor(observation[None])[0]

        # call env.step function
        observation, reward, done, _ = env.step(action)

        """
        since the reward is undiscounted we can simply add
        the reward to the cumulated return
        """
        episode_return += reward

        if done:
            break

    # here an episode is terminated, print the return
    print("Episode return", episode_return)
        """
        here an episode is terminated, print the return
        and the number of steps
        """
    print(f"Episode return {episode_return}, \
Number of steps: {timestep}")
```

If you run the preceding code, you should see the agent's performance on the CartPole task.

You should get, as output, the return for each episode; it should be something similar to the following:

```
Episode return 200.0, Number of steps: 199
Episode return 200.0, Number of steps: 199
Episode return 200.0, Number of steps: 199
Episode return 200.0, Number of steps: 199
Episode return 200.0, Number of steps: 199
```

This means our agent always reaches the maximum possible return for CartPole (**200.0**) and the maximum possible number of steps (**199**).

We can compare the return obtained using a trained DQN agent with respect to the return obtained using a random agent (*Activity 1.01, Measuring the Performance of a Random Agent*). The random agent yields an average return of **20.0**, while DQN obtains the maximum return possible for CartPole, which is **200.0**.

In this section, we presented OpenAI Gym and OpenAI Baselines, the two main frameworks for RL research and experiments. There are many other frameworks for RL, each with their pros and cons. Gym is particularly suited due to its unified interface in the RL loop, while OpenAI Baselines is very useful for understanding how to implement sophisticated state-of-the-art RL algorithms and how to compare new algorithms with existing ones.

In the following section, we will explore some interesting RL applications in order to better understand the possibilities offered by the framework as well as its flexibility.

APPLICATIONS OF REINFORCEMENT LEARNING

RL has exciting and useful applications in many different contexts. Recently, the usage of deep neural networks has augmented the number of possible applications considerably.

When used in a deep learning context, RL can also be referred to as deep RL.

The applications vary from games and video games to real-world applications, such as robotics and autonomous driving. In each of these applications, RL is a game-changer, allowing you to solve tasks that are considered to be almost impossible (or, at least, very difficult) without these techniques.

In this section, we will present some RL applications, describe the challenges of each application, and begin to understand why RL is preferred among other methods, along with its advantages and its drawbacks.

GAMES

Nowadays, RL is widely used in video games and board games.

Games are used to benchmark RL algorithms because, usually, they are very complex to solve yet easy to implement and to evaluate. Games also represent a simulated reality in which the agent can freely move and behave without affecting the real environment:

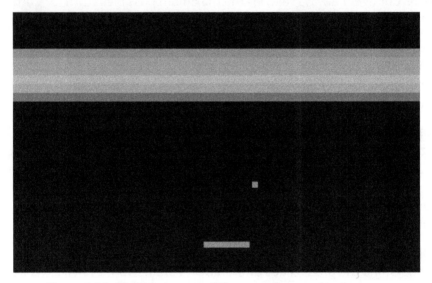

Figure 1.36: Breakout – one of the most famous Atari games

> **NOTE**
>
> The preceding screenshot has been sourced from the official documentation of OpenAI Gym. Please refer to the following link for more examples: https://gym.openai.com/envs/#atari.

Despite appearing to be secondary or relatively limited-use applications, games represent a useful benchmark for RL and, in general, artificial intelligence algorithms. Very often, artificial intelligence algorithms are tested on games due to the significant challenges that arise in these scenarios.

The two main characteristics required to play games are **planning** and **real-time control**.

An algorithm that is not able to plan won't be able to win strategic games. Having a long-term plan is also fundamental in the early stages of a game. Planning is also fundamental in real-world applications in which taken actions may have long-term consequences.

Real-time control is another fundamental challenge that requires an algorithm to be able to respond within a small timeframe. This challenge is similar to one an algorithm has to face when applied to real-world cases such as autonomous driving, robot control, and many others. In these cases, the algorithm can't evaluate all the possible actions or all the possible consequences of these actions; therefore, the algorithm should learn an efficient (and maybe compressed) state representation and should understand the consequences of its actions without simulating all of the possible scenarios.

Recently, RL has been able to exceed human performance in games such as Go, and in video games such as Dota II and StarCraft, thanks to work done by DeepMind and OpenAI.

Go

Go is a very complex, highly strategic board game. In Go, two players are competing against each other. The aim is to use the game pieces, also called stones, to surround more territory than the opponent. At each turn, the player can place its stone in a vacant intersection on the board. At the end of the game, when no player can place a stone, the player surrounding more territories wins.

Go has been studied for many years to understand the strategies and moves necessary to lead a player to victory. Until recently, no algorithm succeeded in producing strong players – even algorithms working very well for similar games, such as chess. This difficulty is due to Go's huge search space, the variety of possible moves, and the average length (in terms of moves) of Go games, which, for example, is longer than the average length of chess games. RL, and in particular **AlphaGo** by DeepMind, succeeded recently in beating a human player on a standard dimension board. AlphaGo is actually a mix of RL, supervised learning, and tree search algorithms trained on an extensive set of games from both human and artificial players. AlphaGo denoted a real milestone in artificial intelligence history, which was made possible mainly due to the advances in RL algorithms and their improved efficiency.

The successor of **AlphaGo** is **AlphaGo Zero**. AlphaGo Zero has been trained fully in a self-play fashion, learning from itself completely with no human intervention (Zero comes from this characteristic). It is currently the world's top player at Go and Chess:

Figure 1.37: The Go board

Both AlphaGo and AlphaGo Zero used a deep **Convolutional Neural Network (CNN)** to learn a suitable game representation starting from the "raw" board. This peculiarity shows that a deep CNN can also extract features starting from a sparse representation such as the Go board. One of the main strengths of RL is that it can use, in a transparent way, machine learning models that are widely studied in other fields or problems.

Deep convolutional networks are usually used for classification or segmentation problems that, at first glance, might seem very different from RL problems. Actually, the way CNNs are used in RL is very similar to a classification or a regression problem. The CNN of AlphaGo Zero, for example, takes the raw board representation and outputs the probabilities for each possible action together with the value of each action. It can be seen as a classification and regression problem at the same time. The difference is that the labels, or actions in the case of RL, are not given in the training set, rather it is the algorithm itself that has to discover the real labels through interaction. AlphaGo, the predecessor of AlphaGo Zero, used two different networks: one for action probabilities and another for value estimates. This technique is called actor-critic. The network tasked with predicting actions is called the actor, and the network that has to evaluate actions is called the critic.

DOTA 2

Dota 2 is a complex, real-time strategy game in which there are two teams of five players competing, with each player controlling a "hero." The characteristics of Dota, from an RL perspective, are as follows:

- **Long-Time Horizon**: A Dota game can have around 20,000 moves and can last for 45 minutes. As a reference, a chess game ends before 40 moves and a Go game ends before 150 moves.

- **Partially Observed State**: In Dota, agents can only see a small portion of the full map, that is, only the portion around them. A strong player should make predictions about the position of the enemies and their actions. As a reference, Go and Chess are fully observable games where agents can see the whole situation and the actions taken by the opponents.

- **High-Dimensional and Continuous Action Space**: Dota has a vast number of actions available to each player at each step. The possible actions have been discretized by researchers in around 170,000 actions, with an average of 1,000 possible actions for each step. In comparison, the number of average actions in chess is 35, and in Go, it is 250. With a huge action space, learning becomes very difficult.

- **High-Dimensional and Continuous Observation Space**: While Chess and Go have a discretized observation space, Dota has a continuous state space with around 20,000 dimensions. The state space, as we will learn later in the book, includes all of the information available to players that must be taken into consideration when selecting an action. In a video game, the state space is represented by the characteristics and position of the enemies, the state of the current player, including its ability, its equipment, and its health status, and other domain-specific features.

OpenAI Five, the RL algorithm able to exceed human performance at Dota, is composed of five neural networks collaborating together. The algorithm learns to play by itself through self-play, playing an equivalent of 180 years per day. The algorithm used for training the five neural networks is called **Proximal Policy Optimization**, representing the current state of the art of RL algorithms.

> **NOTE**
>
> To read more on OpenAI Five, refer to the following link:
> https://openai.com/blog/openai-five/

STARCRAFT

StarCraft has characteristics that make it very similar to Dota, including a huge number of moves per play, imperfect information available to players, and highly dimensional state and action spaces. **AlphaStar**, the player developed by DeepMind, is the first artificial intelligence agent able to reach the top league without any game restrictions. AlphaStar uses machine learning techniques such as neural networks, self-play through RL, multi-agent learning methods, and imitation learning to learn from other human players in a supervised way.

> **NOTE**
>
> For further reading on AlphaStar, refer to the following paper:
> https://arxiv.org/pdf/1902.01724.pdf

ROBOT CONTROL

Robots are starting to become ubiquitous nowadays and are widely used in various industries because of their ability to perform repetitive tasks in a precise and efficient way. RL can be beneficial for robotics applications, by simplifying the development of complex behaviors. At the same time, robotics applications represent a set of benchmark and real-world validations for RL algorithms. Researchers test their algorithm on robotic tasks such as locomotion (for example, learning to move) or grasping (for example, learning how to grasp an object). Robotics offers unique challenges, such as the **curse of dimensionality**, the **effective usage of samples** (also called sample efficiency), the possibility of **transferring knowledge** from similar or simulated tasks, and the **need for safety**:

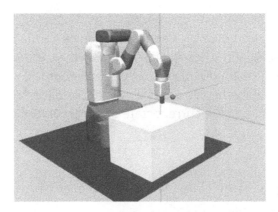

Figure 1.38: A robotic task from the Gym robotics suite

> **NOTE**
>
> The preceding diagram has been sourced from the official documentation for OpenAI Gym: https://gym.openai.com/envs/#robotics
>
> Please refer to the link for more examples of robot control.

The **curse of dimensionality** is a challenge that can also be found in supervised learning applications. Still, in these cases, it is softened by restricting the space of possible solutions to a limited class of functions or by injecting prior knowledge elements in the models through architectural decisions. Robots usually have many degrees of freedom, making the space of possible states and possible actions very large.

Robots interact with the physical environment by definition. The interaction of a real robot with an environment is usually time-consuming, and it can be dangerous. Usually, RL algorithms require millions of samples (or episodes) in order to become efficient. Sample efficiency is a problem in this field, as the required time may be impractical. The usage of collected samples in a smart way is the key to successful RL-based robotics applications. A technique that can be used in these cases is the so-called **sim2real**, in which an initial learning phase is practiced in a simulated environment that is usually safer and faster than the real environment. After this phase, the learned behavior is transferred to the real robot in the real environment. This technique requires a simulated environment that is very similar to the real environment or the generalization capabilities of the algorithm.

AUTONOMOUS DRIVING

Autonomous driving is another exciting application of RL. The main challenge this task presents is the lack of precise specifications. In autonomous driving, it is challenging to formalize what it means to drive well, whether steering in a given situation is good or bad, or whether the driver should accelerate or break. As with robotic applications, autonomous driving can also be hazardous. Testing an RL algorithm, or, in general, a machine learning algorithm, on a driving task, is very problematic and raises many concerns.

Aside from the concerns, the autonomous driving scenario fits very well in the RL framework. As we will explore later in the book, we can think of the driver as the decision-maker. At each step, they receive an observation. The observation includes the road's state, the current velocity, the acceleration, and all of the car's characteristics. The driver, based on the current state, should make a decision corresponding to what to do with the car's commands, steering, brakes, and acceleration. Designing a rule-based system that is able to drive in real situations is complicated, due to the infinite number of different situations to confront. For this reason, a learning-based system would be far more efficient and effective in tasks such as this.

> **NOTE**
>
> There are many simulated environments available for developing efficient algorithms in the context of autonomous driving, listed as follows:
>
> **Voyage Deepdrive:**
> https://news.voyage.auto/introducing-voyage-deepdrive-69b3cf0f0be6
>
> **AWS DeepRacer:** https://aws.amazon.com/fr/deepracer/

In this section, we analyzed some interesting RL applications, the main challenges of them, and the main techniques used by researchers. Games, robotics, and autonomous driving are just some examples of real-world RL applications, but there are many others. In the remainder of this book, we will deep dive into RL; we will understand its components and the techniques presented in this chapter.

SUMMARY

RL is one of the fundamental paradigms under the umbrella of machine learning. The principles of RL are very general and interdisciplinary, and they are not bound to a specific application.

RL considers the interaction of an agent with an external environment, taking inspiration from the human learning process. RL explicitly targets the need to explore efficiently and the exploration-exploitation trade-off appearing in almost all human problems; this is a peculiarity that distinguishes this discipline from others.

We started this chapter with a high-level description of RL, showing some interesting applications. We then introduced the main concepts of RL, describing what an agent is, what an environment is, and how an agent interacts with its environment. Finally, we implemented Gym and Baselines by showing how these libraries make RL extremely simple.

In the next chapter, we will learn more about the theory behind RL, starting with Markov chains and arriving at MDPs. We will present the two functions at the core of almost all RL algorithms, namely the state-value function, which evaluates the goodness of states, and the action-value function, which evaluates the quality of the state-action pair.

2

MARKOV DECISION PROCESSES AND BELLMAN EQUATIONS

OVERVIEW

This chapter will cover more of the theory behind reinforcement learning. We will cover Markov chains, Markov reward processes, and Markov decision processes. We will learn about the concepts of state values and action values along with Bellman equations to calculate previous quantities. By the end of this chapter, you will be able to solve Markov decision processes using linear programming methods.

INTRODUCTION

In the previous chapter, we studied the main elements of **Reinforcement Learning** (**RL**). We described an agent as an entity that can perceive an environment's state and act by modifying the environment state in order to achieve a goal. An agent acts through a policy that represents its behavior, and the way the agent selects an action is based on the environment state. In the second half of the previous chapter, we introduced Gym and Baselines, two Python libraries that simplify the environment representation and the algorithm implementation, respectively.

We mentioned that RL considers problems as **Markov Decision Processes** (**MDPs**), without entering into the details and without giving a formal definition.

In this chapter, we will formally describe what an MDP is, its properties, and its characteristics. When facing a new problem in RL, we have to ensure that the problem can be formalized as an MDP; otherwise, applying RL techniques is impossible.

Before presenting a formal definition of MDPs, we need to understand **Markov Chains** (**MCs**) and **Markov Reward Processes** (**MRPs**). MCs and MRPs are specific cases (simplified) of MDPs. An MC only focuses on state transitions without modeling rewards and actions. Consider the example of the game of snakes and ladders, where the next action is completely dependent on the number displayed on the dice. MRPs also include the reward component in the state transition. MRPs and MCs are useful in understanding the characteristics of MDPs gradually. We will be looking at specific examples of MCs and MRPs later in the chapter.

Along with MDPs, this chapter also presents the concepts of the state-value function and the action-value function, which are used to evaluate how good a state is for an agent and how good an action taken in a given state is. State-value functions and action-value functions are the building blocks of the algorithms used to solve real-world problems. The concepts of state-value functions and action-value functions are highly related to the agent's policy and the environment dynamics, as we will learn later in this chapter.

The final part of this chapter presents two **Bellman equations**, namely the **Bellman expectation equation** and the **Bellman optimality equation**. These equations are helpful in the context of RL in order to evaluate the behavior of an agent and find a policy that maximizes the agent's performance in an MDP.

In this chapter, we will practice with some MDP examples, such as the student MDP and Gridworld. We will implement the solution methods and equations explained in this chapter using Python, SciPy, and NumPy.

MARKOV PROCESSES

In the previous chapter, we described the RL loop as an agent observing a representation of the environment state, interacting with an environment through actions, and receiving a reward based on the action and the environment state. This interaction process is called an MDP. In this section, we will understand what an MDP is, starting with the simplest case of an MDP, an MC. Before describing the various types of MDPs, it is useful to formalize the underlying property of all these processes, the Markov property.

THE MARKOV PROPERTY

Let's start with two examples to help us to understand what the Markov property is. Consider a Rubik's cube. When formalizing the solving of a Rubik's cube as an RL task, we can define the environment state as the state of the cube. The agent can perform actions corresponding to the rotation of the cube's faces. The action results in a state transition that changes the cube. Here, the history is not important – that is, the sequence of actions yielding the current state – in determining the next state. The current state and the present action are the only components that influence the future state:

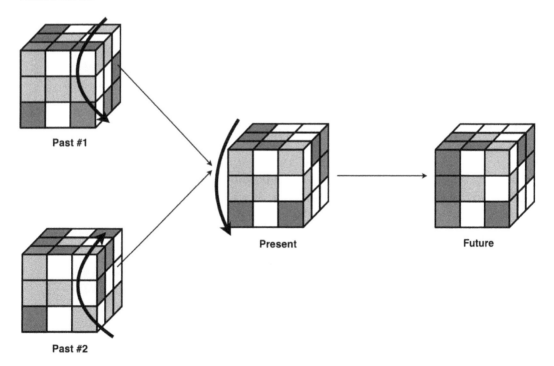

Figure 2.1: Rubik's cube representation

Looking at the preceding figure, suppose the current environment state is the cube with the **Present** label. The current state can be reached by the two states to its left, with the labels **Past #1** and **Past #2**, using two different actions, represented as black arrows. By rotating the face on the left downwards, in the current state, we get the future state on the right, denoted by the label **Future**. The next state, in this case, is independent of the past, in the sense that only the present state and action determine it. It does not matter what the former state was, whether it was **Past #1** or **Past #2**; in both cases, we end up with the same future state.

Let's now consider another classic example: the Breakout game.

Breakout is a classic Atari game. In the game, there is a layer of bricks at the top of the screen; the goal is to break the bricks using a ball, without allowing the ball to touch the bottom of the screen. The player can only move a paddle horizontally. When formalizing the Breakout game as an RL task, we can define the environment state as the image pixels at a certain moment. The agent has at its disposal three possible actions, "Left," "Right," and "None," corresponding to the paddle's movement.

Here, there is a difference with respect to the Rubik's cube example. *Figure 2.2* explains the difference visually. If we represent the environment state using only the current frame, the future is not determined only by the current state and the current action. We can easily visualize this problem by looking at the ball.

In the left part of *Figure 2.2*, we can see two possible past states yielding the same present state. With the arrow, we represent the ball movement. In both cases, the agent's action is "Left."

In the right part of the figure, we have two possible future states, **Future #1** and **Future #2**, starting from the present state and performing the same action (the "Left" action).

By looking only at the current state, it is not possible to decide with certainty which of the two future states will be the next one, as we cannot infer the ball's direction, whether it is going toward the top of the screen or the bottom. We need to know the history, that is, which of the two previous states was the actual previous state, in order to understand what the next state will be.

In this case, the future state is not independent of the past:

> **NOTE**
>
> Notice that the arrow is not actually present in the environment state. We have drawn it in the frame for ease of presentation.

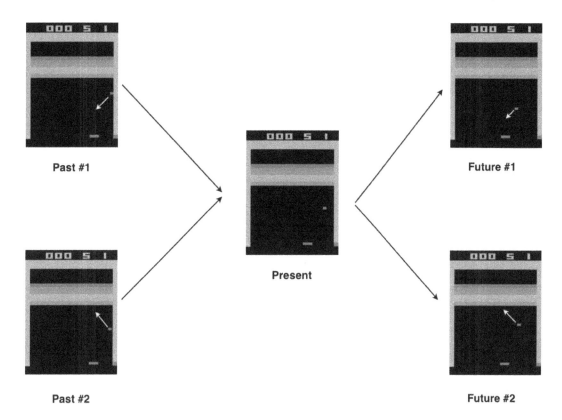

Figure 2.2: Atari game representation

In the Rubik's cube example, the current state contained enough information to determine, together with the current action, the next state. In the Atari example, this is not true. The current state does not contain a crucial piece of information: the movement component. In this case, we need not only the current state but also the past states to determine the next ones.

The Markov property explains exactly the difference between the two examples in mathematical terms. The Markov property states that "the future is independent of the past, given the present."

This means that the future state depends only on the present state, the present state is the only thing influencing the future state, and that we can get rid of the past states. The Markov property can be formalized in the following way:

$$P\Big[\underbrace{s_{t+1}}_{Future}\big|\underbrace{s_t}_{Present}\Big] = P\Big[s_{t+1}\big|\underbrace{s_0, \ldots, s_{t-1}, s_t}_{Past}\Big]$$

Future Present   Past

Figure 2.3: Expression for the Markov property

The probability, $P\big[s_{t+1}\big|s_t\big]$, of the next state, s_{t+1}, given the current one, s_t, is equal to the probability of the next state given the state history, $P\big[s_{t+1}\big|s_0, \ldots, s_t\big]$. This means that the past states, $s_0, \ldots, s_{t-1}$, have no influence over the next state distribution.

In other words, to describe the probability distribution of the next state, we only need the information contained in the current state. Almost all RL environments, being MDPs, assume that the Markov property holds true. We need to remember this property when designing RL tasks; otherwise, the main RL assumptions won't be true anymore, causing the algorithms to fail miserably.

> **NOTE**
>
> In statistical language, the term "given" means that the probability is influenced by some information. In other words, the probability function depends on some other information.

Most of the time, the Markov property holds true; however, there are cases in which we need to design the environment state to ensure the independence of the next state from the past states. This is exactly the case in Breakout. To restore the Markov property, we can define the state as multiple consequent frames so that it is possible to infer the ball direction. Refer to the following figure for a visual representation:

Figure 2.4: Markov state for Breakout

As you can see in the preceding figure, the state is represented by three consequent frames.

> **NOTE**
>
> There are other tricks you can use to restore the Markov property. One of these tricks consists of using policies represented as **Recurrent Neural Networks** (**RNNs**). Using RNNs, the agent can also take into account past states when determining the current action. The usage of RNNs as RL policies will be discussed later on in the book.

In the context of MDPs, the probability of the next state given the current one, $P\big[s_{t+1}\big|s_t\big]$ is referred to as a transition function.

If the state space is finite, composed of *N* states, we can arrange the transition functions evaluated for each couple of states in an N x N matrix, where the sum of all the columns is 1, as we are summing a probability distribution over transition function elements:

$$
P = \begin{array}{cc} & \text{State 1} \ldots \text{State n} \\ \begin{array}{c} \text{State 1} \\ \vdots \\ \text{State n} \end{array} & \begin{pmatrix} P_{11} & \cdots & P_{1n} \\ \vdots & \ddots & \vdots \\ P_{n1} & \cdots & P_{nn} \end{pmatrix} \end{array}
$$

Figure 2.5: Transition probability matrix

In the rows, we have the source states, and in the columns, we have the destination states.

The probability matrix summarizes the transition function. It can be read as follows: P_{ij} is the probability of landing in state j starting from state i.

MARKOV CHAINS

An MC, or, simply, a Markov process, is defined as a tuple of state space s and transition function P. The state space, together with the transition function, defines a memory-less sequence of random states, $s_{0,\ \ldots,}\ s_T$, satisfying the Markov property. A sample from a Markov process is simply a sequence of states, which is also called an episode in the context of RL:

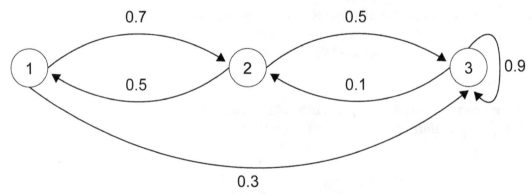

Figure 2.6: MC with three states

Consider the preceding MC. As you can see, we have three states represented by circles. The probability function evaluated for the state pairs is reported on the edges connecting the different states. Looking at the edges starting from each state, we can see that the sum of the probabilities associated with each edge is 1, as it defines a probability distribution. The transition function for a couple of states that are not linked by an edge is 0.

The transition function can be arranged in a matrix, as follows:

$$P = \begin{array}{c} \\ State\ 1 \\ State\ 2 \\ State\ 3 \end{array} \begin{array}{ccc} State\ 1 & State\ 2 & State\ 3 \\ \left[\begin{array}{ccc} 0 & 0.7 & 0.3 \\ 0.5 & 0 & 0.5 \\ 0 & 0.1 & 0.9 \end{array}\right] \end{array}$$

Figure 2.7: Transition matrix for the MC in Figure 2.6

The matrix form of the transition function is very convenient from a programming perspective as it allows us to perform calculations easily.

MARKOV REWARD PROCESSES

An MRP is an MC with values associated with state transitions, called rewards. The reward function evaluates how useful it is to transition from one state to another.

An MRP is a tuple of $\langle s, P, R, \gamma \rangle$ such that the following is true:

- S is a finite set of states.

- P is the transition probability, where $P(s, s') = \mathbb{P}\left[s_{t+1} = s' \middle| s_t = s\right]$ is the probability of transitioning from state s to state s'.

- R is a reward function, where $R(s, s')$ is the reward associated with the transition from state S to state s'.

- γ is the discount factor associated with future rewards, $\gamma \in [0, 1]$:

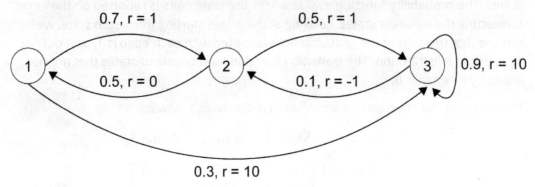

Figure 2.8: An example of an MRP

As you can see in the previous figure, the rewards are represented by **r** and are associated with state transitions.

Let's consider the MRP in *Figure 2.8*. The highest reward (**10**) is associated with transitions *1->3* and the self-loop, *3->3*. The lowest reward is associated with transitions *3->2*, and it is equal to **-1**.

In an MRP, it is possible to calculate the discounted return as the cumulative sum of discounted rewards.

In this context, we use the term "trajectory" or "episode" to denote a sequence of states traversed by the process.

Let's now calculate the discounted return for a given trajectory; for example, the trajectory of 1-2-3-3-3 with discount factor $\gamma = 0.9$.

The discounted return is as follows:

$$G_t = \sum_{k=0}^{T} \gamma^k r_{t+k+1}$$

$$G_t = r(s_1, s_2) + \gamma \cdot r(s_2, s_3) + \gamma^2 \cdot r(s_3, s_3) + \gamma^3 \cdot r(s_3, s_3)$$

$$G_t = 1 + 0.9 \cdot 1 + 0.9^2 \cdot 10 + 0.9^3 \cdot 10 = 17.29$$

Figure 2.9: Discounted return for the trajectory of 1-2-3-3-3

We can also calculate the discounted return for a different trajectory, for example, 1-3-3-3-3:

$$G_t = 10 + 0.9 * 10 + 0.9^2 * 10 + 0.9^3 * 10 = 34.39$$

Figure 2.10: Discounted return for the trajectory of 1-3-3-3-3

In this example, the second trajectory is more convenient than the first one, having a higher return. This means that the associated path is better in comparison to the first one. The return does not represent an absolute feature of a trajectory; it represents the relative goodness with respect to the other trajectories. Trajectory returns of different MRPs are not comparable to each other.

Considering an MRP composed of N states, the reward function can be represented in an N x N matrix, similar to the transition matrix:

$$R = \begin{matrix} R_{11} & \cdots & R_{1n} \\ \vdots & \ddots & \vdots \\ R_{n1} & \cdots & R_{nn} \end{matrix}$$

Figure 2.11: Reward matrix

In the rows, we represent the source states, and in the columns, we represent the destination states.

The reward matrix can be read as follows: R_{ij} is the reward associated with the state transition, $i \rightarrow j$.

For the example in *Figure 2.11*, the reward function arranged in a matrix is as follows:

$$R = \begin{array}{c} \\ State1 \\ State2 \\ State3 \end{array} \begin{array}{ccc} State1 & State2 & State3 \\ \begin{bmatrix} 0 & 1 & 10 \\ 0 & 0 & 1 \\ 0 & -1 & 10 \end{bmatrix} \end{array}$$

Figure 2.12: Reward matrix for the MRP example

When a reward is not specified, we assume that the reward is **0**.

Using Python and NumPy, we can represent the transition matrix in this way:

```
n_states = 3
P = np.zeros((n_states, n_states), np.float)
P[0, 1] = 0.7
P[0, 2] = 0.3
P[1, 0] = 0.5
P[1, 2] = 0.5
P[2, 1] = 0.1
P[2, 2] = 0.9
```

In a similar way, the reward matrix can be represented like this:

```
R = np.zeros((n_states, n_states), np.float)
R[0, 1] = 1
R[0, 2] = 10
R[1, 0] = 0
R[1, 2] = 1
R[2, 1] = -1
R[2, 2] = 10
```

We are now ready to introduce the concepts of value functions and Bellman equations for MRPs.

VALUE FUNCTIONS AND BELLMAN EQUATIONS FOR MRPS

The **value function** in an MRP evaluates the long-term value of a given state, intended as the expected return starting from that state. In this way, the value function expresses a preference over states. A state with a higher value in comparison to another state represents a better state – in other words, a state that it is more rewarding to be in.

Mathematically, the value function is formalized as follows:

$$v(s) = \mathbb{E}\left[G_t \middle| s_t = s\right]$$

Figure 2.13: Expression for the value function

The value function of state s is represented by $v(s)$. The expectation on the right side of the equation is the expected value, represented by $\mathbb{E}$ of the return, G_t, considering the fact that the current state is precisely equal to state s – the state for which we are evaluating the value function. The expectation is taken according to the transition function.

The value function can be decomposed into two parts by considering the immediate reward and the discounted value function of the successor state:

$$v(s) = \mathbb{E}\left[r_{t+1} + \gamma\, r_{t+2} + \gamma^2 r_{t+3} + \cdots \,\middle|\, s_t = s\right]$$

$$v(s) = \mathbb{E}\left[r_{t+1} + \gamma\left(r_{t+2} + \gamma\, r_{t+3} + \cdots\right)\middle|\, s_t = s\right]$$

$$v(s) = \mathbb{E}\left[r_{t+1} + \gamma\, G_{t+1}\,\middle|\, s_t = s\right]$$

$$v(s) = \mathbb{E}\left[r_{t+1} + \gamma\, v(s_{t+1})\,\middle|\, s_t = s\right]$$

Figure 2.14: Decomposition of the value function

The last equation is a recursive equation, known as the **Bellman expectation equation for MRPs**, in which the value function of given states depends on the value function of the successor states.

To highlight the dependency of the equation on the transition function, we can rewrite the expectation as a summation of the possible states weighted by the transition probability. We define with $R_{\mathbb{E}_s}$ the expectation of the reward function in state s, which can also be defined as the average reward.

We can write $R_{\mathbb{E}_s}$ in the following ways:

$$R_{\mathbb{E}_s} = \mathbb{E}\left[R(s_t, s_{t+1})\middle|\, s_t = s\right]$$

$$R_{\mathbb{E}_s} = \sum_{s' \in s} P(s'\,|\,s)\, R(s, s')$$

Figure 2.15: Expression for the expectation of the reward function in state s

We can now rewrite the value function in a more convenient way:

$$v(s) = R_{\mathbb{E}_s} + \gamma \sum_{s' \in S} P(s' | s) v(s')$$

Figure 2.16: Revised expression for the expectation of the value function in state s

This expression can be translated into code, as follows:

```
R_expected = np.sum(P * R, axis=1, keepdims=True)
```

In the preceding code, we calculated the expected reward for each state by multiplying element-wise the probability matrix and the reward matrix. Please note that the **keepdims** parameter is required to obtain a column vector.

This formulation makes it possible to rewrite the Bellman equation using matrix notation:

$$V = R_{\mathbb{E}} + \gamma P V$$

Figure 2.17: Matrix form of the Bellman equation

Here, **V** is a column vector with state values, $R_{\mathbb{E}}$ is the expected reward for each state, and **P** is the transition matrix:

$$V = \begin{bmatrix} V_{s_1} \\ \vdots \\ V_{s_n} \end{bmatrix}, R_{\mathbb{E}} = \begin{bmatrix} R_{\mathbb{E}_{s_1}} \\ \vdots \\ R_{\mathbb{E}_{s_2}} \end{bmatrix}$$

Figure 2.18: Matrix form of the Bellman equation

Using matrix notation, it is also possible to solve the Bellman equation for **V**, finding the value function associated with each state:

$$(I - \gamma P) V = R_{\mathbb{E}}$$

$$V = (I - \gamma P)^{-1} R_{\mathbb{E}}$$

Figure 2.19: Value function using the Bellman equation

Here, **I** is an identity matrix of size N x N, and **N** is the number of states in the MRP.

SOLVING LINEAR SYSTEMS OF AN EQUATION USING SCIPY

SciPy (https://github.com/scipy/scipy) is a Python library used for scientific computing based on NumPy. SciPy offers, inside the **linalg** module (linear algebra), useful methods for solving systems of equations.

In particular, we can use **linalg.solve(A, b)** to solve a system of equations in the form of $Ax = b$. This is precisely the method we can use to solve the system $(I - \gamma P) V = R_{\mathbb{E}}$, where $(I - \gamma P)$ is the matrix, **A**; **V** is the vector of variables, **x**; and $R_{\mathbb{E}}$ is the vector, **b**.

When translated into code, it should look like this:

```
gamma = 0.9
A = np.eye(n_states) - gamma * P
B = R_states
# solve using scipy linalg solve
V = linalg.solve(A, B)
```

As you can see, we have declared the elements of the Bellman equation and are using **scipy.linalg** to calculate the **value** function.

Let's now strengthen our understanding further by completing an exercise.

EXERCISE 2.01: FINDING THE VALUE FUNCTION IN AN MRP

In this exercise, we are going to solve the Bellman expectation equation by finding the value function for the MRP in the following figure. We will use **scipy** and the **linalg** module to solve the linear equation presented in the previous section. We will also demonstrate how to define a transition probability matrix and how to calculate the expected reward for each state:

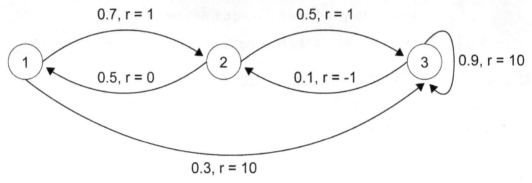

Figure 2.20: Example of an MRP with three states

1. Import the required NumPy and SciPy packages:

```
import numpy as np
from scipy import linalg
```

2. Define the transition probability matrix:

```
# define the Transition Probability Matrix
n_states = 3
P = np.zeros((n_states, n_states), np.float)
P[0, 1] = 0.7
P[0, 2] = 0.3
P[1, 0] = 0.5
P[1, 2] = 0.5
P[2, 1] = 0.1
P[2, 2] = 0.9
print(P)
```

You should obtain the following output:

```
array([[0. , 0.7, 0.3],
       [0.5, 0. , 0.5],
       [0. , 0.1, 0.9]])
```

Let's check the correctness of the matrix. The probability of going from state 1 to state 1 is $P_{1,1}=0$. This is correct as there are no self-loops in state 1. The probability of going from state 1 to state 2 is $P_{1,2}=0.7$ as it's the probability associated with edge *1->2*. This can be done for all elements of the transition matrix. Note that, here, the transition matrix elements are indexed by the state, not by their position in the matrix. This means that with $P_{1,1}$, we refer to element 0,0 of the matrix.

3. Check that the sum of all the columns is exactly equal to **1**, being a probability matrix:

```
# the sum over columns is 1 for each row being a probability matrix
assert((np.sum(P, axis=1) == 1).all())
```

The **assert** function is used to ensure that a particular condition will return **true**. In this case, the **assert** function will make sure that the sum of all the columns is exactly **1**.

4. We can calculate the expected immediate reward for each state using the reward matrix and the transition probability matrix:

```
# define the reward matrix
R = np.zeros((n_states, n_states), np.float)
R[0, 1] = 1
R[0, 2] = 10
R[1, 0] = 0
R[1, 2] = 1
R[2, 1] = -1
R[2, 2] = 10
"""

calculate expected reward for each state by multiplying the
probability matrix for each reward
"""

#keepdims is required to obtain a column vector
R_expected = np.sum(P * R, axis=1, keepdims=True)


# The matrix R_expected
R_expected
```

You should obtain the following column vector:

```
array([[3.7],
       [0.5],
       [8.9]])
```

The **R_expected** vector is the expected immediate reward for each state. State 1 has an expected reward of **3.7**, which is exactly equal to *0.7 * 1 + 0.3*10*. The same logic applies to state 2 and state 3.

5. Now we need to define **gamma**, and we are ready to solve the Bellman equation as a linear equation, $Ax=b$. We have $A=(I-\gamma P)$ and $b=R_E$:

```
# define the discount factor
gamma = 0.9
# Now it is possible to solve the Bellman Equation
A = np.eye(n_states) - gamma * P
B = R_expected
# solve using scipy linalg
V = linalg.solve(A, B)
V
```

You should obtain the following output:

```
array([[65.540732  ],
       [64.90791027],
       [77.5879575 ]])
```

The vector, **V**, represents the value for each state. State 3 has the highest value (**77.58**). This means that state 3 is the state providing the highest expected return. It is the best state in this MRP. Intuitively, state 3 is the best state because, with a high probability (0.9), the transition brings the agent to the same state, and the reward associated with the transition is high (+10).

> **NOTE**
>
> To access the source code for this specific section, please refer to https://packt.live/37o5ZH4.
>
> You can also run this example online at https://packt.live/3dU8cfW.

In this exercise, we solved the Bellman equation for an MRP by finding the state values for our toy problem. The state values describe quantitatively the benefit of being in each state. We described the MRP in terms of a transition probability matrix and a reward matrix. These two matrices permit us to solve the linear system associated with the Bellman equation.

> **NOTE**
>
> The computational complexity of the solution of the Bellman equation is $O(n^3)$; it is cubic in the number of states. Therefore, it is only possible for small MRPs.

In the next section, we will consider an active agent that can perform actions, thus arriving at the description of an MDP.

MARKOV DECISION PROCESSES

An MDP is an MRP with decisions. In this context, we have a set of actions available to an agent that can condition the transition probability to the next state. While, in MRPs, the transition probability depends only on the state of the environment, in MDPs, the agent can perform actions influencing the transition probability. In this way, the agent becomes an active entity in the framework, interacting with the environment through actions.

Formally, an MDP is a tuple, $\langle S, A, R, P, \gamma \rangle$, in which the following is true:

- S is the set of states.

- A is the set of actions.

- R is the reward function, $R(s,a) = \mathbb{E}\left[r_{t+1} | s_t = s, a_t = a\right]$. $R(s,a)$ is the expected reward resulting in action a and state S.

- P is the transition probability function in which $P(s,a,s') = \mathbb{P}\left[s_{t+1} = s' | s_t = s, a_t = a\right]$ is the probability of landing in state S' starting from the current state, S, and performing an action, a.

- γ is the discount factor associated with future rewards, $\gamma \in [0, 1]$.

The difference between an MRP and an MDP is the fact that the agent has at its disposal a set of actions from which it can choose to condition the transition probability to have a higher possibility of landing in good states. If an MRP and MC are only a description of Markov processes without an objective, an MDP contains the concept of a policy and a goal. In an MDP, the agent should take decisions about which action to take, maximizing the discounted return:

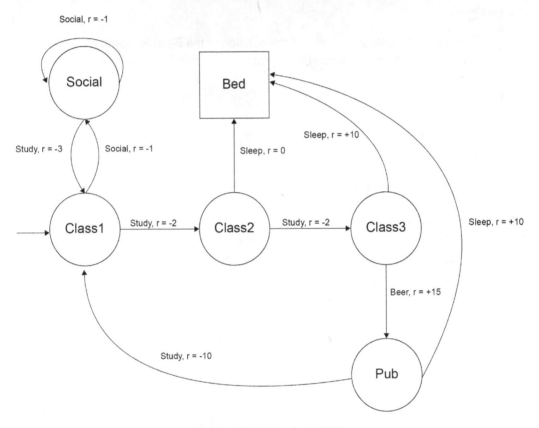

Figure 2.21: A student MDP

Figure 2.21 is an example of an MDP representing the day of a university student. There are six possible states: **Class 1, Class 2, Class 3, Social, Bed,** and **Pub**. The edges between the states represent state transitions. On the edges, we have the action and the reward, denoted by **r**. Possible actions are **Study, Social, Beer,** and **Sleep**. The initial state, represented by the incoming arrow, is **Class 1**. The goal of the student is to select the best actions in each state, maximizing their return.

In the following paragraphs, we will discuss some possible strategies for this MDP.

A student agent starts from **Class 1**. They can decide to study and complete all of the lessons. Each study decision comes with a small negative reward, **-2**. If the student decides to sleep after **Class 3**, they will land in the absorbing state, **Bed**, with a high positive reward of **+10**. This represents a very common situation in daily routines. You have to sacrifice some immediate reward in order to obtain a higher reward in the future. In this case, by deciding to study in **Class 1** and **2**, you obtain a negative reward but are compensated by the positive reward after **Class 3**.

Another possible strategy in this MDP is to select a **Social** action right after the **Class 1** state. This action comes with a small negative reward. The student can continue doing the same action, and each time they get the same reward. The student can also decide to **Study** from the **Social** state (notice that **Social** is both a state and an action) by returning to **Class 1**. Feeling guilty, in **Class 1**, the student can decide to study. After having studied a bit, they may feel tired and decide to sleep for a little while, ending up in the **Bed** state. Having performed the **Social** action, the agent has cumulated a negative return.

Let's evaluate the possible strategies for this example. We will assume a discount factor of $\gamma = 1$, that is, no discount:

- Strategy: Good student. The good student strategy was the first strategy that was described. Supposing the student will end in **Class 1**, they can perform the following actions: **Study, Study**, and **Study**. The associated sequence of states is thus **Class 1, Class 2, Class 3**, and **Sleep**. The associated return is, therefore, the sum of the rewards along the trajectory:

$$G_{\text{good student}} = (-2) + (-2) + (+10) = +6$$

Figure 2.22: Return for the good student

- Strategy: Social student. The social student strategy is the second strategy described. The student can perform the following actions: **Social, Social, Social, Study, Study**, and **Sleep**. The associated sequence of states is **Class 1, Social, Social, Social, Class 1, Class 2**, and **Bed**. The associated return is, in this case, as follows:

$$G_{\text{social student}} = (-1) + (-1) + (-1) + (-3) + (-2) + 0 = -8$$

Figure 2.23: Return for the social student

By looking at the associated return, we can see that the good student strategy is a better strategy in comparison to the social student strategy, having a higher return.

The question you may ask at this point is how can an agent decide which action to take in order to maximize the return? To answer the question, we need to introduce two useful functions: the state-value function and the action-value function.

THE STATE-VALUE FUNCTION AND THE ACTION-VALUE FUNCTION

In the context of MDPs, we can define a function by evaluating how good it is to be in a given state. However, we should take into account the agent's policy, as it defines the agent's decisions and conditions the probability over trajectories, that is, the sequence of future states. So, the value function depends on the agent policy, π.

The **state-value function**, $v_\pi(s)$, of an MDP can be defined as the expected return that starts from state s and follows the policy, π:

$$v_\pi(s) = \mathbb{E}_\pi = \left[G_t \middle| s_t = s \right]$$

Figure 2.24: Definition of the state-value function

In MDPs, we are also interested in defining the benefit of taking an action in a given state. This function is called the action-value function.

The **action-value function**, $q_\pi(s, a)$, (also called the q-function), can be termed as the expected return starting from state s, which takes action a and follows the policy π:

$$q_\pi(s, a) = \mathbb{E}_\pi \left[G_t \middle| s_t = s, a_t = a \right]$$

Figure 2.25: Definition of the action-value function

The state-value function, as we will learn later in the book, provides information that is only useful when it comes to evaluating a policy. The action-value function also provides information about control, that is, for selecting an action in a state.

Suppose that we know the action-value function for an MDP. If we are in given state, S, which action would be the best one?

Well, the best action is the one that yields the highest discounted return. The action-value function measures the discounted return that is obtained by starting from a state and performing an action. In this way, the action-value function provides an ordering (or a preference) over the actions in a state. The best action to perform is the one with the highest q-function:

$$a^*(s) = argmax_a \ q_\pi(s, a)$$

Figure 2.26: Best action using the action-value function

Note that, in this case, we are only doing a one-step optimization of the current policy; that is, we are modifying, possibly, the action in a given state under the assumption that the following actions are taken with the current policy. If we do this, we do not select the best action in this state, but we select the best action under this policy.

Just like in an MRP, in an MDP, the state-value function and the action-value function can be decomposed in a recursive way:

$$v_\pi(s) = \mathbb{E}_\pi \left[r_{t+1} + \gamma v_\pi(s_{t+1}) \middle| s_t = s \right]$$

Figure 2.27: The state-value function in an MDP

$$q_\pi(s, a) = \mathbb{E}_\pi \left[R_{t+1} + \gamma q_\pi(s_{t+1}, A_{t+1}) \middle| s_t = s, A_t = a \right]$$

Figure 2.28: The action-value function in an MDP

These equations are known as the Bellman expectation equations for MDPs.

Bellman expectation equations are recursive as the state-value function of a given state depends on the state-value function of another state. This is also true for the action-value function.

In the action-value function equation, the action, a, for which we are evaluating the function, is an arbitrary action. It is not taken from the action distribution defined by the policy. Instead, the action, A_{t+1}, taken in the following step, is taken according to the action distribution defined in state s_{t+1}.

Let's rewrite the state-value function and the action-value function to highlight the contribution of the agent's policy, π:

$$v_\pi(s) = \sum_{a \in A} \pi(a|s) \left(R(s, a) + \gamma \sum_{s' \in S} P(s, a, s') v_\pi(s') \right)$$

Figure 2.29: The state-value function to highlight the policy contribution

Let's analyze the two terms of the equation:

- $\sum_a \pi(a|s) R(s, a)$: This term is the expectation of the immediate rewards given the action distribution defined by the agent's policy. Each immediate reward for a state-action pair is weighted by the probability of the action given the state, which is defined as $\pi(s, a)$.

- $\gamma \sum \pi(a|s) \sum_{s' \in S} P(s, a, s') v_\pi(s')$ $v_\pi(s')$ is the discounted expected value of the state-value function, given the state distribution defined by the transition function. Note that here the action, **a**, is defined by the agent's policy. Being an expected value, every state value, $v_\pi(s')$, is weighed by the probability of the transition from state s to state s', given the action, a. This is represented by $P(s, a, s') = P(s_{t+1} = s' | s_t = s, a_t = a)$.

The action-value function can be rewritten to highlight the dependency on the transition and value functions:

$$q_\pi(s, a) = R(s, a) + \gamma \sum_{s' \in S} P(s, a, s') v_\pi(s')$$

Figure 2.30: The action-value function, highlighting the dependency on the transition and value functions of the next state

The action-value function, therefore, is given by the summation of the immediate reward and the expected value of the state-value function of the successor state under the environment dynamic (**P**).

By comparing the two equations, we obtain an important relationship between the state value and the action value:

$$v_\pi(s) = \sum_{a \in A} \pi(a|s) \, q_\pi(s, a)$$

Figure 2.31: Expression for the state-value function, in terms of the action-value function

In other words, the state-value state, s, under the policy, π, is the expected value of the action-value function under the actions selected by π. Each action-value function is weighted by the probability of the action given the state.

The state-value function can also be rewritten in matrix form, as in the MRP case:

$$v_\pi = R_\pi + \gamma \, P_\pi \, V_\pi$$

Figure 2.32: Matrix form for the state-value function

There is a direct solution, as follows:

$$V_\pi = \left(1 - \gamma P_\pi \right)^{-1} R_\pi$$

Figure 2.33: Direct solution for the state values

Here, you can see the following:

- R_π (column vector) is the expected value of the immediate reward induced by the policy for each state:

$$R_\pi(s) = \sum_{a \in A} \pi(a|s) \, R(s, a)$$

Figure 2.34: Expected immediate reward

- v_π is the column vector of the state values for each state.

- P_π is the transition matrix based on the action distribution. It is an $|S| \times |S|$ matrix, where $|S|$ is the number of states in the MDP. Given two states, s_i and s_j, we have the following:

$$P_\pi\left(s_i, s_j\right) = \sum_{a \in A} \pi\left(a \mid s_i\right) P\left(s_i, a, s_j\right)$$

Figure 2.35: Transition matrix conditioned on an action distribution

Therefore, the transition matrix is the probability of transitioning from state s_i to state s_j given the actions selected by the policy and the transition function defined by the MDP.

Following the same steps, we can also find the matrix form of the action-value function:

$$Q_\pi = R + \gamma P V_\pi$$

Figure 2.36: Matrix form equation for the action-value function

Here, Q_π is a column vector with $|s| \cdot |A|$ entries. R is the vector of immediate rewards with the same shape of Q_π. P is the transition matrix with a shape of $|s| \cdot |A|$ rows and $|S|$ columns. V_π represents the state values for each state.

The explicit form of Q_π and P is as follows:

$$Q_\pi = \begin{bmatrix} q_\pi\left(s_1, a_1\right) \\ \vdots \\ q_\pi\left(s_1, a_m\right) \\ \vdots \\ q_\pi\left(s_n, a_m\right) \end{bmatrix} \quad p = \begin{bmatrix} P\left(s_1, a_1, s_1\right) & \cdots & P\left(s_1, a_{m_1}, s_n\right) \\ \vdots & \ddots & \vdots \\ P\left(s_n, a_1, s_1\right) & \cdots & P\left(s_n, a_{m_n}, s_n\right) \end{bmatrix}$$

Figure 2.37: Explicit matrix form of the action-value function and the transition function

Here, the number of actions associated with state i is indicated by m_i, thus $|A| = \sum_i m_i$. The number of actions of the MDP is obtained by summing up the actions associated with each state.

Let's now implement our understanding of the state- and action-value functions for our student MDP example. In this example, we will use the calculation of the state-value function and the action-value function for the student MDP in *Figure 2.21*. We will consider the case of an undecided student, that is, a student with a random policy for each state. This means that the probability of each action for each state is exactly 0.5.

We will examine a different case for a myopic student in the following example.

Import the required libraries as follows:

```
import numpy as np
from scipy import linalg
```

Define the environment properties:

```
n_states = 6
# transition matrix together with policy
P_pi = np.zeros((n_states, n_states))
R = np.zeros_like(P_pi)
```

P_pi contains the contribution of the transition matrix and the policy of the agent. **R** is the reward matrix.

We will use the following state encoding:

- **0**: Class 1
- **1**: Class 2
- **2**: Class 3
- **3**: Social
- **4**: Pub
- **5**: Bed

Create the transition matrix by considering a random policy:

```
P_pi[0, 1] = 0.5
P_pi[0, 3] = 0.5
P_pi[1, 2] = 0.5
P_pi[1, 5] = 0.5
P_pi[2, 4] = 0.5
P_pi[2, 5] = 0.5
P_pi[4, 5] = 0.5
P_pi[4, 0] = 0.5
P_pi[3, 0] = 0.5
P_pi[3, 3] = 0.5
P_pi[5, 5] = 1
```

Print **P_pi**:

```
P_pi
```

The output will be as follows:

```
array([[0. , 0.5, 0. , 0.5, 0. , 0. ],
       [0. , 0. , 0.5, 0. , 0. , 0.5],
       [0. , 0. , 0. , 0. , 0.5, 0.5],
       [0.5, 0. , 0. , 0.5, 0. , 0. ],
       [0.5, 0. , 0. , 0. , 0. , 0.5],
       [0. , 0. , 0. , 0. , 0. , 1. ]])
```

Create the reward matrix, **R**:

```
R[0, 1] = -2
R[0, 3] = -1
R[1, 2] = -2
R[1, 5] = 0
R[2, 4] = 15
R[2, 5] = 10
R[4, 5] = 10
R[4, 0] = -10
R[3, 3] = -1
R[3, 0] = -3
```

Print **R**:

```
R
```

The output will be as follows:

```
array([[  0.,  -2.,   0.,  -1.,   0.,   0.],
       [  0.,   0.,  -2.,   0.,   0.,   0.],
       [  0.,   0.,   0.,   0.,  15.,  10.],
       [ -3.,   0.,   0.,  -1.,   0.,   0.],
       [-10.,   0.,   0.,   0.,   0.,  10.],
       [  0.,   0.,   0.,   0.,   0.,   0.]])
```

Being a probability matrix, the sum of all the columns of **P_pi** should be **1**:

```
# check the correctness of P_pi
assert((np.sum(P_pi, axis=1) == 1).all())
```

The assertion should be verified.

We can now calculate the expected reward for each state, using **R** and **P_pi**:

```
# expected reward for each state
R_expected = np.sum(P_pi * R, axis=1, keepdims=True)
R_expected
```

The expected reward, in this case, is as follows:

```
array([[-1.5],
       [-1. ],
       [12.5],
       [-2. ],
       [ 0. ],
       [ 0. ]])
```

The **R_expected** vector contains the expected immediate reward for each state.

We are ready to solve the Bellman equation to find the value for each state. For this, we can use **scipy.linalg.solve**:

```
# Now it is possible to solve the Bellman Equation
gamma = 0.9
A = np.eye(n_states, n_states) - gamma * P_pi
B = R_expected
# solve using scipy linalg
V = linalg.solve(A, B)
V
```

The vector, **V**, contains the following values:

```
array([[-1.78587056],
       [ 4.46226255],
       [12.13836121],
       [-5.09753046],
       [-0.80364175],
       [ 0.        ]])
```

This is the vector of the state values. State **0** has a value of **−1.7**, state **1** has a value of **4.4**, and so on:

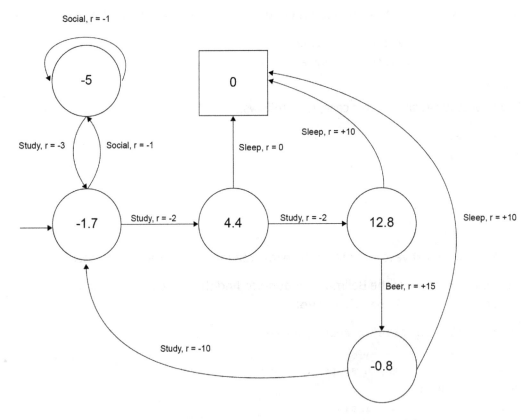

Student MDP γ=0.9

Figure 2.38: State values of the student MDP for $\gamma=0.9$

Let's examine how the results change with $\gamma = 0$, which is the condition assumed for a myopic random student:

```
gamma = 0.
A = np.eye(n_states, n_states) - gamma * P_pi
B = R_expected
# solve using scipy linalg
V_gamma_zero = linalg.solve(A, B)
V_gamma_zero
```

The output will be as follows:

```
array([[-1.5],
       [-1. ],
       [12.5],
       [-2. ],
       [ 0. ],
       [ 0. ]])
```

The visual representation is as follows:

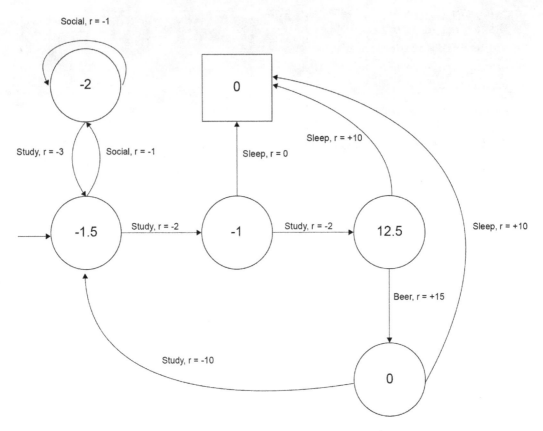

Figure 2.39: State values of the student MDP for γ=0

As you can see, using $\gamma=0$, the value of each state is exactly equal to the expected immediate reward according to the policy.

Now we can calculate the action-value function. We need to use a different form of immediate reward using a matrix with a shape of $(|S| \cdot |A|, 1)$. Each row corresponds to a state-action pair, and the value is the immediate reward for that pair:

$$R(s, a) = \begin{bmatrix} R(\text{Class}1, \text{Study}) \\ R(\text{Class}1, \text{Social}) \\ R(\text{Class}2, \text{Study}) \\ R(\text{Class}2, \text{Sleep}) \\ \vdots \end{bmatrix}$$

Figure 2.40: Immediate rewards

Translate it into code as follows:

```
R_sa = np.zeros((n_states*2, 1))
R_sa[0] = -2 # study in state 0
R_sa[1] = -1 # social in state 0
R_sa[2] = -2 # study in state 1
R_sa[3] = 0 # sleep in state 1
R_sa[4] = 10 # sleep in state 2
R_sa[5] = +15 # beer in state 2
R_sa[6] = -1 # social in state 3 (social)
R_sa[7] = -3 # study in state 3 (social)
R_sa[8] = 10 # sleep in state 4 (pub)
R_sa[9] = -10 # study in state 4 (pub)
R_sa.shape
```

The output will be as follows:

```
(10, 1)
```

We now have to define the transition matrix of the student MDP. The transition matrix contains the probability of landing in a given state, starting from a state and an action. In the rows, we have the source state and action, and in the columns, we have the landing state:

$$
P(s, a, s') =
\begin{array}{c}
\\
\text{Class 1, Study} \\
\text{Class 1, Social} \\
\text{Class 2, Study} \\
\text{Class 2, Sleep} \\
\text{Class 3, Sleep} \\
\text{Class 3, Beer} \\
\text{Social, Social} \\
\text{Social, Study} \\
\text{Pub, Sleep} \\
\text{Pub, Study}
\end{array}
\begin{array}{cccccc}
\text{Class 1} & \text{Class 2} & \text{Class 3} & \text{Social} & \text{Pub} & \text{Bed} \\
& 1 & & & & \\
& & & 1 & & \\
& & 1 & & & \\
& & & & & 1 \\
& & & & & 1 \\
& & & & 1 & \\
& & & 1 & & \\
1 & & & & & \\
& & & & & 1 \\
1 & & & & &
\end{array}
$$

Figure 2.41: Transition matrix of the student MDP

When translating the probability transition matrix into code, you should see the following:

```
# Transition Matrix (states x action, states)
P = np.zeros((n_states*2, n_states))
P[0, 1] = 1 # study in state 0 -> state 1
P[1, 3] = 1 # social in state 0 -> state 3
P[2, 2] = 1 # study in state 1 -> state 2
```

```
P[3, 5] = 1 # sleep in state 1 -> state 5 (bed)
P[4, 5] = 1 # sleep in state 2 -> state 5 (bed)
P[5, 4] = 1 # beer in state 2 -> state 4 (pub)
P[6, 3] = 1 # social in state 3 -> state 3 (social)
P[7, 0] = 1 # study in state 3 -> state 0 (Class 1)
P[8, 5] = 1 # sleep in state 4 -> state 5 (bed)
P[9, 0] = 1 # study in state 4 -> state 0 (class 1)
```

We can now calculate the action-value function using $\gamma = 0.9$:

```
gamma = 0.9
Q_sa_pi = R_sa + gamma * P @ V
Q_sa_pi
```

The action-value vector contains the following values:

```
array([[  2.01603629],
       [ -5.58777741],
       [  8.92452509],
       [  0.        ],
       [ 10.        ],
       [ 14.27672242],
       [ -5.58777741],
       [ -4.60728351],
       [ 10.        ],
       [-11.60728351]])
```

Q_sa_pi is the action-value vector. For each state-action pair, we have the value of the action in that state. The action-value function is represented in the following figure. Action values are represented with q_π:

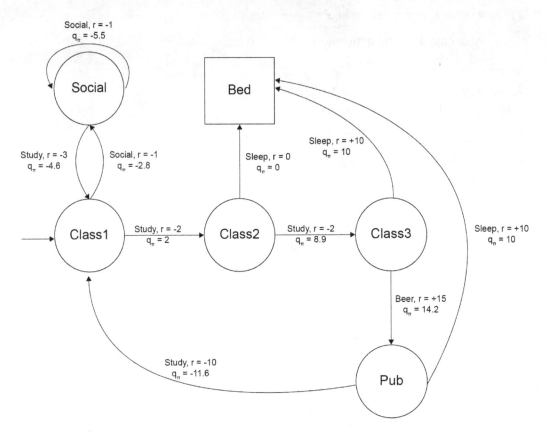

Student MDP γ=0.9

Figure 2.42: Action values for the student MDP

We are now interested in extracting the best action for each state:

```
"""
reshape the column so that we obtain a vector with shape (n_states, n_
actions)
"""
n_actions = 2
Q_sa_pi2 = np.reshape(Q_sa_pi, (-1, n_actions))
Q_sa_pi2
```

The output will be as follows:

```
array([[  2.01603629,  -5.58777741],
       [  8.92452509,   0.        ],
       [ 10.        ,  14.27672242],
       [ -5.58777741,  -4.60728351],
       [ 10.        , -11.60728351]])
```

In this way, performing the **argmax** function, we obtain the index of the best action in each state:

```
best_actions = np.reshape(np.argmax(Q_sa_pi2, -1), (-1, 1))
best_actions
```

The **best_actions** vector contains the following values:

```
array([[0],
       [0],
       [1],
       [1],
       [0]])
```

The best actions can be visualized as follows:

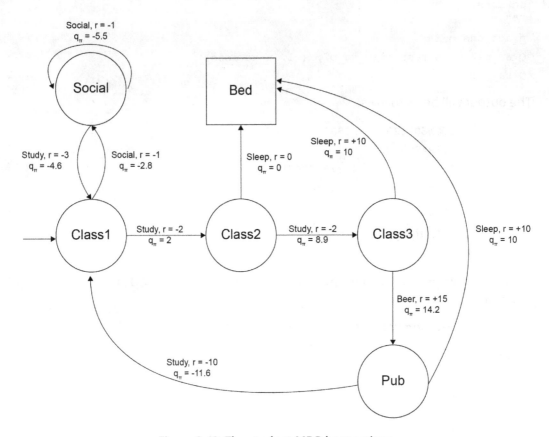

Figure 2.43: The student MDP best actions

In *Figure 2.43*, the dotted arrows are the best actions in each state. We can easily find them by looking at the action maximizing the q function in each state.

From the action-value calculation, we can see that when $\gamma=0$, the action-value function is equal to the expected immediate reward:

```
Q_sa_pi_gamma_zero = R_sa
Q_sa_pi_gamma_zero
```

The output will be as follows:

```
array([[ -2.],
       [ -1.],
       [ -2.],
       [  0.],
       [ 10.],
       [ 15.],
       [ -1.],
       [ -3.],
       [ 10.],
       [-10.]])
```

Reshape the columns with **n_actions = 2**, as follows:

```
n_actions = 2
Q_sa_pi_gamma_zero2 = np.reshape(Q_sa_pi_gamma_zero, \
                                 (-1, n_actions))
Q_sa_pi_gamma_zero2
```

The output will be as follows:

```
array([[ -2.,   -1.],
       [ -2.,    0.],
       [ 10.,   15.],
       [ -1.,   -3.],
       [ 10.,  -10.]])
```

By performing the **argmax** function, we obtain the index of the best action in each state as follows:

```
best_actions_gamma_zero = np.reshape(np.argmax\
                                     (Q_sa_pi_gamma_zero2, -1), \
                                     (-1, 1))
best_actions_gamma_zero
```

The output will be as follows:

```
array([[1],
       [1],
       [1],
       [0],
       [0]])
```

The state diagram can be visualized as follows:

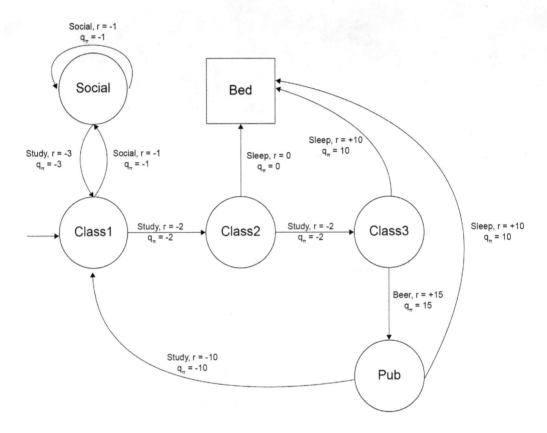

Figure 2.44: The best actions and the action-value function for the student MDP when $\gamma = 0$

It is interesting to note how the best actions are changed by only modifying the discount factor. Here, the best action the agent can take, starting from **Class 1**, is **Social** as it provides a bigger immediate reward compared to the **Study** action. The **Social** action brings the agent to state **Social**. Here, the best the agent can do is to repeat the **Social** action, cumulating negative rewards.

In this example, we learned how to calculate the state-value function using `scipy.linalg.solve` and how to calculate the action-value function using the matrix form. We noticed that both the state values and the action values depend on the discount factor.

In the next section, we will illustrate the Bellman optimality equation, which makes it possible to solve MDPs by finding the best policy and the best state values.

BELLMAN OPTIMALITY EQUATION

It is natural to ask whether it is possible to define an order for policies that determines whether one policy is better than another one. It turns out that the value function provides ordering over policies.

Policy π can be considered better than or equal to ($\geqslant$) policy π' if the expected return from that policy is greater than or equal to the expected return of π' for all states:

$$\pi \geqslant \pi' \Leftrightarrow v_\pi(s) \geq v_{\pi'}(s) \; \forall s \in S$$

Figure 2.45: Preference over policies

In this example, we substituted the expected return in a state with the state-value function, using the state-value function definition.

Following the previous definition, an optimal policy is a policy that is better than or equal to all other policies in all states. The optimal state-value function, v^*, and the optimal action-value function, q^*, are simply the ones associated with the best policy:

$$v^*(s) = \max_\pi v_\pi(s)$$

Figure 2.46: Optimal state-value function

$$q^*(s, a) = \max_\pi q_\pi(s, a)$$

Figure 2.47: Optimal action-value function

Some important properties of MDPs are as follows:

- There is always at least one optimal (deterministic) policy maximizing the state-value function in every state.

- All optimal policies share the same optimal state-value function.

An MDP is solved if we know the optimal state-value function and the optimal action-value function.

Knowing the optimal value function, q^*, makes it possible to find the optimal policy of the MDP by maximizing over $q^*(s, a)$. We can define the optimal policy associated with the optimal action-value function as follows:

$$\pi(a|s) = \begin{cases} 1 \ \textit{if} \ a = \textit{argmax}_a \ q^*(s, a) \\ 0 \ \textit{otherwise} \end{cases}$$

Figure 2.48: Optimal policy associated with the optimal action-value function

As you can see, this policy is simply telling us to perform the action, a, with a probability of 1 (essentially, in a deterministic way) if the action, a, maximizes the action-value function in this state. In other words, we need to take the action that guarantees the highest discounted return following the optimal policy. All other actions, being suboptimal, are taken with probability 0; therefore, they are, essentially, never taken. Notice that the policy obtained in this way is deterministic, not stochastic.

Analyzing this result, we uncover two essential facts:

- There is always a deterministic optimal policy for any MDP.

- The optimal policy is determined by the knowledge of the optimal action-value function, $q^*(s, a)$.

The optimal value functions are related to the Bellman optimality equation. The Bellman optimality equation states that the optimal state-value function in a state is equal to the overall maximum actions of the optimal action-value function in the same state:

$$v^*(s) = \textit{max}_a \ q_*(s, a)$$

Figure 2.49: The Bellman optimality equation

Using the definition of the action-value function, we can expand the previous equation to a more explicit form:

$$v^* (s) = \max_a \left[R(s, a) + \gamma \sum_{s' \in S} P(s' | s, a) v^* (s') \right]$$

Figure 2.50: The Bellman optimality equation in terms of the action-value function

The previous equation tells us that the optimal value function of a state is equal to the maximum over actions of the immediate reward, $R(s, a)$, plus the discounted (γ), expected optimal value of the successor state, $v^* (s')$, where the expected value is determined by the transition function.

Also, the optimal action-value function has an explicit formulation, known as the Bellman optimality equation, for q:

$$q^* (s, a) = R(s, a) + \gamma \sum_{s' \in S} P(s' | s, a) v^* (s')$$

Figure 2.51: The Bellman optimality equation for q

This can be rewritten only in terms of q^* by using the relationship between q^* and v^*:

$$q^* (s, a) = R(s, a) + \gamma \sum_{s' \in S} P(s' | s, a) \max_{a'} q^* (s', a')$$

Figure 2.52: The Bellman optimality equation, using the relationship between q^* and v^*

The Bellman optimality equation for v^* expresses the fact that the optimal state-value function must equal the expected return for the best action in that state. Similarly, the Bellman optimality equation for q^* expresses the fact that the optimal q-function must equal the immediate reward plus the discounted return of the best action in the next state according to the environment dynamic.

SOLVING THE BELLMAN OPTIMALITY EQUATION

The presence of a maximization makes the Bellman optimality equation non-linear. This means that we do not have a closed-form solution for these equations in the general case. However, there are many iterative solution methods that we will analyze in the next sections and chapters.

The main methods include value iteration, policy iteration, Q learning, and SARSA, which we will study in later chapters.

SOLVING MDPS

Now that we have gained a fair understanding of all the important concepts and equations, let's move on to solving actual MDPs.

ALGORITHM CATEGORIZATION

Before considering the different algorithms for solving MDPs, it is beneficial for us to understand the family of algorithms along with their pros and cons. Knowing the main family of algorithms makes it possible for us to select the correct family based on our task:

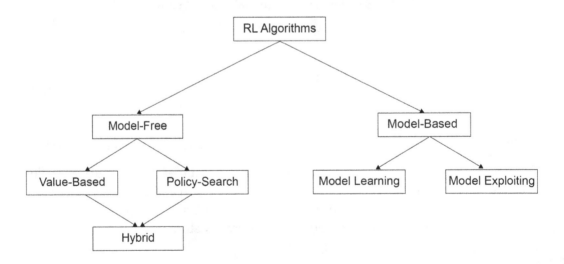

Figure 2.53: Taxonomy of RL algorithms

The first main distinction is between model-based algorithms and model-free algorithms:

- A model-based algorithm requires knowledge of the environment dynamic (model). This is a strong requirement, as the environment model is usually unknown. Let's consider an autonomous driving problem. Here, knowing the environment dynamic means that we should know exactly how the agent's actions influence the environment and the next state distribution. This depends on many factors: the street state, weather conditions, car characteristics, and much more. For many problems, the dynamic is unknown, too complex, or too inaccurate to be used successfully. Nonetheless, the dynamic provides beneficial information for solving the task.

  When the model is known (**Model Exploiting**), model-based algorithms are preferred over their counterparts for their sample efficiency, as they require fewer samples to learn good policies.

  The environment model in these cases can also be unknown; the algorithm itself explicitly learns an environment model (**Model Learning**) and uses it to plan its actions. Dynamic programming algorithms use this model knowledge to perform bootstrapping, which uses a previous estimation for the estimate of another quantity.

- Model-free algorithms do not require a model of the environment. These types of algorithms are, therefore, preferred for real-world applications. Note that these algorithms may build an environment representation internally, taking into account the environment dynamic. However, usually, this process is implicit, and the users just don't care about these aspects.

Model-free algorithms can also be classified as value-based algorithms or policy-search algorithms.

VALUE-BASED ALGORITHMS

A value-based algorithm focuses on learning the action-value function and the state-value function. Learning the value functions is done by using the Bellman equations presented in the previous sections. An example of a value-based algorithm is Q learning, where the objective is to learn the action-value function, which, in turn, is used for control. A deep Q network is an extension of Q learning in which a neural network is used to approximate the q-function. Value-based algorithms are usually off-policy, which means they can reuse previous samples collected with a different policy with respect to the policy being optimized at the moment. This is a very powerful property as it allows us to obtain more efficient algorithms in terms of samples. We will learn about Q learning and deep Q networks in more detail in *Chapter 9, What Is Deep Q-Learning?*.

POLICY SEARCH ALGORITHMS

Policy Search (**PS**) methods explore the policy space directly. In PS, the RL problem is formalized as the maximization of the performance measure depending on the policy parameters. You will study PS methods and policy gradients in more detail in *Chapter 11, Policy-Based Methods for Reinforcement Learning*.

LINEAR PROGRAMMING

Linear programming is an optimization technique that is used for problems with linear constraints and linear objective functions. The objective function describes the quantity to be optimized. In the case of RL, this quantity is the expected discounted return of all the states weighted by the initial state distribution, which is the probability of starting an episode in that state.

When the starting state is precisely one, this simplifies to the optimization of the expected discounted return starting from the initial state.

Linear programming is a model-based, model-exploiting technique. Solving an MDP with linear programming, therefore, requires perfect knowledge of the environment dynamics, which translates into knowledge of the transition probability matrix, P. Using linear programming, we can solve MDPs by finding the best state values for each state. From our knowledge of state values, we can derive knowledge of the optimal action-value function. In this way, we can find a control policy for our agent and maximize its performance in the given task.

The basic idea follows on from the definition of ordering over policies; we want to find the state-value function by maximizing the value of each state weighted by the initial state distribution, $\mu(s)$, subject to a feasibility constraint:

$$\min_{v} \sum_{s \in S} \mu(s)\, v(s)$$

$$subject\ to\colon v(s) \geq R(s, a) + \gamma \sum_{s' \in S} P(s' | s, a)\, v(s'), \quad \forall s \in S, \forall a \in A$$

Figure 2.54: Linear programming formulation for solving MDPs

Here, we have $|S|$ variables and $|S| \cdot |A|$ constraints. The variables are the values, $v(s)$, for each state, **s**, in the state space, **S**.

Note that the maximization role is taken by the constraints, while we need to minimize the objective function because, otherwise, an optimal solution would have infinite values for all variables, $v(s)$.

The constraints are based on the idea that the value of a state must be greater than or equal to the immediate reward plus the discounted expected value of the successor states. This must be true for all states and all actions.

The huge number of variables and constraints makes it possible to use linear programming techniques for only finite-state and finite-action MDPs.

We will be using the following notation:

$$\min_{x} c^T x$$

$$such\ that\ A_{ub}\, x \leq b_{ub},$$

$$A_{eq}\, x \leq b_{eq},$$

$$l \leq x \leq u$$

Figure 2.55: Linear programming notation

In the preceding notation, **c** is the vector of coefficients of the objective function, A_{ub} is the matrix of the upper bound constraints, and b_{ub} is the associated coefficient vector.

In Python, SciPy offers the **linprog** function (inside the **optimize** module, **scipy.optimize.linprog**), which optimizes linear programs given the objective function and the constraints.

The signature of the function is **scipy.optimize.linprog(c, A_ub, b_ub)**.

To rephrase the problem using upper bounds, we have the following:

$$V \geq R + \gamma PV \rightarrow (\gamma P - I) V \leq -R$$

Figure 2.56: Linear programming constraints using upper bounds

> **NOTE**
>
> For further reading on linear programming for MDPs, refer to the following paper from *de Farias, D. P. (2002): The Linear Programming Approach to Approximate Dynamic Programming: Theory and Application*: http://www.mit.edu/~pucci/discountedLP.pdf.

Let's now solve a quick exercise to strengthen our understanding of linear programming.

EXERCISE 2.02: DETERMINING THE BEST POLICY FOR AN MDP USING LINEAR PROGRAMMING

The goal of this exercise is to solve the MDP in the following figure using linear programming. In this MDP, the environment model is straightforward and the transition function is deterministic, determined uniquely by the action. We will be finding the best action (the one with the maximum reward) taken by the agent, which determines the best policy of the environment, using linear programming:

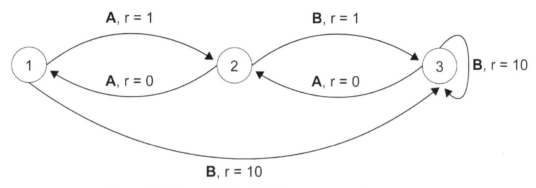

Figure 2.57: Simple MDP with three states and two actions

The variables of the linear program are the state values. The coefficients are given by the initial state distribution, which, in our case, is a deterministic function, as state 1 is the initial state. Therefore, the coefficients of the objective function are **[1, 0, 0]**.

We are now ready to tackle our problem:

1. As always, import the required libraries:

```
import numpy as np
import scipy.optimize
```

2. Define the number of states and actions and the discount factor for this problem:

```
# number of states and number of actions
n_states = 3
n_actions = 2
```

3. Define the initial state distribution. In our case, it is a deterministic function:

```
# initial state distribution
mu = np.array([[1, 0, 0]]).T # only state 1
mu
```

The output will be as follows:

```
array([[1],
       [0],
       [0]])
```

4. Now we need to build the upper bound coefficients for action **A**:

```
# Build the upper bound coefficients for the action A
# define the reward matrix for action A
R_A = np.zeros((n_states, 1), np.float)
R_A[0, 0] = 1
R_A[1, 0] = 0
R_A[2, 0] = 0
R_A
```

The output will be as follows:

```
array([[1.],
       [0.],
       [0.]])
```

5. Define the transition matrix for action **A**:

```
# Define the transition matrix for action A
P_A = np.zeros((n_states, n_states), np.float)
P_A[0, 1] = 1
P_A[1, 0] = 1
P_A[2, 1] = 1
P_A
```

The output will be as follows:

```
array([[0., 1., 0.],
       [1., 0., 0.],
       [0., 1., 0.]])
```

6. We are ready to build the upper bound matrix for action **A**:

```
gamma = 0.9
# Upper bound A matrix for action A
A_up_A = gamma * P_A - np.eye(3,3)
A_up_A
```

The output will be as follows:

```
array([[-1. ,  0.9,  0. ],
       [ 0.9, -1. ,  0. ],
       [ 0. ,  0.9, -1. ]])
```

7. We need to do the same for action **B**:

```
# The same for action B
# define the reward matrix for action B
R_B = np.zeros((n_states, 1), np.float)
R_B[0, 0] = 10
R_B[1, 0] = 1
R_B[2, 0] = 10
# Define the transition matrix for action B
P_B = np.zeros((n_states, n_states), np.float)
P_B[0, 2] = 1
P_B[1, 2] = 1
P_B[2, 2] = 1
# Upper bound A matrix for action B
A_up_B = gamma * P_B - np.eye(3,3)
A_up_B
```

The output will be as follows:

```
array([[-1. ,  0. ,  0.9],
       [ 0. , -1. ,  0.9],
       [ 0. ,  0. , -0.1]])
```

8. We are ready to concatenate the results for the two actions:

```
# Upper bound matrix for all actions and all states
A_up = np.vstack((A_up_A, A_up_B))
"""
verify the shape: number of constraints are equal to |actions| *
|states|
"""
assert(A_up.shape[0] == n_states * n_actions)
# Reward vector is obtained by stacking the two vectors
R = np.vstack((R_A, R_B))
```

9. The only thing we have to do now is to solve the linear program using **scipy. optimize.linprog**:

```
c = mu
b_up = -R
# Solve the linear program
res = scipy.optimize.linprog(c, A_up, b_up)
```

10. Let's collect the results:

```
# Obtain the results: state values
V_ = res.x
V_
V = V_.reshape((-1, 1))
V
np.savetxt("solution/V.txt", V)
```

Let's analyze the results. We can see that the value of state 2 is the lowest one, as expected. The values of states 1 and 3 are very close to each other and are approximately equal to 1e+2:

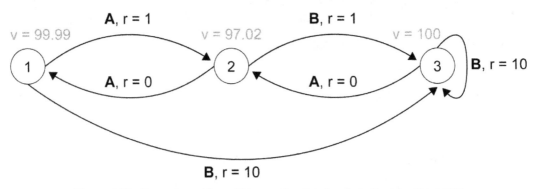

Figure 2.58: Representation of the optimal value function for the MDP

11. Now we can calculate the optimal policy by calculating the optimal action-value function for each state-action pair:

```
"""
transition matrix. On the rows, we have states and actions, and on
the columns, we have the next states
"""
P = np.vstack((P_A, P_B))
P
```

The output will be as follows:

```
array([[0., 1., 0.],
       [1., 0., 0.],
       [0., 1., 0.],
       [0., 0., 1.],
       [0., 0., 1.],
       [0., 0., 1.]])
```

12. Use the action-value formula to calculate the action values for each state-action pair:

```
"""
Use the action value formula to calculate the action values for each
state action pair.
"""
Q_sa = R + gamma * P.dot(V)
"""
The first three rows are associated to action A, the last three are
associated # to action B
"""
Q_sa
```

The output is as follows:

```
array([[ 88.32127683],
       [ 89.99999645],
       [ 87.32127683],
       [100.00000622],
       [ 91.00000622],
       [100.00000622]])
```

13. Reshape and use the **argmax** function to better understand the best actions:

```
Q_sa_2 = np.stack((Q_sa[:3, 0], Q_sa[3:, 0]), axis=1)
Q_sa_2
```

The output will be as follows:

```
array([[ 88.32127683, 100.00000622],
       [ 89.99999645,  91.00000622],
       [ 87.32127683, 100.00000622]])
```

Use the following code to better understand the best actions:

```
best_actions = np.reshape(np.argmax(Q_sa_2, axis=1), (3, 1))
best_actions
```

The output will be as follows:

```
array([[1],
       [1],
       [1]])
```

By visually inspecting the result, we can see that action **B** is the best action for all states, having acquired the highest q values for all states. Thus, the optimal policy decides to always take action **B**. Doing this, we will land in state **3**, and we will follow the self-loop, cumulating high positive rewards:

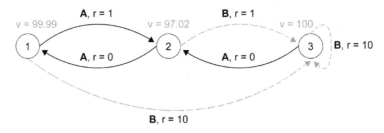

Figure 2.59: Representation of the optimal policy and the optimal value function for the MDP

The optimal policy is represented in *Figure 2.59*. The dotted arrows represent the best action for each state.

> **NOTE**
>
> To access the source code for this specific section, please refer to https://packt.live/2Arr9rO.
>
> You can also run this example online at https://packt.live/2Ck6neR.

In this exercise, we used linear programming techniques to solve a simple MDP with finite states and actions. By using the correspondence between the state-value function and the action-value function, we extracted the value of each state-action pair. From this knowledge, we extracted the optimal policy for this environment. In this case, the best policy is always just to take action **B**.

In the next activity, we will use the Bellman expectation equation to evaluate a policy for a more complex task. Before that, let's explore the environment that we are going to use in the activity, Gridworld.

GRIDWORLD

Gridworld is a classical RL environment with many variants. The following figure displays the visual representation of the environment:

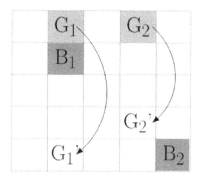

Figure 2.60: The Gridworld environment

As you can see, the states are represented by cells, and there are 25 states arranged in a 5 x 5 grid. There are four available actions: left, right, up, and down. These actions move the current state in the direction of the action, and the associated reward is 0 for all actions. The exceptions are as follows:

- Border cells: If an action takes the agent outside of the grid, the agent state does not change, and the agent receives a reward of -1.

- Good cells: G_1 and G_2 are good cells. For these cells, each action brings the agent to states G_1' and G_2', respectively. The associated reward is +10 for going outside state G_1 and +5 for going outside state G_2.

- Bad cells: B_1 and B_2 are bad cells. For these cells, the associated reward is -1 for all actions.

Now that we have an understanding of the environment, let's attempt an activity that implements it.

ACTIVITY 2.01: SOLVING GRIDWORLD

In this activity, we will be working on the Gridworld environment. The goal of the activity is to calculate and visually represent the state values for a random policy, in which the agent selects each action with an equal probability (1/4) in all states. The discount factor is assumed to be equal to 0.9.

The following steps will help you to complete the activity:

1. Import the required libraries. Import **Enum** and **auto** from **enum**, **matplotlib.pyplot**, **scipy**, and **numpy**, and import **tuple** from **typing**.

2. Define the visualization function and the possible actions for the agent.

3. Write a policy class that returns the action probability in a given state; for a random policy, the state can be ignored.

4. Write an **Environment** class with a step function that returns the next state and the associated reward given the current state and action.

5. Loop for all states and actions and build a transition matrix (width*height, width*height) and a reward matrix of the same dimension. The transition matrix contains the probability of going from one state to another, so the sum of the first axis should be equal to 1 for all rows.

6. Use the matrix form of the Bellman expectation equation to compute the state values for each state. You can use **scipy.linalg.solve** or directly compute the inverse matrix and solve the system.

The output will be as follows:

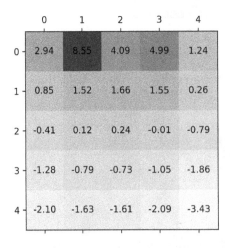

Figure 2.61: State values of Gridworld

> **NOTE**
>
> It is useful to visualize the state values and the expected reward, so write a function visually representing the calculated matrices.
>
> The solution to this activity can found on page 689.

SUMMARY

In this chapter, we learned the differences between MCs, MRPs, and MDPs. An MC is the most straightforward description of a generic process that is composed of states and a probability function that describes the transition between states. An MRP includes the concept of rewards as a measure of how good a transition is. The MDP is what we are most interested in; it includes the concept of actions, policies, and goals.

In the context of Markov processes, we introduced Bellman equations in different forms and also analyzed the relationship between the state-value function and the action-value function.

We discussed various methods for solving MDPs, categorizing algorithms based on the information they require and on the methods they use. These algorithms will be presented in more detail in the following chapters. We focused on linear programming, showing how it is possible to solve MDPs using these techniques.

In the next chapter, you will learn how to use TensorFlow 2 to implement deep learning algorithms and machine learning models.

3

DEEP LEARNING IN PRACTICE WITH TENSORFLOW 2

OVERVIEW

This chapter will introduce you to TensorFlow and Keras and provide an overview of their key features and applications, as well as how they work in synergy. You will be able to implement a deep neural network with TensorFlow by addressing the main topics, from model creation, training, and validation, to testing. You will perform a regression task and solve a classification problem, thereby gaining hands-on experience with the frameworks. Finally, you will build and train a model to classify clothes images with high accuracy. By the end of this chapter, you will be able to design, build, and train deep learning models using the most advanced machine learning frameworks available.

INTRODUCTION

In the previous chapter, we covered the theory behind Reinforcement Learning (RL), explaining topics such as Markov chains and Markov Decision Processes (MDPs), Bellman equations, and a number of techniques we can use to solve MDPs. In this chapter, we will be looking at deep learning methods, all of which will play a primary role in building approximate functions for reinforcement learning. Specifically, we will look at different families of deep neural networks: fully connected, convolutional, and recurrent networks. These algorithms have the key capability of encoding knowledge that's been learned through examples in a compact and effective representation. In RL, they are typically used to approximate the so-called policy functions and value functions, which encode how the RL agent chooses its action, given the current state and the value associated with the current state, respectively. We will study the policy and value functions in the upcoming chapters.

Data is the new oil: This famous quote is being heard more and more frequently these days, especially in tech and economic industries. With the great amount of data available today, techniques to leverage such enormous quantities of information, thereby creating value and opportunities, are becoming key competitive factors and skills to have. All products and platforms that are provided to users for free (from social networks to apps related to wearable devices) use data that is provided by the users to generate revenues: think about the huge quantity of information they collect every day relating to our habits, preferences, or even body weight trends. These provide high-value insights that can be leveraged by advertisers, insurance companies, and local businesses to improve their offers so that they fit the market.

Thanks to the relevant increase in computational power availability and theory breakthroughs such as backpropagation-based training, deep learning has seen an explosion in the last 10 years, achieving unprecedented results in many fields, from image processing to speech recognition to natural language processing and understanding. In fact, it is now possible to successfully train large and deep neural networks by leveraging huge amounts of data and overcoming practical roadblocks that impeded their adoption in past decades. These models demonstrated the capability to exceed human performances in terms of both speed and accuracy. This chapter will teach you how to adopt deep learning to solve real-world problems by taking advantage of the top machine learning frameworks. TensorFlow and Keras, are the de facto production standards in the industry. Their success is mainly related to two aspects: TensorFlow's unrivaled performance in production environments in terms of both speed and scalability, and Keras' ease of use, which provides a very powerful, high-level interface that can be used to create deep learning models.

Now, let's take a look at the frameworks.

AN INTRODUCTION TO TENSORFLOW AND KERAS

In this section, both frameworks will be presented, thus providing you with a general overview of their architecture, the fundamental elements they are composed of, and listing some of their typical applications.

TENSORFLOW

TensorFlow is an open source numerical computation software library that leverages data flow computational graphs. Its architecture allows users to run it on a wide variety of hardware: from CPUs to **Tensor Processing Units** (**TPUs**), including GPUs as well as mobile and embedded platforms. The main difference between the three is the speed and the type of data they are able to perform computations with (multiplications and additions), which, of course, is of primary importance when aiming for maximum performance.

> **NOTE**
>
> We will be looking at various code implementation examples for TensorFlow in the *Keras* section of this chapter.
>
> You can refer to the official documentation of TensorFlow for more information here: https://www.tensorflow.org/
>
> The following article is a very good reference if you wish to find out more about the differences between GPUs and TPUs: https://iq.opengenus.org/cpu-vs-gpu-vs-tpu/

TensorFlow is based on a high-performance core implemented in C++ that's provided by a distributed execution engine that works as an abstraction toward the many devices it supports. We will be using TensorFlow 2, which has recently been released. It represents a major milestone for TensorFlow. Its main differences with respect to version 1 are related to its greater ease of use, in particular for model building. In fact, Keras has become the lead tool that's used to easily create models and experiment with them. TensorFlow 2 uses eager execution by default. This allowed the creators of TensorFlow to eliminate the previous complex workflow, which was based on the construction of a computational graph that's then run in a session. With eager execution, this is no longer required. Finally, the data pipeline has been simplified by means of the TensorFlow dataset, which is a common interface that's used to ingest standard or custom datasets with no need to define placeholders.

The execution engine is then interfaced with Python and C++ frontends, which, in turn, are the basis for the Layers API, which provides a simple interface for common layers in deep learning models. This hierarchical structure continues with higher-level APIs, including Keras (which we will describe later in this section). Finally, a set of common models are provided and can be used out of the box.

The following diagram provides an overview of how different TensorFlow modules are hierarchically organized, starting from the low level (bottom) up to the highest level (top):

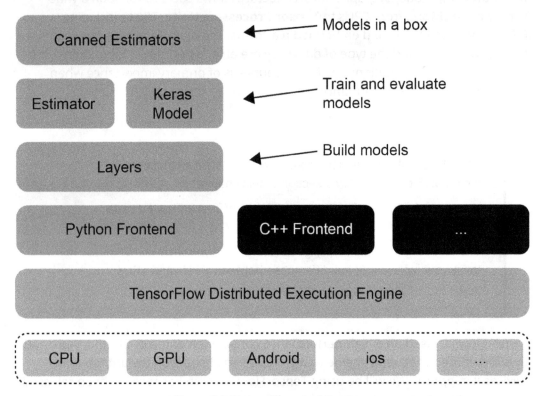

Figure 3.1: TensorFlow architecture

The historical execution model of TensorFlow was based on computational graphs. Using this approach, the first step when building a model is to create a computation graph that fully describes the calculations we want to perform. The second step is to execute it. This approach has the drawback of being less intuitive with respect to common implementations, where the graph doesn't have to be completed before it can be executed. At the same time, it provides several advantages, making the algorithm highly portable, deployable on different types of hardware platforms, and capable of running in parallel on multiple instances.

In the latest version of TensorFlow (starting with v. 1.7), a new execution model called "eager execution" has been introduced. This is an imperative style for writing code. With eager execution enabled, all algorithmic operations can be run immediately, with no need to build a graph first and then execute it. This new approach has been greeted with enthusiasm and has some very important pros: first, it is much simpler to inspect and debug algorithms and access intermediate values; it is possible to directly use a Python control flow inside TensorFlow APIs; and it makes building and training complex algorithms very easy.

In addition, once the model that has been created using eager execution satisfies requirements, it is possible to automatically convert it into a graph, which makes it possible to leverage all the advantages we looked at previously, such as saving, porting, and distributing models optimally.

Like other machine learning frameworks, TensorFlow provides a large number of ready-to-use models and for many of them, it also provides trained model weights along with the model graph, meaning we can run such models out of the box, and even tune them for a specific use case to take advantage of techniques such as transfer learning with fine tuning. We will cover these in the following sections.

The models provided cover a wide range of different applications, for example:

- **Image classification**: Able to classify images into categories.

- **Object detection**: Capable of detecting and localizing multiple objects in images.

- **Language understanding and translation**: Performing natural language processing for tasks such as word prediction and translation.

- **Patch harmonization and style transfer**: The algorithm is able to apply a given style (represented, for example, through a painting) to a given photo (refer to the following example).

As we mentioned previously, many of the models include trained weights and examples explaining how to use them. Thus, it is very straightforward to adopt "transfer learning," that is, to take advantage of these pretrained models by creating new ones, retraining only a part of the network on a new dataset. This can be significantly smaller with respect to the one used to train the entire network from scratch.

TensorFlow models can also be deployed on mobile devices. After being trained on large systems, they are optimized to reduce their footprint, which cannot be too big to meet platform limitations. For example, the TensorFlow project known as **MobileNet** is developing a set of computer vision models specifically designed with optimal speed/accuracy trade-offs in mind. These are typically considered for embedded devices and mobile applications.

The following image represents a typical example of an object detection application where the input image is processed and three objects have been detected, localized, and classified:

Figure 3.2: Object detection

The following image shows how style transfer works: the style of the famous painting "*The Great Wave off Kanagawa*" has been applied to a photo of the Seattle skyline. The results keep the key parts of the picture (the majority of the buildings are there, mountains, and so on), but it is represented through stylistic elements that have been extrapolated from the reference image:

Figure 3.3: Style transfer

Now, let's learn about Keras.

KERAS

Building deep learning models is quite complex, especially when we have to deal with all the typical low-level aspects of major frameworks, and this is one of the most relevant barriers for newcomers in the machine learning field. As an example, the following code shows how to create a simple neural network (one hidden layer with an input size of **100** and an output size of **10**) with a low-level TensorFlow API.

In the following code snippet, two functions are being defined. The first builds the weights matrix of a network layer, while the second one creates the bias vector:

```
def weight_variable(shape):
    shape = tf.TensorShape(shape)
    initial_values = tf.truncated_normal(shape, stddev=0.1)
    return tf.Variable(initial_values)

def bias_variable(shape):
    initial_values = tf.zeros(tf.TensorShape(shape))
    return tf.Variable(initial_values)
```

Next, the placeholders for the input (**X**) and labels (**y**) are created. They will contain the training samples that will be used to fit the model:

```
# Define placeholders
X = tf.placeholder(tf.float32, shape=[None, 100])
y = tf.placeholder(tf.int32, shape=[None, 10])
```

Two matrices and two vectors are created, one couple for each of the two hidden layers of the network to be created, with the functions previously defined. These will contain trainable parameters (network weights):

```
# Define variables
w1 = weight_variable([X_input.shape[1], 64])
b1 = bias_variable([64])
w2 = weight_variable([64, 10])
b2 = bias_variable([10])
```

The two network layers are defined via their mathematical definition: matrix multiplication, plus the bias sum and activation function applied to the result:

```
# Define network
# Hidden layer
z1 = tf.add(tf.matmul(X, w1), b1)
a1 = tf.nn.relu(z1)

# Output layer
z2 = tf.add(tf.matmul(a1, w2), b2)
y_pred = tf.nn.softmax(z2)
y_one_hot = tf.one_hot(y, 10)
```

The **loss** function is defined, the optimizer is initialized, and the training metrics are chosen. Finally, the graph is run to perform training:

```
# Define loss function
loss = tf.losses.softmax_cross_entropy(y, y_pred, \
        reduction=tf.losses.Reduction.MEAN)

# Define optimizer
optimizer = tf.train.AdamOptimizer(0.01).minimize(loss)

# Metric
accuracy = tf.reduce_mean(tf.cast(tf.equal(tf.argmax(y, axis=1), \
        tf.argmax(y_pred, axis=1)), tf.float32))

for _ in range(n_epochs):
    sess.run(optimizer, feed_dict={X: X_train, y: y_train})
```

As you can see, we need to manually manage many different aspects: variable declaration, weights initialization, layer creation, layer-related mathematical operations, and the definition of the loss function, optimizers, and metrics. For comparison, the same neural network will be created using Keras later in this section.

> **NOTE**
>
> The preceding code snippet is an example that demonstrates how to implement a simple fully connected neural network with a TensorFlow low-level API. In *Exercise 3.01, Building a Sequential Model with the Keras High-Level API*, you will see how much more straightforward it is to do the same job using a Keras high-level API.

Among many different proposals, Keras has become one of the main references for high-level APIs, especially the context of those targeted at creating neural networks. It is written in Python and can be interfaced with different backend computation engines, one of which is, of course, TensorFlow.

> **NOTE**
>
> You can refer to the official documentation for further reading on Keras here: https://keras.io/.

Keras' conception has been driven by some clear principles, in particular, modularity, user friendliness, easy extendibility, and its straightforward integration with Python. Its aim is to favor adoption by newcomers and non-experienced users, and it presents a very gentle learning curve. It provides many different standalone modules, ranging from neural network layers to optimizers, from initialization schemes to cost functions. These can be easily created to create deep learning models quickly and to code them directly in Python, with no need to use separate configuration files. Given these features, its wide adoption, the fact that it can be interfaced with a large number of different backend engines (for example, TensorFlow, CNTK, Theano, MXNet, and PlaidML) and its wide choice of deployment options, it has risen to become the standard choice in the field.

Since it doesn't have its own low-level implementation, Keras needs to rely on an external element. This can be easily modified by editing (for Linux users) the **$HOME/.keras/keras.json** file, where it is possible to specify the backend name. It is also possible to specify it by means of the **KERAS_BACKEND** environment variable.

Keras' fundamental class is **Model**. There are two different types of model available: The sequential model (which we will use extensively), and the **Model** class, which is used with the functional API.

The sequential model can be seen as a linear stack of layers, piled one after the other in a very simple way, and these layers can be described very easily. The following exercise shows how short a Python script in Keras that builds a deep neural network using **model.add()** can be in order to define two dense layers in a sequential model.

EXERCISE 3.01: BUILDING A SEQUENTIAL MODEL WITH THE KERAS HIGH-LEVEL API

This exercise shows how to easily build a sequential model, composed of two dense layers, with the Keras high-level API, step by step:

1. Import the TensorFlow module and print its version:

```
import tensorflow as tf
from __future__ import absolute_import, division, \
print_function, unicode_literals

import tensorflow as tf

print("TensorFlow version: {}".format(tf.__version__))
```

This outputs the following line:

```
TensorFlow version: 2.1.0
```

2. Build the model using Keras' **sequential** and **add** methods and print a network summary. To continue in parallel with a low-level API, the same activation functions are used. We are using **ReLu** here, which is a typical activation function that's used for hidden layers. It is a key element that provides nonlinearity to the model thanks to its nonlinear shape. We also use **Softmax**, which is the activation function typically used for output layers in classification problems. It receives the output values (so-called "logits") from the previous layer and performs a weighting of them, defining all the probabilities of the output classes. The **input_dim** is the dimension of the input feature vector; it is assumed to have a dimension of **100**:

```
model = tf.keras.Sequential()

model.add(tf.keras.layers.Dense(units=64, \
                              activation='relu', input_dim=100))
model.add(tf.keras.layers.Dense(units=10, activation='softmax'))
```

3. Print the standard model architecture:

```
model.summary()
```

In our case, the network model summary is as follows:

```
Model: "sequential_1"
```

Layer (type)	Output Shape	Param #
dense_2 (Dense)	(None, 64)	6464
dense_3 (Dense)	(None, 10)	650

```
Total params: 7,114
Trainable params: 7,114
Non-trainable params: 0
```

The preceding output is a useful visualization that gives us a clear understanding of layers, their type and shape, and the number of network parameters.

> **NOTE**
>
> To access the source code for this specific section, please refer to https://packt.live/30A9Dw9.
>
> You can also run this example online at https://packt.live/3cT0cKL.

As anticipated, this exercise showed us how to create a sequential model and how to add two layers to it in a very straightforward way.

We will deal with the remaining aspects later on, but it is still worth noting that training the model we just created and performing inference only requires very few lines of code, as presented in the following snippet, which needs to be appended to the snippet of *Exercise 3.01, Building a Sequential Model with the Keras High-Level API*:

```
model.compile(loss='categorical_crossentropy', optimizer='sgd', \
              metrics=['accuracy'])

model.fit(x_train, y_train, epochs=5, batch_size=32)

loss_and_metrics = model.evaluate(x_test, y_test, batch_size=128)

classes = model.predict(x_test, batch_size=128)
```

If more complex models are required, the sequential API is too limited. For these needs, Keras provides the functional API, which allows us to create models that are able to manage complex networks graphs, such as networks with multiple inputs and/or multiple outputs, recurrent neural networks where data processing is not sequential but instead is cyclic, and context, where layers' weights are shared among different parts of the network. For this purpose, Keras allows us to leverage the same set of layers as the sequential model, but provides more flexibility in putting them together. First, we have to define the layers and put them together. An example is presented in the following snippet.

First, after importing TensorFlow, an input layer of dimension **784** is created:

```
import tensorflow as tf

inputs = tf.keras.layers.Input(shape=(784,))
```

Inputs are processed by the first hidden layer. They go through the ReLu activation function and are returned as output. This output then becomes the input for the second hidden layer, which is exactly the same as the first one, and returns another output, again stored in the **x** variable:

```
x = tf.keras.layers.Dense(64, activation='relu')(inputs)
x = tf.keras.layers.Dense(64, activation='relu')(x)
```

Finally, the **x** variable goes as input to the final output layer, which has a **softmax** activation function, and returns predictions:

```
predictions = tf.keras.layers.Dense(10, activation='softmax')(x)
```

Once all the passages have been completed, the model can be created by telling Keras where it starts (input variable) and where it ends (predictions variable):

```
model = tf.keras.models.Model(inputs=inputs, outputs=predictions)
```

After the model has been built, it is compiled by specifying the optimizer, the loss, and the metrics. Finally, it is fitted onto the training data:

```
model.compile(optimizer='rmsprop', \
              loss='categorical_crossentropy', \
              metrics=['accuracy'])
model.fit(data, labels)  # starts training
```

Keras provides a large number of predefined layers, as well as the possibility to code custom ones. Among those, the following are the already available layers:

- Dense layers, which are typically used for fully connected neural networks. They consist of a matrix of weights and a bias.

- Convolution layers are filters that are defined by specific kernels, which are then convolved with the inputs they are applied to. There are layers available for different input dimensions, from 1D to 3D, including the possibility to embed in them complex operations, such as cropping or transposition.

- Locally connected layers are similar to convolution layers in the sense that they act only on a subgroup of the input features, but, unlike convolution layers, they don't share weights.

- Pooling layers are layers that are used to downscale the input. As convolutional layers, they are available for inputs with dimensionality ranging from 1D to 3D. They include most of the common variants, such as max and average pooling.

- Recurrent layers are used for recurrent neural networks, where the output of a layer is also fed backward in the network. They support state-of-the-art units such as **Gated Recurrent Units** (**GRUs**), **Long Short-Term Memory** (**LSTM**) units, and others.

- Activation functions are also available in the form of layers. These are functions that are applied to layer outputs, such as `ReLu`, `Elu`, `Linear`, `Tanh`, and `Softmax`.

- Lambda layers are layers for embedding arbitrary, user-defined expressions.

- Dropout layers are special objects that randomly set a fraction of the input units to **0** at each training update to avoid overfitting (more on this later).

- Noise layers are additional layers, such as dropout, that are used to avoid overfitting.

Keras also provides common datasets, as well as famous models. For image-related applications, many networks are available, such as Xception, VGG16, VGG19, ResNet50, InceptionV3, InceptionResNetV2, MobileNet, DenseNet, NASNet, and MobileNetV2TK, all of which are pretrained on ImageNet. Keras also provides text and sequences and generative models, making a total of more than 40 algorithms.

As we saw for TensorFlow, Keras models have a vast choice of deployment platforms, including iOS, via CoreML (supported by Apple); Android, via the TensorFlow Android runtime; in a browser, via Keras.js and WebDNN; on Google Cloud, via TensorFlow-Serving; in a Python webapp backend; on the JVM, via DL4J model import; and on a Raspberry Pi.

Now that we've looked at both TensorFlow and Keras, from the next section onward, our main focus will be on how to use them in combination to create deep neural networks. Keras will be used as a high-level API, given its user-friendliness, including TensorFlow, which will be the backend.

HOW TO IMPLEMENT A NEURAL NETWORK USING TENSORFLOW

In this section, we will look at the most important aspects to consider when implementing a deep neural network. Starting with the very basic concepts, we will go through all the steps that lead up to the creation of a state-of-the-art deep learning model. We will cover the network architecture's definition, training strategies, and performance improvement techniques, understanding how they work, and preparing you so that you can tackle the next section's exercises, where these concepts will be applied to solve real-world problems.

To successfully implement a deep neural network in TensorFlow, we have to complete a given number of steps. These can be summarized and grouped as follows:

1. **Model creation**: Network architecture definition, input features encoding, embeddings, output layers

2. **Model training**: Loss function definition, optimizer choice, features normalization, backpropagation

3. **Model validation**: Strategies and key elements

4. **Model improvement**: Overfitting countermeasures

5. **Model test and inference**: Performance evaluation and online predictions

Let's look at each of these steps in detail.

MODEL CREATION

The very first step is to create a model. Choosing an architecture is hardly something that can be done *a priori* on paper. It is a typical process that requires experimentation, going back and forth between model design and field validation and testing. This is the phase where all network layers are created and properly linked to generate a complete processing operation set that goes from inputs to outputs.

The very first layer is the one that is interfaced with input data, specifically, the so-called "input features." In the case of images, for example, input features are image pixels. Depending on the nature of the layer, the input features' dimensionality needs to be taken into account. You will learn how to choose layer dimensions, depending on the layer's nature, in the upcoming sections.

The very last layer is called the output layer. It generates model predictions, so its dimensions depend on the nature of the problem. For example, in classification problems, where the model has to predict in which of the, say, 10 classes a given instance falls, the model will have 10 neurons in the output layer providing 10 scores (one per class). In the upcoming sections, we will illustrate how to create output layers with the correct dimensions.

Between the first and last layers, there are intermediate layers, called hidden layers. These layers constitute the network architecture, and they are responsible for the core processing capabilities of the model. At the time of writing, a rule that can be used to choose the best network architecture doesn't exist; this is a process that requires a lot of experimentation, under the guidance of some general principles.

A very powerful and common approach is to leverage proven models from academic papers, using them as a starting point, and then adjusting the architecture appropriately to fit and fine-tune it to the custom problem. When pretrained literature models are used and fine-tuned, the procedure is called "transfer learning," meaning we are leveraging an already trained model and transferring its knowledge to the new model, which then won't start from scratch.

Once the model has been created, all its parameters (weights/biases) must be initialized (for all non-pretrained layers). You might be tempted to set them all equal to zero, but this is hardly a good choice. There are many different initialization schemes available, and again, which one to choose requires experience and experimentation. This aspect will become clearer in the following sections. Implementation will rely on default initialization to be performed by Keras/ TensorFlow, which is usually a good and safe starting point.

A typical code example for model creation can be seen in the following snippet, which we studied in the previous section:

```
inputs = tf.keras.layers.Input(shape=(784,))

x = tf.keras.layers.Dense(64, activation='relu')(inputs)
x = tf.keras.layers.Dense(64, activation='relu')(x)
predictions = tf.keras.layers.Dense(10, activation='softmax')(x)

model = tf.keras.models.Model(inputs=inputs, outputs=predictions)
```

MODEL TRAINING

When a model is initialized and applied to input data without undergoing a training phase, it outputs random values. In order to improve its performance, we need to adjust its parameters (weights) to minimize its errors. This is the aim of the model training stage, which requires the following steps:

1. First, we have to evaluate how "wrong" the model is with a given parameter configuration by computing a so-called "loss," which is a measure of model prediction error.

2. Second, a hyperdimensional gradient is computed, which tells us how (in which direction) the model needs to change its parameters in order to improve current performance, thereby minimizing the loss function (it is indeed an optimization process).

3. Finally, the model parameters are updated by taking a "step" in the negative gradient direction (following some precise rules) and the whole process restarts from the loss evaluation stage.

This procedure is repeated as many times as needed until the system converges and the model reaches its maximum performance (minimum loss).

A typical code example for model training is shown in the following snippet, which we studied in the previous sections:

```
model.compile(optimizer='rmsprop', \
              loss='categorical_crossentropy', \
              metrics=['accuracy'])
model.fit(data, labels)   # starts training
```

LOSS FUNCTION DEFINITION

Model error can be measured by means of different loss functions. How to choose the best one requires experience. For complex applications, we often need to carefully adapt the loss function in order to drive training in directions we are interested in. As an example, let's look at how to define a typical loss that's used for classification problems: the sparse categorical cross entropy. To create it in Keras, we can use the following instruction:

```
loss_CatCrossEntropy = tf.keras.losses\
                        .SparseCategoricalCrossentropy()
```

This function operates on two inputs: true labels and predicted labels. Based on their values, it computes the loss associated with the model:

```
loss_CatCrossEntropy(y_true=groundTruth, y_pred=predictions)
```

OPTIMIZER CHOICE

The second and third steps, estimating the gradient and updating the parameters, respectively, are addressed by optimizers. These objects calculate gradients and perform update steps in the gradient's direction to minimize model loss. There are many optimizers available, from the simplest ones to the most advanced (refer to the following diagram). They provide different performances, and which one to select is, again, a matter of experience and a trial-and-error process. As an example, the following code selects the **Adam** optimizer, assigning a specific learning rate of **0.01**. This parameter regulates how "large" the step taken will be along the gradient direction:

```
optimizer = tf.keras.optimizers.Adam(learning_rate=0.01)
optimizer = tf.keras.optimizers.Adadelta(learning_rate=0.01)
optimizer = tf.keras.optimizers.Adagrad(learning_rate=0.01)
optimizer = tf.keras.optimizers.Adamax(learning_rate=0.01)
optimizer = tf.keras.optimizers.Ftrl(learning_rate=0.01)
optimizer = tf.keras.optimizers.Nadam(learning_rate=0.01)
optimizer = tf.keras.optimizers.RMSprop(learning_rate=0.01)
optimizer = tf.keras.optimizers.SGD(learning_rate=0.01)
```

The following diagram is an instantaneous snapshot comparing different optimizers. It shows how *quickly* they move toward the minimum, starting all at the same time. We can see how some of them are faster than others:

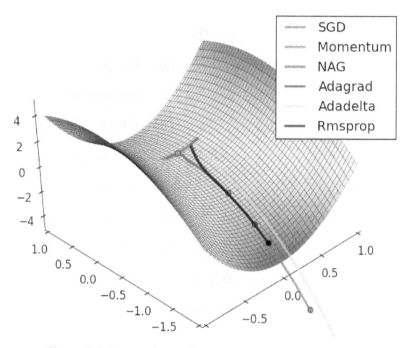

Figure 3.4: Comparison of optimizer minimization steps

NOTE

The preceding diagram was created by Alec Radford (https://twitter.com/alecrad).

LEARNING RATE SCHEDULING

In most cases, and for most deep learning models, the best results are achieved if the learning rate is gradually reduced during training. The reason for this can be seen in the following diagram:

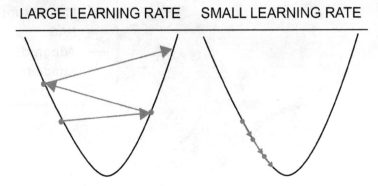

LARGE LEARNING RATE   SMALL LEARNING RATE

Figure 3.5: Optimization behavior when using different learning rate values

When approaching the minimum of the loss function, we want to take smaller and smaller steps to efficiently reach the very bottom of the hyperdimensional concavity.

With Keras, it is possible to prescribe many different decreasing functions for the learning rate trend over epochs by means of a scheduler. One common choice is **InverseTimeDecay**. This can be implemented as follows:

```
lr_schedule = tf.keras.optimizers.schedules\
             .InverseTimeDecay(0.001,\
                        decay_steps=STEPS_PER_EPOCH*1000,\
                        decay_rate=1, staircase=False)
```

The preceding code sets a decreasing function through **InverseTimeDecay** to hyperbolically decrease the learning rate to 1/2 of the base rate at 1,000 epochs, 1/3 at 2,000 epochs, and so on. This can be seen in the following graph:

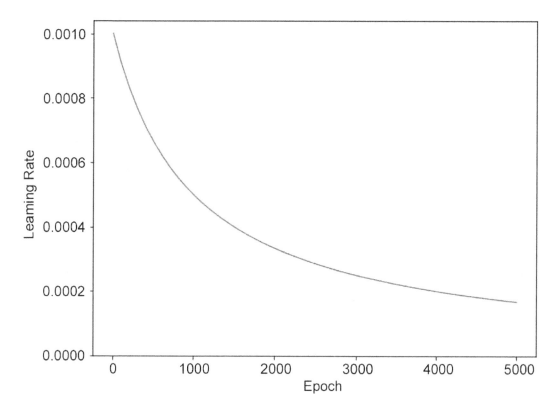

Figure 3.6: Inverse time decay learning rate scheduling

Then, it is applied to an optimizer as an argument, as shown in the following snippet for the **Adam** optimizer:

```
tf.keras.optimizers.Adam(lr_schedule)
```

Each optimization step makes the loss drop, thereby improving the model. It is then possible to repeat the same process over and over until convergence is reached and the loss stops decreasing. The number of optimization steps performed is usually called the number of epochs.

FEATURE NORMALIZATION

The broad applications for deep neural networks favor their usage on very different types of inputs, from image pixels to credit card transaction history, from social account profile habits to audio recordings. From this, it is clear that raw input features cover very different numerical scales. As mentioned previously, training these models requires solving an optimization problem using a loss gradient calculation. For this reason, numerical aspects are of paramount importance, resulting in a speeding up of the process, as well as making it more robust. One of the most important practices, in this context, is feature normalization or standardization. The most common approach consists of performing the following steps for each feature:

1. Calculating the mean and standard deviation using all the training set instances.

2. Subtracting the mean and dividing by standard deviation. Values calculated on the training set must be applied to the training, validation, and test sets.

This way, all the features will have zero mean and standard deviation equal to **1**. Different, but similar, approaches scale feature values between a user-defined minimum-maximum range (for example, between –1 and 1) or apply similar transformations (for example, log scaling). As usual, in the field, which approach works better is hardly predictable and requires experience and a trial-and-error approach.

The following code snippet shows how data normalization is performed, wherein the mean and standard deviation of the original values are calculated, the mean is then subtracted from the original values, and the result is then divided by the standard deviation:

```
train_stats = train_dataset.describe()
train_stats = train_stats.transpose()

def norm(x):
    return (x - train_stats['mean']) / train_stats['std']

normed_train_data = norm(train_dataset)
```

MODEL VALIDATION

As stated in the previous subsections, a large portion of choices require experimentation, meaning we have to select a given configuration and evaluate how the corresponding model performs. In order to compute this performance measure, the candidate model must be applied to a set of instances and its output compared against ground truth values. This step can be repeated many times, depending on how many alternative configurations we want to compare. In the long run, these configuration choices can suffer an excessive influence of the set of instances used to measure model performance. For this reason, in order to have a final accurate performance measure of the model of choice, it has to be tested on a new set of instances that have never been seen before. The first set of instances is called a "validation set," while the final one is called a "test set."

There are different choices we can adopt when defining training, validation, and test sets, such as the following:

- 70:20:10: The initial dataset is decomposed into three chunks, that is, the training, validation, and test sets, with the proportion 70:20:10, respectively.

- 80:20 + k-Folding: The initial dataset is decomposed into two chunks, 80% training and 20% testing, respectively. Validation is performed using k-Folding on the training dataset: it is divided into 'k' folds and, in turn, training is carried out in 'k-1' folds, while validation is performed on the k-th piece. 'K' varies from 1 to k and metrics are averaged to obtain a global measure.

Many variants of the preceding methods can be used. The choices are strictly related to the problem and the available dataset.

The following code snippet shows how to prescribe an 80:20 split for validation when fitting a model on a training dataset:

```
model.fit(normed_train_data, train_labels, epochs=epochs, \
          validation_split = 0.2, verbose=2)
```

PERFORMANCE METRICS

In order to measure performances, beside the loss functions, other metrics are usually adopted. There is a very wide set of metrics available, and the question as to which you should use depends on many factors, including the type of problem, dataset characteristics, and so on. The following is a list of the most common ones:

- **Mean Squared Error** (**MSE**): Used for regression problems.

- **Mean Absolute Error** (**MAE**): Used for regression problems.

- Accuracy: Number of correct predictions divided by the number of total tested instances. This is used for classification problems.

- **Receiver Operating Characteristic Area Under Curve** (**ROC AUC**): Used for binary classification, especially in the presence of highly unbalanced data.

- Others: Fβ score, precision, and recall.

MODEL IMPROVEMENT

In this section, we will look at a few techniques that can be used to improve the performance of a model.

OVERFITTING

A common problem we may typically encounter when training deep neural networks is a critical drop in model performance (measured, of course, on the validation or test set) when the number of training epochs passes a given threshold, even if, at the same time, the training loss continues to decrease. This phenomenon is called **overfitting**. It can be defined as follows: a highly representative model, a model with the relevant number of degrees of freedom (for example, a neural network with many layers and neurons), if trained "*too much*," bends itself to adhere to the training data, with the intent to minimize the training loss. This results in poor generalization performances, making validation and/or test errors higher. Deep learning models, thanks to their high-dimensional parameter space, are usually very good at fitting the training data, but the actual aim of building a machine learning model is being able to generalize what has been learned, not merely fit a dataset.

At this point, we might be tempted to significantly reduce the number of model parameters to avoid overfitting. But this would cause different problems. In fact, a model with an insufficient number of parameters would incur **underfitting**. Basically, it would not be able to properly fit the data, again resulting in poor performance, this time on both the training and validation/test sets.

The correct solution is the one that finds a proper balance between having a large number of parameters that would perfectly fit training data and having too small a number of model degrees of freedom, resulting in it being able to capture important information from data. It is currently not possible to identify the right size for a model so that it won't face overfitting or underfitting problems. Experimentation is a key element in this regard, thereby requiring the data engineer to build and test different architectures. A good rule is to start with models with a relatively small number of parameters and then increase them until generalization performance grows.

The best solution against overfitting is to enrich the training dataset with new data. Aim for complete coverage of the full range of inputs that are supported and expected by the model. New data should also contain additional information with respect to starting the dataset in order to effectively contrast overfitting and to result in a better generalization error. When collecting additional data is not possible or too expensive, it is necessary to adopt specific, very powerful techniques. The most important ones will be described here.

REGULARIZATION

Regularization is one of the most powerful tools used to contrast overfitting. Given a network architecture and a set of training data, there is an entire space of possible weights that produce the same results. Every combination of weights in this space defines a specific model. As we saw in the preceding section, we have to prefer, as a general principle, simple models over complex ones. A common way to reach this goal is to force network weights to assume small values, thereby regularizing the distribution of weights. This can be achieved through "weight regularization". This consists of shaping the loss function so that it can take weight values into consideration, adding a new term to it that is directly proportional to their magnitude. Two approaches are usually encountered:

- **L1 regularization**: The term that's added to the loss function is proportional to the absolute value of the weight coefficients, commonly referred to as the "L1 norm" of the weights.

- **L2 regularization**: The term that's added to the loss function is proportional to the square of the value of the weight coefficients, commonly referred to as the "L2 norm" of the weights.

Both of these have the effect of limiting the magnitude of the weights, but while L1 regularization tends to drive weights toward exactly zero, L2 regularization penalizes weights with a less strict constraint since the additional loss term grows at a higher rate. L2 is, in general, more common.

Keras contains pre-built L1 and L2 regularization objects. The user has to pass them as arguments to the network layers that they want to apply the technique to. The following code shows how to apply it to a common dense layer:

```
tf.keras.layers.Dense(512, activation='relu', \
                  kernel_regularizer=tf.keras\
                              .regularizers.l2(0.001))
```

The parameter that was passed to the L2 regularizer (`0.001`) shows that an additional loss term equal to `0.001 * weight_coefficient_value**2` will be added to the total loss of the network for every coefficient in the weight matrix.

EARLY STOPPING

Early stopping is a specific form of regularization. The idea is to keep track of both training and validation errors during training and to continue training the model until both training and validation losses decrease. This allows us to spot the epochs threshold, after which the training loss' decrease would come as an expense of increased generalization error, so that we can stop training when validation/test performances have reached their maximum. One typical parameter the user has to choose when adopting this technique is the number of epochs the system should wait for and monitor before stopping the iterations if no improvement in the validation error is shown. This parameter is commonly named "patience."

DROPOUT

One of the most popular and effective reliable regularization techniques for neural networks is Dropout. It was developed at the University of Toronto by Prof. Hinton and his research group.

When Dropout is applied to a layer, a certain percentage of the layer output features during training are randomly set to zero (they drop out). For example, if the output of a given layer would normally have been [0.3, 0.4, 1.2, 0.1, 1.5] for a given set of input features during training, when dropout is applied, the same output vector will have some zero entries randomly distributed; for example, [0.3, 0, 1.2, 0.1, 0].

The idea behind dropout is to encourage each node to output values that are highly informative and meaningful on their own, without relying on its neighboring ones.

The parameter to be set when inserting a dropout layer is called the **dropout rate**: this represents the fraction of features that are being set to zero and is usually chosen in a range between **0.2** and **0.5**. When performing inference, dropout is deactivated, and an additional operation needs to be executed to take into account the fact that more units will be active with respect to training time. To re-establish a balance between these two situations, the layer's output values are multiplied by a factor equal to the dropout rate, resulting in a scaling-down operation. In Keras, dropout can be introduced in a network using the dropout layer, which is applied to the output of the layer immediately before it. Consider the following code snippet:

```
dropout_model = tf.keras.Sequential([
    #[...]
    tf.keras.layers.Dense(512, activation='relu'), \
    tf.keras.layers.Dropout(0.5), \
    tf.keras.layers.Dense(256, activation='relu'), \
    #[...]
    ])
```

As you can see, dropout is applied to the layer with **512** neurons, setting 50% of their values to 0.0 at training time, and multiplying their values by 0.5 at inference time.

DATA AUGMENTATION

Data augmentation is particularly useful when the number of instances available for training is limited. It is super easy to understand how it is implemented and works in the context of image processing. Suppose we want to train a network to classify images of different breeds of a specific species and we only have a limited number of examples for each breed. How can we enlarge the dataset to help the model generalize better? Data augmentation plays a major role in this context: the idea is to create new training instances, starting from those we already have and tweaking them appropriately. In the case of images, we can act on them by doing the following:

- Random rotations with respect to a point in the vicinity of the center
- Random crops
- Random affine transformations (shear, resize, and so on)
- Random horizontal/vertical flips
- White noise superimposition
- Salt and pepper noise superimposition

These are a few examples of data augmentation techniques that can be used for images, which, of course, have counterparts in other domains. This approach makes the model way more robust and improves its generalization performance, allowing it to abstract notions and knowledge about the specific problem it is facing in a more general way by giving privilege to the most informative input features.

BATCH NORMALIZATION

Batch normalization is a technique that consists of applying a normalization transform to every batch of data. For example, in the context of training a deep network with a batch size of 128, meaning the system will process 128 training samples at a time, the batch normalization layer works this way:

1. It calculates the mean and variance for each feature using all the samples of the given batch.

2. It subtracts the corresponding feature mean that was previously calculated from each feature of every batch sample.

3. It divides each feature of every batch sample by the square root of the corresponding feature variance.

Batch normalization has many benefits. It was initially proposed to solve *internal covariate shift*. While training deep networks, the layer's parameters continuously change, causing internal layers to constantly adapt and readjust to new distributions they see as inputs coming from the preceding layers. This is particularly critical for deep networks, where small changes in the first layers are amplified through the network. Normalizing the layer's output helps in bounding these shifts, speeding up training and generating more reliable models.

In addition, using batch normalization, we can do the following:

- We can adopt a higher learning rate without the risk of incurring the problem of vanishing or exploding gradients.

- We can favor network regularization by making its generalization better and mitigating overfitting.

- We can make the model become more robust to different initialization schemes and learning rates.

MODEL TESTING AND INFERENCE

Once the model has been trained and its validation performances are satisfactory, we can move on to the final stage. As already stated, a final, accurate, model performance estimation requires that we test the model on a set of instances it has never seen before: the test set. After performance has been confirmed, the model can be moved to production for online inference, where it will serve as designed: new instances will be provided to the model and it will output predictions, leveraging the knowledge it has been designed and trained to have.

In the following subsections, three types of neural networks with specific elements/ layers will be described. They will provide straightforward examples of different technologies that are widely encountered in the field.

STANDARD FULLY CONNECTED NEURAL NETWORKS

The term *fully connected neural network* is commonly used to indicate deep neural networks that are only composed of fully connected layers. Fully connected layers are the layers whose neurons are connected to all the neurons of the previous layer, as well as all the neurons of the next one, as shown in the following diagram:

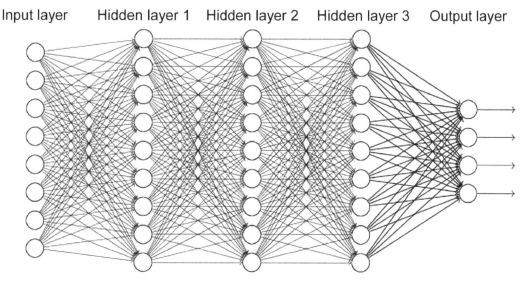

Figure 3.7: A fully connected neural network

This chapter will mainly deal with fully connected networks. They map inputs to outputs through a series of intermediate hidden layers. These architectures are capable of handling a wide variety of problems, but they are limited in terms of the input dimensions they can handle, as well as the number of layers and number of neurons, due to the rapid growth of the number of parameters, which is strictly dependent on these variables.

An example of a fully connected neural network that will be encountered later on is the one presented as follows, built with the Keras API. It connects an input layer who dimension is equal to **len(train_dataset.keys())** to an output layer of dimension **1**, by means of two hidden layers with **64** neurons each:

```
model = tf.keras.Sequential([tf.keras.layers.Dense\
        (64, activation='relu',\
        input_shape=[len(train_dataset.keys())]),\
        tf.keras.layers.Dense(64, activation='relu'),\
        tf.keras.layers.Dense(1)])
```

Now, let's quickly solve an exercise in order to aid our understanding of fully connected neural networks.

EXERCISE 3.02: BUILDING A FULLY CONNECTED NEURAL NETWORK MODEL WITH THE KERAS HIGH-LEVEL API

In this exercise, we will build a fully connected neural network with an input dimension of **100**, 2 hidden layers, and an output layer of **10** neurons. The following are the steps to complete this exercise:

1. Import the **TensorFlow** module and print its version:

```
from __future__ import absolute_import, division, \
print_function, unicode_literals

import tensorflow as tf
print("TensorFlow version: {}".format(tf.__version__))
```

This prints out the following line:

```
TensorFlow version: 2.1.0
```

2. Create the network using the Keras **sequential** module. This allows us to build a model by stacking a series of layers, one after the other. In this specific case, we're using two hidden layers and an output layer:

```
INPUT_DIM = 100
OUTPUT_DIM = 10

model = tf.keras.Sequential([tf.keras.layers.Dense\
        (128, activation='relu', \
        input_shape=[INPUT_DIM]), \
        tf.keras.layers.Dense(256, activation='relu'), \
        tf.keras.layers.Dense(OUTPUT_DIM, activation='softmax')])
```

3. Print the summary to look at the model description:

```
model.summary()
```

The output will be as follows:

```
Model: "sequential"
```

Layer (type)	Output Shape	Param #
dense (Dense)	(None, 128)	12928
dense_1 (Dense)	(None, 256)	33024
dense_2 (Dense)	(None, 10)	2570

```
Total params: 48,522
Trainable params: 48,522
Non-trainable params: 0
```

As you can see, the model has been created and the summary provides us with a clear understanding of the layers, their types and shapes, and the number of parameters of the network, which is very useful when building neural networks in real life.

> **NOTE**
>
> To access the source code for this specific section, please refer to https://packt.live/37s1M5w.
>
> You can also run this example online at https://packt.live/3f9WzSq.

Now, let's move on and understand convolutional neural networks.

CONVOLUTIONAL NEURAL NETWORKS

The term **Convolutional Neural Network** (**CNN**) usually identifies a deep neural network composed of a combination of the following:

- Convolutional layers

- Pooling layers

- Fully connected layers

One of the most successful applications of CNNs is in image and video processing tasks. In fact, they are way more capable, with respect to fully connected ones, of handling high-dimensional inputs such as images. They are also widely used for anomaly detection tasks, being used in autoencoders, as well as encoders for reinforcement learning algorithms, specifically for policy and value networks.

Convolutional layers can be thought of as a series of filters applied (convolved) to layer inputs to generate layer outputs. The main parameters of these layers are the number of filters they have and the dimension of the convolution kernel.

Pooling layers reduce the dimensions of the data; they combine the outputs of neuron clusters at one layer into a single neuron in the next layer. Pooling layers may compute a max (**MaxPooling**), which uses the maximum value from each cluster of neurons at the prior layer, or an average (**AveragePooling**), which uses the average value from each cluster of neurons at the prior layer.

These convolution/pooling operations encode input information in a compressed representation, up to a point where these new deep features, also called embeddings, are typically provided as inputs to standard fully connected layers at the very end of the network. A classic convolutional neural network schematization is represented in the following figure:

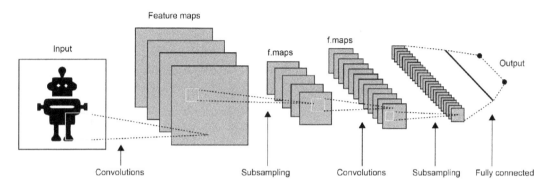

Figure 3.8: Convolutional neural network scheme

The following exercise shows how to create a convolutional neural network using the Keras high-level API.

EXERCISE 3.03: BUILDING A CONVOLUTIONAL NEURAL NETWORK MODEL WITH THE KERAS HIGH-LEVEL API

This exercise will show you how to build a convolutional neural network with three convolutional layers (number of filters equal to **16**, **32**, and **64**, respectively, and a kernel size of **3**), alternated with three **MaxPooling** layers, and, at the end, two fully connected layers with **512** and **1** neurons, respectively. Here is the step-by-step procedure:

1. Import the **TensorFlow** module and print its version:

```
from __future__ import absolute_import, division, \
print_function, unicode_literals

import tensorflow as tf
print("TensorFlow version: {}".format(tf.__version__))
```

This prints out the following line:

```
TensorFlow version: 2.1.0
```

2. Create the network using the Keras sequential module:

```
IMG_HEIGHT = 480
IMG_WIDTH = 680

model = tf.keras.Sequential([tf.keras.layers.Conv2D\
        (16, 3, padding='same',\
        activation='relu',\
        input_shape=(IMG_HEIGHT, IMG_WIDTH, 3)),\
        tf.keras.layers.MaxPooling2D(),\
        tf.keras.layers.Conv2D(32, 3, padding='same',\
        activation='relu'),\
        tf.keras.layers.MaxPooling2D(),\
        tf.keras.layers.Conv2D(64, 3, padding='same',\
        activation='relu'),\
        tf.keras.layers.MaxPooling2D(),\
        tf.keras.layers.Flatten(),\
        tf.keras.layers.Dense(512, activation='relu'),\
        tf.keras.layers.Dense(1)])

model.summary()
```

The preceding code allows us to build a model by stacking a series of layers, one after the other. In this specific case, three series of convolutional layers and max pooling layers are followed by a flattening layer and two dense layers.

This outputs the following model description:

```
Model: "sequential"
```

Layer (type)	Output Shape	Param #
conv2d (Conv2D)	(None, 480, 680, 16)	448
max_pooling2d (MaxPooling2D)	(None, 240, 340, 16)	0
conv2d_1 (Conv2D)	(None, 240, 340, 32)	4640
max_pooling2d_1 (MaxPooling2	(None, 120, 170, 32)	0
conv2d_2 (Conv2D)	(None, 120, 170, 64)	18496

```
max_pooling2d_2 (MaxPooling2 (None, 60, 85, 64)          0

flatten (Flatten)            (None, 326400)              0

dense (Dense)                (None, 512)                 167117312

dense_1 (Dense)              (None, 1)                   513
=================================================================
Total params: 167,141,409
Trainable params: 167,141,409
Non-trainable params: 0
```

Thus, we have successfully created a CNN using Keras. The preceding summary gives us significant information about the layers and the different parameters of the network.

> **NOTE**
>
> To access the source code for this specific section, please refer to https://packt.live/2AZJqwn.
>
> You can also run this example online at https://packt.live/37p1OuX.

Now that we've dealt with convolutional neural networks, let's focus on another important architecture family: recurrent neural networks.

RECURRENT NEURAL NETWORKS

Recurrent neural networks are models composed of particular units that, in the same way as feedforward networks, are able to process data from input to output, but, unlike them, are also able to process data in the opposite direction using feedback loops. They are basically designed so that the output of a layer is redirected and becomes the input of the same layer using specific internal states capable of "remembering" previous states.

This specific feature makes them particularly suited for solving tasks characterized by temporal/sequential development. It can be useful to compare CNNs and RNNs to understand which problems one is more suited to than the other. CNNs are the best fit for problems where local coherence is strongly enhanced and is particularly the case for images/video. Local coherence is exploited to drastically reduce the number of weights needed to process high-dimensional inputs. RNNs, on the other hand, perform best on problems characterized by temporal development, which means tasks where data can be represented by time series. This is the case for natural language processing or speech recognition, where words and sounds are meaningful if they're considered in a specific sequence.

Recurrent architectures can be thought of as sequences of operations, and they are perfectly designed to keep track of historical data:

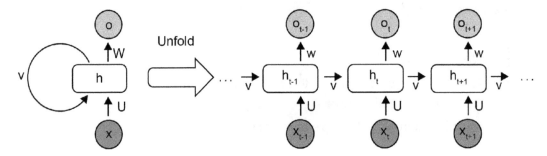

Figure 3.9: Recurrent neural network block diagram

The most important components they are based on are GRUs and LSTMs. These blocks have internal elements and states explicitly dedicated to keeping track of important information for the task they aim to solve. They both address the issue of learning long-term dependencies successfully when training machine learning algorithms on temporal data. They tackle this problem by storing "memory" from data seen in the past in order to help the network make predictions in the future.

The main differences between GRUs and LSTMs are the number of gates, the inputs the unit has, and the cell states, which are the internal elements the make up the unit's memory. GRUs have one gate, while LSTMs have three gates, called the input, forget, and output gates. LSTMs are more flexible than GRUs since they have more parameters, which, on the other hand, makes them less efficient in terms of both memory and time.

These networks have been responsible for the great advancements in fields such as speech recognition, natural language processing, text-to-speech, machine translation, language modeling, and many other similar tasks.

The following is a block diagram of a typical GRU:

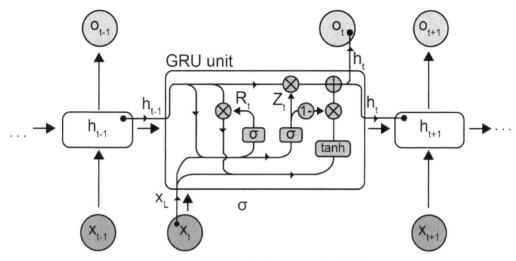

Figure 3.10: Block diagram of a GRU

The following is a block diagram of a typical LSTM:

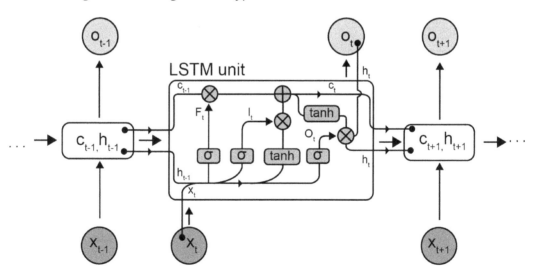

Figure 3.11: Block diagram of an LSTM

The following exercise shows how a recurrent network with LSTM units can be created using the Keras API.

EXERCISE 3.04: BUILDING A RECURRENT NEURAL NETWORK MODEL WITH THE KERAS HIGH-LEVEL API

In this exercise, we will create a recurrent neural network using the Keras high-level API. It will have the following architecture: the very first layer is simply a layer that encodes, using certain rules, the input features, thereby producing a given set of embeddings. The second layer is a layer where **64** LSTM units are added to it. They are added inside a bidirectional wrapper, which is a specific layer that's used to improve and speed up learning by doubling the units it acts on and training the first ones with the input as-is, and the second ones with the input reversed (for example, words in a sentence read from right to left). Then, the outputs are concatenated. This technique has been proven to generate faster and better learning. Finally, two dense layers are added that have **64** and **1** neurons, respectively. Perform the following steps to complete this exercise:

1. Import the **TensorFlow** module and print its version:

```
from __future__ import absolute_import, division, \
print_function, unicode_literals
import tensorflow as tf
print("TensorFlow version: {}".format(tf.__version__))
```

This outputs the following line:

```
TensorFlow version: 2.1.0
```

2. Build the model using the Keras **sequential** method and print the network summary:

```
EMBEDDING_SIZE = 8000

model = tf.keras.Sequential([\
        tf.keras.layers.Embedding(EMBEDDING_SIZE, 64),\
        tf.keras.layers.Bidirectional(tf.keras.layers.LSTM(64)),\
        tf.keras.layers.Dense(64, activation='relu'),\
        tf.keras.layers.Dense(1)])

model.summary()
```

In the preceding code, the model is simply built by stacking up consecutive layers. First, there is the embedding layer, then the bidirectional one, which operates on the LSTM layer, and finally two dense layers at the end of the model.

The model summary will be as follows:

```
Model: "sequential"
_____
Layer (type)                  Output Shape              Param #
=================================================================
embedding (Embedding)         (None, None, 64)          512000

bidirectional (Bidirectional  (None, 128)               66048

dense (Dense)                 (None, 64)                8256

dense_1 (Dense)               (None, 1)                 65
=================================================================
Total params: 586,369
Trainable params: 586,369
Non-trainable params: 0
```

> **NOTE**
>
> To access the source code for this specific section, please refer to https://packt.live/3cX01QO.
>
> You can also run this example online at https://packt.live/37nw1ud.

With this overview of how to implement a neural network using TensorFlow, the following sections will show you how to combine all these notions to tackle typical machine learning problems, including regression and classification problems.

SIMPLE REGRESSION USING TENSORFLOW

This section will explain, step by step, how to successfully tackle a regression problem. You will learn how to take a preliminary look at the dataset to understand its most important properties, as well as how to prepare it to be used during training, validation, and inference. Then, a deep neural network will be built from a clean sheet using TensorFlow via the Keras API. This model will then be trained and its performance will be evaluated.

In a regression problem, the aim is to predict the output of a continuous value, such as a price or a probability. In this exercise, the classic Auto MPG dataset will be used and a deep neural network will be trained on it to accurately predict car fuel efficiency, using no more than the following seven features: Cylinders, Displacement, Horsepower, Weight, Acceleration, Model Year, and Origin.

The dataset can be thought of as a table with eight columns (seven features, plus one target value) and as many rows as instances the dataset has. As per the best practices we looked at in the previous sections, it will be divided as follows: 20% of the total number of instances will create the test set, while the remaining ones will be split again into training and validation sets with an 80:20 ratio.

As a first step, the training set will be inspected for missing values, and cleaned if needed. Then, a chart showing variable correlation will be plotted. The only categorical variable present will be converted into numerical form via one-hot encoding. Finally, all the features will be normalized.

The deep learning model will then be created. A three-layered fully connected architecture will be used: the first and the second layer will have 64 nodes, while the last one, being the output layer of a regression problem, will have only one node.

Standard choices for the loss function (mean squared error) and optimizer (RMSprop) will be applied. Training will then be performed with and without early stopping to highlight the different effects they have on training and validation loss.

Finally, the model will be applied to the test set to evaluate performances and make predictions.

EXERCISE 3.05: CREATING A DEEP NEURAL NETWORK TO PREDICT THE FUEL EFFICIENCY OF CARS

In this exercise, we will build, train, and measure performances of a deep neural network model that predicts car fuel efficiency using only seven car features: **Cylinders**, **Displacement**, **Horsepower**, **Weight**, **Acceleration**, **Model Year**, and **Origin**.

The step-by-step procedure for this is as follows:

1. Import all the required modules and print the versions of the most important ones:

```
from __future__ import absolute_import, division, \
print_function, unicode_literals

import matplotlib.pyplot as plt
import numpy as np
import pandas as pd
import seaborn as sns

import tensorflow as tf

print("TensorFlow version: {}".format(tf.__version__))
```

The output will be as follows:

```
TensorFlow version: 2.1.0
```

2. Import the Auto MPG dataset, read it with pandas, and show the last five rows:

```
dataset_path = tf.keras.utils.get_file("auto-mpg.data", \
                "https://raw.githubusercontent.com/"\
                "PacktWorkshops/"\
                "The-Reinforcement-Learning-Workshop/master/"\
                "Chapter03/Dataset/auto-mpg.data")

column_names = ['MPG','Cylinders','Displacement','Horsepower',\
                'Weight', 'Acceleration', 'Model Year', 'Origin']
```

```
raw_dataset = pd.read_csv(dataset_path, names=column_names,\
                          na_values = "?", comment='\t',\
                          sep=" ", skipinitialspace=True)

dataset = raw_dataset.copy()
dataset.tail()
```

> **NOTE**
>
> Watch out for the slashes in the string below. Remember that the
> backslashes (\ ) are used to split the code across multiple lines, while the
> forward slashes (/) are part of the URL.

The output will be as follows:

	MPG	Cylinders	Displacement	Horsepower	Weight	Acceleration	Model Year	Origin
393	27.0	4	140.0	86.0	2790.0	15.6	82	1
394	44.0	4	97.0	52.0	2130.0	24.6	82	2
395	32.0	4	135.0	84.0	2295.0	11.6	82	1
396	28.0	4	120.0	79.0	2625.0	18.6	82	1
397	31.0	4	119.0	82.0	2720.0	19.4	82	1

Figure 3.12: Last five rows of the dataset imported in pandas

3. Let's clean the data from unknown values. Check how much **Not available** data is present and where:

```
dataset.isna().sum()
```

This produces the following output:

```
MPG              0
Cylinders        0
Displacement     0
Horsepower       6
Weight           0
Acceleration     0
Model Year       0
Origin           0
dtype: int64
```

4. Given the small number of rows with unknown values, simply drop them:

```
dataset = dataset.dropna()
```

5. Use one-hot encoding for the **Origin** variable, which is categorical:

```
dataset['Origin'] = dataset['Origin']\
                    .map({1: 'USA', 2: 'Europe', 3: 'Japan'})
dataset = pd.get_dummies(dataset, prefix='', prefix_sep='')
dataset.tail()
```

The output will be as follows:

	MPG	Cylinders	Displacement	Horsepower	Weight	Acceleration	Model Year	Europe	Japan	USA
393	27.0	4	140.0	86.0	2790.0	15.6	82	0	0	1
394	44.0	4	97.0	52.0	2130.0	24.6	82	1	0	0
395	32.0	4	135.0	84.0	2295.0	11.6	82	0	0	1
396	28.0	4	120.0	79.0	2625.0	18.6	82	0	0	1
397	31.0	4	119.0	82.0	2720.0	19.4	82	0	0	1

Figure 3.13: Last five rows of the dataset imported into pandas using one-hot encoding

6. Split the data into training and test sets with an 80:20 ratio:

```
train_dataset = dataset.sample(frac=0.8,random_state=0)
test_dataset = dataset.drop(train_dataset.index)
```

7. Now, let's take a look at some training data statistics, that is, the joint distributions of some pairs of features from the training set, using the **seaborn** module. The **pairplot** command takes in the features of the dataset as input to evaluate them, couple by couple. Along the diagonal (where the couple is composed of two instances of the same feature), it shows the distribution of the variable, while in the off-diagonal terms, it shows the scatterplot of the two features. This is useful if we wish to highlight correlations:

```
sns.pairplot(train_dataset[["MPG", "Cylinders", "Displacement", \
                    "Weight"]], diag_kind="kde")
```

This generates the following image:

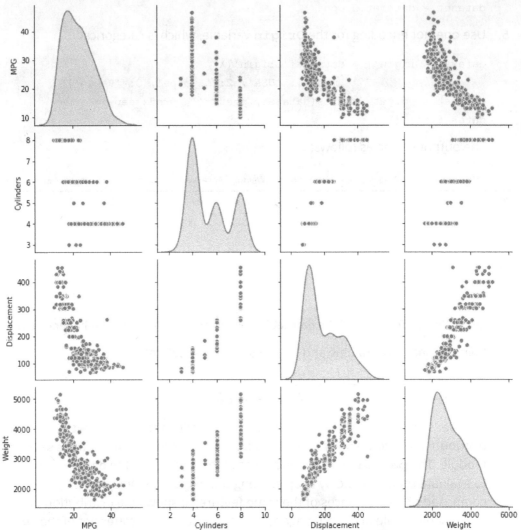

Figure 3.14: Joint distributions of some pairs of features from the training set

8. Let's now take a look at the overall statistics:

```
train_stats = train_dataset.describe()
train_stats.pop("MPG")
train_stats = train_stats.transpose()
train_stats
```

The output will be as follows:

	count	mean	std	min	25%	50%	75%	max
Cylinders	314.0	5.477707	1.699788	3.0	4.00	4.0	8.00	8.0
Displacement	314.0	195.318471	104.331589	68.0	105.50	151.0	265.75	455.0
Horsepower	314.0	104.869427	38.096214	46.0	76.25	94.5	128.00	225.0
Weight	314.0	2990.251592	843.898596	1649.0	2256.50	2822.5	3608.00	5140.0
Acceleration	314.0	15.559236	2.789230	8.0	13.80	15.5	17.20	24.8
Model Year	314.0	75.898089	3.675642	70.0	73.00	76.0	79.00	82.0
Europe	314.0	0.178344	0.383413	0.0	0.00	0.0	0.00	1.0
Japan	314.0	0.197452	0.398712	0.0	0.00	0.0	0.00	1.0
USA	314.0	0.624204	0.485101	0.0	0.00	1.0	1.00	1.0

Figure 3.15: Overall training set statistics

9. Split the features from the labels and normalize the data:

```
train_labels = train_dataset.pop('MPG')
test_labels = test_dataset.pop('MPG')

def norm(x):
    return (x - train_stats['mean']) / train_stats['std']

normed_train_data = norm(train_dataset)
normed_test_data = norm(test_dataset)
```

10. Now, let's look at the model's creation and a summary of the same:

```
def build_model():
    model = tf.keras.Sequential([
            tf.keras.layers.Dense(64, activation='relu',\
                            input_shape=[len\
                            (train_dataset.keys())]),\
            tf.keras.layers.Dense(64, activation='relu'),\
            tf.keras.layers.Dense(1)])

    optimizer = tf.keras.optimizers.RMSprop(0.001)
```

```
        model.compile(loss='mse', optimizer=optimizer,\
                    metrics=['mae', 'mse'])
    return model

model = build_model()
model.summary()
```

This generates the following output:

```
Model: "sequential"
```

Layer (type)	Output Shape	Param #
dense (Dense)	(None, 64)	640
dense_1 (Dense)	(None, 64)	4160
dense_2 (Dense)	(None, 1)	65

```
Total params: 4,865
Trainable params: 4,865
Non-trainable params: 0
```

Figure 3.16: Model summary

11. Use the **fit** model function to train the network for 1,000 epochs by using a validation set of 20%:

```
epochs = 1000

history = model.fit(normed_train_data, train_labels,\
                    epochs=epochs, validation_split = 0.2, \
                    verbose=2)
```

This will produce a very long output. We will only report the last few lines here:

```
Epoch 999/1000251/251 - 0s - loss: 2.8630 - mae: 1.0763
- mse: 2.8630 - val_loss: 10.2443 - val_mae: 2.3926
- val_mse: 10.2443

Epoch 1000/1000251/251 - 0s - loss: 2.7697 - mae: 0.9985
- mse: 2.7697 - val_loss: 9.9689 - val_mae: 2.3709 - val_mse: 9.9689
```

12. Visualize the training and validation metrics by plotting the MAE and MSE.

The following snippet plots the MAE:

```
hist = pd.DataFrame(history.history)
hist['epoch'] = history.epoch

plt.plot(hist['epoch'],hist['mae'])
plt.plot(hist['epoch'],hist['val_mae'])
plt.ylim([0, 10])
plt.ylabel('MAE [MPG]')
plt.legend(["Training", "Validation"])
```

The output will be as follows:

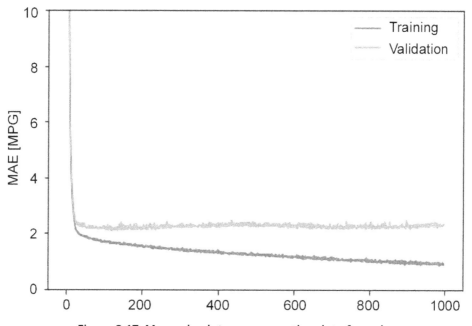

Figure 3.17: Mean absolute error over the plot of epochs

The preceding figure shows how increasing the training epochs causes the validation error to grow, meaning the system is experiencing an overfitting problem.

13. Now, let's visualize the MSE using a plot:

```
plt.plot(hist['epoch'],hist['mse'])
plt.plot(hist['epoch'],hist['val_mse'])
plt.ylim([0, 20])
plt.ylabel('MSE [MPG^2]')
plt.legend(["Training", "Validation"])
```

The output will be as follows:

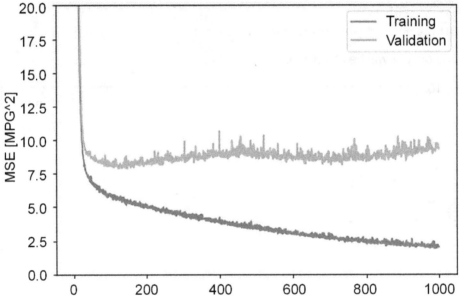

Figure 3.18: Mean squared error over the plot of epochs

Also, in this case, the figure shows how increasing the training epochs causes the validation error to grow, meaning the system is experiencing an overfitting problem.

14. Use Keras callbacks to add early stopping (with the patience parameter equal to 10 epochs) to avoid overfitting. First of all, build the model:

```
model = build_model()
```

15. Then, define an early stopping callback. This entity will be passed to the **`model.fit`** function and will be called every fit step to check whether the validation error stops decreasing for more than **10** consecutive epochs:

```
early_stop = tf.keras.callbacks\
            .EarlyStopping(monitor='val_loss', patience=10)
```

16. Finally, call the **`fit`** method with the early stop callback:

```
early_history = model.fit(normed_train_data, train_labels,\
                          epochs=epochs, validation_split=0.2,\
                          verbose=2, callbacks=[early_stop])
```

The last few lines of the output are as follows:

```
Epoch 42/1000251/251 - 0s - loss: 7.1298 - mae: 1.9014
- mse: 7.1298 - val_loss: 8.1151 - val_mae: 2.1885
- val_mse: 8.1151


Epoch 43/1000251/251 - 0s - loss: 7.0575 - mae: 1.8513
- mse: 7.0575 - val_loss: 8.4124 - val_mae: 2.2669
- val_mse: 8.4124
```

17. Visualize the train and validation metrics for early stopping. Firstly, collect all the training history data and put it into a pandas DataFrame, for both the metric and epoch values:

```
early_hist = pd.DataFrame(early_history.history)
early_hist['epoch'] = early_history.epoch
```

18. Then, plot the training and validation MAE against the epochs, limiting the max **`y`** values to **10**:

```
plt.plot(early_hist['epoch'],early_hist['mae'])
plt.plot(early_hist['epoch'],early_hist['val_mae'])
plt.ylim([0, 10])
plt.ylabel('MAE [MPG]')
plt.legend(["Training", "Validation"])
```

The preceding code will produce the following output:

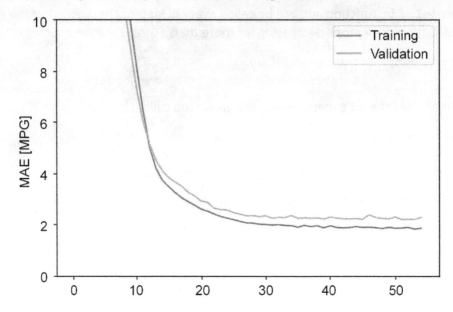

Figure 3.19: Mean absolute error over the plot of epochs (early stopping)

As demonstrated by the preceding figure, training is stopped as soon as the validation error stops decreasing, thereby avoiding overfitting.

19. Evaluate the model accuracy on the test set:

```
loss, mae, mse = model.evaluate(normed_test_data, \
                           test_labels, verbose=2)

print("Testing set Mean Abs Error: {:5.2f} MPG".format(mae))
```

The output will be as follows:

```
78/78 - 0s - loss: 6.3067 - mae: 1.8750 - mse: 6.3067
Testing set Mean Abs Error:  1.87 MPG
```

NOTE

The accuracy may show slightly different values due to random sampling with a variable random seed.

20. Finally, perform model inference by predicting all the MPG values for all test instances. Then, plot these values with respect to their true values so that you have a visual estimation of the model error:

```
test_predictions = model.predict(normed_test_data).flatten()

a = plt.axes(aspect='equal')
plt.scatter(test_labels, test_predictions)
plt.xlabel('True Values [MPG]')
plt.ylabel('Predictions [MPG]')
lims = [0, 50]
plt.xlim(lims)
plt.ylim(lims)
_ = plt.plot(lims, lims)
```

The output will be as follows:

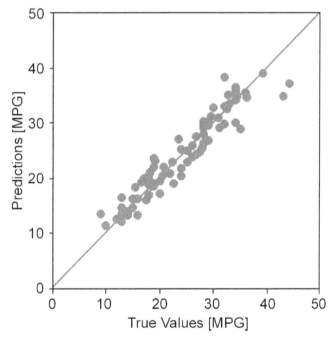

Figure 3.20: Predictions versus ground truth scatterplot

The scatterplot puts predicted values versus true values in correspondence with one another, which means that the closer the points are to the diagonal line, the more accurate the predictions will be. It is evident how clustered the points are, meaning predictions are fairly accurate.

> **NOTE**
>
> To access the source code for this specific section, please refer to https://packt.live/3feCLNN.
>
> You can also run this example online at https://packt.live/37n5WeM.

This section has shown how to successfully tackle a regression problem. The selected dataset has been imported, cleaned, and subdivided into training, validation, and test sets. Then, a brief exploratory data analysis was carried out before a three-layered fully connected deep neural network was created. The network has been successfully trained and its performance has been evaluated on the test set.

Now, let's study classification problems using TensorFlow.

SIMPLE CLASSIFICATION USING TENSORFLOW

This section will help you understand and solve a typical supervised learning problem that falls under the category conventionally named **classification**.

Classification tasks, in their simplest generic form, aim to associate one category, among a predefined set, with instances. An intuitive example of a classification task that's often used for introductory courses is classifying the images of domestic pets in the correct category they belong to, such as "cat" or "dog." Classification plays a fundamental role in many everyday activities and can easily be encountered in different contexts. The previous example is a specific case of classification called **image classification**, and many similar applications can be found in this category.

However, classification extends beyond images. The following are some examples:

- Customer classification for video recommendation systems (answering the question, "In which market segment this user falls?")

- Spam filters ("What are the chances this email is spam?")

- Malware detection ("Is this program a cyber threat?")

- Medical diagnosis ("Is this patient sick?")

For image classification tasks, images are fed to the classification algorithm as inputs, and it returns the class they belong to as output. Images are three-dimensional arrays of numbers representing per-pixel brightness (height x width x number of channels, where color images have three channels – red, green, blue (RGB) – and grayscale images only have one), and these numbers are the features that the algorithm uses to determine the class images belong to.

When dealing with other types of inputs, features can be different. For example, in the case of a medical diagnosis classification system, blood test parameters, age, sex, and suchlike can be features that are used by the algorithm to identify the class the instance belongs to, that is, "sick" or "not sick."

In the following exercise, we will create a deep neural network by building upon what we described in the previous sections. This will be able to achieve an accuracy of around 70% when classifying signals that have been detected inside a simulated ATLAS experiment, distinguishing between background noise and Higgs Boson Tau-Tau decay using a set of 28 features: yes, machine learning applied to particle physics!

> **NOTE**
>
> For additional information on the dataset, visit the official website:
> http://archive.ics.uci.edu/ml/datasets/HIGGS.

Given the huge size of the dataset, to keep the exercise easy to run and still meaningful, it will be subsampled: 10,000 rows will be used for training and 1,000 rows each for validation and test. Three different models will be trained: a small model that will be a reference (two layers with 16 and 1 neurons each), a large model with no overfit countermeasures (five layers; four with 512 neurons and the last one with 1 neuron) to demonstrate problems that may be encountered in this scenario, and then regularization and dropout will be added to the large model, effectively limiting overfitting and improving performance.

EXERCISE 3.06: CREATING A DEEP NEURAL NETWORK TO CLASSIFY EVENTS GENERATED BY THE ATLAS EXPERIMENT IN THE QUEST FOR HIGGS BOSON

In this exercise, we will build, train, and measure the performance of a deep neural network in order to improve the discovery significance of the ATLAS experiment by using simulated data with features for characterizing events. The task is to classify events into two categories: "tau decay of a Higgs Boson" versus "background."

This dataset can be found in the TensorFlow dataset (https://www.tensorflow.org/datasets), which is a collection of ready-to-use datasets. It is available to download and interface via the processing pipeline. In our case, the original dataset is too big for our purposes, so we will postpone dataset usage until we get to this chapter's activity. For now, we will use a subgroup of the dataset that's directly available through the repository.

> **NOTE**
>
> You can find the dataset in this book's GitHub repository here: https://packt.live/3dUfYq8.

The step-by-step procedure is described in detail as follows:

1. Import all the required modules and print the versions of the most important ones:

```
from __future__ import absolute_import, division, \
print_function, unicode_literals

from  IPython import display
from matplotlib import pyplot as plt
from scipy.ndimage.filters import gaussian_filter1d
import pandas as pd
import numpy as np

import tensorflow as tf
print("TensorFlow version: {}".format(tf.__version__))
```

The output will be as follows:

```
TensorFlow version: 2.1.0
```

2. Import the dataset and prepare the data for preprocessing.

For this exercise, we will download a custom-made smaller subset that's been pulled from the original dataset:

```
higgs_path = tf.keras.utils.get_file('HIGGSSmall.csv.gz', \
            'https://github.com/PacktWorkshops/'\
            'The-Reinforcement-Learning-Workshop/blob/'\
            'master/Chapter03/Dataset/HIGGSSmall.csv.gz?raw=true')
```

3. Read the CSV dataset into a TensorFlow dataset class and repack it so that it has tuples (**features**, **labels**):

```
N_TEST = int(1e3)
N_VALIDATION = int(1e3)
N_TRAIN = int(1e4)
BUFFER_SIZE = int(N_TRAIN)
BATCH_SIZE = 500
STEPS_PER_EPOCH = N_TRAIN//BATCH_SIZE


N_FEATURES = 28

ds = tf.data.experimental\
     .CsvDataset(higgs_path,[float(),]*(N_FEATURES+1), \
                compression_type="GZIP")


def pack_row(*row):
    label = row[0]
    features = tf.stack(row[1:],1)
    return features, label


packed_ds = ds.batch(N_TRAIN).map(pack_row).unbatch()
```

Take a look at the value distribution of the features:

```
for features,label in packed_ds.batch(1000).take(1):
    print(features[0])
    plt.hist(features.numpy().flatten(), bins = 101)
```

The output will be as follows:

```
tf.Tensor(
[ 0.8692932  -0.6350818   0.22569026   0.32747006 -0.6899932
  0.7542022  -0.2485731  -1.0920639   0.          1.3749921
 -0.6536742   0.9303491   1.1074361    1.1389043  -1.5781983
 -1.0469854   0.          0.65792954  -0.01045457 -0.04576717
  3.1019614   1.35376     0.9795631    0.97807616  0.92000484
  0.72165745 0.98875093 0.87667835], shape=(28,), dtype=float32)
```

The plot will be as follows:

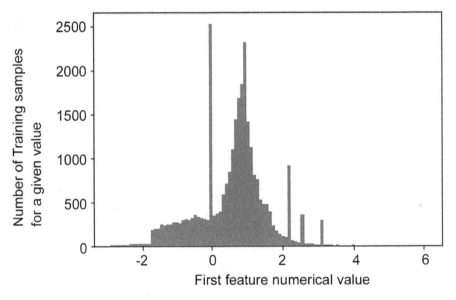

Figure 3.21: First feature value distribution

In the preceding graph, the *x* axis represents the number of training samples for a given value, while the *y* axis denotes the first feature's numerical value.

4. Create training, validation, and test sets:

```
validate_ds = packed_ds.take(N_VALIDATION).cache()
test_ds = packed_ds.skip(N_VALIDATION).take(N_TEST).cache()
train_ds = packed_ds.skip(N_VALIDATION+N_TEST)\
        .take(N_TRAIN).cache()
```

5. Define feature, label, and class names:

```
feature_names = ["lepton pT", "lepton eta", "lepton phi",\
                "missing energy magnitude", \
                "missing energy phi",\
                "jet 1 pt", "jet 1 eta", "jet 1 phi",\
                "jet 1 b-tag",\
                "jet 2 pt", "jet 2 eta", "jet 2 phi",\
                "jet 2 b-tag",\
                "jet 3 pt", "jet 3 eta", "jet 3 phi",\
                "jet 3 b-tag",\
                "jet 4 pt", "jet 4 eta", "jet 4 phi",\
                "jet 4 b-tag",\
                "m_jj", "m_jjj", "m_lv", "m_jlv", "m_bb",\
                "m_wbb", "m_wwbb"]
label_name = ['Measure']
class_names = ['Signal', 'Background']

print("Features: {}".format(feature_names))
print("Label: {}".format(label_name))
print("Class names: {}".format(class_names))
```

The output will be as follows:

```
Features: ['lepton pT', 'lepton eta', 'lepton phi',
'missing energy magnitude', 'missing energy phi',
'jet 1 pt', 'jet 1 eta', 'jet 1 phi', 'jet 1 b-tag',
'jet 2 pt', 'jet 2 eta', 'jet 2 phi', 'jet 2 b-tag',
'jet 3 pt', 'jet 3 eta', 'jet 3 phi', 'jet 3 b-tag',
'jet 4 pt', 'jet 4 eta', 'jet 4 phi', 'jet 4 b-tag',
'm_jj', 'm_jjj', 'm_lv', 'm_jlv', 'm_bb', 'm_wbb', 'm_wwbb']
Label: ['Measure']
Class names: ['Signal', 'Background']
```

6. Show a sample of a training instance for features and labels:

```
features, labels = next(iter(train_ds))
print("Features =")
print(features.numpy())
print("Labels =")
print(labels.numpy())
```

The output will be as follows:

```
Features =
[ 0.3923715    1.3781117    1.5673449    0.17123567   1.6574531
0.86394763    0.88821083   1.4797885    2.1730762    1.2008675
0.9490923 -0.30092147    2.2148721    1.277294     0.4025028
0.50748837  0.           0.50555664
 -0.55428815 -0.7055601   0.           0.94152564   0.9448251
0.9839765     0.7801499    1.4989641    0.91668195   0.8027126 ]
Labels = 0.0
```

7. Assign a batch size to the datasets:

```
test_ds = test_ds.batch(BATCH_SIZE)
validate_ds = validate_ds.batch(BATCH_SIZE)
train_ds = train_ds.shuffle(BUFFER_SIZE).repeat()\
           .batch(BATCH_SIZE)
```

8. Now, let's start creating the model and training it. Create a decaying learning rate:

```
lr_schedule = tf.keras.optimizers.schedules\
              .InverseTimeDecay(0.001,\
                                decay_steps=STEPS_PER_EPOCH*1000, \
                                decay_rate=1,   staircase=False)
```

9. Define a function that will compile a model with an **Adam** optimizer, use binary cross entropy as the **loss** function, and fit it on training data by using early stopping on the validation dataset.

    The function takes in the model as input, chooses the **Adam** optimizer, and compiles the model with it, as well as with the binary cross entropy loss and the accuracy metrics:

```
def compile_and_fit(model, name, max_epochs=3000):
    optimizer = tf.keras.optimizers.Adam(lr_schedule)
    model.compile(optimizer=optimizer,\
    loss=tf.keras.losses.BinaryCrossentropy(from_logits=True),\
    metrics=[tf.keras.losses.BinaryCrossentropy(from_logits=True,\
            name='binary_crossentropy'),'accuracy'])
```

A summary of the model is then printed, as follows:

```
model.summary()
```

10. The model is then fitted on the training dataset using a validation dataset and the early stopping callback. The training **history** is saved and returned as output:

```
history = model.fit(train_ds, \
            steps_per_epoch = STEPS_PER_EPOCH,\
            epochs=max_epochs, validation_data=validate_ds, \
            callbacks=[tf.keras.callbacks\
                    .EarlyStopping\
                    (monitor='val_binary_crossentropy',\
                    patience=200)],verbose=2)
    return history
```

11. Create a small model with just two layers with 16 and 1 neurons, respectively, and compile it and fit it on the dataset:

```
small_model = tf.keras.Sequential([\
            tf.keras.layers.Dense(16, activation='elu',\
                                input_shape=(N_FEATURES,)),\
            tf.keras.layers.Dense(1)])

size_histories = {}

size_histories['small'] = compile_and_fit(small_model, 'sizes/small')
```

This will produce a long output, where the last two lines will be similar to the following:

```
Epoch 1522/3000
20/20 - 0s - loss: 0.5693 - binary_crossentropy: 0.5693
- accuracy: 0.6846 - val_loss: 0.5841
- val_binary_crossentropy: 0.5841 - val_accuracy: 0.6640

Epoch 1523/3000
20/20 - 0s - loss: 0.5695 - binary_crossentropy: 0.5695
- accuracy: 0.6822 - val_loss: 0.5845
- val_binary_crossentropy: 0.5845 - val_accuracy: 0.6600
```

12. Check the model's performance on the test set:

```
test_accuracy = tf.keras.metrics.Accuracy()

for (features, labels) in test_ds:
    logits = small_model(features)
    probabilities = tf.keras.activations.sigmoid(logits)
    predictions = 1*(probabilities.numpy() > 0.5)
    test_accuracy(predictions, labels)
    small_model_accuracy = test_accuracy.result()

print("Test set accuracy:{:.3%}".format(test_accuracy.result()))
```

The output will be as follows:

```
Test set accuracy: 68.200%
```

> **NOTE**
>
> The accuracy may show slightly different values due to random sampling with a variable random seed.

13. Create a large model with five layers – four with **512** neurons and the last one with **1** neuron, respectively – and compile and fit it:

```
large_model = tf.keras.Sequential([\
            tf.keras.layers.Dense(512, activation='elu',\
                                input_shape=(N_FEATURES,)),\
            tf.keras.layers.Dense(512, activation='elu'),\
            tf.keras.layers.Dense(512, activation='elu'),\
            tf.keras.layers.Dense(512, activation='elu'),\
            tf.keras.layers.Dense(1)])
size_histories['large'] = compile_and_fit(large_model, "sizes/large")
```

This will produce a long output, where the last two lines will be similar to the following:

```
Epoch 221/3000
20/20 - 0s - loss: 1.0285e-04 - binary_crossentropy: 1.0285e-04
- accuracy: 1.0000 - val_loss: 2.5506
- val_binary_crossentropy: 2.5506 - val_accuracy: 0.6660
Epoch 222/3000
20/20 - 0s - loss: 1.0099e-04 - binary_crossentropy: 1.0099e-04
- accuracy: 1.0000 - val_loss: 2.5586
- val_binary_crossentropy: 2.5586 - val_accuracy: 0.6650
```

14. Check the model's performance on the test set:

```
test_accuracy = tf.keras.metrics.Accuracy()

for (features, labels) in test_ds:
    logits = large_model(features)
    probabilities = tf.keras.activations.sigmoid(logits)
    predictions = 1*(probabilities.numpy() > 0.5)
    test_accuracy(predictions, labels)
    large_model_accuracy = test_accuracy.result()
    regularization_model_accuracy = test_accuracy.result()

print("Test set accuracy: {:.3%}"\
      . format(regularization_model_accuracy))
```

The output will be as follows:

```
Test set accuracy: 65.200%
```

NOTE

The accuracy may show slightly different values due to random sampling with a variable random seed.

15. Create the same large model as before, but add regularization items such as L2 regularization and dropout. Then, compile it and fit the model to the set:

```
regularization_model = tf.keras.Sequential([\
                    tf.keras.layers.Dense(512,\
                    kernel_regularizer=tf.keras.regularizers\
                                    .l2(0.0001),\
                    activation='elu', \
                    input_shape=(N_FEATURES,)),\
                    tf.keras.layers.Dropout(0.5),\
                    tf.keras.layers.Dense(512,\
                    kernel_regularizer=tf.keras.regularizers\
                                    .l2(0.0001),\
                    activation='elu'),\
                    tf.keras.layers.Dropout(0.5),\
                    tf.keras.layers.Dense(512,\
                    kernel_regularizer=tf.keras.regularizers\
                                    .l2(0.0001),\
                    activation='elu'),\
                    tf.keras.layers.Dropout(0.5),\
                    tf.keras.layers.Dense(512,\
                    kernel_regularizer=tf.keras.regularizers\
                                    .l2(0.0001),\
                    activation='elu'),\
                    tf.keras.layers.Dropout(0.5),\
                    tf.keras.layers.Dense(1)])


size_histories['regularization'] = compile_and_fit\
                            (regularization_model,\
                            "regularizers/regularization",\
                            max_epochs=9000)
```

This will produce a long output, where the last two lines will be similar to the following:

```
Epoch 1264/9000
20/20 - 0s - loss: 0.5873 - binary_crossentropy: 0.5469
- accuracy: 0.6978 - val_loss: 0.5819
- val_binary_crossentropy: 0.5416 - val_accuracy: 0,7030
Epoch 1265/9000
20/20 - 0s - loss: 0.5868 - binary_crossentropy: 0.5465
- accuracy: 0.7024 - val_loss: 0.5759
- val_binary_crossentropy: 0.5356 - val_accuracy: 0.7100
```

16. Check the model's performance on the test set:

```
test_accuracy = tf.keras.metrics.Accuracy()

for (features, labels) in test_ds:
    logits = regularization_model (features)
    probabilities = tf.keras.activations.sigmoid(logits)
    predictions = 1*(probabilities.numpy() > 0.5)
    test_accuracy(predictions, labels)

print("Test set accuracy: {:.3%}".format(test_accuracy.result()))
```

The output will be as follows:

```
Test set accuracy: 69.300%
```

> **NOTE**
>
> The accuracy may show slightly different values due to random sampling with a variable random seed.

17. Compare the binary cross entropy trend of the three models over epochs:

```
histSmall = pd.DataFrame(size_histories["small"].history)
histSmall['epoch'] = size_histories["small"].epoch

histLarge = pd.DataFrame(size_histories["large"].history)
histLarge['epoch'] = size_histories["large"].epoch

histReg = pd.DataFrame(size_histories["regularization"].history)
histReg['epoch'] = size_histories["regularization"].epoch

trainSmoothSmall = gaussian_filter1d\
                    (histSmall['binary_crossentropy'], sigma=3)
testSmoothSmall = gaussian_filter1d\
                    (histSmall['val_binary_crossentropy'], sigma=3)

trainSmoothLarge = gaussian_filter1d\
                    (histLarge['binary_crossentropy'], sigma=3)
testSmoothLarge = gaussian_filter1d\
                    (histLarge['val_binary_crossentropy'], sigma=3)

trainSmoothReg = gaussian_filter1d\
                    (histReg['binary_crossentropy'], sigma=3)
testSmoothReg = gaussian_filter1d\
                    (histReg['val_binary_crossentropy'], sigma=3)

plt.plot(histSmall['epoch'], trainSmoothSmall, '-', \
         histSmall['epoch'], testSmoothSmall, '--')
plt.plot(histLarge['epoch'], trainSmoothLarge, '-', \
         histLarge['epoch'], testSmoothLarge, '--')
plt.plot(histReg['epoch'], trainSmoothReg, '-', \
         histReg['epoch'], testSmoothReg, '--',)
plt.ylim([0.5, 0.7])
plt.ylabel('Binary Crossentropy')
plt.legend(["Small Training", "Small Validation", \
            "Large Training", "Large Validation", \
            "Regularization Training", \
            "Regularization Validation"])
```

This will produce the following graph:

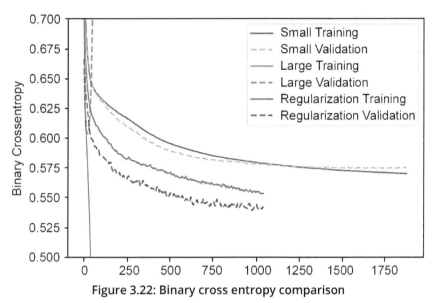

Figure 3.22: Binary cross entropy comparison

The preceding graph shows a comparison of the different models, in terms of both training and validation errors, to demonstrate how overfitting works. The training error goes down for each of them as the number of training epochs increases. The validation error for the large model, on the other hand, rapidly increases after a certain number of epochs. In the small model, it goes down, following the training error closely and reaching a final performance that is worse than the one obtained by the model with regularization, which avoids overfitting and has the best performance among the three.

18. Compare the accuracy trend of the three models over epochs:

```
trainSmoothSmall = gaussian_filter1d\
                   (histSmall['accuracy'], sigma=6)
testSmoothSmall = gaussian_filter1d\
                  (histSmall['val_accuracy'], sigma=6)

trainSmoothLarge = gaussian_filter1d\
                   (histLarge['accuracy'], sigma=6)
testSmoothLarge = gaussian_filter1d\
                  (histLarge['val_accuracy'], sigma=6)

trainSmoothReg = gaussian_filter1d\
                 (histReg['accuracy'], sigma=6)
```

```
testSmoothReg = gaussian_filter1d\
                (histReg['val_accuracy'], sigma=6)

plt.plot(histSmall['epoch'], trainSmoothSmall, '-', \
         histSmall['epoch'], testSmoothSmall, '--')
plt.plot(histLarge['epoch'], trainSmoothLarge, '-', \
         histLarge['epoch'], testSmoothLarge, '--')
plt.plot(histReg['epoch'], trainSmoothReg, '-', \
         histReg['epoch'], testSmoothReg, '--',)

plt.ylim([0.5, 0.75])
plt.ylabel('Accuracy')
plt.legend(["Small Training", "Small Validation", \
            "Large Training", "Large Validation",\
            "Regularization Training", \
            "Regularization Validation",])
```

This will produce the following graph:

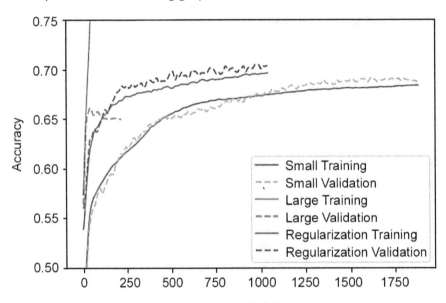

Figure 3.23: Accuracy comparison

In a specular way with respect to the previous one, this graph shows, once again, a comparison of the different models, but in terms of accuracy. The training accuracy grows for each model when the number of training epochs increases. The validation accuracy for the large model, on the other hand, stops growing after a certain number of epochs. In the small model, it goes up, following the training one closely and reaching a final performance that is worse than the one obtained by the model with regularization, which avoids overfitting and attains the best performance among the three.

> **NOTE**
>
> To access the source code for this specific section, please refer to https://packt.live/37m9huu.
>
> You can also run this example online at https://packt.live/3hhIDaZ.

In this section, we solved a fancy classification problem, resulting in the creation of a deep learning model able to achieve about 70% accuracy when classifying Higgs boson-related signals using simulated ATLAS experiment data. After a first general overview of the dataset, where we understood how it is arranged and the nature of its features and labels, a set of three deep fully connected neural networks were created using the Keras API. These models were trained and tested, and their performances in terms of loss and accuracy over epochs have been compared, thereby giving us a firm grasp of the overfitting problem and which techniques help in solving it.

TENSORBOARD – HOW TO VISUALIZE DATA USING TENSORBOARD

TensorBoard is a web-based tool embedded in TensorFlow. It provides a suite of methods that we can use to get insights into TensorFlow sessions and graphs, thus allowing the user to inspect, visualize, and understand them deeply. It provides access to many functionalities in a straightforward way, as follows:

- It allows us to explore the details of TensorFlow model graphs, making the user able to zoom in to specific blocks and subsections.

- It can generate plots of typical quantities of interest that we can take a look at during training, such as loss and accuracy.

- It gives us access to histogram visualizations that show tensors changing over time.

- It provides trends of layer weights and bias over epochs.

- It stores runtime metadata for a run, such as total memory usage.

- It visualizes embeddings.

TensorBoard reads TensorFlow log files containing summary information about the training process at hand. These are generated with the appropriate callbacks, which are then passed to TensorFlow jobs.

The following screenshot shows some typical visualizations that are provided by TensorBoard. The first one is the "Scalars" section, which shows scalar quantities associated with the training stage. In this example, accuracy and binary cross entropy are being represented:

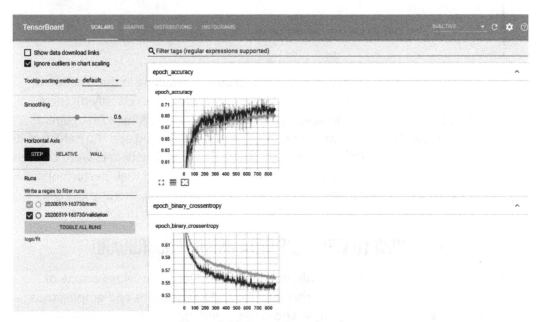

Figure 3.24: TensorBoard scalars

The second view provides a block diagram visualization of the computational graph, where all the layers are reported together with their relations, as shown in the following screenshot:

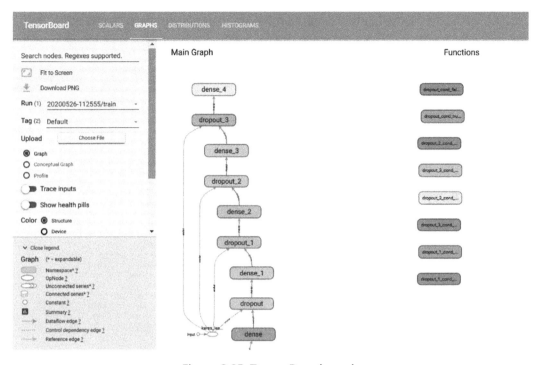

Figure 3.25: TensorBoard graph

The **DISTRIBUTIONS** tab provides an overview of how the model parameters are distributed across epochs, as shown in the following figure:

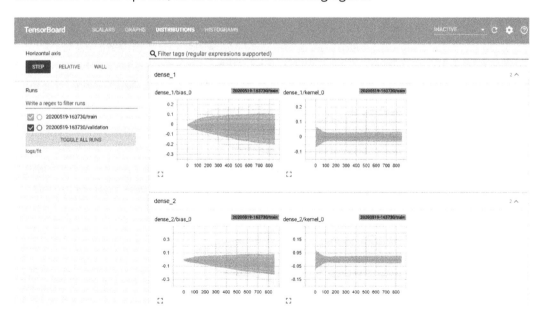

Figure 3.26: TensorBoard distributions

Finally, the **HISTOGRAMS** tab provides similar information to the **DISTRIBUTIONS** tab, but is unfolded in 3D, as shown in the following screenshot:

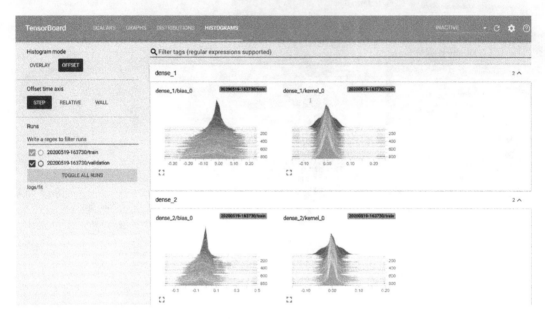

Figure 3.27: TensorBoard histograms

In this section and, in particular, in the following exercise, TensorBoard will be leveraged to easily visualize metrics in terms of trends, tensor graphs, distributions, and histograms.

In order to focus only on TensorBoard, the very same classification exercise we performed in the previous section will be used. Only the large model will be trained. All we need is to import TensorBoard and activate it, as well as a definition of the log file directory.

A TensorBoard callback is then created and passed to the **fit** method of the model. This will generate all TensorBoard files inside the log directory. Once training is complete, this log directory path is passed to TensorBoard as an argument. This will open a web-based visualization where the user is able to gain deep insights into its model and training-related aspects.

EXERCISE 3.07: CREATING A DEEP NEURAL NETWORK TO CLASSIFY EVENTS GENERATED BY THE ATLAS EXPERIMENT IN THE QUEST FOR THE HIGGS BOSON USING TENSORBOARD FOR VISUALIZATION

In this exercise, we will build, train, and measure the performance of a deep neural network with the same goal of *Exercise 3.06, Creating a Deep Neural Network to Classify Events Generated by the ATLAS Experiment in the Quest for Higgs Boson* in mind, but instead, we will leverage TensorBoard so that we can gain additional training insights.

The following steps need to be implemented in order to complete this exercise:

1. Import all the required modules:

```
from __future__ import absolute_import, division, \
print_function, unicode_literals

from  IPython import display
from matplotlib import pyplot as plt
from scipy.ndimage.filters import gaussian_filter1d
import pandas as pd
import numpy as np
import datetime

import tensorflow as tf

!rm -rf ./logs/

# Load the TensorBoard notebook extension
%load_ext tensorboard
```

2. Download the custom smaller subset of the original dataset:

```
higgs_path = tf.keras.utils.get_file('HIGGSSmall.csv.gz', \
            'https://github.com/PacktWorkshops/'\
            'The-Reinforcement-Learning-Workshop/blob/master/'\
            'Chapter03/Dataset/HIGGSSmall.csv.gz?raw=true')
```

3. Read the CSV dataset into the TensorFlow dataset class and repack it so that it has tuples (**features**, **labels**):

```
N_TEST = int(1e3)
N_VALIDATION = int(1e3)
N_TRAIN = int(1e4)
BUFFER_SIZE = int(N_TRAIN)
BATCH_SIZE = 500
STEPS_PER_EPOCH = N_TRAIN//BATCH_SIZE

N_FEATURES = 28

ds = tf.data.experimental.CsvDataset\
    (higgs_path,[float(),]*(N_FEATURES+1), \
     compression_type="GZIP")

def pack_row(*row):
    label = row[0]
    features = tf.stack(row[1:],1)
    return features, label

packed_ds = ds.batch(N_TRAIN).map(pack_row).unbatch()
```

4. Create training, validation, and test sets and assign them the **BATCH_SIZE** parameter:

```
validate_ds = packed_ds.take(N_VALIDATION).cache()
test_ds = packed_ds.skip(N_VALIDATION).take(N_TEST).cache()
train_ds = packed_ds.skip(N_VALIDATION+N_TEST)\
            .take(N_TRAIN).cache()

test_ds = test_ds.batch(BATCH_SIZE)
validate_ds = validate_ds.batch(BATCH_SIZE)
train_ds = train_ds.shuffle(BUFFER_SIZE)\
            .repeat().batch(BATCH_SIZE)
```

5. Now, let's start creating the model and training it. Create a decaying learning rate:

```
lr_schedule = tf.keras.optimizers.schedules\
              .InverseTimeDecay(0.001, \
                                decay_steps=STEPS_PER_EPOCH*1000,\
                                decay_rate=1, staircase=False)
```

6. Define a function that will compile a model with the **Adam** optimizer and use binary cross entropy as the **loss** function. Then, fit it on the training data using early stopping by using the validation dataset, as well as a TensorBoard callback:

```
log_dir = "logs/fit/" + datetime.datetime.now()\
          .strftime("%Y%m%d-%H%M%S")

def compile_and_fit(model, name, max_epochs=3000):

    optimizer = tf.keras.optimizers.Adam(lr_schedule)

    model.compile(optimizer=optimizer, \
    loss=tf.keras.losses.BinaryCrossentropy(from_logits=True),\
    metrics=[tf.keras.losses.BinaryCrossentropy\
            (from_logits=True, name='binary_crossentropy'),\
             'accuracy'])

    model.summary()

    tensorboard_callback = tf.keras.callbacks.TensorBoard\
                           (log_dir=log_dir, \
                            histogram_freq=1,\
                            profile_batch=0)
```

```
    history = model.fit\
            (train_ds,\
            steps_per_epoch = STEPS_PER_EPOCH,\
            epochs=max_epochs,\
            validation_data=validate_ds,\
            callbacks=[tf.keras.callbacks.EarlyStopping\
                        (monitor='val_binary_crossentropy',\
                        patience=200),\
                        tensorboard_callback],
            verbose=2)
    return history
```

7. Create the same large model as before with regularization items such as L2 regularization and dropout, and then compile it and fit it on the dataset:

```
regularization_model = tf.keras.Sequential([\
                    tf.keras.layers.Dense(512,\
                    kernel_regularizer=tf.keras.regularizers\
                                    .12(0.0001),\
                    activation='elu', \
                    input_shape=(N_FEATURES,)),\
                    tf.keras.layers.Dropout(0.5),\
                    tf.keras.layers.Dense(512,\
                    kernel_regularizer=tf.keras.regularizers\
                                    .12(0.0001),\
                    activation='elu'),\
                    tf.keras.layers.Dropout(0.5),\
                    tf.keras.layers.Dense(512,\
                    kernel_regularizer=tf.keras.regularizers\
                                    .12(0.0001),\
                    activation='elu'),\
                    tf.keras.layers.Dropout(0.5),\
                    tf.keras.layers.Dense(512,\
                    kernel_regularizer=tf.keras.regularizers\
                                    .12(0.0001),\
                    activation='elu'),\
                    tf.keras.layers.Dropout(0.5),\
                    tf.keras.layers.Dense(1)])
compile_and_fit(regularization_model,\
            "regularizers/regularization", max_epochs=9000)
```

The last output line will be as follows:

```
Epoch 1112/9000
20/20 - 1s - loss: 0.5887 - binary_crossentropy: 0.5515
- accuracy: 0.6949 - val_loss: 0.5831
- val_binary_crossentropy: 0.5459 - val_accuracy: 0.6960
```

8. Check the model's performances on the test set:

```
test_accuracy = tf.keras.metrics.Accuracy()

for (features, labels) in test_ds:
    logits = regularization_model(features)
    probabilities = tf.keras.activations.sigmoid(logits)
    predictions = 1*(probabilities.numpy() > 0.5)
    test_accuracy(predictions, labels)

print("Test set accuracy: {:.3%}".format(test_accuracy.result()))
```

The output will be as follows:

```
Test set accuracy: 69.300%
```

NOTE

The accuracy may show slightly different values due to random sampling with a variable random seed.

9. Visualize the variables with TensorBoard:

```
%tensorboard --logdir logs/fit
```

This command starts the web-based visualization tool. Four main windows are represented in the following figure, displaying information about loss and accuracy, model graphs, histograms, and distributions in a clockwise order, starting from the top left:

Figure 3.28: TensorBoard visualization

The advantages of using TensorBoard are quite evident: all the training information is collected in a single place, allowing the user to easily navigate through it. The top-left panel, the **SCALARS** tab, allows the user to monitor loss and accuracy so that they are able to check the same chart we saw previously in an easier way.

In the top right, the model graph is shown, so it is possible to visualize how input data flows into the computational graph by going through each block.

The two views at the bottom show the same information in two different representations: all the model parameter (networks weights and biases) distributions are shown across training epochs. On the left, the **DISTRIBUTIONS** tab displays the parameters in 2D, while the **HISTOGRAMS** tab unfolds the parameters in 3D. They both allow the user to monitor how trainable parameters vary during the training step.

> **NOTE**
>
> To access the source code for this specific section, please refer to https://packt.live/2AWGjFv.
>
> You can also run this example online at https://packt.live/2YrWl2d.

In this section, we focused on providing some insights into how to use TensorBoard to visualize training-related model parameters. We saw how, starting with an already familiar problem, it is super easy to add the TensorBoard web-based visualization tool and navigate through all of its plugins directly inside a Python notebook.

Now, let's complete an activity to put all our knowledge to the test.

ACTIVITY 3.01: CLASSIFYING FASHION CLOTHES USING A TENSORFLOW DATASET AND TENSORFLOW 2

Suppose you need to code an image processing algorithm for a company that owns a clothes warehouse. They want to autonomously classify clothes based on a camera output, thereby allowing them to group clothes together with no human intervention.

In this activity, we will create a deep fully connected neural network capable of doing such a task, meaning that it will accurately classify images of clothes by assigning them to the class they belong to.

The following steps will help you to complete this activity:

1. Import all the required modules, such as **numpy**, **matplotlib.pyplot**, **tensorflow**, and **tensorflow_datasets**, and print out their main module versions.

2. Import the Fashion MNIST dataset using TensorFlow datasets and split it into train and test sets.

3. Explore the dataset to get familiar with the input features, that is, shapes, labels, and classes.

4. Visualize some instances of the training set.

5. Perform data normalization by building the classification model.

6. Train the deep neural network.

7. Test the model's accuracy. You should obtain an accuracy in excess of 88%.

8. Perform inference and check the predictions against the ground truth.

By the end of this activity, the trained model should be able to classify all the fashion items (clothes, shoes, bags, and so on) with an accuracy greater than 88%, thus producing a result similar to the one shown in the following image:

Figure 3.29: Clothes classification with a deep neural network output

> **NOTE**
>
> The solution to this activity can be found on page 696.

SUMMARY

In this chapter, we were introduced to practical deep learning with TensorFlow 2 and Keras, their key features and applications, and how they work together. We became familiar with the differences between low- and high-level APIs, as well as how to leverage the most advanced modules to ease the creation of deep models. Then, we discussed how to implement a deep neural network with TensorFlow and addressed some major topics: from model creation, training, validation, and testing, we highlighted the most important aspects to consider so as to avoid pitfalls. We saw how to build different types of deep learning models, such as fully connected, convolutional, and recurrent neural networks, via the Keras API. We solved a regression task and a classification problem, which gave us hands-on experience with this. We learned how to leverage TensorBoard to visualize many different training trends regarding metrics and model parameters. Finally, we built and trained a model that is able to classify fashion item images with high accuracy, an activity that shows that a possible real-world problem can be solved with the help of the most advanced deep learning techniques.

In the next chapter, we will be studying the OpenAI Gym environment and how to use TensorFlow 2 for reinforcement learning.

4

GETTING STARTED WITH OPENAI AND TENSORFLOW FOR REINFORCEMENT LEARNING

OVERVIEW

This chapter introduces you to some key technologies and concepts to get started with reinforcement learning. You will become familiar with and use two OpenAI tools: Gym and Universe. You will learn how to deal with the interfaces of these environments and how to create a custom environment for a specific problem. You will build a policy network with TensorFlow, feed it with environment states to retrieve corresponding actions, and save the policy network weights. You will also learn how to use another OpenAI resource, Baselines, and use it to train a reinforcement learning agent to solve a classic control problem. By the end of this chapter, you will be able to use all the elements we will introduce to build and train an agent to play a classic Atari video game, thus achieving better-than-human performance.

INTRODUCTION

In the previous chapter, you were introduced to TensorFlow and Keras, along with an overview of their key features and applications and how they work in synergy. You learned how to implement a deep neural network with TensorFlow, addressing all major topics, that is, model creation, training, validation, and testing, using the most advanced machine learning frameworks available. In this chapter, we will use this knowledge to build models that are able to solve some classical reinforcement learning problems.

Reinforcement learning is a branch of machine learning that comes closest to the idea of artificial intelligence. The goal of training an artificial system to learn a given task, without any prior information, and only by means of experiences of an environment, represents the ambitious aim of replicating human learning. Applying deep learning techniques to the field has recently led to a great increase in performance, thus allowing us to solve problems in very different domains, from classic control problems to video games and even robotic locomotion. This chapter will introduce the various resources, methods, and tools you can use to become familiar with the context and problems typically encountered when getting started in the field. In particular, we will look at **OpenAI Gym** and **OpenAI Universe**, two libraries that allow us to easily create environments where we can train Reinforcement Learning (RL) agents, and OpenAI Baselines, a tool that provides a clear and simple interface for state-of-the-art reinforcement learning algorithms. By the end of this chapter, you will be able to leverage top libraries and modules to easily train a state-of-the-art reinforcement learning agent to solve classic control problems, as well as to achieve better-than-human performance on classic video games.

Now, let's begin our journey, starting with the very first important concept: how to correctly model a proper reinforcement learning environment where we can train an agent. For this, we will be using OpenAI Gym and Universe.

OPENAI GYM

In this section, we will study the OpenAI Gym tool. We will go through the motivations behind its creation and its main elements, learning how to interact with them to properly train a reinforcement learning algorithm to tackle state-of-the-art benchmark problems. Finally, we will build a custom environment with the same set of standardized interfaces.

The role of shared standard benchmarks for machine learning algorithms is of paramount importance to measure performance and state-of-the-art improvements. While for supervised learning there have been many different examples since the early days of the discipline, the same is not true for the reinforcement learning field.

With the aim of fulfilling this need, in 2016, OpenAI released OpenAI Gym (https://gym.openai.com/). It was conceived to be to reinforcement learning what standardized datasets such as ImageNet and COCO are to supervised learning: a standard, shared context in which the performance of RL methods can be directly measured and compared, both to identify the highest-achieving ones as well as to monitor current progress.

OpenAI Gym acts as an interface between the typical Markov decision process formulation of the reinforcement learning problem and a variety of environments, covering different types of problems the agent has to solve (from classic control to Atari video games), as well as different observations and action spaces. Gym is completely independent of the structure of the agent that will be interfaced with, as well as the machine learning framework used to build and run it.

Here is the list of environment categories Gym offers, ranging from easy to difficult and involving many different kinds of data:

- **Classic control and toy text**: Small-scale easy tasks, frequently found in reinforcement learning literature. These environments are the best place to start in order to gain confidence with Gym and to familiarize yourself with agent training.

The following figure shows an example of classic control problem of CartPole:

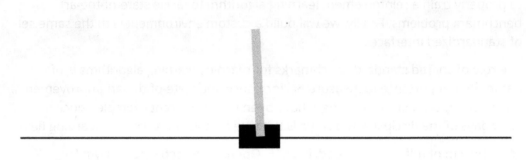

Episode 12

Figure 4.1: Classic control problem- CartPole

The following figure shows an example of classic control problem of MountainCar:

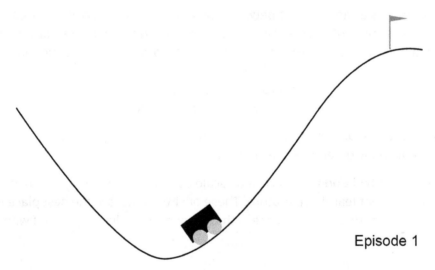

Episode 1

Figure 4.2: Classic control problem - Mountain Car

- **Algorithmic**: In these environments, the system has to learn, autonomously and purely from examples, to perform computations ranging from multi-digit additions to alphanumeric-character sequence reversal.

The following figure shows screenshot representing instances of the algorithmic problem set:

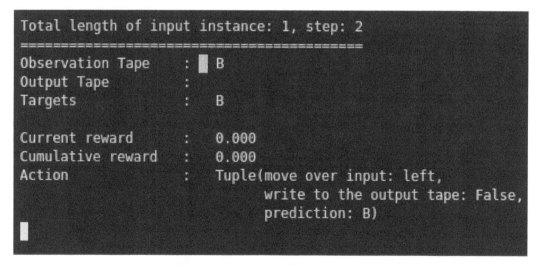

```
Total length of input instance: 2, step: 2
==========================================
Observation Tape      :    E
Output Tape           :
Targets               :    E

Current reward        :    0.000
Cumulative reward     :    0.000
Action                :    Tuple(move over input: right,
                               write to the output tape: False,
                               prediction: A)
```

Figure 4.3: Algorithmic problem – copying multiple instances of the input sequence

The following figure shows screenshot representing instances of the algorithmic problem set:

```
Total length of input instance: 1, step: 2
==========================================
Observation Tape      :    B
Output Tape           :
Targets               :    B

Current reward        :    0.000
Cumulative reward     :    0.000
Action                :    Tuple(move over input: left,
                               write to the output tape: False,
                               prediction: B)
```

Figure 4.4: Algorithmic problem - copying instance of the input sequence

- **Atari**: Gym integrates the **Arcade Learning Environment** (**ALE**), a software library that provides an interface we can use to train an agent to play classic Atari video games. It played a major role in helping reinforcement learning research achieve outstanding results.

The following figure shows Atari video game Breakout, provided by ALE:

Figure 4.5: Atari video game of Breakout

The following figure shows Atari video game Pong, provided by ALE:

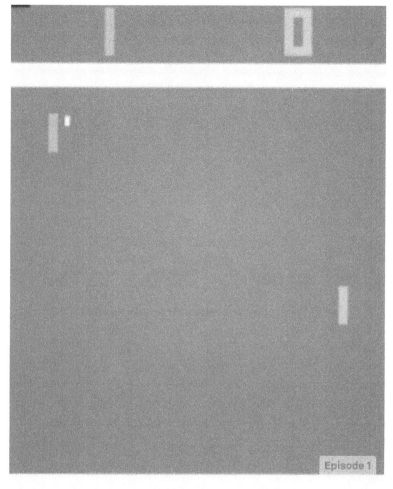

Figure 4.6: Atari video game of Pong

NOTE

The preceding figures have been sourced from the official documentation for OpenAI Gym. Please refer to the following link for more visual examples of Atari games: https://gym.openai.com/envs/#atari.

- **MuJoCo and Robotics**: These environments expose typical challenges that are encountered in the field of robot control. Some of them take advantage of the MuJoCo physics engine, which was designed for fast and accurate robot simulation and offers free licenses for trial.

The following figure shows three MuJoCo environments, all of which provide a meaningful overview of robotic locomotion tasks:

Figure 4.7: Three MuJoCo-powered environments – Ant (left), Walker (center), and Humanoid (right)

> **NOTE**
>
> The preceding images have been sourced from the official documentation for OpenAI Gym. Please refer to the following link for more visual examples of MuJoCo environments: https://gym.openai.com/envs/#mujoco.

- The following figure shows two environments contained in the "Robotics" category, where RL agents are trained to perform robotic manipulation tasks:

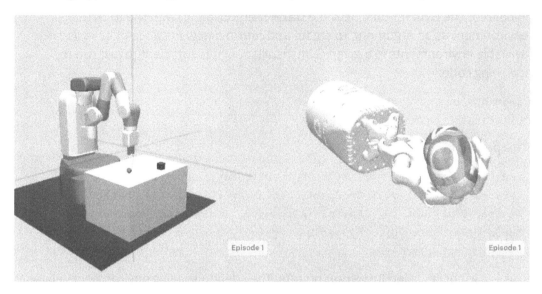

Figure 4.8: Two robotics environments – FetchPickAndPlace (left) and HandManipulateEgg (Right)

NOTE

The preceding images have been sourced from the official documentation for OpenAI Gym. Please refer to the following link for more visual examples of Robotics environments: https://gym.openai.com/envs/#robotics.

- **Third-party environments**: Environments developed by third parties are also available with a very broad landscape of applications, complexity, and data types (https://github.com/openai/gym/blob/master/docs/environments.md#third-party-environments).

HOW TO INTERACT WITH A GYM ENVIRONMENT

In order to interact with a Gym environment, it has to, first of all, be created and initialized. The Gym module uses the **make** method, along with the ID of the environment as an argument, to create and return a new instance of it. To list all available environments in a given Gym installation, it is sufficient to just run the following code:

```
from gym import envs
print(envs.registry.all())
```

This prints out the following:

```
[EnvSpec(DoubleDunk-v0), EnvSpec(InvertedDoublePendulum-v0),
EnvSpec(BeamRider-v0), EnvSpec(Phoenix-ram-v0), EnvSpec(Asterix-v0),
EnvSpec(TimePilot-v0), EnvSpec(Alien-v0), EnvSpec(Robotank-ram-v0),
EnvSpec(CartPole-v0), EnvSpec(Berzerk-v0), EnvSpec(Berzerk-ram-v0),
EnvSpec(Gopher-ram-v0), ...
```

This is a list of so-called **EnvSpec** objects. They define specific environment-related parameters, such as the goal to be achieved, the reward threshold defining when the task is considered solved, and the maximum number of steps allowed for a single episode.

One interesting thing to note is that it is possible to easily add custom environments, as we will see later on. Thanks to this, a user can implement a custom problem using standard interfaces, making it straightforward for it to be tackled by standardized, off-the-shelf, reinforcement learning algorithms.

The fundamental elements of an environment are as follows:

- **Observation** (object): An environment-specific object representing what can be observed of the environment; for example, the kinematic variables (that is, velocities and positions) of a mechanical system, pawn positions in a chess game, or the pixel frames of a video game.

- **Actions** (object): An environment-specific object representing actions the agent can perform in the environment; for example, joint rotations and/or joint torques for a robot, a legal move in a board game, or buttons being pressed in combination for a video game.

- **Reward** (float): The amount of reward achieved by executing the last step with the prescribed action. The reward range differs between different tasks, but in order to solve the environment, the aim is always to increase it, since this is what the RL agent tries to maximize.

- **Done** (bool): This indicates whether the episode has finished. If true, the environment needs to be reset. Most, but not all, tasks are divided into well-defined episodes, where a terminated episode may represent that the robot has fallen on the ground, the board game has reached a final state, or the agent lost its last life in a video game.

- **Info** (dict): This contains diagnostic information on environment internals and is useful for both debugging purposes and for an RL agent training, even if it's not allowed for standard benchmark comparisons.

The fundamental methods of an environment are as follows:

- `reset()`: Input: none, output: observation. Resets the environment, bringing it to the starting point. It takes no input and outputs the corresponding observation. It has to be called right after environment creation and every time a final state is reached (**done** flag equal to `True`).

- `step(action)`: Input: action, output: observation – reward – done – info. Advances the environment by one step, applying the selected input action. Returns the observation of the newly reached state, which is a reward associated with the transition from the previous to the new state under the selected action. The **done** flag is used to indicate whether the new state is a terminal one or not (`True/False`, respectively), as well as the `Info` dict with environment internals.

- `render()`: Input: none, output: environment rendering. Renders the environment and is used for visualization/presentation purposes only. It is not used during agent training, which only needs observations to know the environment's state. For example, it presents robot movements via animation graphics or outputs a video game video stream.

- `close()`: Input: none, output: none. Shuts down the environment gracefully.

These elements allow us to have complete interaction with the environment simply by executing it with random inputs, training an agent, and running it. It is, in fact, an implementation of the standard reinforcement learning contextualization, which is described by the agent-environment interaction. For each timestep, the agent executes an action. This interaction with the environment causes a transition from the current state to a new state, resulting in an observation of the new state and a reward, which are returned as results. As a preliminary step, the following exercise shows how to create a CartPole environment, reset it, run it for 1,000 steps while randomly sampling one action for each step, and finally close it.

EXERCISE 4.01: INTERACTING WITH THE GYM ENVIRONMENT

In this exercise, we will familiarize ourselves with the Gym environment by looking at a classic control example, CartPole. Follow these steps to complete this exercise:

1. Import the OpenAI Gym module:

```
import gym
```

2. Instantiate the environment and reset it:

```
env = gym.make('CartPole-v0')
env.reset()
```

The output will be as follows:

```
array([ 0.03972635,  0.00449595,  0.04198141, -0.01267544])
```

3. Run the environment for **1000** steps, rendering it and resetting it if a terminal state is encountered. After all steps are completed, close the environment:

```
for _ in range(1000):
    env.render()
    # take a random action
    _, _, done, _ = env.step(env.action_space.sample())

    if done:
        env.reset()

env.close()
```

It renders the environment and plays it for 1,000 steps. The following figure shows one frame that was extracted from step number 12 of the entire sequence:

Episode 12

Figure 4.9: One frame of the 1,000 rendered steps for the CartPole environment

> **NOTE**
>
> To access the source code for this specific section, please refer to https://packt.live/30yFmOi.
>
> This section does not currently have an online interactive example, and will need to be run locally.

This shows that the black cart can move along its rail (the horizontal line), with its pole fixed on the cart with a hinge that allows it to rotate freely. The goal is to control the cart while pushing it left and right in order to maintain the pole's vertical equilibrium, as seen in the preceding figure.

ACTION AND OBSERVATION SPACES

In order to appropriately interact with an environment and train an agent on it, a fundamental initial step is to familiarize yourself with its action and observation spaces. For example, in the preceding exercise, the action was randomly sampled from the environment's action space.

Every environment is characterized by **action_space** and **observation_space**, which are instances of the **Space** class that describe the actions and observations required by Gym. The following snippet prints them out for the **CartPole** environment:

```
import gym
env = gym.make('CartPole-v0')
print("Action space =", env.action_space)
print("Observation space =", env.observation_space)
```

This outputs the following two rows:

```
Action space = Discrete(2)
Observation space = Box(4,)
```

The **Discrete** space represents the set of non-negative integer numbers (natural numbers plus 0). Its dimension defines which numbers represent valid actions. For example, in the **CartPole** case, it is of dimension **2** because the agent can only push the cart left and right, so the admissible values are 0 or 1. The **Box** space can be thought of as an n-dimensional array. In the **CartPole** case, the system state is defined by four variables: cart position and velocity, and pole angle with respect to the vertical and angular velocity. So, the "box observation" space dimension is equal to 4, and valid observations will be an array of four real numbers. In the latter case, it is useful to check their upper and lower bounds. This can be done as follows:

```
print("Observations superior limit =", env.observation_space.high)
print("Observations inferior limit =", env.observation_space.low)
```

This prints out the following:

```
Observations superior limit = array([ 2.4, inf, 0.20943951, inf])
Observations inferior limit = array([-2.4, -inf,-0.20943951, -inf])
```

With these new elements, it is possible to write a more complete snippet to interact with the environment, using all the previously presented interfaces. The following code shows a complete loop, executing **20** episodes, each for **100** steps, rendering the environment, retrieving observations, and printing them out while taking random actions and resetting once it reaches a terminal state:

```
import gym
env = gym.make('CartPole-v0')
for i_episode in range(20):
    observation = env.reset()
    for t in range(100):
        env.render()
        print(observation)
        action = env.action_space.sample()
        observation, reward, done, info = env.step(action)
        if done:
            print("Episode finished after {} timesteps".format(t+1))
            break
env.close()
```

The preceding code runs the environment for **20** episodes of **100** steps each, also rendering the environment, as we saw in *Exercise 4.01, Interacting with the Gym Environment.*

> **NOTE**
>
> In the preceding case, we run each episode for 100 steps instead of 1,000, as we did previously. There is no particular reason for doing so, but we are running 20 different episodes, not a single one, so we opted for 100 steps to keep the code execution time short enough.

In addition to that, this code also prints out the sequence of observations, as returned by the environment, for each step performed. The following are a few lines that are received as output:

```
[-0.061586   -0.75893141  0.05793238  1.15547541]
[-0.07676463 -0.95475889  0.08104189  1.46574644]
[-0.0958598  -1.15077434  0.11035682  1.78260485]
[-0.11887529 -0.95705275  0.14600892  1.5261692 ]
[-0.13801635 -0.7639636   0.1765323   1.28239155]
[-0.15329562 -0.57147373  0.20218013  1.04977545]
Episode finished after 14 timesteps
[-0.02786724  0.00361763 -0.03938967 -0.01611184]
[-0.02779488 -0.19091794 -0.03971191  0.26388759]
[-0.03161324  0.00474768 -0.03443415 -0.04105167]
```

Looking at the previous code example, we can see how, for now, the action choice is completely random. It is right here that a trained agent would make a difference: it should choose actions based on environment observations, thus appropriately responding to the state it finds itself in. So, revising the previous code by substituting a trained agent in place of a random action choice looks as follows:

1. Import the OpenAI Gym and CartPole modules:

```
import gym
env = gym.make('CartPole-v0')
```

2. Run **20** episodes of **100** steps each:

```
for i_episode in range(20):
    observation = env.reset()
    for t in range(100):
```

3. Render the environment and print the observation:

```
env.render()
print(observation)
```

4. Use the agent's knowledge to choose the action, given the current environment state:

```
action = RL_agent.select_action(observation)
```

5. Step the environment:

```
observation, reward, done, info = env.step(action)
```

6. If successful, break the inner loop and start a new episode:

```
if done:
    print("Episode finished after {} timesteps"\
            .format(t+1))
    break
env.close()
```

With a trained agent, actions will be chosen optimally since a function of the state that the agent is in, is used to maximize the expected reward. This code would result in an output similar to the previous one.

But how do we proceed and train an agent from scratch? As you will learn throughout this book, there are many different approaches and algorithms we can use to achieve this quite complex task. In general, they all need the following tuple of elements: current state, chosen action, reward obtained by performing the chosen action, and new state reached by performing the chosen action.

So, elaborating again on the previous code snippet to introduce the agent training step, it would look like this:

```
import gym
env = gym.make('CartPole-v0')
for i_episode in range(20):
    observation = env.reset()
    for t in range(100):
        env.render()
        print(observation)
        action = RL_agent.select_action(observation)
        new_observation, reward, done, info = env.step(action)
        RL_agent.train(observation, action, reward, \
                    new_observation)
        observation = new_observation
        if done:
            print("Episode finished after {} timesteps"\
                    .format(t+1))
            break
env.close()
```

The only difference in this code with respect to the previous block is the following line:

```
RL_agent.train(observation, action, reward, new_observation)
```

This refers to the agent training step. The purpose of this code is to give us a high-level idea of all the steps involved in training an RL agent in a given environment.

This is the high-level idea behind the method adopted to carry out reinforcement learning agent training with the Gym environment. It provides access to all the required details through a very clean standard interface, thus giving us access to an extremely large set of different problems against which measuring algorithms and techniques can be used.

HOW TO IMPLEMENT A CUSTOM GYM ENVIRONMENT

All the environments that are available through Gym are perfect for learning purposes, but eventually, you will need to train an agent to solve a custom problem. One good way to achieve this is to create a custom environment, specific to the problem domain.

In order to do so, a class derived from **gym.Env** must be created. It will implement all the objects and methods described in the previous section so that it supports the agent-world interaction cycle that's typical of any reinforcement learning setting.

The following snippet represents a frame guiding a custom environment's development:

```
import gym
from gym import spaces

class CustomEnv(gym.Env):
    """Custom Environment that follows gym interface"""
    metadata = {'render.modes': ['human']}

    def __init__(self, arg1, arg2, ...):
        super(CustomEnv, self).__init__()
        # Define action and observation space
        # They must be gym.spaces objects
        # Example when using discrete actions:
        self.action_space = spaces.Discrete(N_DISCRETE_ACTIONS)
        # Example for using image as input:
        self.observation_space = spaces.Box\
```

```
                               (low=0, high=255, \
                                 shape=(HEIGHT, WIDTH, \
                                        N_CHANNELS), \
                                 dtype=np.uint8)

    def step(self, action):
      # Execute one time step within the environment
      ...
      # Compute reward
      ...
      # Check if in final state
      ...
      return observation, reward, done, info

    def reset(self):
      # Reset the state of the environment to an initial state
      ...
      return observation

    def render(self, mode='human', close=False):
      # Render the environment to the screen
      ...
      return
```

In the constructor, **action_space** and **observation_space** are defined. As mentioned previously, they will contain all possible actions the agent can take in the environment and all environment data observable by the agent. They are to be attributed to the specific problem: in particular, **action_space** will reflect elements the agent can control to interact with the environment, while **observation_space** will contain all the variables we want the agent to consider when choosing the action.

The **reset** method will be called to periodically reset the environment to an initial state, typically after the first initialization and every time after the end of an episode. It will return the observation.

The **step** method receives an action as input and executes it. This will result in an environment transitioning from the current state to a new state. The observation related to the new state is returned. This is also the method where the reward is calculated as a result of the state transition generated by the action. The new state is checked to determine whether it is a terminal one, in which case, the **done** flag that's returned is set to **true**. As the last step, all useful internals are returned in the **info** dictionary.

Finally, the **render** method is the one in charge of rendering the environment. Its complexity may range from being as simple as a print statement to being as complicated as rendering a 3D environment using OpenGL.

In this section, we studied the OpenAI Gym tool. We had an overview that explained the context and motivations behind its conception, provided details about its main elements, and saw how to interact with the elements to properly train a reinforcement learning algorithm to tackle state-of-the-art benchmark problems. Finally, we saw how to build a custom environment with the same set of standardized interfaces.

OPENAI UNIVERSE – COMPLEX ENVIRONMENT

OpenAI Universe was released by OpenAI a few months after Gym. It's a software platform for measuring and training artificial general intelligence on different applications, ranging from video games to websites. It makes an AI agent able to use a computer as a human does: the environment state is represented by screen pixels and the actions are all operations that can be performed by operating a virtual keyboard and mouse.

With Universe, it is possible to adapt any program, thus transforming the program into a Gym environment. It executes the program using **Virtual Network Computing (VNC)** technology, a software technology that allows the remote control of a computer system via graphical desktop-sharing over a network, transmitting keyboard and mouse events and receiving screen frames. By mimicking execution behind a remote desktop, it doesn't need to access program memory states, customized source code, or have a set of APIs.

The following snippet shows how to use Universe in a simple Python program, where a scripted action is always executed in every step:

1. Import the OpenAI Gym and OpenAI Universe modules:

```python
import gym
# register Universe environments into Gym
import universe
```

2. Instantiate the OpenAI Universe environment and reset it:

```python
# Universe env ID here
env = gym.make('flashgames.DuskDrive-v0')
observation_n = env.reset()
```

3. Execute a prescribed action to interact with the environment and render it:

```
while True:
    # agent which presses the Up arrow 60 times per second
    action_n = [[('KeyEvent', 'ArrowUp', True)] \
            for _ in observation_n]
    observation_n, reward_n, done_n, info = env.step(action_n)
    env.render()
```

The preceding code successfully runs a Flash game in the browser.

The goal behind Universe is to favor the development of an AI agent that's capable of applying its past experience to master complex new environments, which would represent a fundamental step in the quest for artificial general intelligence.

Despite the great success of AI in recent years, all developed systems can still be considered "Narrow AI." This is because they can only achieve better-than-human performance in a limited domain. Building something with a general problem-solving ability on a par with human common sense requires overcoming the goal of carrying agent experience along when shifting to a completely new task. This would allow an agent to avoid training from scratch, randomly going through tens of millions of trials.

Now, let's take a look at the infrastructure of OpenAI Universe.

OPENAI UNIVERSE INFRASTRUCTURE

The following diagram effectively describes how OpenAI Universe works: it exposes all its environments, which will be described in detail later, through a common interface: by leveraging VNC technology, it makes the environment act as a server and the agent as a client so that the latter operates a remote desktop by observing the pixels of a screen (observations of the environment) and producing keyboard and mouse commands (actions of the agent). VNC is a well-established technology and is the standard for interacting with computers remotely through the network, as in the case of cloud computing systems or decentralized infrastructures:

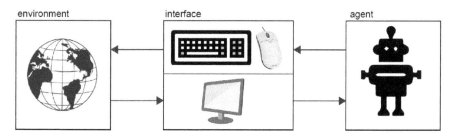

Figure 4.10: VNC server-client Universe infrastructure

Universe's implementation has some notable properties, as follows:

- **Generality**: By adopting the VNC interface, it doesn't require emulators or access to a program's source code or memory states, thus opening a relevant number of opportunities in fields such as computer games, web browsing, CAD software usage, and much more.

- **Familiarity to humans**: It can be easily used by humans to provide baselines for AI algorithms, which are useful to initialize agents with human demonstrations recorded in the form of VNC traffic. For example, a human can solve one of the tasks provided by OpenAI Universe by using it through VNC and recording the corresponding traffic. Then, it can use it to train an agent, providing good examples of policies to learn from.

- **Standardization**. Leveraging VNC technology ensures portability in all major operating systems that have VNC software by default.

- **Easiness of debugging**: It is super easy to observe the agent during training or evaluation by simply connecting a client for visualization to the environment's VNC shared server. Saving VNC traffic also helps.

ENVIRONMENTS

In this section, we will look at the most important categories of problems that are already available inside Universe. Each environment is composed of a Docker image and hosts a VNC server. The server has the role of the interface and is in charge of the following:

- Sending observations (screen pixels)

- Receiving actions (keyboard/mouse commands)

- Providing information for reinforcement learning tasks (reward signal, diagnosis elements, and so on) through a Web Socket server

Now, let's take a look at each of the different categories of environments.

ATARI GAMES

These are the classic Atari 2600 games from the ALE. Already encountered in OpenAI Gym, they are also part of Universe.

FLASH GAMES

The landscape of Flash games offers a large number of games with more advanced graphics with respect to Atari, but still with simple mechanics and goals. Universe's initial release contained 1,000 Flash games, 100 of which also provided reward as a function.

With the Universe approach, there is a major aspect to be addressed: how the agent knows how well it performed, which is related to the rewards returned by interacting with the environment. If you don't have access to an application's internal states (that is, its RAM addresses), the only way to do so is to extract such information from the onscreen pixels. Many games have a score associated with them, and this is printed out on each frame so that it can be parsed via some image processing algorithm. For example, Atari Pong shows both players' scores in the top part of the frame, so it is possible to parse those pixels to retrieve it. Universe developed a high-performing image-to-text model based on convolutional neural networks that's embedded into the Python controller and runs inside a Docker container. On the environments where it can be applied, it retrieves the user's score from the frame buffer and provides this information through the Web Socket's score from the frame buffer, thus providing this information through the Web Socket.

BROWSER TASKS

Universe adds a unique set of tasks based on the usage of a web browser. These environments put the AI agent in front of a common web browser, presenting it with problems that require the use of the web: reading content, navigating through pages and clicking buttons while observing only pixels, and using the keyboard and mouse. Depending on the complexity, these tasks can, conceptually, be grouped into two categories: Mini World of Bits and real-world browser tasks:

- **Mini World of Bits**:

  These environments are to browser-based tasks as what the MNIST dataset is to image recognition: they are basic building blocks that can be found on complex browsing problems on which training is easier but also insightful. They are environments of differing difficulty levels, for example, that you click on a specific button or reply to a message using an email client.

- **Real-world browser tasks**:

  With respect to the previous category, these environments require the agent to solve more realistic problems, usually in the form of an instruction expressed to the agent, which has to perform a sequence of actions on a website. An example could be a request for an agent to book a specific flight that would require it to interact with the platform in order to find the right answer.

RUNNING AN OPENAI UNIVERSE ENVIRONMENT

Being a large collection of tasks that can be accessed via a common interface, running an environment requires performing only a few steps:

- Install Docker and Universe, which can be done with the following command:

```
git clone https://github.com/openai/universe && pip install -e
universe
```

- Start a runtime, which is a server that groups a collection of similar environments into a "runtime" exposing two ports: **5900** and **15900**. Port **5900** is used for the VNC protocol to exchange pixel information or keyboard/mouse actions, while **15900** is used to maintain the **WebSocket** control protocol. The following snippet shows how to boot a runtime from a PC console (for example, a Linux shell):

```
# -p 5900:5900 and -p 15900:15900
# expose the VNC and WebSocket ports
# --privileged/--cap-add/--ipc=host
# needed to make Selenium work
$ docker run --privileged --cap-add=SYS_ADMIN --ipc=host \
    -p 5900:5900 -p 15900:15900 quay.io/openai/universe.flashgames
```

With this command, the Flash game's Docker container will be downloaded. You can then use a VNC viewer to view and control the created remote desktop. The target port is **5900**. It is also possible to use the browser-based VNC client through the web server using port **15900** and the password **openai**.

The following snippet is the very same as the one we saw previously, except it only adds the VNC connection step. This means that the output is also the same, so it is not reported here. As we saw, writing a custom agent is quite straightforward. Observations include a NumPy pixel array, and actions are a list of VNC events (mouse/keyboard interactions):

```
import gym
import universe # register Universe environments into Gym

# Universe [environment ID]
env = gym.make('flashgames.DuskDrive-v0')
"""
If using docker-machine, replace "localhost" with specific Docker IP
"""
env.configure(remotes="vnc://localhost:5900+15900")
observation_n = env.reset()

while True:
    # agent which presses the Up arrow 60 times per second
    action_n = [[('KeyEvent', 'ArrowUp', True)] \
                for _ in observation_n]
    observation_n, reward_n, done_n, info = env.step(action_n)
    env.render()
```

Exploiting the same VNC connection, the user is able to watch the agent in action and also send action commands using the keyboard and mouse. The VNC interface, managing environments as server processes, allows us to run them on remote machines, thus allowing us to leverage in-house computation clusters or even cloud solutions. For more information, refer to the OpenAI Universe website (https://openai.com/blog/universe/).

VALIDATING THE UNIVERSE INFRASTRUCTURE

One of the intrinsic problems of Universe is the associated lag in observations and execution of actions that comes with the choice of architecture. In fact, agents must operate in real time and are accountable for fluctuating action and observation delays. Most environments can't be solved with current techniques, but the creators of Universe performed tests to guarantee that it is actually possible for an RL agent to learn. During these tests, the reward trends during training for Atari games, Flash games, and browser tasks confirm that it is actually possible to obtain results even in such a complex setting.

Now that we've introduced the OpenAI tools for reinforcement learning, we can now move on and learn how to use TensorFlow in this context.

TENSORFLOW FOR REINFORCEMENT LEARNING

In this section, we will learn how to create, run, and save a policy network using TensorFlow. Policy networks are one of the fundamental pieces, if not the most important one, of reinforcement learning. As will be shown throughout this book, they are a very powerful implementation of containers for the knowledge the agent has to learn, which tells them how to choose actions based on environment observations.

IMPLEMENTING A POLICY NETWORK USING TENSORFLOW

Building a policy network is not too different from building a common deep learning model. Its goal is to output the "optimal" action, given the input it receives, that represents the environment's observation. So, it acts as a link between the environment state and the optimal agent behavior associated with it. Being optimal here means doing what maximizes the cumulative expected reward of the agent.

To make things as clear as possible, we will focus on a specific problem here, but the same approach can be adopted to solve other tasks, such as controlling a robotic arm or teaching locomotion to a humanoid robot. We will see how to create a policy network for a classic control problem that will also be at the core of an exercise later in this chapter. This problem is the "CartPole" problem: the goal is to maintain the balance of the vertical pole so that it is upright at all times. Here, the only way to do this is by moving the cart along either direction of the x-axis. The following figure shows a frame from this problem:

Episode 12

Figure 4.11: CartPole control problem

As we mentioned previously, the policy network links the observations of the environment with the actions that the agent can take. So, they act as the input and output, respectively.

As we saw in the previous chapter, this is the first information that you need in order to build a neural network. To retrieve the input and output dimensions, you have to instantiate the environment (in this case, this is done via OpenAI Gym) and print out information about the observation and action spaces.

Let's perform this first task by completing the following exercise.

EXERCISE 4.02: BUILDING A POLICY NETWORK WITH TENSORFLOW

In this exercise, we will learn how to build a policy network with TensorFlow for a given Gym environment. We will learn how to take its observation space and action space into account, which constitute the input and output of the network, respectively. We will then create a deep learning model that is able to generate actions for the agent in the environment in response to environment observations. This network is the piece that needs to be trained and is the final goal of every RL algorithm. Follow these steps to complete this exercise:

1. Import the required modules:

    ```
    import numpy as np
    import gym
    import tensorflow as tf
    ```

2. Instantiate the environment:

    ```
    env = gym.make('CartPole-v0')
    ```

3. Print out the action and observation spaces:

```
print("Action space =", env.action_space)
print("Observation space =", env.observation_space)
```

This prints out the following:

```
Action space = Discrete(2)
Observation space = Box(4,)
```

4. Print out the action and observation space dimensions:

```
print("Action space dimension =", env.action_space.n)
print("Observation space dimension =", \
    env.observation_space.shape[0])
```

The output will be as follows:

```
Action space dimension = 2
Observation space dimension = 4
```

As you can see from the preceding output, the action space is a discrete space of dimension **2**, meaning it can take the value **0** or **1**. The observation space is of the **Box** type with a dimension of **4**, meaning it consists of four real numbers inside the lower and upper boundaries, which, as we already saw for the CartPole environment, are [±2.4, ± inf, ±0.20943951, ±inf].

With this information, it is now possible to build a policy network that can be interfaced with the CartPole environment. The following code block shows one of many possible choices: it uses two hidden layers with **64** neurons each and an output layer with **2** neurons (as this is the action space's dimension) with a **softmax** activation function. The model summary prints out the outline of the model.

5. Build the policy network and print its summary:

```
model = tf.keras.Sequential\
        ([tf.keras.layers.Dense(64, activation='relu', \
          input_shape=[env.observation_space.shape[0]]), \
          tf.keras.layers.Dense(64, activation='relu'), \
          tf.keras.layers.Dense(env.action_space.n, \
          activation="softmax")])

model.summary()
```

The output will be as follows:

```
Model: "sequential_2"

_____
Layer (type)                 Output Shape              Param #
=================================================================
dense (Dense)                (None, 64)                320
_____
dense_1 (Dense)              (None, 64)                4160
_____
dense_2 (Dense)              (None, 2)                 130
=================================================================
Total params: 4,610
Trainable params: 4,610
Non-trainable params: 0
```

As you can see, the model has been created and we also have an elaborate summary of it, which gives us significant information about the model, regarding the layers, the parameters of the network, and so on.

> **NOTE**
>
> To access the source code for this specific section, please refer to https://packt.live/3fkxfce.
>
> You can also run this example online at https://packt.live/2XSXHnF.

Once the policy network has been built and initialized, it is possible to feed it. Of course, since the network hasn't been trained, it will generate random outputs, but still, it can be used, for example, to run a random agent in an environment of choice. This is what we will implement in the following exercise: the neural network model will be fed with the observation provided by the environment step or **reset** function through the **predict** method. This outputs the action probabilities. The action with the highest probability is chosen and used to step through the environment until the episode ends.

EXERCISE 4.03: FEEDING THE POLICY NETWORK WITH ENVIRONMENT STATE REPRESENTATION

In this exercise, we will be feeding information to the policy network with the environment state representation. This exercise is a continuation of *Exercise 4.02, Building a Policy Network with TensorFlow*, so in order to carry it out, you need to perform all the steps of the preceding exercise and then begin this one right after. Follow these steps to complete this exercise:

1. Reset the environment:

```
t = 1
observation = env.reset()
```

2. Start a loop that will run until the episode is complete. Render the environment and print the observations:

```
while True:

    env.render()

    # Print the observation
    print("Observation = ", observation)
```

3. Feed the network with the environment observations, let it choose the appropriate actions, and print it:

```
action_probabilities =model.predict\
                    (np.expand_dims(observation, axis=0))
action = np.argmax(action_probabilities)
print("Action = ", action)
```

4. Step through the environment with the selected action. Print the received reward and close the environment if the terminal state has been reached:

```
observation, reward, done, info = env.step(action)

# Print received reward
print("Reward = ", reward)

# If terminal state reached, close the environment
```

```
     if done:
         print("Episode finished after {} timesteps".format(t+1))
         break

     t += 1

env.close()
```

This produces the following output (only the last few lines have been shown):

```
Observation =   [-0.00324467 -1.02182257  0.01504633  1.38740738]
Action =   0
Reward =   1.0
Observation =   [-0.02368112 -1.21712879  0.04279448  1.684757  ]
Action =   0
Reward =   1.0
Observation =   [-0.0480237  -1.41271906  0.07648962  1.99045154]
Action =   0
Reward =   1.0
Observation =   [-0.07627808 -1.60855467  0.11629865  2.30581208]
Action =   0
Reward =   1.0
Observation =   [-0.10844917 -1.80453455  0.16241489  2.63191088]
Action =   0
Reward =   1.0
Episode finished after 11 timesteps
```

> **NOTE**
>
> To access the source code for this specific section, please refer
> to https://packt.live/2AmwUHw.
>
> You can also run this example online at https://packt.live/3kvuhVQ.

By completing this exercise, we've built a policy network and used it to guide an agent's behavior in a Gym environment. At the moment, it behaves randomly, but apart from policy network training, which will be explained in the following chapters, every other piece of the big picture is already in place.

HOW TO SAVE A POLICY NETWORK

The goal of reinforcement learning is to effectively train the network so that it learns how to perform the optimal action for every given environment state. RL theory deals with how to achieve this goal and, as we will see, different approaches have been successful. Supposing one of them has been applied to the previous network, the trained model needs to be saved so that it can be loaded every time it needs to run the agent on the environment.

To save the policy network, we need to follow the very same steps of saving a common neural network, where all the weights of all the layers are dumped into a save file to be loaded again in the network at a later stage. The following code is an example of this implementation:

```
save_dir = "./"
model_name = "modelName"

print("Saving best model to {}".format(save_dir))
model.save_weights(os.path.join(save_dir,\
                              'model_{}.h5'.format(model_name)))
```

This produces the following output:

```
Saving best model to ./
```

In this section, we learned how to create, run, and save a policy network using TensorFlow. Once the inputs (environment states/observations) and outputs (actions the agent can perform) are clear, there is no big difference with respect to standard deep neural networks. The model has also been used to run the agent. When fed with the environment state, it produced actions for the agent to take. Being an untrained network, the agent behaved randomly. The only missing piece in this section is how to effectively train the policy network, which is the goal of reinforcement learning and will be covered in detail in this book and, partially, in the following sections.

Now that we've learned how to build a policy network with TensorFlow, let's dive into another OpenAI resource that will allow us to easily train an RL agent.

OPENAI BASELINES

So far, we have studied the two different frameworks that allow us to solve reinforcement learning problems (OpenAI Gym and OpenAI Universe). We also studied how to create the "brain" of the agent, known as the policy network, with TensorFlow.

The next step is to train the agent and make it learn how to act optimally, only through experience. Learning how to train an RL agent is the ultimate goal of this book. We will see how most advanced methods work and find out about all their internal elements and algorithms. But even before we find out all the details of how these approaches are implemented, it is possible to rely on some tools that make the task more straightforward.

OpenAI Baselines is a Python-based tool, built on TensorFlow, that provides a library of high-quality, state-of-the-art implementations of reinforcement learning algorithms. It can be used as an out-of-the-box module, but it can also be customized and expanded. We will be using it to solve a classic control problem and a classic Atari video game by training a custom policy network.

> **NOTE**
>
> Please make sure you have installed OpenAI Baselines by using the instructions mentioned in the preface, before moving on.

PROXIMAL POLICY OPTIMIZATION

It is worth providing a high-level idea of what **Proximal Policy Optimization** (**PPO**) is. We will remain at the highest level when describing this state-of-the-art RL algorithm because, in order to deeply understand how it works, you will need to become familiar with the topics that will be presented in the following chapters, thereby preparing you to study and build other state-of-the-art RL methods by the end of this book.

PPO is a reinforcement learning method that is part of the policy gradient family. Algorithms in this category aim to directly optimize the policy, instead of building a value function to then generate a policy. To do so, they instantiate a policy (in our case, in the form of a deep neural network) and build a method to calculate a gradient that defines where to move the policy function's approximator parameters (the weights of our deep neural network, in our case) to directly improve the policy. The word "proximal" suggests a specific feature of these methods: in the policy update step, when adjusting policy parameters, the update is constrained, thus preventing it from moving "too far" from the starting policy. All these aspects will be transparent to the user, thanks to the OpenAI Baselines tool, which will take care of carrying out the job under the hood. You will learn about these aspects in the upcoming chapters.

> **NOTE**
>
> Please refer to the following paper to learn more about PPO:
> https://arxiv.org/pdf/1707.06347.pdf.

COMMAND-LINE USAGE

As stated earlier, OpenAI Baselines allows us to train state-of-the-art RL algorithms easily for OpenAI Gym problems. The following code snippet, for example, trains a PPO algorithm for 20 million steps in the Pong Gym environment:

```
python -m baselines.run --alg=ppo2 --env=PongNoFrameskip-v4
    --num_timesteps=2e7 --save_path=./models/pong_20M_ppo2
    --log_path=./logs/Pong/
```

It saves the model in the user-defined save path so that it is possible to reload the weights on the policy network and deploy the trained agent in the environment with the following command-line instruction:

```
python -m baselines.run --alg=ppo2 --env=PongNoFrameskip-v4
    --num_timesteps=0 --load_path=./models/pong_20M_ppo2 --play
```

You can easily train every available method on every OpenAI Gym environment by changing only the command-line arguments, without knowing anything about how they work internally.

METHODS IN OPENAI BASELINES

OpenAI Baselines gives us access to the following RL algorithm implementations:

- **A2C**: Advantage Actor-Critic
- **ACER**: Actor-Critic with Experience Replay
- **ACKTR**: Actor-Critic using Kronecker-factored Trust Region
- **DDPG**: Deep Deterministic Policy Gradient
- **DQN**: Deep Q-Network
- **GAIL**: Generative Adversarial Imitation Learning
- **HER**: Hindsight Experience Replay
- **PPO2**: Proximal Policy Optimization
- **TRPO**: Trust Region Policy Optimization

For the upcoming exercise and activity, we will be using PPO.

CUSTOM POLICY NETWORK ARCHITECTURE

Despite its out-of-the-box usability, OpenAI Baselines can also be customized and expanded. In particular, as something that will also be used in the next two sections of this chapter, it is possible to provide a custom definition to the module for the policy network architecture.

One aspect that needs to be clear is the fact that the network will be used as an encoder of the environment state or observation. OpenAI Baselines will then take care of creating the final layer, which is in charge of linking the latent space (space of embeddings) to the proper output layer. The latter is chosen depending on the type of the action space (is it discrete or continuous? How many available actions are there?) of the selected environment.

First of all, the user needs to import the Baselines register, which allows them to define a custom network and register it with a user-defined name. Then, they can define a custom deep learning model in the form of a function using a custom architecture. In this way, we are able to change the policy network architecture at will, testing different solutions to find the best one for a specific problem. A practical example will be presented in the exercise in the following section.

Now, we are ready to train our first RL agent and solve a classic control problem.

TRAINING AN RL AGENT TO SOLVE A CLASSIC CONTROL PROBLEM

In this section, we will learn how to train a reinforcement learning agent capable of solving a classic control problem named CartPole by building upon all the concepts explained previously. OpenAI Baselines will be leveraged and, following the steps highlighted in the previous section, we will use a custom fully connected network as a policy network, which is provided as input for the PPO algorithm.

Let's have a quick recap of the CartPole control problem. It is a classic control problem with a continuous four-dimensional observation space and a discrete two-dimensional action space. The observations that are recorded are the position and velocity of the cart along its line of movement, as well as the angle and angular velocity of the pole. The actions are the left/right movement of the cart along its rail. The reward is +1.0 for every step that does not result in a terminal state, which is the case if the pole moves more than 15 degrees from the vertical or if the cart moves outside the rail boundary placed at +/- 2.4. The environment is considered solved if it does not end before having completed 200 steps.

Now, let's put all these concepts together by completing an exercise.

EXERCISE 4.04: SOLVING A CARTPOLE ENVIRONMENT WITH THE PPO ALGORITHM

The CartPole problem in this exercise will be solved using the PPO algorithm. We will use two slightly different approaches so that we will learn about both approaches to using OpenAI Baselines. The first approach will take advantage of Baselines' infrastructure but will adopt a custom path where a user-defined network is used as the policy network. It will be trained and run in the environment after being trained in a "manual" way, without relying on Baselines' automation. This will give you the chance to take a look at what is happening under the hood. The second approach will be simpler, wherein we will be directly adopting Baselines' pre-defined command-line interface.

A custom deep network will be built that will encode environment states and create embeddings in the latent space. The OpenAI Baselines module will then take care of creating the remaining layer of the policy (and value) network for linking the embedding space with action spaces.

We will also create a specific function, which is created by customizing an OpenAI Baselines function, with the specific aim of building the Gym environment, as expected by the infrastructure. There is no particular value in it, but this is required in order to then leverage all Baselines modules.

> **NOTE**
>
> In order to properly run this exercise, you will need to install OpenAI Baselines. Please refer to the preface for the installation instructions.
>
> Also, in order to properly train the RL agent, many episodes are needed, so the training phase may take several hours to complete. A set of weights for the pretrained agent will be provided at the end of this exercise so that you can see the trained agent in action.

Follow these steps to complete this exercise:

1. Open a new Jupyter Notebook and import all the required modules from OpenAI Baselines and TensorFlow to use the PPO algorithm:

```
from baselines.ppo2.ppo2 import learn
from baselines.ppo2 import defaults
from baselines.common.vec_env import VecEnv, VecFrameStack
from baselines.common.cmd_util import make_vec_env, make_env
from baselines.common.models import register
import tensorflow as tf
```

2. Define and register a custom multi-layer perceptron for the policy network. Here, some arguments have also been defined so that you can easily control network architecture, making the user able to specify the number of hidden layers, the number of neurons for the hidden layers, and their activation functions:

```
@register("custom_mlp")
def custom_mlp(num_layers=2, num_hidden=64, activation=tf.tanh):
    """
    Stack of fully-connected layers to be used in a policy /
    q-function approximator
    Parameters:
    ----------
    num_layers: int    number of fully-connected layers (default: 2)
    num_hidden: int    size of fully-connected layers (default: 64)
    activation:        activation function (default: tf.tanh)
```

```
Returns:
-------

function that builds fully connected network with a
given input tensor / placeholder
"""

def network_fn(input_shape):
    print('input shape is {}'.format(input_shape))
    x_input = tf.keras.Input(shape=input_shape)
    h = x_input
    for i in range(num_layers):
        h = tf.keras.layers.Dense\
            (units=num_hidden, \
            name='custom_mlp_fc{}'.format(i),\
            activation=activation)(h)

    network = tf.keras.Model(inputs=[x_input], outputs=[h])
    network.summary()
    return network

return network_fn
```

3. Create a function that will build the environment in the format required by OpenAI Baselines:

```
def build_env(env_id, env_type):

    if env_type in {'atari', 'retro'}:
        env = make_vec_env\
            (env_id, env_type, 1, None, gamestate=None,\
            reward_scale=1.0)
        env = VecFrameStack(env, 4)

    else:
        env = make_vec_env\
            (env_id, env_type, 1, None,\
            reward_scale=1.0, flatten_dict_observations=True)

    return env
```

4. Build the **CartPole-v0** environment, choose the necessary policy network parameters, and train it using the specific PPO **learn** function that has been imported:

```
env_id = 'CartPole-v0'
env_type = 'classic_control'
print("Env type = ", env_type)


env = build_env(env_id, env_type)


hidden_nodes = 64
hidden_layers = 2


model = learn(network="custom_mlp", env=env, \
              total_timesteps=1e4, num_hidden=hidden_nodes, \
              num_layers=hidden_layers)
```

While training, the model will produce an output similar to the following:

```
Env type =  classic_control
Logging to /tmp/openai-2020-05-11-16-00-34-432546
input shape is (4,)
Model: "model"
```

Layer (type)	Output Shape	Param #
input_1 (InputLayer)	[(None, 4)]	0
custom_mlp_fc0 (Dense)	(None, 64)	320
custom_mlp_fc1 (Dense)	(None, 64)	4160

```
Total params: 4,480
Trainable params: 4,480
Non-trainable params: 0
```

```
------------------------------------------
| eplenmean         | 22.3            |
| eprewmean         | 22.3            |
| fps               | 696             |
| loss/approxkl     | 0.00013790815 |
```

```
| loss/clipfrac           | 0.0          |
| loss/policy_entropy     | 0.6929994    |
| loss/policy_loss        | -0.0029695872 |
| loss/value_loss         | 44.237858    |
| misc/explained_variance | 0.0143       |
| misc/nupdates           | 1            |
| misc/serial_timesteps   | 2048         |
| misc/time_elapsed       | 2.94         |
| misc/total_timesteps    | 2048         |
```

This shows the policy network architecture, as well as the bookkeeping of some quantities related with the training process, where the first two are, for example, the mean episode length and the mean episode reward.

5. Run the trained agent in the environment and print the cumulative reward:

```
obs = env.reset()
if not isinstance(env, VecEnv):
    obs = np.expand_dims(np.array(obs), axis=0)


episode_rew = 0

while True:
    actions, _, state, _ = model.step(obs)
    obs, reward, done, info = env.step(actions.numpy())
    if not isinstance(env, VecEnv):
        obs = np.expand_dims(np.array(obs), axis=0)
    env.render()
    print("Reward = ", reward)
    episode_rew += reward

    if done:
        print('Episode Reward = {}'.format(episode_rew))
        break

env.close()
```

The output should be similar to the following:

```
#[...]
Reward =   [1.]
Reward =   [1.]
Reward =   [1.]
Reward =   [1.]
Reward =   [1.]
Reward =   [1.]
Reward =   [1.]
Reward =   [1.]
Reward =   [1.]
Reward =   [1.]
Episode Reward = [28.]
```

6. Use the built-in OpenAI Baselines run script to train PPO on the **CartPole-v0** environment:

```
!python -m baselines.run --alg=ppo2 --env=CartPole-v0
--num_timesteps=1e4 --save_path=./models/CartPole_2M_ppo2
--log_path=./logs/CartPole/
```

The last few lines of the output should be similar to the following:

```
-------------------------------------------
| eplenmean              | 20.8             |
| eprewmean              | 20.8             |
| fps                    | 675              |
| loss/approxkl          | 0.00041882397 |
| loss/clipfrac          | 0.0              |
| loss/policy_entropy    | 0.692711         |
| loss/policy_loss       | -0.004152138 |
| loss/value_loss        | 42.336742        |
| misc/explained_variance | -0.0112         |
| misc/nupdates          | 1                |
| misc/serial_timesteps  | 2048             |
| misc/time_elapsed      | 3.03             |
| misc/total_timesteps   | 2048             |
-------------------------------------------
```

7. Use the built-in OpenAI Baselines run script to run the trained model on the **CartPole-v0** environment:

```
!python -m baselines.run --alg=ppo2 --env=CartPole-v0
--num_timesteps=0
    --load_path=./models/CartPole_2M_ppo2 --play
```

The last few lines of the output should be similar to the following:

```
episode_rew=27.0
episode_rew=27.0
episode_rew=11.0
episode_rew=11.0
episode_rew=13.0
episode_rew=29.0
episode_rew=28.0
episode_rew=14.0
episode_rew=18.0
episode_rew=25.0
episode_rew=49.0
episode_rew=26.0
episode_rew=59.0
```

8. Use the pretrained weights provided to see the trained agent in action:

```
!wget -O cartpole_1M_ppo2.tar.gz \
https://github.com/PacktWorkshops/The-Reinforcement-Learning-\
Workshop/blob/master/Chapter04/cartpole_1M_ppo2.tar.gz?raw=true
```

The output will be similar to the following:

```
Saving to: 'cartpole_1M_ppo2.tar.gz'

cartpole_1M_ppo2.ta 100%[====================>]   53,35K   --.-KB/s
in 0,05s

2020-05-11 15:57:07 (1,10 MB/s) - 'cartpole_1M_ppo2.tar.gz' saved
[54633/54633]
```

You can read the **.tar** file using the following command:

```
!tar xvzf cartpole_1M_ppo2.tar.gz
```

The last few lines of the output should be similar to the following:

```
cartpole_1M_ppo2/ckpt-1.index
cartpole_1M_ppo2/ckpt-1.data-00000-of-00001
cartpole_1M_ppo2/
cartpole_1M_ppo2/checkpoint
```

9. Use the built-in OpenAI Baselines run script to train PPO on the CartPole environment:

```
!python -m baselines.run --alg=ppo2 --env=CartPole-v0
--num_timesteps=0 --load_path=./cartpole_1M_ppo2 -play
```

The output will be similar to the following:

```
episode_rew=16.0
episode_rew=200.0
episode_rew=200.0
episode_rew=200.0
episode_rew=26.0
episode_rew=176.0
```

This step will show you how a trained agent behaves so that it can solve the CartPole environment. It uses a set weights for the policy network that were ready to be used. The output will be similar to the one shown in *Step 5*, confirming that the environment has been solved.

> **NOTE**
>
> To access the source code for this specific section, please refer to https://packt.live/2XS69n8.
>
> This section does not currently have an online interactive example, and will need to be run locally.

In this exercise, we learned how to train a reinforcement learning agent capable of solving the CartPole classic control problem. We successfully used a custom fully connected network as a policy network. This allowed us to take a look at what happens behind the automation provided by OpenAI Baselines' command-line interface. In this hands-on exercise, we have also familiarized ourselves with OpenAI Baselines' out-of-the-box method, confirming that it is a straightforward resource that can be easily used to train a reinforcement learning agent.

ACTIVITY 4.01: TRAINING A REINFORCEMENT LEARNING AGENT TO PLAY A CLASSIC VIDEO GAME

In this activity, the challenge is to adopt the same approach we used in *Exercise 4.04, Solving the CartPole Environment with the PPO Algorithm*, to create a reinforcement learning bot that's able to achieve better-than-human performance on a classic Atari video game, Pong. The game is represented in the following way: two paddles, one per user, can move up and down. The goal is to make the white ball pass the opposite paddle to score one point. The game ends when one of the two players reaches a score equal to **21**.

An approach similar to the one we saw in *Exercise 4.04, Solving the CartPole Environment with the PPO Algorithm*, has to be adopted, with a custom convolutional neural network, which will work as the encoder for the environment's observation (the pixels frame):

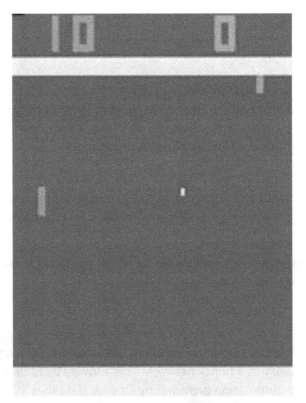

Figure 4.12: One frame of the Pong game

OpenAI Gym will be used to create the environment, while the OpenAI Baselines module will be used to train a custom policy network using the PPO algorithm.

As we saw in *Exercise 4.04, Solving the CartPole Environment with the PPO Algorithm*, both the custom approach, that is, using specific OpenAI modules, and the simple one, that is, using the built-in general command-line interface, will be implemented (in *steps 1* to *5* and *step 6*, respectively).

> **NOTE**
>
> In order to run this exercise, you will need to install OpenAI Baselines. Please refer to the preface for the installation instructions.
>
> In order to properly train the RL agent, many episodes are needed, so the training phase may take several hours to complete. A set of weights you can use for a pretrained agent has been provided at this address: https://packt.live/2XSY4yz. Use them to see the trained agent in action.

The following steps will help you complete this activity:

1. Import all the required modules from OpenAI Baselines and TensorFlow in order to use the **PPO** algorithm.

2. Define and register a custom convolutional neural network for the policy network.

3. Create a function to build the environment in the format required by OpenAI Baselines.

4. Build the **PongNoFrameskip-v4** environment, choose the required policy network parameters, and train it.

5. Run the trained agent in the environment and print the cumulative reward.

6. Use the built-in OpenAI Baselines run script to train PPO on the **PongNoFrameskip-v0** environment.

7. Use the built-in OpenAI Baselines run script to run the trained model on the **PongNoFrameskip-v0** environment.

8. Use the pretrained weights provided to see the trained agent in action.

   At the end of this activity, the agent is expected to easily win most of the time.

   The final score of the agent should be like the one represented in the following frame most of the time:

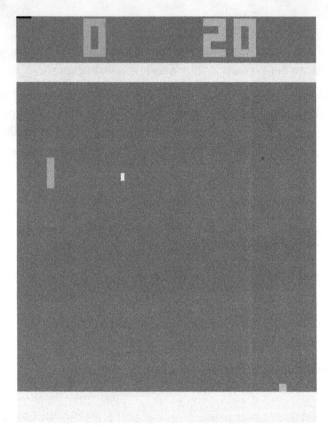

Figure 4.13: One frame of the real-time environment, after rendering

---

NOTE

The solution to this activity can be found on page 704.

SUMMARY

This chapter introduced us to the key technologies and concepts we can use to get started with reinforcement learning. The first two sections described two OpenAI Tools, OpenAI Gym and OpenAI Universe. These are collections that contain a large number of control problems that cover a broad spectrum of contexts, from classic tasks to video games, from browser usage to algorithm deduction. We learned how the interfaces of these environments are formalized, how to interact with them, and how to create a custom environment for a specific problem. Then, we learned how to build a policy network with TensorFlow, how to feed it with environment states to retrieve corresponding actions, and how to save the policy network weights. We also studied another OpenAI resource, Baselines. We solved problems that demonstrated how to train a reinforcement learning agent to solve a classic control task. Finally, using all the elements introduced in this chapter, we built an agent and trained it to play a classic Atari video game, thus achieving better-than-human performance.

In the next chapter, we will be delving deep into dynamic programming for reinforcement learning.

5

DYNAMIC PROGRAMMING

OVERVIEW

In this chapter, you will be introduced to the driving principles of dynamic programming. You will be introduced to the classic coin-change problem as an application of dynamic programming. Furthermore, you will learn how to implement policy evaluation, policy iteration, and value iteration and learn the differences between them. By the end of the chapter, you will be able to implement dynamic programming to solve problems in **Reinforcement Learning (RL)**.

INTRODUCTION

In the previous chapter, we were introduced to the OpenAI Gym environment and also learned how to implement custom environments, depending on the application. You also learned the basics of TensorFlow 2, how to implement a policy using the TensorFlow 2 framework, and how to visualize learning using TensorBoard. In this chapter, we will see how **Dynamic Programming** (**DP**) works in general, from a computer science perspective. Then, we'll go over how and why it is used in RL. Next, we will dive deep into classic DP algorithms such as policy evaluation, policy iteration, and value iteration and compare them. Lastly, we will implement the algorithms in the classic coin-change problem.

DP is one of the most fundamental and foundational topics in computer science. Furthermore, RL algorithms such as **Value Iteration**, **Policy Iteration**, and others, as we will see, use the same basic principle: avoid repeated computations to save time, which is what DP is all about. The philosophy of DP is not new; it is self-evident and commonplace once you learn the ways to solve it. The hard part is identifying whether a problem can be solved using DP.

The basic principle can be explained to a child as well. Imagine counting the number of candies in a box. If you know there are 100 candies in a box, and the shopkeeper offers 5 extra candies, you don't start counting the candies all over again. You use the prior information to add 5 to the original count and say, "I have 105 candies." That's the core of DP: saving intermediate information and reusing it, if required, to avoid re-computation. While it sounds simple, as mentioned before, the hard part is identifying whether a problem can be solved using DP. As we will see later, in the *Identifying Dynamic Programming Problems* section, a problem must satisfy a specific prerequisite, such as optimal substructure and overlapping subproblems, to be solved using DP, which we will study in the *Identifying Dynamic Programming Problems* section. Once a problem qualifies, there are some well-known techniques such as top-down memoization, that is, saving intermediate states in an unordered fashion, and bottom-up tabulation, which is saving the states in an ordered array or matrix.

Combining these techniques can achieve a considerable performance boost over solving them using brute force. Furthermore, the difference in time increases with an increase in the number of operations. Mathematically speaking, solutions solved using DP usually run in $O(n^2)$, while those using brute force execute in $O(2^n)$ time, where the notation "O" (Big-O) can be loosely thought of as the number of operations performed. So, for instance, if N=500, which is a reasonably small number, a DP algorithm will roughly execute 500^2 steps, compared to a brute force algorithm, which will use 2^{500} steps. For reference, there are 2^{80} hydrogen atoms in the sun, which is undoubtedly a much smaller number than 2^{500}.

The following figure depicts the difference in the number of operations executed for both algorithms:

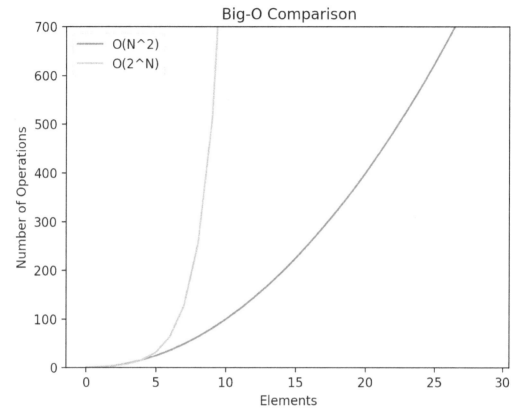

Figure 5.1: Visualizing Big-O values

Let's now move toward studying the approach of solving DP problems.

SOLVING DYNAMIC PROGRAMMING PROBLEMS

There are two popular ways to solve DP problems: the tabular method and memoization. In the tabular method, we build a matrix that stores the intermediate values one by one in the lookup table. On the other hand, in the memoization method, we store the same values in an unstructured way. Here, unstructured way refers to the fact that the lookup table may be filled all at once.

Imagine you're a baker and are selling cakes to shops. Your job is to sell cakes and make the maximum profit out of it. For simplicity, we will assume that all other costs are fixed, and the highest price offered for your product is the only indicator of profits earned, which is a fair assumption for most business cases. So, naturally, you'd wish to sell all your cakes to the shop offering the highest price, but there's a decision to make as there are multiple shops that offer different prices on different sizes of cakes. So, you have two choices: how much to sell, and which shop to trade with. For this example, we'll forget other variables and assume there are no additional hidden costs. We'll tackle the problem using the tabular method, as well as memoization.

Phrasing the problem formally, you have a cake with weight W, and an array of prices that different shops are willing to offer, and you have to find out the optimal configuration that yields the highest price (and by the assumptions stated previously, highest profit).

> **NOTE**
>
> In the code examples, which will be listed further in this section, we have used profit and price interchangeably. So, for example, if you encounter a variable such as **best_profit**, it would also be an indicator of best price and vice-versa.

For instance, say W = 5, meaning we have a cake that weighs 5 kilograms and the prices, indicated in the following table, are what are offered by restaurants:

Weight in KGs	Price Offered
1	9
2	40
3	50
4	70
5	80

Figure 5.2: Different prices offered for different weights of cakes

Now consider restaurant A pays $10 for a 1 kg cake, but $40 for a 2 kg cake. So, the question is: should I sell a 5 kg cake and partition it into 5 x 1 kg slices, which will yield $45, or should I sell the 5 kg cake as a whole to restaurant B, which is offering $80. In this case, the most optimal configuration is to partition the cake into a 3 kg part that yields $50 and a 2 kg part that generates $40, which yields a total of $90. The following table indicates various ways of partitioning and the corresponding price that we'll get:

Partition	Price Offered
1 + 1 + 1 + 1 + 1	45
2 + 1 + 1 + 1	67
2 + 2 + 1	89
2 + 3	90
4 + 1	79
5	80

Figure 5.3: Different combinations for cake partitioning

Now, from the preceding table, it is quite evident that the best price is provided by the combination of 2 kg + 3 kg. But to really understand the limitation of the brute force approach, we'll assume that we don't know the best combination for yielding the maximum price. We'll try to implement the brute force approach in code. In reality, the number of observations for an actual business problem may be too large for you to arrive at an answer as quickly as you may have done here. The preceding table is just an example to help you understand the limitations of the brute force approach.

So, let's try to solve this problem using brute force. We can rephrase the question slightly differently: at every junction, we have a choice – partition or not. If we choose to partition the cake into two unequal parts first, the left side, for instance, becomes one part of the cake, and the right side can be treated as an independent partition. In the next iteration, we'll only concentrate on the right side / the other part. Now, again, we can partition it, and the right side becomes a part of the cake that is divided further. This paradigm is also called **recursion**.

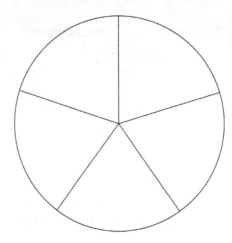

Figure 5.4: Cake partitioned into several pieces

In the preceding figure, we can see a cake being partitioned into multiple pieces. For a cake that weighs 5 kg (and assuming you can partition the cake in a manner that the minimum weight of each partition is 1 kg, and thus the partitions can only be integral multiples of 1), we are presented with "partition or not" a total of 32 times; here's how:

```
2 x 2 x 2 x 2 x 2 = 2⁵= 32
```

So, for starters, let's do this: for each of the 32 possible combinations, calculate the total price, and in the end, report the combination with the highest amount of price. We've defined the price in a list, where the index tells us the weight of a slice of cake:

```
PRICES = ["NA", 9, 40, 50, 70, 80]
```

For instance, selling a whole 1 kg cake yields a price of $9; whereas selling a 2 kg cake/slice yields a price of $40. The price on the zero[th] index is NA because we won't ever have a cake that weighs 0 kg. Here is pseudo-code formulated to implement the preceding scenario:

```
def partition(cake_size):
    """

    Partitions a cake into different sizes, and calculates the
    most profitable cut configuration
    Args:
        cake_size: size of the cake

    Returns:
        the best profit possible
    """
    if cake_size == 0:
        return 0
    best_profit = -1
    for i in range(1, cake_size + 1):
        best_profit = max(best_profit, PRICES[i] \
                        + partition(cake_size - i))
    return best_profit
```

The preceding function partition, **cake_size**, will take an integer input: the size of the cake. Then, in the **for** loop, we will cut the cake in every possible way and calculate the best profit. Given that we are taking a partition/no partition decision for every single place, the code runs in $O(2^n)$ time. Now let's call the function using the following code. The **if __name__** block will make sure that your code runs only when you run the script (and not when you import it):

```
if __name__ == '__main__':
    size = 5
    best_profit_result = partition(size)
    print(f"Best profit: {best_profit_result}")
```

Upon running it, we can see the best possible profit for a cake of size **5**:

```
Best profit: 90
```

The preceding method solves the problem of calculating maximum profit, but it has a huge flaw: it is very slow. We are performing unnecessary computations, and exploring the entire search tree (all possible combinations). Why is this a bad idea? Well, imagine you're traveling from point A to point C, and it costs $10. Would you ever consider traveling from A to B to D to F and then to C, which might cost, say, $150? Of course not, right? The idea is similar: if I know the current path is not the most optimal one, why bother exploring that way?

To solve this problem more efficiently, we will look at two great techniques: the tabular method and memoization. Both are based on the same principle: avoid unproductive exploration. But each uses a slightly fundamentally different approach to solving the problem, as you will see.

Let's explore memoization in the following section.

MEMOIZATION

The **memoization** method refers to a method in which we save the results of the intermediate outputs for further use in a dictionary, also known as memo. Hence the name "memoization."

Coming back to our cake partition example, if we modify the **partition** function and print the value of **cake_size** and the best solution for the size, there's a new pattern to be found. Using the same code as was used in the brute force approach before, we add a **print** statement to display the cake size and the corresponding profit:

```
def partition(cake_size):
    """
    Partitions a cake into different sizes, and calculates the
    most profitable cut configuration
    Args:
        cake_size: size of the cake

    Returns:
        the best profit possible
    """
    if cake_size == 0:
        return 0
```

```
    best_profit = -1
    for i in range(1, cake_size + 1):
        best_profit = max(best_profit, PRICES[i] \
                      + partition(cake_size - i))
    print(f"Best profit for size {cake_size} is {best_profit}")
    return best_profit
```

Call the function using the **main** block:

```
if __name__ == '__main__':
    size = 5
    best_profit_result = partition(size)
    print(f"Best profit: {best_profit_result}")
```

We then see the output as follows:

```
Best profit for size 1 is 9
Best profit for size 2 is 40
Best profit for size 1 is 9
Best profit for size 3 is 50
Best profit for size 1 is 9
Best profit for size 2 is 40
Best profit for size 1 is 9
Best profit for size 4 is 80
Best profit for size 1 is 9
Best profit for size 2 is 40
Best profit for size 1 is 9
Best profit for size 3 is 50
Best profit for size 1 is 9
Best profit for size 2 is 40
Best profit for size 1 is 9
Best profit for size 5 is 90
Best profit: 90
```

As you can see in the preceding output, there is a pattern here – the best profit for a given size remains the same, but we calculate it many times. Especially pay attention to the size and the order of the calculations. It calculates the profit for size 1, and then 2, and now when it wants to calculate it for size 3, it does so by starting from scratch again by calculating the answer for 1, and then 2, and then finally 3. This happens repeatedly since it doesn't store any intermediate results. An obvious improvement would be to store the profit in a memo and then use it later.

We add a small modification here: if the **best_profit** for a given **cake_size** is already calculated, we just use it right away without calculating it, as shown in the following code:

```
if cake_size == 0:
    return 0
if cake_size in memo:
    return memo[cake_size]
```

Let's now look at the complete code snippet:

```
def memoized_partition(cake_size, memo):
    """
        Partitions a cake into different sizes, and calculates the
        most profitable cut configuration using memoization.
        Args:
            cake_size: size of the cake
            memo: a dictionary of 'best_profit' values indexed
                by 'cake_size'

        Returns:
            the best profit possible
    """
    if cake_size == 0:
        return 0
    if cake_size in memo:
        return memo[cake_size]
    else:
        best_profit = -1
        for i in range(1, cake_size + 1):
            best_profit = max(best_profit, \
                            PRICES[i] + memoized_partition\
                                    (cake_size - i, memo))
        print(f"Best profit for size {cake_size} is {best_profit}")
        memo[cake_size] = best_profit
        return best_profit
```

Now if we run this program, we get the following output:

```
Best profit for size 1 is 9
Best profit for size 2 is 40
Best profit for size 3 is 50
Best profit for size 4 is 80
Best profit for size 5 is 90
Best profit: 90
```

Here, instead of running the calculations 2^n times, we're running it just **n** times. That's a vast improvement. And all we had to do was save the result of the output in a dictionary, or memo, hence the name **memoization**. In this method, we essentially save the intermediate solution in a dictionary to avoid re-computation. This method is also called the top-bottom method as we follow natural ordering analogous to searching in a binary tree, for instance.

Next, we will be looking at the tabular method.

THE TABULAR METHOD

Using memoization, we arbitrarily store the intermediate computation. The tabular method does almost the same thing, but slightly differently: it goes in a predetermined order, which is almost always fixed – from small to large. This means that to obtain the most profitable cuts, we will first get the most profitable cut in a 1 kg cake, then a 2 kg cake, then 3 kg, and so on. This is usually done using a matrix and is called the bottom-up method as we solve the smaller problems first.

Consider the following code snippet:

```python
def tabular_partition(cake_size):
    """
    Partitions a cake into different sizes, and calculates the
    most profitable cut configuration using tabular method.
    Args:
        cake_size: size of the cake

    Returns:
        the best profit possible

    """
    profits = [0] * (cake_size + 1)
    for i in range(1, cake_size + 1):
        best_profit = -1
```

```
    for current_size in range(1, i + 1):
        best_profit = max(best_profit, \
                          PRICES[current_size] \
                          + profits[i - current_size])
    profits[i] = best_profit
return profits[cake_size]
```

The output will be as follows:

```
Best profit: 90
```

In the preceding code, we are iterating over the sizes first and then cuts. A good exercise would be to run the code in an IDE using a debugger to see how the **profits** array is updated. First, it would find the most profit in the cake of size 1, and then it would find the most profit in the cake of size 2. But here, the second **for** loop would try both the configurations: one cut (two cakes of size 1), and no cuts (one cake of size 2) indicated by **profits[i - current_size]**. Now, similarly, for every size, it would try to cut the cake in all the possible configurations, without recalculating the profits on the smaller part. For instance, **profits[i - current_size]** would return the best possible configuration, without recalculating it.

EXERCISE 5.01: MEMOIZATION IN PRACTICE

In this exercise, we will try to solve a DP problem using the memoization method. The problem is as follows:

Given a number **n**, print the nth Tribonacci number. The Tribonacci sequence is similar to the Fibonacci sequence but uses three numbers instead of two. This means that the nth Tribonacci number is the sum of the prior three numbers. The following is an example:

Fibonacci sequence 0, 1, 2, 3, 5, 8... is defined as follows:

$$F(n) = F(n-1) + F(n-2)$$

Figure 5.5: Fibonacci sequence

Tribonacci sequence 0, 0, 1, 1, 2, 4, 7.... is defined as follows:

$$T(n) = T(n-1) + T(n-2) + T(n-3)$$

Figure 5.6: Tribonacci sequence

The generalized formula for the Tribonacci sequence is as follows:

```
Fibonacci(n) = Fibonacci(n - 1) + Fibonacci(n - 2)
Tribonacci(n) = Tribonacci(n - 1) \
                + Tribonacci(n - 2) + Tribonacci(n - 3)
```

The following steps will help you complete the exercise:

1. Now that we know the formula, the first step is to create a simple recursive implementation in Python. Use the formulas in the description and convert them into a Python function. You can choose to do it in a Jupyter notebook, or just a simple **.py** Python file:

```python
def tribonacci_recursive(n):
    """
    Uses recursion to calculate the nth tribonacci number
    Args:
        n: the number

    Returns:
        nth tribonacci number
    """
    if n <= 1:
        return 0
    elif n == 2:
        return 1
    else:
        return tribonacci_recursive(n - 1) \
                + tribonacci_recursive(n - 2) \
                + tribonacci_recursive(n - 3)
```

In the preceding code, we are recursively calculating the value of the Tribonacci number. Furthermore, if the number is less than or equal to 1, we know the answer is going to be 0, and for 2 it's going to be 1, so we add the **if-else** condition to take care of the edge cases. To test the preceding code, just call it in the **main** block and check the output is as expected:

```python
if __name__ == '__main__':
    print(tribonacci_recursive(6))
```

2. As we've learned, this implementation is quite slow and grows exponentially with higher values of **n**. Now, using the principle of memoization, store the intermediate results so they are not recomputed. Create a dictionary that will check whether the answer to that n[th] tribonacci number is already added to the dictionary. If yes, just return that; otherwise, try to compute it:

```
def tribonacci_memo(n, memo):
    """
    Uses memoization to calculate the nth tribonacci number
    Args:
        n: the number
        memo: the dictionary that stores intermediate results
    Returns:
        nth tribonacci number
    """
    if n in memo:
        return memo[n]
    else:
        ans1 = tribonacci_memo(n - 1, memo)
        ans2 = tribonacci_memo(n - 2, memo)
        ans3 = tribonacci_memo(n - 3, memo)
        res = ans1 + ans2 + ans3
        memo[n] = res
        return res
```

3. Now, using the previous code snippet, calculate the nth Tribonacci number without using recursion. Run the code and make sure the output matches the expectation by running it in the **main** block:

```
if __name__ == '__main__':
    memo = {0: 0, 1: 0, 2: 1}
    print(tribonacci_memo(6, memo))
```

The output will be as follows:

As you can see in the output, the sum is **7**. We have learned how to convert a simple recursive function into memoized DP code.

> **NOTE**
>
> To access the source code for this specific section, please refer to https://packt.live/3dghMJ1.
>
> You can also run this example online at https://packt.live/3fFE7RK.

Next, we will try to do the same with the tabular method.

EXERCISE 5.02: THE TABULAR METHOD IN PRACTICE

In this exercise, we will solve a DP problem using the tabular method. The goal of the exercise is to identify the length of the longest common substring between two strings. For instance, if the two strings are **BBBABDABAA** and **AAAABDABBAABB**, then the longest common substring is **ABDAB**. Other common substrings are **AA**, **BB**, and **BA**, and **BAA** but they're not the longest:

1. Import the **numpy** library:

    ```
    import numpy as np
    ```

2. Implement the brute force method to calculate the longest common substring of two strings first. Imagine we have two variables, **i** and **j**, that indicate the start and end of the substring. Use these pointers to indicate the start and end of the substring for both strings. You can use the **==** operator in Python to see whether the strings match:

    ```
    def lcs_brute_force(first, second):
        """
        Use brute force to calculate the longest common
        substring of two strings
        Args:
            first: first string
            second: second string
        Returns:
            the length of the longest common substring
        """
        len_first = len(first)
        len_second = len(second)
    ```

```
    max_lcs = -1
    lcs_start, lcs_end = -1, -1
    # for every possible start in the first string
    for i1 in range(len_first):
        # for every possible end in the first string
        for j1 in range(i1, len_first):
            # for every possible start in the second string
            for i2 in range(len_second):
                # for every possible end in the second string
                for j2 in range(i2, len_second):
                    """
                    start and end position of the current
                    candidates
                    """
                    slice_first = slice(i1, j1)
                    slice_second = slice(i2, j2)
                    """
                    if the strings match and the length is the
                    highest so far
                    """
                    if first[slice_first] == second[slice_second] \
                        and j1 - i1 > max_lcs:
                        # save the lengths
                        max_lcs = j1 - i1
                        lcs_start = i1
                        lcs_end = j1

    print("LCS: ", first[lcs_start: lcs_end])
    return max_lcs
```

3. Call the function using the **main** block:

```
if __name__ == '__main__':
    a = "BBBABDABAA"
    b = "AAAABDABBAABB"
    lcs_brute_force(a, b)
```

We can verify that the output is correct:

```
LCS:  ABDAB
```

4. Let's implement the tabular method. Now that we have a simple solution, we can proceed to optimize it. Look at the main loop, which nests four times. Meaning the solution runs in **O(N^4)**. It performs the same calculations irrespective of whether we have the longest common substring or not. Use the tabular method to come up with more solutions:

```python
def lcs_tabular(first, second):
    """
    Calculates the longest common substring using memoization.
    Args:
        first: the first string
        second: the second string
    Returns:
        the length of the longest common substring.
    """
    # initialize the table using numpy
    table = np.zeros((len(first), len(second)), dtype=int)
    for i in range(len(first)):
        for j in range(len(second)):
            if first[i] == second[j]:
                table[i][j] += 1 + table[i - 1][j - 1]
    print(table)
    return np.max(table)
```

The problem has a nice matrix structure inherent to it. Consider the length of one string to be the rows and the length of the other string as the columns of the matrix. Initialize this matrix with **0**. The values in the matrix at position **i**, **j** will indicate whether the **i**th character in the first string is the same as the **j**th character in the second string.

Now the longest common substring will have the highest number of ones in a diagonal. Use this fact to increment the maximum length of the substring by 1 if there's a match at the current position and there's a **1** in the **i-1** and **j-1** positions. This will essentially indicate that there are two subsequent matches. Return the **max** element in the matrix using **np.max(table)**. We can also look at the diagonally increasing sequence until the value reaches **5**.

5. Call the function using the **main** block:

```
if __name__ == '__main__':
    a = "BBBABDABAA"
    b = "AAAABDABBAABB"
    lcs_tabular(a, b)
```

The output will be as follows:

```
[[0 0 0 0 1 0 0 1 1 0 0 1 1]
 [0 0 0 0 1 0 0 1 2 0 0 1 2]
 [0 0 0 0 1 0 0 1 2 0 0 1 2]
 [3 1 1 1 0 0 1 0 0 3 1 0 0]
 [0 0 0 0 2 0 0 2 1 0 0 2 1]
 [0 0 0 0 0 3 0 0 0 0 0 0 0]
 [1 1 1 1 0 0 4 0 0 1 1 0 0]
 [0 0 0 0 2 0 0 5 1 0 0 2 1]
 [2 1 1 1 0 0 1 0 0 2 1 0 0]
 [1 3 2 2 0 0 1 0 0 1 3 0 0]]
```

Figure 5.7: Output for LCS

As you can see, there is a direct mapping between the rows (the first string) and the columns (the second string), so the LCS string would just be the diagonal elements counted backward from the LCS length. In the preceding output, you can see that the highest element is 5 and hence you know that the length is 5. The LCS string would be the elements going diagonally upward from the element **5**. The direction of the string will always be diagonally upward since the columns always run from left to right. Note that the solution involves just calculating the length of the LCS and not finding the actual LCS.

> **NOTE**
>
> To access the source code for this specific section, please refer to https://packt.live/3fD79BC.
>
> You can also run this example online at https://packt.live/2UYVlfK.

Now that we have learned how to solve DP problems, we should next learn how to identify them.

IDENTIFYING DYNAMIC PROGRAMMING PROBLEMS

While it is easy to solve a DP problem once you identify how it recurses, it is difficult to determine whether a problem can be solved using DP. For instance, the traveling salesman problem, where you are given a graph and wish to cover all the vertices in the least possible time, is something that can't be solved using DP. Every DP problem must satisfy two prerequisites: it should have an optimal substructure and should have overlapping subproblems. We'll look into exactly what they mean and how to solve them in the subsequent section.

OPTIMAL SUBSTRUCTURES

Recall the best path example we discussed earlier. If you want to go from point A to point C through B, and you know that's the best path, there's no point in exploring others. Rephrasing this: If I want to go from A to D and I know the best path from A to C, then the best route from A to D will include the path from A to C. This is called the optimal substructure. Essentially, what it means is the optimal solution to the problem contains optimal solutions to subproblems. Remember how we didn't care to recalculate the best profit for a cake of size n once we knew it? Because we know the best profit for the cake of size $n + 1$ will include n while considering making a cut and dividing the cake into size n and 1. To reiterate, the property of optimal substructure would be a requirement if we were to solve the problem using DP.

OVERLAPPING SUBPROBLEMS

Remember when we were initially designing a brute force solution for the cake partition example, and later using memoization. Initially, it required 32 steps for the brute force approach to arrive at the solution, while memoization took only 5. This was because the brute force approach performed the same computation repeatedly: the optimal solution for size three would call for size two and then one. Then, for size 4, it would again call for three, and then two, and then one. This recursive re-computation is due to the nature of the problem: the overlapping subproblems. This is the reason we could save the answer in a memo and later use the same solution without recomputing it. The overlapping subproblem is another requirement that a problem must have to be solved using DP.

THE COIN-CHANGE PROBLEM

The coin-change problem is one of the most commonly asked interview questions in software engineering interviews. The statement is simple: given a list of coin denominations, and a sum value N, identify the number of unique ways to arrive at the sum. For instance, if N = 3 and D, the coin denomination, = {1, 2} the answer is 2. That is, there are two ways to arrive at 3 by having coins of denomination 1 and 2, which are {1, 1, 1} and {2, 1}:

1. To solve the problem, you would need to prepare the recursion formula that will calculate the number of ways to arrive at a sum. To do this, you might want to start with a simple version that solves just a single number and then try to convert it to a more general solution.

2. The end output could be a table as shown in the following figure, which can be used to summarize the result. In the following table, the first row represents the denominations, and the first column represents the sum. More specifically, the first row, 0, 1, 2, 3, 4, 5, represents the sum. And the first column represents the available denominations. We initialize the base cases with 1 and not 0 because if the denomination is less than the sum, then we just copy the previous combinations over.

    The following table represents how to count the number of ways to get to 5 using coins [1, 2]:

X	0	1	2	3	4	5
1	1	1	1	1	1	1
2	1	1	2	2	3	3

    Figure 5.8: Counting the number of ways to get to sum 5 using denominations of 1, 2

3. So, we can see the number of ways to arrive at the sum of 5 using coins of denominations 1 and 2 is 3, which is basically 1+1+1+1+1, 2+1+1+1, and 2+2+1. Remember we're looking for only unique ways, meaning, 2+2+1 is the same as 1+2+2.

Let's execute an exercise to solve the coin-change problem.

EXERCISE 5.03: SOLVING THE COIN-CHANGE PROBLEM

In this exercise, we will be solving the classic and very popular coin-change problem. Our goal is to find the number of permutations, in which the coins can be used to arrive at a sum, 5, using the coin denominations of 1, 2, and 3. The following steps will help you complete the exercise:

1. Import the **numpy** and **pandas** libraries:

```
import numpy as np
import pandas as pd
```

2. Let's now try to identify the overlapping subproblem. As previously, there's one common thing: we have to search for all possible denominations and check whether they sum to a certain number. Furthermore, it's a little more complicated than the cake example since we have got two things to iterate on: firstly, the denomination, and secondly the total sum (in the cake example, it was only one variable, the cake size). So, we need a 2D array, or a matrix.

    On the columns, we will have the sum we are trying to reach, and on the rows, we will consider various denominations available. As we loop over the denominations (columns), we will calculate the number of ways to sum up to **n** by first adding the number of ways to reach the sum without considering the current denomination, and then by considering it. This is analogous to the cake example, where we first performed the cut, calculated the profit, and then didn't perform the cut and calculate the profit. The difference, however, is this time the previous best configuration would be fetched from the row above, and also, we would add the two numbers instead of selecting the maximum out of it since we are interested in the total number of ways to reach the sum. For example, the number of ways to sum up to 4 using {1, 2} would be first to use {2} and then add the number of ways to sum up to 4 – 2 = 2. We could fetch it from the same row and the index would be 2. We will also initiate the first row with 1s as they are either invalid (the number of ways to reach zeros using 1) or valid with one solution:

	0	1	2	3	4	5
1	1	1	1	1	1	1
2						

Figure 5.9: Initial setup of the algorithm

This logic can be translated into code as follows:

```python
def count_changes(N, denominations):
    """
    Counts the number of ways to add the coin denominations
    to N.
    Args:
        N: number to sum up to
        denominations: list of coins
    Returns:
    """
    print(f"Counting number of ways to get to {N} using coins:\
{denominations}")
```

3. Next, we will initialize a table with the dimension **len(denomination)** x **(N + 1)**. The number of columns is **N + 1** since the index includes zero as well:

```python
table = np.ones((len(denominations), N + 1)).astype(int)
# run the loop from 1 since the first row will always 1s
for i in range(1, len(denominations)):
    for j in range(N + 1):
        if j < denominations[i]:
            """
            If the index is less than the denomination
            then just copy the previous best
            """
            table[i, j] = table[i - 1, j]
        else:
            """
            If not, the add two things:
            1. The number of ways to sum up to
               N *without* considering
               the existing denomination.
            2. And, the number of ways to sum up to N minus
               the value of the current denomination
               (by considering the current and the
               previous denominations)
            """
            table[i, j] = table[i - 1, j] \
                        + table[i, j - denominations[i]]
```

4. Now, in the end, we will print the table:

```
# print the table
print_table(table, denominations)
```

5. Create a Python script with the following utility, which pretty prints a table. This will be useful for debugging. Pretty printing is essentially used to present data in a more legible and comprehensive way. By setting the denominations as the index, we will see the output more clearly:

```
def print_table(table, denominations):
    """
    Pretty print a numpy table
    Args:
        table: table to print
        denominations: list of coins

    Returns:

    """
    df = pd.DataFrame(table)
    df = df.set_index(np.array(denominations))
    print(df)
```

> **NOTE**
>
> For more details on pretty printing, you can refer to the official documentation at the following link: https://docs.python.org/3/library/pprint.html.

6. Initialize the script with the following configuration:

```
if __name__ == '__main__':
    N = 5
    denominations = [1, 2]
    count_changes(N, denominations)
```

The output will be as follows:

```
Counting number of ways to get to 5 using coins: [1, 2]
    0  1  2  3  4  5
 1  1  1  1  1  1  1
 2  1  1  2  2  3  3
```

As we can see in the entry in the last row and column, the number of ways to get a 5 using [1, 2] is 3. We have now learned about the concept of DP in detail.

> **NOTE**
>
> To access the source code for this specific section, please refer to https://packt.live/2NeU4lT.
>
> You can also run this example online at https://packt.live/2YUd6DD.

Next, let's see how it is used to solve problems in RL.

DYNAMIC PROGRAMMING IN RL

DP plays an important role in RL as the number of choices you have at a given time is too large. For instance, whether the robot should take a left or right turn given the current state of the environment. To solve such a problem, it's infeasible to find the outcome of every state using brute force. We can do that, however, using DP, using the methods we learned in the previous section.

We have seen the Bellman equation in previous chapters. Let's reiterate the basics and see how the Bellman equation has both of the required properties for using DP.

Assuming the environment is a finite **Markov Decision Process** (**MDP**), let's define the state of the environment by a finite set of states, S. This indicates the state configuration, for instance, the current position of the robot. A finite set of actions, A, gives the action space, and a finite set of rewards, R. Let's denote the discounting rate using γ, which is a value between 0 and 1.

Given a state, S, the algorithm chooses one of the actions in A using a deterministic policy, π. The policy is nothing but a mapping between state S and action A, for instance, a choice a robot would make such as go left or right. And a deterministic policy allows us to choose an action in a non-random fashion (as opposed to a stochastic policy, which has a significant random component).

To concretize our understanding, let's take an example of a simple autonomous car. To make it simple, we will make some reasonable assumptions here. The action space can be defined as {left, right, straight, reverse}. A deterministic policy is: if there's a hole in the ground, take a left or right turn to avoid it. A stochastic policy, however, would say: if here's a hole in the ground, take a left turn with 80% probability, which means there's a small chance that the car would purposely enter the hole. While this move might not make sense at the moment, we will see later, in the *Chapter 7, Temporal Difference Learning*, that this is a rather important thing to do and addresses one of the critical concepts in RL: the exploration versus exploitation dilemma.

Coming back to the original point of using DP in RL, the following is the **simplified** Bellman equation:

$$V(s) = \max_a \; (R(s,a) + \gamma * V(s'))$$

Figure 5.10: Simplified Bellman equation

The only difference with the complete equation is we are not summing over $P(s,a,s')$, which is valid in the case that we have a non-deterministic environment. Here is the complete Bellman equation:

$$V(s) = \max_a \left(R(s,a) + \gamma \sum_{s'} P(s,a,s') V(s') \right)$$

Figure 5.11: Complete Bellman equation

In the preceding equation, $V(s)$ is the value function, the reward for being in a particular state. We will look more deeply into it later. $R(s,a)$ is the reward of taking action **a** and $\gamma * V(s)$ is the reward of the next state. Two things you can observe are the following:

- The recursive nature between $V(s)$ and $V(s')$, meaning $V(s)$ has an optimal substructure.

- The computation of $V(s)$ will have to be recomputed at some point meaning it has overlapping subproblems. Both conditions of DP are qualified so we can use it to speed up our solutions.

As we will see later, the structure of the value function is similar to the one we saw before in the coin denomination problem. Instead of saving the number of ways to reach the sum, we are going to save the best $V(s)$, that is, the best value of the value function that yields the highest return. Next, we will look at policy and value iteration, which are the basic algorithms that help us solve RL problems.

POLICY AND VALUE ITERATION

The main idea of solving a RL problem is to search for the best policies (a way to make decisions) using value functions. This method works well for simple RL problems as we need information on the entire environment: the number of states and the action space. We can use this method even in a continuous space, but the exact solution is not possible in every case. During the updating process, we will have to iterate over all the possible scenarios, and that's the reason using this method becomes infeasible when the state and action space is too high:

1. Policy Iteration: start with a random policy and iteratively converge to the best one.

2. Value Iteration: state with random values and iteratively update them toward convergence.

STATE-VALUE FUNCTIONS

The state-value function is an array that represents the reward for being in that state. Imagine having four possible states in a particular game: **S1**, **S2**, **S3**, and **S4**, with **S4** being the terminal (end) state. The state-value table can be represented by an array, as indicated in the following table. Please note that the values are simply examples. Every state has a "value," hence state-value function. This table can be used to make decisions later on in the game:

S1	S2	S3	S4
0.82	0.88	0.91	1

Figure 5.12: Sample table for the state-value function

For instance, if you're in state **S3**, you have two possible choices, **S4** and **S2**; you'd go to **S4** since the value of being in that state is higher than that of **S2**.

ACTION-VALUE FUNCTIONS

The action-value function is a matrix that represents the reward for every state-action pair. This again can be used to select the best action to take in a particular state. Unlike the previous state-action table, this time, we have rewards associated with every action as well, as depicted in the following table:

	A1	A2	A3	A4
S1	0.45	0.55	0.1	0.01
S2	0.55	0.7	0.12	0.32
S3	0.67	0.44	0.56	0.34
S4	0.87	0.83	0.15	0.66

Figure 5.13: Sample table for the action-value function

Note these are just example values and will be calculated using a specific update policy. We will be looking at more specific examples of updating policies in the *Policy Improvement* section. The table will be later used in the value iteration algorithm so we can update the table iteratively and not wait till the very end. More on this is in the *Value Iteration* section.

OPENAI GYM: TAXI-V3 ENVIRONMENT

We saw what an OpenAI Gym environment is in previous chapters, but we'll be playing a different game this time: Taxi-v3. In this game, we will teach our agent taxi driver to pick up and drop off passengers. The yellow block represents the taxi. There are four possible locations that are labeled with different characters: R, G, B, and Y for Red, Green, Blue, and Yellow, as you can see in the following figure. The agent has to pick up the passenger at a location and drop them off at a second location. Moreover, there are walls in the environment depicted by a |. Whenever there's a wall, the number of possible actions is limited as the taxi is not allowed to pass through a wall. This makes the problem interesting as the agent has to smartly navigate through the grid while avoiding the walls and finding the best possible (shortest) solution:

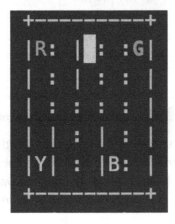

Figure 5.14: Taxi-v3 environment

The following is the list of rewards offered for every action:

- **+20**: On a successful drop-off.

- **-1**: On every step you take. This is important since we are interested in finding the shortest path.

- **-10**: On an illegal drop-off or pickup.

Policy

Every state in the environment is encoded by a number. For instance, the state in the previous photo can be represented by **54**. There are 500 such unique states in this game. For every such state, we have a corresponding policy (that is, which action to perform).

Let's now try the game ourselves.

Initialize the environment and print the possible number of states and the action space, which are 500 and 6 currently. In real-world problems, this number will be huge (in the billions) and we can't use discrete agents. But let's make these assumptions for the sake of simplicity and solve it:

```
def initialize_environment():
    """initialize the OpenAI Gym environment"""
    env = gym.make("Taxi-v3")
    print("Initializing environment")
    # reset the current environment
    env.reset()
    # show the size of the action space
    action_size = env.action_space.n
    print(f"Action space: {action_size}")
    # Number of possible states
    state_size = env.observation_space.n
    print(f"State space: {state_size}")
    return env
```

The preceding code will print the following output:

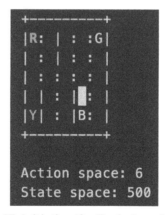

Figure 5.15: Initiating the Taxi-v3 environment

As you can see, the grid represents the current (initial) state of the environment. The yellow box represents the taxi. The six possible choices are: left, right, up, down, pickup, and drop. Let's go ahead and see how we can control the taxi.

Using the following code, we will randomly step through the environment and look at the output. The **env.step** function is used to go from one state to another. The argument it accepts is one of the valid actions in its action space. On stepping, it returns a few values as follows:

- **new_state**: The new state (an integer denoting the next state)

- **reward**: The reward obtained from transitioning to the next state

- **done**: If the environment needs to be reset (meaning you've reached a terminal state)

- **info**: Debug info that indicates transition probabilities

Since we're using a deterministic environment, we will always have transition probabilities that are **1.0**. There are other environments that have non-1 transition probability that indicate if you take a certain decision; for instance, if you take a right turn, the environment will take a right turn with said probability, meaning there's a chance that you will stay in the same place even after taking a specific action. The agent is not allowed to learn this information as it interacts with the environment as, otherwise, it would be unfair if the agent knows the environment information:

```python
def random_step(n_steps=5):
    """

    Steps through the taxi v3 environment randomly
    Args:
        n_steps: Number of steps to step through
    """
    # reset the environment
    env = initialize_environment()
    state = env.reset()
    for i in range(n_steps):
        # choose an action at random
        action = env.action_space.sample()
        env.render()

        new_state, reward, done, info = env.step(action)
        print(f"New State: {new_state}\n"\
              f"reward: {reward}\n"\
              f"done: {done}\n"\
              f"info: {info}\n")\
        print("*" * 20)
```

Using this code, we will take random (but valid) steps in the environment and stop when we've reached the terminal state. If we execute the code, we will see the following output:

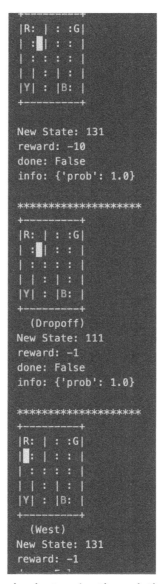

Figure 5.16: Randomly stepping through the environment

Looking at the output, we can see the new state that is stepped through after taking an action and the reward received for taking the action; done will indicate that we've arrived at a terminal stage; and some environment information such as transition probabilities. Next, we will look at our first RL algorithm: policy iteration.

POLICY ITERATION

As the name suggests, in policy iteration, we iterate over multiple policies and then optimize them. The policy iteration algorithm works in two steps:

1. Policy evaluation

2. Policy improvement

Policy evaluation calculates the value function for the current policy, which is initialized randomly. We then use the Bellman optimality equation to update the values for every single state. Then, once we have a new value function, we update the policy to maximize the rewards and update the policy, which is also called policy improvement. Now if the policy is updated (that is, even if a single decision in the policy is changed), this newer policy is guaranteed to be better than the older once. If the policy doesn't update, it means that the current policy is already the most optimal one (otherwise, it would have updated and found a better one).

The following are the steps in which the policy iteration algorithm works:

1. Start with a random policy.

2. Compute the value function for all the states.

3. Update the policy to choose the action that maximizes the rewards (Policy Improvement).

4. Stop when the policy doesn't change. This indicates the optimal policy has been obtained.

Let's take a dry run through the algorithm manually and see how it is updated, using a simple example:

1. Start with a random policy. The following table lists the possible actions for an agent to take in a given position in the Taxi-v3 environment:

↑	↓	→	↓	←
→	→	↑	↑	↑
←	↑	↓	←	↓
→	←	←	↓	←
←	↑	↑	→	↓

Figure 5.17: Possible actions for an agent

In the preceding figure, the table is the environment and the boxes represent the choices. The arrows indicate the action to take if the agent were in that position.

2. Calculate the value function for all the unique states. The following table lists the sample state values for each state of the agent. The values are initiated with zeros (some variations of the algorithm also use small random values close to 0):

0.43	0.64	0.12	0.51	0.77
0.84	0.55	0.34	0.53	0.66
0.85	0.52	0.234	0.53	0.75
0.67	0.82	0.57	0.13	0.72
0.37	0.26	0.73	0.35	0.57

Figure 5.18: Reward values for each state

To understand the update rule visually, let's use an extremely simple example:

Figure 5.19: Sample policy to understand the update rule

Starting from the blue position, the policy will end in the green (terminal) position after the first `policy_evaluation` step. The values will be updated the following way (one diagram for every iteration):

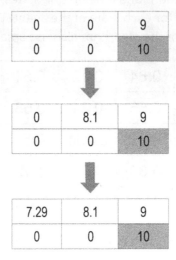

Figure 5.20: Reward multiplying at every step

At every step, the reward is multiplied by gamma (**0.9** in this case). Also, in this example, we already started out with an optimal policy, so the updated policy will look exactly the same as the current one.

3. Update the policy. Let's look at the update rule with a small example. Consider the following as the current value function and the corresponding policy:

0.33	0.33
0.33	0.33

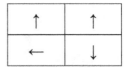

Figure 5.21: The sample value function and the corresponding policy.

As you can see in the preceding figure, the left table indicates the values, and the right table indicates the policy (decision).

Once we perform an update, imagine the value function changes to the following:

0.33	0.67
0.55	0.81

Figure 5.22: Updated values of the sample value function

Now, the policy, in every cell, will update so that the action will take the agent to the state that yields the highest reward and thus the corresponding policy will look something like the following:

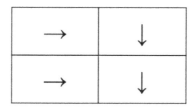

Figure 5.23: Corresponding policy to the updated value function

4. Repeat steps 1-3 until the policy no longer changes.

    We will train the algorithm to iteratively approximate the true value function and do that in episodes, which will give us the most optimal policy. One episode is a series of actions until the agent reaches the terminal state. This can be the goal (drop-off, for instance, in the Taxi-v3 environment) state or it can be a number that defines the maximum number of steps the agent can take to avoid infinite loops.

    Let's use the following code to initialize the environment and the value function table. We will save the value function in the variable **V**. Furthermore, following the first step in the algorithm, we will start out with a random policy using the **env.action_space.sample()** method, which will return a random action every time it's called:

```
def policy_iteration(env):
    """

    Find the most optimal policy for the Taxi-v3 environment
    using Policy Iteration

    Args:
        env: Taxi=v3 environment
```

```
    Returns:
        policy: the most optimal policy
    """
    V = dict()
```

5. Now, in the next section, we will define the variables and initialize them:

```
    """
    initially the value function for all states
    will be random values close to zero
    """
    state_size = env.observation_space.n
    for i in range(state_size):
        V[i] = np.random.random()


    # when the change is smaller than this, stop
    small_change = 1e-20
    # future reward coefficient
    gamma = 0.9
    episodes = 0
    # train for this many episodes
    max_episodes = 50000


    # initially we will start with a random policy
    current_policy = dict()
    for s in range(state_size):
        current_policy[s] = env.action_space.sample()
```

6. Now comes the main loop, which will perform the iteration:

```
    while episodes < max_episodes:
        episodes += 1
        # policy evaluation
        V = policy_evaluation(V, current_policy, \
                            env, gamma, small_change)
        # policy improvement
        current_policy, policy_changed = policy_improvement\
                                        (V, current_policy, \
                                        env, gamma)
        # if the policy didn't change, it means we have converged
        if not policy_changed:
            break
```

```
print(f"Number of episodes trained: {episodes}")
return current_policy
```

7. Now that we have the basic setup ready, we will first do the policy evaluation step using the following code:

```
def policy_evaluation(V, current_policy, env, gamma, \
                      small_change):
    """
    Perform policy evaluation iterations until the smallest
    change is less than
    'smallest_change'

    Args:
        V: the value function table
        current_policy: current policy
        env: the OpenAI Tax-v3 environment
        gamma: future reward coefficient
        small_change: how small should the change be for the
            iterations to stop

    Returns:
        V: the value function after convergence of the evaluation
    """
    state_size = env.observation_space.n
```

8. In the following code, we will loop through the states and update $V(s)$:

```
    while True:
        biggest_change = 0
        # loop through every state present
        for state in range(state_size):
            old_V = V[state]
            # take the action according to the current policy
            action = current_policy[state]
            prob, new_state, reward, done = env.env.P[state]\
                                            [action][0]
```

9. Next, we will use the Bellman optimality equation to update $V(s)$:

```
            V[state] = reward + gamma * V[new_state]
            """
            if the biggest change is small enough then it means
            the policy has converged, so stop.
            """
            biggest_change = max(biggest_change, \
                                abs(V[state] - old_V))
        if biggest_change < small_change:
            break
    return V
```

10. Once we do the policy evaluation step, we will perform policy improvement with the following code:

```
def policy_improvement(V, current_policy, env, gamma):
    """
    Perform policy improvement using the
    Bellman Optimality Equation.

    Args:
        V: the value function table
        current_policy: current policy
        env: the OpenAI Tax-v3 environment
        gamma: future reward coefficient

    Returns:
        current_policy: the updated policy
        policy_changed: True, if the policy was changed,
        else, False
    """
```

11. Let's start by defining all the required variables:

```
state_size = env.observation_space.n
action_size = env.action_space.n
policy_changed = False
for state in range(state_size):
    best_val = -np.inf
    best_action = -1
    # loop over all actions and select the best one
    for action in range(action_size):
        prob, new_state, reward, done = env.env.P[state]\
                                            [action][0]
```

12. Now, here, we will calculate the future reward by taking this action. Note that we're using a simplified equation because we don't have non-one transition probabilities:

```
        future_reward = reward + gamma * V[new_state]
        if future_reward > best_val:
            best_val = future_reward
            best_action = action
    """
    using assert statements we can avoid getting
    into unwanted situations
    """
    assert best_action != -1
    if current_policy[state] != best_action:
        policy_changed = True
    # update the best action for this current state
    current_policy[state] = best_action
# if the policy didn't change, it means we have converged
return current_policy, policy_changed
```

13. Once the optimal policy is learned, we will test it on a fresh environment. Now that both the parts are ready. Let's call them using the **main** block of code:

```
if __name__ == '__main__':
    env = initialize_environment()
    policy = value_iteration(env)
    play(policy, render=True)
```

14. Next, we will add a **play** function that will test the policy on a
    fresh environment:

```python
def play(policy, render=False):
    """

    Perform a test pass on the Taxi-v3 environment

    Args:
        policy: the policy to use
        render: if the result should be rendered at every step.
            False by default

    """
    env = initialize_environment()
    rewards = []
```

15. Next, let's define **max_steps**. This is essentially the maximum number of steps
    the agent is allowed to take. If it doesn't reach a solution in this time, then we call
    it an episode and proceed:

```python
max_steps = 25
test_episodes = 2
for episode in range(test_episodes):
    # reset the environment every new episode
    state = env.reset()
    total_rewards = 0
    print("*" * 100)
    print("Episode {}".format(episode))

    for step in range(max_steps):
```

Here, we will take the action that we saved in the policy earlier:

```python
        action = policy[state]
        new_state, reward, done, info = env.step(action)
        if render:
            env.render()
        total_rewards += reward
        if done:
            rewards.append(total_rewards)
            print("Score", total_rewards)
            break
```

```
                state = new_state
        env.close()
        print("Average Score", sum(rewards) / test_episodes)
```

After running the main block, we see the following output:

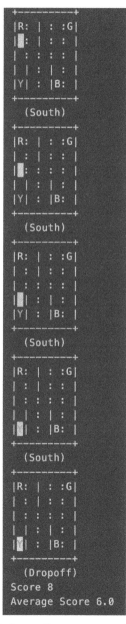

Figure 5.24: The agent drops the passenger in the correct location

As you can see, the agent drops the passenger in the right location. Note that the output is truncated for presentation purposes.

VALUE ITERATION

As you saw in the previous section, we arrived at the optimal solution after a few iterations, but policy iteration has one disadvantage: we get to improve the policy only once after multiple iterations of evaluation.

The simplified Bellman equation can be updated in the following way. Note that this is similar to the policy evaluation step, but the only addition is taking the max value of the value function over all the possible actions:

$$V(s) = \max_a \left[r + \gamma V(s') \right]$$

Figure 5.25: Updated Bellman equation

The equation can be comprehended as follows:

"*For a given state, take all the possible actions and then store the one with the highest V[s] value.*"

It's as simple as that. Using this technique, we can combine both evaluation and improvement in a single step as you will see now.

We will start off as usual by defining the important variables, such as **gamma**, **state_size**, and **policy**, and the value function dictionary:

```
def value_iteration(env):
    """
    Performs Value Iteration to find the most optimal policy for the
    Tax-v3 environment

    Args:
        env: Taxiv3 Gym environment

    Returns:
        policy: the most optimum policy
    """
    V = dict()
    gamma = 0.9
    state_size = env.observation_space.n
```

```
action_size = env.action_space.n
policy = dict()
# initialize the value table randomly
# initialize the policy randomly
for x in range(state_size):
    V[x] = 0
    policy[x] = env.action_space.sample()
```

And using the equation defined before, we will take the same loop and make the change in the $V(s)$ calculation part. We are now using the updated Bellman equation, which was defined earlier:

```
"""
this loop repeats until the change in value function
is less than delta
"""
while True:
    delta = 0
    for state in reversed(range(state_size)):
        old_v_s = V[state]
        best_rewards = -np.inf
        best_action = None
        # for all the actions in current state
        for action in range(action_size):
            # check the reward obtained if we were to perform
            # this action
            prob, new_state, reward, done = 
              env.env.P[state][action][0]
            potential_reward = reward + gamma * V[new_state]
            # select the one that has the best reward
            # and also save the action to the policy
            if potential_reward > best_rewards:
                best_rewards = potential_reward
                best_action = action
        policy[state] = best_action
        V[state] = best_rewards
        # terminate if the change is not high
        delta = max(delta, abs(V[state] - old_v_s))
    if delta < 1e-30:
        break
```

```
if __name__ == '__main__':
    env = initialize_environment()
    # policy = policy_iteration(env)
    policy = value_iteration(env)
    play(policy, render=True)
```

Thus, we have successfully implemented the policy iteration and value iteration for the Taxi-v3 environment.

In the next activity, we will be using the very popular FrozenLake-v0 environment for policy and value iteration. Before we begin, let's quickly explore the basics of the environment.

THE FROZENLAKE-V0 ENVIRONMENT

The environment is based on a scenario in which there is a frozen lake, except for some parts where the ice has melted. Suppose that a group of friends is playing frisbee near the lake and one of them made a wild throw that landed the frisbee right in the middle of the lake. The goal is to navigate across the lake and get the frisbee back. Now, the fact that has to be considered here is that the ice is slippery, and you cannot always move in the intended direction. The surface is described using a grid as follows:

```
SFFF        (S: starting point, safe)
FHFH        (F: frozen surface, safe)
FFFH        (H: hole, fall to your doom)
HFFG        (G: goal, where the frisbee is located)
```

Note that the episode ends when one of the players reaches the goal or falls in the hole. The player is rewarded with a 1 or 0 respectively.

Now, in the Gym environment, the agent is supposed to control the movement of the player accordingly. As you know, some tiles in the grid can be stepped upon and some may land you directly into the hole where the ice has melted. Hence, the movement of the player is highly unpredictable and is partially dependent on the direction that the agent has chosen.

> ### NOTE
>
> For more information on the FrozenLake-v0 environment, please refer to the following link: https://gym.openai.com/envs/FrozenLake-v0/

Let's now implement the policy and value iteration techniques to solve the problem and retrieve the frisbee.

ACTIVITY 5.01: IMPLEMENTING POLICY AND VALUE ITERATION ON THE FROZENLAKE-V0 ENVIRONMENT

In this activity, we will solve FrozenLake-v0 using policy and value iteration. The goal of the activity is to define a safe path through the frozen lake and retrieve the frisbee. The episode ends when the goal is achieved or when the agent falls into the hole. The following steps will help you complete the activity:

1. Import the required libraries: **numpy** and **gym**.

2. Initialize the environment and reset the current one. Set **is_slippery=False** in the initializer. Show the size of the action space and the number of possible states.

3. Perform policy evaluation iterations until the smallest change is less than **smallest_change**.

4. Perform policy improvement using the Bellman optimality equation.

5. Find the most optimal policy for the FrozenLake-v0 environment using policy iteration.

6. Perform a test pass on the FrozenLake-v0 environment.

7. Take steps through the FrozenLake-v0 environment randomly.

8. Perform value iteration to find the most optimal policy for the FrozenLake-v0 environment. Note that the aim here is to make sure the reward value for each action should be one (or close to one) to ensure maximum rewards.

The output should be similar to the following:

```
SFFF
FHFH
FFFH
HF G
  (Right)
SFFF
FHFH
FFFH
HFF
Score 1.0
Average Score 1.0
```

Figure 5.26: Expected output average score (1.0)

> **NOTE**
>
> The solution to this activity can be found on page 711.

Thus, with this activity, we have successfully implemented the policy and value iteration methods in the FrozenLake-v0 environment.

With this, we have reached the end of the chapter, and you can now confidently implement the techniques learned in this chapter for various environments and scenarios.

SUMMARY

In this chapter, we looked at the two most commonly used techniques to solve DP problems. The first method, memoization, also called the top-bottom method uses a dictionary (or HashMap-like structure) to store intermediate results in a natural (unordered) manner. While the second method, the tabular method, also called the bottom-up method, sequentially solves problems from small to large and usually saves the result in a matrix-like structure.

Next, we also looked at how to use DP to solve RL problems using policy and value iteration, and how we overcome the disadvantage of policy iteration by using the modified Bellman equation. We implemented policy and value iteration in two very popular environments: Taxi-v3 and FrozenLake-v0.

In the next chapter, we will be studying Monte Carlo methods, which are used to simulate real-world scenarios and are some of the most widely used tools in domains such as finance, mechanics, and trading.

6

MONTE CARLO METHODS

OVERVIEW

In this chapter, you will learn about the various types of Monte Carlo methods, including the first visit and every visit techniques. In the case, if the model of the environment is not known, you can use Monte Carlo methods to learn the environment by generating experience samples or by simulation. This chapter teaches you importance sampling and how to apply Monte Carlo methods to solve the frozen lake problem. By the end of this chapter, you will be able to identify problems where Monte Carlo methods of reinforcement learning can be applied. You will be able to solve prediction, estimation, and control problems using Monte Carlo reinforcement learning.

INTRODUCTION

In the previous chapter, we learned about dynamic programming. Dynamic programming is a way of doing reinforcement learning where the model of the environment is known beforehand. Agents in reinforcement learning can learn a policy, value function, and/or model. Dynamic programming helps solve a known **Markov Decision Process (MDP)**. The probabilistic distribution for all possible transitions is known in an MDP and is required for dynamic programming.

But what happens when the model of the environment is not known? In many real-life situations, the model of the environment is not known beforehand. Can the algorithm learn the model of the environment? Can the agents in reinforcement learning still learn to make good decisions?

Monte Carlo methods are a way of learning when the model of the environment is not known and so they are called model-free learning. We can make a model-free prediction that estimates the value function of an unknown MDP. We can also use model-free control, which optimizes the value functions of an unknown MDP. Monte Carlo methods can also handle non-Markovian domains too.

The transition probabilities between one state and another are not known in many cases. You need to play around and get a sense of the environment before learning how to play the game well. Monte Carlo methods can learn a model of an environment from experiencing the environment. Monte Carlo methods take actual or stochastically simulated scenarios and get an average of the sample returns. By using the sample sequence of states, actions, and rewards from actual or simulated interactions with the environment, Monte Carlo methods can learn from experience. A well-defined set of rewards is needed for Monte Carlo methods to work. This criterion is met only for episodic tasks, where experience is divided into clearly defined episodes, and episodes eventually terminate irrespective of the action selected. An example application is AlphaGo, which is one of the most complex games; the number of possible moves in any state is over 200. One of the key algorithms used to solve it was a tree search based on Monte Carlo.

In this chapter, we will first understand Monte Carlo methods of reinforcement learning. We will apply them to the Blackjack environment in OpenAI. We will learn about various methods, such as the first visit method and every visit method. We will also learn about importance sampling and, later in the chapter, revisit the frozen lake problem. In the next section, we will introduce the basic workings of Monte Carlo methods.

THE WORKINGS OF MONTE CARLO METHODS

Monte Carlo methods solve reinforcement problems by averaging the sample returns for each state-action pair. Monte Carlo methods work only for episodic tasks. This means the experience is split into various episodes and all episodes finally terminate. Only after the episode is complete are the value functions recalculated. Monte Carlo methods can be incrementally optimized episode by episode but not step by step.

Let's take the example of a game like Go. This game has millions of states; it is going to be difficult to learn all of those millions of states and their transition probabilities beforehand. The other approach would be to play the game of Go repeatedly and assign a positive reward for winning and a negative reward for losing.

As we don't have information about the policy of the model, we need to use experience samples to learn. This technique is also a sample-based model. We call this direct sampling of episodes in Monte Carlo.

Monte Carlo is model-free. As no knowledge of MDP is required, the model is inferred from the samples. You can perform model-free prediction or model-free estimation. We can perform an evaluation, also called a prediction, on a policy. We can also evaluate and improve a policy, which is often called control or optimization. Monte Carlo reinforcement learning can learn only from episodes that terminate.

For example, if you have a game of chess, played by a set of rules or policies, that would be playing several episodes according to those rules or policies and evaluating the success rate of the policy. If we are playing a game according to a policy and modifying the policy based on the game, then it would be a policy improvement, optimization, or control.

UNDERSTANDING MONTE CARLO WITH BLACKJACK

Blackjack is a simple card game that is quite popular in casinos. It is a great game, as it is simple to simulate and take samples, and lends itself to Monte Carlo methods. Blackjack is also available as part of the OpenAI framework. Players and the dealer are dealt two cards each. The dealer shows one card face up and lays the other card face down. The players and the dealer have a choice of whether to be dealt additional cards or not:

- **The aim of the game**: To obtain cards whose sum is close to or equal to 21 but not greater than 21.

- **Players**: There are two players, called the player and the dealer.

- **The start of the game**: The player is dealt with two cards. The dealer is also dealt with two cards, and the rest of the cards are pooled into a stack. One of the dealer's cards is shown to the player.

- **Possible actions – stick or hit**: "Stick" is to stop asking for more cards. "Hit" is to ask for more cards. The player will choose "Hit" if the sum of their cards is less than 17. If the sum of their cards is greater than or equal to 17, the player will stick. This threshold of 17 to decide whether to hit or stick can be changed if needed in various versions of Blackjack. In this chapter, we will consistently keep the threshold at 17 to decide whether to hit or stick.

- **Rewards**: +1 for a win, -1 for a loss, and 0 for a draw.

- **Strategy**: The player has to decide whether to stick or hit by looking at the dealer's cards. The ace can be considered to be 1 or 11, based on the value of the other cards.

We will explain the game of Blackjack in the following table. The table has the following columns:

- **Game**: The game number and the sub-state of the game: i, ii, or iii

- **Player Cards**: The cards the player has; for example, K♣, 8♦ means the player has the King of clubs and the eight of diamonds.

- **Dealer Cards**: The cards the dealer gets. For example, 8♠, Xx means the dealer has the eight of spades and a hidden card.

- **Action**: This is the action the player decides to choose.

- **Result**: The result of the game based on the player's actions and the cards the dealer gets.

- **Sum of Player Cards**: The sum of the player's two cards. Please note that the King (K), Queen (Q), and Jack (J) face cards are scored as 10.

- **Comments**: An explanation of why a particular action was taken or a result was declared.

In game 1, the player decided to stick as the sum of the cards was 18. "Stick" means the player will no longer receive cards. Now the dealer shows the hidden card. It is a draw as both the dealer's and player's cards sum 18. In game 2, the player's cards sum 15, which is less than 17. The player hits and gets another card, which takes the sum to 17. The player then sticks, which means the player will no longer receive cards. The dealer shows the cards and as the sum of the cards is less than 17, gets another card. With the dealer's new card, the sum is 25, which is greater than 21. The game aims to get close to or equal to 21 without the score becoming greater than 21. The dealer loses and the player wins the second game. The following figure presents a summary of this game:

Game	Player Cards	Dealer Cards	Action	Result	Sum of Player Cards	Comments
1 (i)	K♣, 8♦	8♠, Xx	Stick		18	Player's cards >=17
1 (ii)	K♣, 8♦	8♠, 10♥		Draw	18	The dealer shows and it is a draw
2 (i)	5♥, Q♣	5♠, Xx	Hit		15	Player's cards < 17
2 (ii)	5♥, Q♣, 2♦	5♠, Xx	Stick		17	Player's cards >=17
2 (iii)	5♥, Q♣, 2♦	5♠, Q♦			17	The dealer shows
2 (iii)	5♥, Q♣, 2♦	5♠, Q♦, 10♥		Win	17	The dealer gets a card and the sum > 21

Figure 6.1: Explanation of a Blackjack game

Next, we will be implementing the game of Blackjack using the OpenAI framework. This will serve as a foundation for the simulation and application of Monte Carlo methods.

EXERCISE 6.01: IMPLEMENTING MONTE CARLO IN BLACKJACK

We will learn how to use the OpenAI framework for Blackjack, and get to know about observation space, action space, and generating an episode. The goal of this exercise is to implement Monte Carlo techniques in the game of Blackjack.

Perform the following steps to complete the exercise:

1. Import the necessary libraries:

```
import gym
import numpy as np
from collections import defaultdict
from functools import partial
```

gym is the OpenAI framework, **numpy** is the framework for data processing, and **defaultdict** is for dictionary support.

2. We start the **Blackjack** environment with **gym.make()** and assign it to **env**:

```
#set the environment as blackjack
env = gym.make('Blackjack-v0')
```

Find the number of observation spaces and action spaces:

```
#number of observation space value
print(env.observation_space)
#number of action space value
print(env.action_space)
```

You will get the following output:

```
Tuple(Discrete(32), Discrete(11), Discrete(2))
Discrete(2)
```

The number of observation spaces is the number of states. The number of action spaces is the number of actions possible in each state. The output shows as discrete, as the observation and action space in a Blackjack game is not continuous. For example, there are other games in OpenAI, such as balancing a CartPole and pendulum where the observation and action spaces are continuous.

3. Write a function to play the game. If the sum of the player's cards is more than or equal to 17, stick (don't choose more cards); otherwise, hit (choose more cards), as shown in the following code:

```
def play_game(state):
    player_score, dealer_score, usable_ace = state
    #if player_score is greater than 17, stick
    if (player_score >= 17):
        return 0 # don't take any cards, stick
    else:
        return 1 # take additional cards, hit
```

Here, we are initializing the episode, choosing the initial state, and assigning it to **player_score**, **dealer_score**, and **usable_ace**.

4. Add a dictionary, **action_text**, that has a key-value mapping for two action integers to action text. Here's the code to convert the integer value of the action into text format:

```
for game_num in range(100):
    print('***Start of Game:', game_num)
    state = env.reset()
    action_text = {1:'Hit, Take more cards!!', \
                   0:'Stick, Dont take any cards' }
    player_score, dealer_score, usable_ace = state
    print('Player Score=', player_score,', \
        Dealer Score=', dealer_score, ', \
        Usable Ace=', usable_ace)
```

5. Play the game in batches of 100 and calculate **state**, **reward**, and **action**:

```
    for i in range(100):
        action = play_game(state)
        state, reward, done, info = env.step(action)
        player_score, dealer_score, usable_ace = state
        print('Action is', action_text[action])
        print('Player Score=', player_score,', \
            Dealer Score=', dealer_score, ', \
            Usable Ace=', usable_ace, ', Reward=', reward)
        if done:
            if (reward == 1):
                print('***End of Game:', game_num, \
                    ' You have won Black Jack!\n')
```

```
            elif (reward == -1):
                print('***End of Game:', game_num, \
                        ' You have lost Black Jack!\n')
            elif (reward ==0):
                print('***End of Game:', game_num, \
                        ' The game is a Draw!\n')
            break
```

You will get the following output:

```
***Start of Game: 0
Player Score= 20 , Dealer Score= 4 , Usable Ace= False
Action is Stick, Dont take any cards
Player Score= 20 , Dealer Score= 4 , Usable Ace= False , Reward= 1.0
***End of Game: 0  You have won Black Jack!

***Start of Game: 1
Player Score= 13 , Dealer Score= 2 , Usable Ace= False
Action is Hit, Take more cards!!
Player Score= 23 , Dealer Score= 2 , Usable Ace= False , Reward= -1
***End of Game: 1  You have lost Black Jack!

***Start of Game: 2
Player Score= 14 , Dealer Score= 10 , Usable Ace= False
Action is Hit, Take more cards!!
Player Score= 20 , Dealer Score= 10 , Usable Ace= False , Reward= 0
Action is Stick, Dont take any cards
Player Score= 20 , Dealer Score= 10 , Usable Ace= False , Reward= 1.0
***End of Game: 2  You have won Black Jack!
```

Figure 6.2: The output is the episode of the Blackjack game in progress

NOTE

The Monte Carlo technique is based on generating random samples. As such, two executions of the same code will not match in values. So, you might have a similar output but not the same for all the exercises and activities.

In the code, **done** has the value of **True** or **False**. If **done** is **True**, the game stops, we note the value of the rewards and print the game result. In the output, we simulated the game of Blackjack using the Monte Carlo method and noted the various actions, states, and game completion. We were also able to simulate the rewards when the game ends.

> **NOTE**
>
> To access the source code for this specific section, please refer to https://packt.live/2XZssYh.
>
> You can also run this example online at https://packt.live/2Ys0cMJ.

Next, we will describe the different types of Monte Carlo methods, namely, the first visit and every visit method, which will be used to estimate the value function.

TYPES OF MONTE CARLO METHODS

We have implemented the game of Blackjack using Monte Carlo. Typically, a trajectory of Monte Carlo is a sequence of state, action, and reward. In several episodes, it is possible that the state repeats. For example, the trajectory could be S0, S1, S2, S0, S3. How do we handle the calculation of the reward function when we have multiple visits to the states?

Broadly, this highlights that there are two types of Monte Carlo methods – first visit and every visit. We will understand the implications of both methods.

As stated previously, in Monte Carlo methods, we approximate the value function by averaging the rewards. In the first visit Monte Carlo method, only the first visit to a state in an episode is included to calculate the average reward. For example, in a given game of traversing a maze, you could make several visits to the sample place. In the first visit Monte Carlo method, only the first visit is used for the calculation of the reward. When the agent revisits the same state in the episode, the reward is not included for the calculation of the average reward.

In every visit Monte Carlo, every time the agent visits the same state, the rewards are included in the calculation of the average return. For example, let's use the same game of maze. Every time the agent comes to the same point in the maze, we include the rewards earned in that state for the calculation of the reward function.

Both first visit and every visit converge to the same value function. For a smaller number of episodes, the choice between the first visit and every visit is based on the particular game and the rules of the game.

Let's understand the pseudocode for first visit Monte Carlo prediction.

FIRST VISIT MONTE CARLO PREDICTION FOR ESTIMATING THE VALUE FUNCTION

In the pseudocode for first visit Monte Carlo prediction for estimating the value function, the key is to calculate the value function *V(s)*. Gamma is the discount factor. The discount factor is used to reward future rewards less than immediate rewards:

- Input a *Pi* policy to be evaluated
- Initialize the `v(s)` and `R(s)`
- Loop forever for every episode
  - ○ Generate an `episode` of `state`, `action`, and `rewards` by following the *Pi* policy
  - ○ Set `G` as `0`
  - ○ For each step in episode
    - ▪ Set *G = gamma * G + Rewards*
    - ▪ For the first visit to the state
      - • Append `G` to `Returns`
      - • `V(s)` is the average of the `Returns`

Figure 6.3: Pseudocode for first visit Monte Carlo prediction

What we have done in the first visit is to generate an episode, calculate the result value, and append the result to the rewards. We then calculate the average returns. In the upcoming exercise, we will apply the first visit Monte Carlo prediction to estimate the value function by following the steps detailed in the pseudocode. The key block of code for the first visit algorithm is navigating the states only through the first visit:

```
if current_state not in states[:i]:
```

Consider the **states** that have not been visited. We increase the count for the number of **states** by **1**, calculate the value function with the incremental method, and return the value function. This is implemented as follows:

```
"""
only include the rewards of the states that have not been visited before
"""
        if current_state not in states[:i]:
            #increasing the count of states by 1
```

```
            num_states[current_state] += 1


            #finding the value_function by incremental method
            value_function[current_state] \
            += (total_rewards - value_function[current_state]) \
            / (num_states[current_state])
        return value_function
```

Let's understand it better through the next exercise.

EXERCISE 6.02: FIRST VISIT MONTE CARLO PREDICTION FOR ESTIMATING THE VALUE FUNCTION IN BLACKJACK

This exercise aims to understand how to apply first visit Monte Carlo prediction to estimate the value function in the game of Blackjack. We will apply the steps outlined in the pseudocode step by step.

Perform the following steps to complete the exercise:

1. Import the necessary libraries:

```
import gym
import numpy as np
from collections import defaultdict
from functools import partial
```

gym is the OpenAI framework, **numpy** is the framework for data processing, and **defaultdict** is for dictionary support.

2. Select the environment as **Blackjack** in OpenAI:

```
env = gym.make('Blackjack-v0')
```

3. Write the **policy_blackjack_game** function, which takes the state as input and returns the action **0** or **1** based on **player_score**:

```
def policy_blackjack_game(state):
    player_score, dealer_score, usable_ace = state
    if (player_score >= 17):
        return 0 # don't take any cards, stick
    else:
        return 1 # take additional cards, hit
```

In the function, if the player score is greater than or equal to **17**, it does not take more cards. But if **player_score** is less than 17, it takes additional cards.

4. Write a function to generate a Blackjack episode. Initialize **episode**, **states**, **actions**, and **rewards**:

```
def generate_blackjack_episode():
    #initializing the value of episode, states, actions, rewards
    episode = []
    states = []
    actions = []
    rewards = []
```

5. Reset the environment and set the **state** value to **player_score, dealer_score**, and **usable_ace**:

```
#starting the environment
state = env.reset()


"""
setting the state value to player_score,
dealer_score and usable_ace
"""

player_score, dealer_score, usable_ace = state
```

6. Write a function that generates the action from the state. We then step through the action and find **next_state** and **reward**:

```
while (True):
    #finding the action by passing on the state
    action = policy_blackjack_game(state)
    next_state, reward, done, info = env.step(action)
```

7. Create a list of **episode**, **state**, **action**, and **reward** by appending them to the existing list:

```
#creating a list of episodes, states, actions, rewards
episode.append((state, action, reward))
states.append(state)
actions.append(action)
rewards.append(reward)
```

If the episode is complete (**done** is true), we **break** the loop. If not, we update **state** to **next_state** and repeat the loop:

```
if done:
    break
state = next_state
```

8. We return **episodes**, **states**, **actions**, and **rewards** from the function:

```
return episode, states, actions, rewards
```

9. Write the function for calculating the value function for Blackjack. The first step is to initialize the value of **total_rewards**, **num_states**, and **value_function**:

```
def black_jack_first_visit_prediction(policy, env, num_episodes):
    """
    initializing the value of total_rewards,
    number of states, and value_function
    """
    total_rewards = 0
    num_states = defaultdict(float)
    value_function = defaultdict(float)
```

10. Generate an **episode**, and for an **episode**, we find the total **rewards** for all the **states** in reverse order in the **episode**:

```
for k in range (0, num_episodes):
    episode, states, actions, rewards = \
    generate_blackjack_episode()
    total_rewards = 0
    for i in range(len(states)-1, -1,-1):
        current_state = states[i]
        #finding the sum of rewards
        total_rewards += rewards[i]
```

11. Consider the **states** that have not been visited. We increase the count for the number of **states** by **1** and calculate the value function using the incremental method, and return the value function:

```
    """
    only include the rewards of the states that
    have not been visited before
```

```
        """

            if current_state not in states[:i]:
                #increasing the count of states by 1
                num_states[current_state] += 1

                #finding the value_function by incremental method
                value_function[current_state] \
                += (total_rewards \
                - value_function[current_state]) \
                / (num_states[current_state])
        return value_function
```

12. Now, execute first visit prediction 10,000 times:

```
black_jack_first_visit_prediction(policy_blackjack_game, env, 10000)
```

You will get the following output:

```
defaultdict(float,
            {(14, 10, False): -0.49732620320855625,
             (10, 10, False): -0.16666666666666669,
             (12, 6, False): -0.29761904761904767,
             (19, 10, False): 0.013698630136986337,
             (11, 10, False): 0.13978494623655924,
             (21, 7, False): 0.9268292682926828,
             (11, 7, False): 0.35185185185185197,
             (14, 8, False): -0.45192307692307704,
             (9, 8, False): 0.055555555555555558,
             (19, 5, False): 0.425,
             (13, 5, False): -0.27272727272727276,
             (20, 5, False): 0.7540983606557375,
             (20, 10, False): 0.45719844357976686,
             (20, 4, False): 0.6953124999999999,
             (15, 4, False): -0.4666666666666665,
             (16, 3, False): -0.625,
             (12, 10, False): -0.38387096774193563,
             (17, 10, False): -0.56479217603912,
```

Figure 6.4: First visit value function

The value function for the first visit is printed. For all the states, a combination of **player_score**, **dealer_score**, and **usable_space** has a value function value from the first visit evaluation. Take the example output of **(16, 3, False): -0.625**. This means that the value function for the state with player score **16**, dealer score **3**, and a reusable ace as **False** is **-0.625**. The number of episodes and batches are configurable.

> ### NOTE
>
> To access the source code for this specific section, please refer to https://packt.live/37zbza1.
>
> You can also run this example online at https://packt.live/2AYnhyH.

We have covered the first visit Monte Carlo in this section. In the next section, we will understand every visit Monte Carlo prediction for estimating the value function.

EVERY VISIT MONTE CARLO PREDICTION FOR ESTIMATING THE VALUE FUNCTION

In every visit Monte Carlo prediction, every visit to the state is used for the reward calculation. We have a gamma factor that is the discount factor, which enables us to discount the rewards in the far future relative to rewards in the immediate future:

- Input a policy *Pi* to be evaluated
- Initialize the v(s) and R(s)
- Loop forever for every episode
  - Generate an episode of state, action, and rewards by following the *Pi* policy
  - Set G as 0
  - For each step in episode
    - Set *G = gamma * G + Rewards*
    - For every visit to the state
      - Append G to Returns
      - V(s) is the average of the Returns

Figure 6.5: Pseudocode for every visit Monte Carlo prediction

The difference is primarily visiting every step instead of just the first to calculate the rewards. The code remains similar to the first visit exercise, except for the Blackjack prediction function where the rewards are calculated.

The following line in the first visit implementation checks if the current state has not been traversed before. This line is no longer in every visit algorithm:

```
if current_state not in states[:i]:
```

The code for the calculation of the value function is as follows:

```
            #all the state values of every visit are considered
            #increasing the count of states by 1
            num_states[current_state] += 1

            #finding the value_function by incremental method
            value_function[current_state] \
            += (total_rewards - value_function[current_state]) \
            / (num_states[current_state])


    return value_function
```

In this exercise, we will use every visit Monte Carlo method to estimate the value function.

EXERCISE 6.03: EVERY VISIT MONTE CARLO PREDICTION FOR ESTIMATING THE VALUE FUNCTION

This exercise aims to understand how to apply every visit Monte Carlo prediction to estimate the value function. We will apply the steps outlined in the pseudocode step by step. Perform the following steps to complete the exercise:

1. Import the necessary libraries:

```
import gym
import numpy as np
from collections import defaultdict
from functools import partial
```

2. Select the environment as **Blackjack** in OpenAI:

```
env = gym.make('Blackjack-v0')
```

3. Write the **policy_blackjack_game** function that takes the state as input and returns the action **0** or **1** based on **player_score**:

```
def policy_blackjack_game(state):
    player_score, dealer_score, usable_ace = state
    if (player_score >= 17):
        return 0 # don't take any cards, stick
    else:
        return 1 # take additional cards, hit
```

In the function, if the player score is greater than or equal to **17**, it does not take more cards. But if **player_score** is less than **17**, it takes additional cards.

4. Write a function to generate a Blackjack episode. Initialize **episode**, **states**, **actions**, and **rewards**:

```
def generate_blackjack_episode():
    #initializing the value of episode, states, actions, rewards
    episode = []
    states = []
    actions = []
    rewards = []
```

5. We reset the environment and set the value of **state** to **player_score**, **dealer_score**, and **usable_ace**, as shown in the following code:

```
    #starting the environment
    state = env.reset()
    """
    setting the state value to player_score, dealer_score and
    usable_ace
    """
    player_score, dealer_score, usable_ace = state
```

6. Write a function that generates **action** from **state**. We then step through **action** and find **next_state** and **reward**:

```
    while (True):
        #finding the action by passing on the state
        action = policy_blackjack_game(state)
        next_state, reward, done, info = env.step(action)
```

7. Create a list of **episode**, **state**, **action**, and **reward** by appending them to the existing list:

```
#creating a list of episodes, states, actions, rewards
episode.append((state, action, reward))
states.append(state)
actions.append(action)
rewards.append(reward)
```

8. If the episode is complete (**done** is true), we **break** the loop. If not, we update **state** to **next_state** and repeat the loop:

```
if done:
    break
state = next_state
```

9. We return **episodes**, **states**, **actions**, and **rewards** from the function:

```
return episode, states, actions, rewards
```

10. Write the function for calculating the value function for Blackjack. The first step is to initialize the values of **total_rewards**, **num_states**, and **value_function**:

```
def black_jack_every_visit_prediction\
(policy, env, num_episodes):

    """
    initializing the value of total_rewards, number of states,
    and value_function
    """
    total_rewards = 0
    num_states = defaultdict(float)
    value_function = defaultdict(float)
```

11. Generate an **episode** and for the **episode**, we find the total **rewards** for all the **states** in reverse order in the **episode**:

```
for k in range (0, num_episodes):
    episode, states, actions, rewards = \
    generate_blackjack_episode()
    total_rewards = 0
    for i in range(len(states)-1, -1,-1):
        current_state = states[i]
```

```
#finding the sum of rewards
total_rewards += rewards[i]
```

12. Consider every **state** visited. We increase the count for the number of **states** by **1** and calculate the value function with the incremental method and return the value function:

```
#all the state values of every visit are considered
#increasing the count of states by 1
num_states[current_state] += 1

#finding the value_function by incremental method
value_function[current_state] \
+= (total_rewards - value_function[current_state]) \
/ (num_states[current_state])
```

```
return value_function
```

13. Now, execute every visit prediction 10,000 times:

```
black_jack_every_visit_prediction(policy_blackjack_game, \
                                  env, 10000)
```

You will get the following output:

```
defaultdict(float,
            {(21, 10, False): 0.9304812834224596,
             (11, 10, False): 0.21839080459770124,
             (18, 3, False): 0.1666666666666667,
             (8, 3, False): -0.3043478260869565,
             (17, 10, False): -0.4124700239808153,
             (13, 10, False): -0.4823848238482386,
             (16, 10, False): -0.5656836461126005,
             (12, 6, False): -0.18478260869565222,
             (16, 2, False): -0.5714285714285714,
             (13, 2, False): -0.3925233644859814,
             (21, 1, True): 0.6279069767441859,
             (15, 10, False): -0.5040650406504064,
             (8, 10, False): -0.27272727272727293,
             (19, 1, True): -0.4444444444444444,
```

Figure 6.6: Every visit value function

The value function for every visit is printed. For all the states, a combination of **player_score**, **dealer_score,** and **usable_space** has a value function value from every visit evaluation. We can increase the number of episodes and run this again too. As the number of episodes is made larger and larger, the first visit and every visit functions will converge.

> **NOTE**
>
> To access the source code for this specific section, please refer to https://packt.live/2C0wAP4.
>
> You can also run this example online at https://packt.live/2zqXsH3.

In the next section, we will talk about a key concept of Monte Carlo reinforcement learning, which is the need to balance exploration and exploitation. This is also the basis of the greedy epsilon technique of the Monte Carlo method. Balancing exploration and exploitation helps us to improve the policy function.

EXPLORATION VERSUS EXPLOITATION TRADE-OFF

Learning happens by exploring new things and exploiting or applying what has been learned before. The right combination of these is the essence of any learning. Similarly, in the context of reinforcement learning, we have exploration and exploitation. **Exploration** is trying out different actions, while **exploitation** is following an action that is known to have a good reward.

Reinforcement learning has to balance between exploration and exploitation. Every agent can learn only from the experience of trying an action. Exploration helps try new actions that might enable the agent to make better decisions in the future. Exploitation is choosing actions that yield good rewards based on experience. The agent needs to trade off gaining rewards by exploitation by experimenting in exploration. If an agent exploits more, the agent might miss learning about other policies with even greater rewards. If the agent explores more, the agent might miss the opportunity to exploit a known path and lose out on rewards.

For example, think of a student who is trying to maximize their grades in college. The student can either "explore" by taking courses with new subjects or "exploit" by taking courses in their favorite subjects. If the student indexes towards "exploitation," the student might miss out on both getting good grades in new unexplored courses and the overall learning. If the student explores too many diverse subjects by taking courses in them, this might impact their grades and might make the learning too broad.

Similarly, if you choose to read books, you could exploit by reading books belonging to the same genre or author or explore by reading books across different genres and authors. Similarly, while driving from one place to another, you could exploit by following the same known route based on past experience or explore by taking different routes. In the next section, we will get an understanding of the techniques of on-policy and off-policy learning. We will then get an understanding of a key factor called importance sampling, for off-policy learning.

Exploration and exploitation are techniques used in reinforcement learning. In off-policy learning, you can have an exploitation technique as the target policy and an exploration technique as the behavior policy. We could have a greedy policy as the exploitation technique and a random policy as the exploration technique.

IMPORTANCE SAMPLING

Monte Carlo methods can be on-policy or off-policy. In **on-policy** learning, we learn from the agent experience of the following policy. In **off-policy** learning, we learn how to estimate a target policy from the experience of following a different behavioral policy. Importance sampling is a key technique for off-policy learning. The following figure compares on-policy and off-policy learning:

	On-Policy	Off-Policy
1	Learn by playing	Learn by watching
2	Variance is lower	Variance is higher
3	Self-learning	Transfer learning
4	Simpler	Complex

Figure 6.7: On-Policy versus Off-Policy comparison

You might think that on-policy learning is learning while playing, while off-policy learning is learning by watching someone else play. You could improve your cricket game by playing cricket yourself. This will help you learn from your mistakes and best actions. That would be on-policy learning. You could also learn by watching others play the game of cricket and learning from their mistakes and best actions. That would be off-policy learning.

Human beings typically do both on-policy and off-policy learning. For example, cycling is primarily on-policy learning. We learn to cycle by learning to balance ourselves while cycling. Dancing is a kind of off-policy learning; you watch someone else dance and learn the dance steps.

On-policy methods are simple compared to off-policy methods. Off-policy methods are more powerful due to the "transfer learning" effect. In off-policy methods, you are learning from a different policy, the convergence is slower, and the variance is higher.

The advantage of off-policy learning is that the behavior policy can be very exploratory in nature, while the target policy can be deterministic and greedily optimize the rewards.

Off-policy reinforcement methods are based on a concept called importance sampling. This methodology helps estimate values under one policy probability distribution given samples from another policy probability distribution. Let's understand Monte Carlo off-policy evaluation by detailing the pseudocode. We'll then apply it to the Blackjack game in the OpenAI framework.

THE PSEUDOCODE FOR MONTE CARLO OFF-POLICY EVALUATION

What we see in the following figure is that we are estimating $Q(s,a)$ by learning from behavior policy **b**.

- Initialize the value of $Q(s,a)$
- Initialize the value of $C(s,a)$ as 0
- *Pi(S) <= argmax Q(s,a)*, set the target policy as greedy policy
- Loop forever for every episode
  - Behavior policy b is any soft policy
  - Generate episode using behavior policy b: sequence of state, action, and reward
  - Assign *G <= 0*
  - Assign *W <= 1*
  - Loop for each step of the episode in reverse
    - *G <= gamma * G + Rewards*
    - *C(s,a) <= C(s,a) + W*
    - *Q(s,a) <= Q(s,a) + W/C (s,a) [G - Q(s ,a)]*
    - *Pi (S) <= argmax Q(s,a)*
    - If action not equal to *Pi(s)* proceed to next episode
    - *W <= W * 1 / b(a/s)*

Figure 6.8: Pseudocode for Monte Carlo off-policy evaluation

The target policy is a greedy policy; hence we choose the action with the maximum rewards by using **argmax Q(s,a)**. Gamma is the discount factor that allows us to discount rewards in the distant future compared to immediate rewards in the future. The cumulative value function **C(s,a)** is calculated by incrementing it with weight **W**. Gamma is used to discount the rewards.

The essence of off-policy Monte Carlo is the loop through every episode:

```
for step in range(len(episode))[::-1]:
        state, action, reward = episode[step]

        #G <- gamma * G + Rt+1
        G = discount_factor * G + reward

        # C(St, At) = C(St, At) + W
        C[state][action] += W
```

```
#Q (St, At) <- Q (St, At) + W / C (St, At)
Q[state][action] += (W / C[state][action]) \
* (G - Q[state][action])


"""
If action not equal to argmax of target policy
proceed to next episode
"""
if action != np.argmax(target_policy(state)):
    break
# W <- W * Pi(At/St) / b(At/St)
W = W * 1./behaviour_policy(state)[action]
```

Let's understand the implementation of the off-policy method of Monte Carlo by using importance sampling. This exercise will help us learn how to set the target policy and the behavior policy, and learn the target policy from the behavior policy.

EXERCISE 6.04: IMPORTANCE SAMPLING WITH MONTE CARLO

This exercise will aim to do off-policy learning by using a Monte Carlo method. We have chosen a greedy target policy. We also have a behavior policy, which is any soft, non-greedy policy. By learning from the behavior policy, we will estimate the value function of the target policy. We will apply this technique of importance sampling to the Blackjack game environment. We will apply the steps outlined in the pseudocode step by step.

Perform the following steps to complete the exercise:

1. Import the necessary libraries:

```
import gym
import numpy as np
from collections import defaultdict
from functools import partial
```

2. Select the environment as **Blackjack** in OpenAI by using **gym.make**:

```
env = gym.make('Blackjack-v0')
```

3. Create two policy functions. One of them is a random policy. The random policy chooses a random action, which is a list of size n with 1/n probability where n is the number of actions:

```
"""
creates a random policy which is a linear probability distribution
num_Action is the number of Actions supported by the environment
"""
def create_random_policy(num_Actions):
#Creates a list of size num_Actions, with a fraction 1/num_Actions.
#If 2 is numActions, the array value would [1/2, 1/2]
    Action = np.ones(num_Actions, dtype=float)/num_Actions
    def policy_function(observation):
        return Action
    return policy_function
```

4. Write a function to create a greedy policy:

```
#creates a greedy policy,
"""
sets the value of the Action at the best_possible_action,
that maximizes the Q, value to be 1, rest to be 0
"""
def create_greedy_policy(Q):
    def policy_function(state):
        #Initializing with zero the Q
        Action = np.zeros_like(Q[state], dtype = float)
        #find the index of the max Q value
        best_possible_action = np.argmax(Q[state])
        #Assigning 1 to the best possible action
        Action[best_possible_action] = 1.0
        return Action
    return policy_function
```

The greedy policy chooses an action that maximizes the rewards. We first identify the **best_possible_action**, that is, the maximum value of **Q** across states. We then assign a value to the **Action** corresponding to the **best_possible_action**.

5. Define a function for Blackjack importance sampling that takes
    env, num_episodes, behaviour_policy, and **discount_factor**
    as arguments:

```
def black_jack_importance_sampling\
(env, num_episodes, behaviour_policy, discount_factor=1.0):

        #Initialize the value of Q
        Q = defaultdict(lambda: np.zeros(env.action_space.n))

        #Initialize the value of C
        C = defaultdict(lambda: np.zeros(env.action_space.n))

        #target policy is the greedy policy
        target_policy = create_greedy_policy(Q)
```

We initialize the value of **Q** and **C**, and set the target policy as a greedy policy.

6. We loop for the number of episodes, initialize the episodes list, and state the
    initial set by doing **env.reset()**:

```
for i_episode in range(1, num_episodes + 1):
    episode = []
    state = env.reset()
```

7. For a batch of 100, apply the behavior policy on a state to calculate
    the probability:

```
for i in range(100):
    probability = behaviour_policy(state)
    action = np.random.choice\
             (np.arange(len(probability)), p=probability)
    next_state, reward, done, info = env.step(action)
    episode.append((state, action, reward))
```

We pick a random action from that list. The step is taken with the random action,
returning **next_state** and **reward**. The episode list is appended with the
state, action, and **reward**.

8. If the **episode** is completed, we break the loop and assign **next_state** to **state**:

```
if done:
    break
state = next_state
```

9. Initialize **G**, the results as **0** and **W**, and the weight as **1**:

```
# G <- 0
    G = 0.0
    # W <- 0
    W = 1.0
```

10. Perform the steps detailed in the pseudocode using a **for** loop, as shown in the following code:

```
"""
Loop for each step of episode t=T-1, T-2,...,0
while W != 0
"""
for step in range(len(episode))[::-1]:
    state, action, reward = episode[step]
    #G <- gamma * G + Rt+1
    G = discount_factor * G + reward
    # C(St, At) = C(St, At) + W
    C[state][action] += W
    #Q (St, At) <- Q (St, At) + W / C (St, At)
    Q[state][action] += (W / C[state][action]) \
    * (G - Q[state][action])


    """
    If action not equal to argmax of target policy
    proceed to next episode
    """
    if action != np.argmax(target_policy(state)):
        break
    # W <- W * Pi(At/St) / b(At/St)
    W = W * 1./behaviour_policy(state)[action]
```

11. Return **Q** and **target_policy**:

```
return Q, target_policy
```

12. Create a random policy:

```
#create random policy
random_policy = create_random_policy(env.action_space.n)
"""

using importance sampling evaluates the target policy
by learning from the behaviour policy
"""

Q, policy = black_jack_importance_sampling\
            (env, 50000, random_policy)
```

The random policy is used as a behavior policy. We pass the behavior policy and using the importance sampling method, get the **Q** value function or the target policy.

13. Iterate through the items in **Q** and then find the action that has the maximum value. This is then stored as the value function for the corresponding state:

```
valuefunction = defaultdict(float)
for state, action_values in Q.items():
    action_value = np.max(action_values)
    valuefunction[state] = action_value
    print("state is", state, "value is", valuefunction[state])
```

You will get the following output:

```
state is (21, 2, False) value is 0.9043478260869562
state is (12, 2, False) value is -0.25263157894736843
state is (19, 2, True) value is 0.3846153846153846
state is (20, 1, False) value is 0.07167235494880556
state is (18, 5, False) value is 0.1361256544502617
state is (14, 5, False) value is -0.09452736318407957
state is (12, 5, False) value is -0.06214689265536725
state is (17, 1, False) value is -0.5369458128078817
state is (19, 5, False) value is 0.4191616766467066
state is (13, 5, False) value is -0.15454545454545454
state is (20, 10, False) value is 0.453049370764763
state is (17, 6, False) value is -0.06060606060606062
state is (14, 7, False) value is -0.2565445026178012
state is (11, 9, False) value is 0.26785714285714296
state is (15, 3, False) value is -0.2972972972972972
state is (8, 7, False) value is 0.28571428571428575
state is (11, 3, False) value is 0.29787234042553185
state is (19, 4, False) value is 0.4595959595959597
state is (21, 10, False) value is 0.8541666666666662
state is (19, 10, False) value is -0.016620498614958432
```

Figure 6.9: Output of off-policy Monte Carlo evaluation

The off-policy evaluation has calculated and returned the value function for every state-action pair. In this exercise, we have applied the concept of importance sampling using a behavior policy and applied the learning to a target policy. The output is provided for every combination state and action pair. It has helped us understand off-policy learning. We had two policies – a behavior policy and a target policy. We learned the target policy by following the behavior policy.

> **NOTE**
>
> To access the source code for this specific section, please refer to https://packt.live/3hpOOKa.
>
> You can also run this example online at https://packt.live/2B1GQGa.

In the next section, we will learn how to solve the frozen lake problem, available in the OpenAI framework, using Monte Carlo techniques.

SOLVING FROZEN LAKE USING MONTE CARLO

Frozen Lake is another simple game found in the OpenAI framework. This is a classic game where you can do sampling and simulations for Monte Carlo reinforcement learning. We have already described and used the Frozen Lake environment in *Chapter 05, Dynamic Programming*. Here we shall quickly revise the basics of the game so that we can solve it using Monte Carlo methods in the upcoming activity.

We have a 4x4 grid of cells, which is the entire frozen lake. It contains 16 cells (a 4x4 grid). The cells are marked as **S** – Start, **F** – Frozen, **H** – Hole, and **G** – Goal. The player needs to move from the Start cell, **S**, to the Goal cell, along with the Frozen areas (**F** cells), without falling into Holes (**H** cells). The following figure visually presents the aforementioned information:

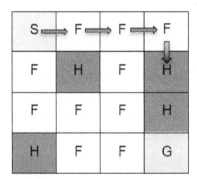

A) Frozen Lake Grid

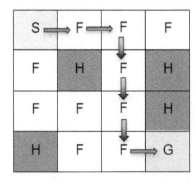

B) Failed Game ending by agent falling into a hole

C) Winning the Game by agent reaching the goal

Figure 6.10: The Frozen Lake game

Here are some basic details of the game:

- **The aim of the game**: The aim of the game is to move from the Start (cell **S**) to the Goal (cell **G**).

- **States** = 16

- **Actions** = 4

- **Total state-action pairs** = 64

- **Strategy**: The player should move along the Frozen cells (cells **F**) without falling into the Holes in the lake (cells **H**). Reaching the Goal (cell **G**) or falling into any Hole (cells **H**) ends the game.

- **Actions**: The actions that can be performed in any cell are left, down, right, and up.

- **Players**: It is a single-player game.

- **Rewards**: The reward for moving into a Frozen cell is 0 (cells **F**), +1 for reaching the Goal (cell **G**), and 0 for falling into a Hole (cells **H**).

- **Configuration**: You can configure the frozen lake to be slippery or non-slippery. If the frozen lake is slippery, then the intended action and the actual action can vary, so if someone wants to move left, they might end up moving right or down or up. If the frozen lake is non-slippery, the intended action and the actual action are always aligned. The grid has 16 possible cells where the agent can be at any point in time. The agent can take 4 possible actions in each of these cells. So, there are 64 possibilities in the game, whose likelihood is updated based on the learning. In the next activity, we will learn more about the Frozen Lake game, and understand the various steps and actions.

ACTIVITY 6.01: EXPLORING THE FROZEN LAKE PROBLEM – THE REWARD FUNCTION

Frozen Lake is a game in OpenAI Gym that's helpful to apply learning and reinforcement techniques. In this activity, we will solve the Frozen Lake problem and determine the various states and actions using Monte Carlo methods. We will track the success rate through batches of episodes.

Perform the following steps to complete the activity:

1. We import the necessary libraries: **gym** for the OpenAI Gym framework, **numpy**, and **defaultdict** is required to process dictionaries.

2. The next step is to select the environment as **FrozenLake**. **is_slippery** is set to **False**. The environment is reset with the line **env.reset()** and rendered with the line **env.render()**.

3. The number of possible values in the observation space is printed with **print(env.observation_space)**. Similarly, the number of action values is printed with the **print(env.action_space)** command.

4. The next step is to define a function to generate a frozen lake **episode**. We initialize the episodes and the environment.

5. We simulate various episodes by using a Monte Carlo method. We then navigate step by step and store **episode** and return **reward**. The action is obtained with **env.action_space.sample()**. **next_state**, **action**, and **reward** are obtained by calling the **env_step(action)** function. They are then appended to an episode. The episode is now a list of states, actions, and rewards.

6. The key is now to calculate the success rate, which is the likelihood of success for a batch of episodes. The way we do this is by calculating the total number of attempts in a batch of episodes. We calculate how many of them successfully reached the goal. The ratio of the agent successfully reaching the goal to the number of attempts made by the agent is the success ratio. First, we initialize the total rewards.

7. We generate **episode** and **reward** for every iteration and calculate the total **reward**.

8. The success ratio is calculated by dividing **total_reward** by **100** and is printed.

9. The frozen lake prediction is calculated using the **frozen_lake_prediction** function. The final output will demonstrate the default success ratio of the game without any reinforcement learning when the game is played randomly.

    You will get the following output:

    ```
    Episode 0 Policy Win Rate=> 1.0 %
    Episode 100 Policy Win Rate=> 1.0 %
    Episode 200 Policy Win Rate=> 1.0 %
    Episode 300 Policy Win Rate=> 2.0 %
    Episode 400 Policy Win Rate=> 1.0 %
    Episode 500 Policy Win Rate=> 0.0 %
    Episode 600 Policy Win Rate=> 1.0 %
    Episode 700 Policy Win Rate=> 1.0 %
    Episode 800 Policy Win Rate=> 2.0 %
    Episode 900 Policy Win Rate=> 0.0 %
    Episode 1000 Policy Win Rate=> 0.0 %
    Episode 1100 Policy Win Rate=> 2.0 %
    Episode 1200 Policy Win Rate=> 0.0 %
    Episode 1300 Policy Win Rate=> 0.0 %
    ```

    Figure 6.11: Output of Frozen Lake without learning

> **NOTE**
>
> The solution to this activity can be found on page 719.

In the next section, we detail how we can enable improvement by balancing exploration and exploitation, by using the epsilon soft policy and greedy policy. This ensures that we balance exploration and exploitation.

THE PSEUDOCODE FOR EVERY VISIT MONTE CARLO CONTROL FOR EPSILON SOFT

We have previously implemented every visit Monte Carlo algorithm for estimating the value function. In this section, we will briefly describe every visit Monte Carlo control for epsilon soft so that we can use this in our final activity of this chapter. The following figure shows the pseudo-code for every visit for Epsilon soft by balancing exploration and exploitation:

- Initialize the epsilon value as a small fraction
- Initialize the policy *Pi*, `Q(s,a)` and `Returns(s,a)`
- Loop for every episode
    - ○ Generate an episode following *Pi* with `State, Action`, and `Reward`
    - ○ *G <= 0*
    - ○ Within each step of the episode loop
        - ▪ Choose an action based on *epsilon soft - greedy policy*
        - ▪ Calculate *G <= gamma * G + Rewards*
        - ▪ Append `G` to `Returns(s,a)`
        - ▪ `Q(s,a)` is the average of `Returns(s,a)`

Figure 6.12: Pseudocode for Monte Carlo every visit for epsilon soft

The following code picks a random action with epsilon probability and picks an action that has a maximum **Q(s,a)** with 1-epsilon probability. So, we can choose between exploration with epsilon probability and exploitation with 1-epsilon probability:

```
while not done:


        #random action less than epsilon
        if np.random.rand() < epsilon:
            #we go with the random action
            action = env.action_space.sample()
        else:
            """
            1 - epsilon probability, we go with the greedy algorithm
            """


            action = np.argmax(Q[state, :])
```

In the next activity, we will evaluate and improve the policy for Frozen Lake by implementing the Monte Carlo control every visit for the epsilon soft method.

ACTIVITY 6.02 SOLVING FROZEN LAKE USING MONTE CARLO CONTROL EVERY VISIT EPSILON SOFT

The activity aims to evaluate and improve the policy for the Frozen Lake problem by using every visit epsilon soft method.

You can launch the Frozen Lake game by importing **gym** and executing **gym.make()**:

```
import gym
env = gym.make("FrozenLake-v0", is_slippery=False)
```

Perform the following step to complete the activity:

1. Import the necessary libraries.

2. Select the environment as **FrozenLake**. **is_slippery** is set to **False**.

3. Initialize the **Q** value and **num_state_action** to zeros.

4. Set the value of **num_episodes** to **100000** and create **rewardsList**. Set epsilon to **0.30**.

5. Run the loop till **num_episodes**. Initialize the environment, **results_List**, and **result_sum** to zero. Also, reset the environment.

6. We need to now have both exploration and exploitation. Exploration will be a random policy with epsilon probability and exploitation will be a greedy policy with 1-epsilon. We start a **while** loop and check whether we need to pick a random value with probability epsilon or a greedy policy with a probability of 1-epsilon.

7. Step through the **action** and get **new_state** and **reward**.

8. The result list is appended, with the state and action pair. **result_sum** is incremented by the value of the result.

9. **new_state** is assigned to **state** and **result_sum** is appended to **rewardsList**.

10. Calculate **Q[s,a]** using the incremental method, as **Q[s,a] + (result_sum - Q[s,a]) / N(s,a)**.

11. Print the value of the success rate in batches of **1000**.

12. Print the final success rate.

You will get the following output initially:

```
Frozen Lake Success rate=> 0.0 %
Frozen Lake Success rate=> 0.0 %
Frozen Lake Success rate=> 0.0 %
Frozen Lake Success rate=> 0.0 %
Frozen Lake Success rate=> 0.0 %
Frozen Lake Success rate=> 0.0 %
Frozen Lake Success rate=> 8.4 %
Frozen Lake Success rate=> 15.0625 %
Frozen Lake Success rate=> 20.27777777777778 %
Frozen Lake Success rate=> 24.53 %
Frozen Lake Success rate=> 28.318181818181817 %
Frozen Lake Success rate=> 31.525 %
Frozen Lake Success rate=> 34.06153846153846 %
Frozen Lake Success rate=> 36.121428571428574 %
Frozen Lake Success rate=> 38.02 %
Frozen Lake Success rate=> 39.5375 %
Frozen Lake Success rate=> 40.81764705882353 %
Frozen Lake Success rate=> 41.87777777777778 %
```

Figure 6.13: Initial output of the Frozen Lake success rate

You will get the following output finally:

```
Frozen Lake Success rate=> 60.38953488372093 %
Frozen Lake Success rate=> 60.44712643678161 %
Frozen Lake Success rate=> 60.492045454545455 %
Frozen Lake Success rate=> 60.53370786516854 %
Frozen Lake Success rate=> 60.59222222222222 %
Frozen Lake Success rate=> 60.62967032967033 %
Frozen Lake Success rate=> 60.69565217391305 %
Frozen Lake Success rate=> 60.75268817204301 %
Frozen Lake Success rate=> 60.82127659574468 %
Frozen Lake Success rate=> 60.87684210526316 %
Frozen Lake Success rate=> 60.95104166666667 %
Frozen Lake Success rate=> 61.00309278350515 %
Frozen Lake Success rate=> 61.039795918367346 %
Frozen Lake Success rate=> 61.07575757575758 %
Frozen Lake Success rate=> 61.117999999999995 %
```

Figure 6.14: Final output of the Frozen Lake success rate

NOTE

The solution to this activity can be found on page 722.

SUMMARY

Monte Carlo methods learn from experience in the form of sample episodes. Without having a model of the environment, by interacting with the environment, the agent can learn a policy. In several cases of simulation or sampling, an episode is feasible. We learned about the first visit and every visit evaluation. Also, we learned about the balance between exploration and exploitation. This is achieved by having an epsilon soft policy. We then learned about on-policy and off-policy learnings, and how importance sampling plays a key role in off-policy methods. We learned about the Monte Carlo methods by applying them to Blackjack and the Frozen Lake environment available in the OpenAI framework.

In the next chapter, we will learn about temporal learning and its applications. Temporal learning combines the best of dynamic programming and the Monte Carlo methods. It can work where the model is not known, like the Monte Carlo methods, but can provide incremental learning instead of waiting for the episode to end.

7

TEMPORAL DIFFERENCE LEARNING

OVERVIEW

In this chapter, we will be introduced to **Temporal Difference** (**TD**) learning and focus on how it develops the ideas of the Monte Carlo methods and dynamic programming. TD learning is one of the key topics in the field and studying it allows us to have a deep understanding of reinforcement learning and how it works at the most fundamental level. A new perspective will allow us to see MC methods as a particular case of TD ones, unifying the approach and extending their applicability to non-episodic problems. By the end of this chapter, you will be able to implement the **TD(0)**, **SARSA**, **Q-learning**, and **TD(λ)** algorithms and use them to solve environments with both stochastic and deterministic transition dynamics.

INTRODUCTION TO TD LEARNING

After having studied dynamic programming and Monte Carlo methods in the previous chapters, in this chapter, we will focus on temporal difference learning, one of the main stepping stones of reinforcement learning. We will start with their simplest formulation, that is, the one-step methods, and we will build on them to create their most advanced formulation, which is based on the eligibility traces concept. We will see how this new approach allows us to frame TD and MC methods under the same derivation idea, giving us the ability to compare the two. Throughout this chapter, we will implement many different flavors of TD methods and apply them to the FrozenLake-v0 environment under both the deterministic and the stochastic environment dynamics. Finally, we will solve the stochastic version of FrozenLake-v0 with an off-policy TD method known as Q-learning.

Temporal difference learning, whose name derives from the fact that it uses differences in state (or state-actions pairs) values between subsequent timesteps to learn, can be considered a central idea in the field of reinforcement learning algorithms. It shares some important aspects with the methods we studied in previous chapters – in fact, just like those methods, it learns through experience, with no need to have a model (like Monte Carlo methods do), and it "bootstraps," meaning it learns how to use information it's acquired before reaching the end of the episode (like dynamic programming methods do).

These differences are strictly related to the advantages that TD methods offer with respect to MC and DP ones: it doesn't need a model of the environment and it can be applied with a greater generality with respect to DP methods. Its ability to bootstrap, on the other hand, makes TD more suited for tasks with very long episodes and the only solution for non-episodic ones – to which Monte Carlo methods cannot be applied. As an example of a long-term or non-episodic task, think of an algorithm that is used to grant user access to a server that's rewarded every time the first user in the queue is assigned to a resource and receiving zero reward if user access is not granted. The queue typically never ends, so this is a continuing task that has no episodes.

As seen in previous chapters, the exploration versus exploitation trade-off is a very important subject, and again also in the case of temporal difference algorithms. They fall into two main classes: on-policy and off-policy methods. As we saw in the previous chapters, in on-policy methods, the same policy that is learned is used to explore the environment, while in off-policy ones, the two can be different: one if used for exploration, while the other one is the target to be learned. In the following sections, we will address the general problem of estimating the state value function for a given policy. Then, we will see how, by building upon it, we can obtain a complete RL algorithm to train both on-policy and off-policy methods to find the optimal policy for a given problem.

Let's start with our first steps in the temporal difference methods world.

TD(0) – SARSA AND Q-LEARNING

TD methods are model-free, meaning they do not need a model of the environment to learn a state value representation. For a given policy, π, they accumulate experience associated with it and update their estimate of the value function for every state encountered during the corresponding experience. In doing so, TD methods update a given state value, visited at time **t**, using the value of state (or states) encountered at the next few time steps, so for time **t+1**, **t+2**, ..., **t+n**. An abstract example is as follows: an agent is initialized in the environment and starts interacting with it by following a given policy, without any knowledge of what results are generated by which action. Following a certain number of steps, the agent will eventually reach a state associated with a reward. This reward signal is used to increment the values of previously visited states (or action-state pairs) with the TD learning rule. In fact, those states have allowed the agent to reach the goal, so they are to be associated with a high value. Repeating this process over and over will allow the agent to build a complete and meaningful value map of all states (or state-action pairs) so that it will exploit this acquired knowledge to select the best actions, thereby leading to states associated with a reward.

This means that TD methods do not have to wait until the end of the episode to improve their policy; instead, they can build upon values of states they encounter, and the learning process can start right after initialization.

In this section, we will focus on the so-called one-step method, also named TD(0). In this method, the only value considered to build the update for a given state value function is the one found at the next time step, nothing else. So, for example, the value function update for a state at time **t** looks as follows:

$$V(S_t) \leftarrow V(S_t) + \alpha \left[R_{t+1} + \gamma V(S_{t+1}) - V(S_t) \right]$$

Figure 7.1: Value function update for a state at time 't'

Here, (S_{t+1}) is the next state where the environment transitioned, R_{t+1} is the reward obtained in the transition, α is the learning rate, and γ is the discount factor. It is clear how TD methods "bootstrap": in order to update the value function for a state **(t)**, they use the current value function for the next state **(t+1)** without waiting until the end of the episode. It is worth noting that the quantity between square brackets in the previous equation can be interpreted as an error term. This error term measures the difference between the estimated value of state S_t and the new, better estimate, $R_{t+1} + \gamma V(S_{t+1})$. This quantity is called the TD error, and we will encounter it many times in RL theory:

$$\delta_t \overset{def}{=} R_{t+1} + \gamma V(S_{t+1}) - V(S_t)$$

Figure 7.2: TD error at time 't'

This error is specific for the given time it has been calculated at, and it depends on the values at the next time step (that is, the error at time t depends on the values at time **t+1**).

One important theory result for TD methods is their proof of convergence: in fact, it has been demonstrated that, for any fixed policy, π, the algorithm TD(0) described in the preceding equation converges to the state (or action-state pair) value function, V_π. Convergence is reached for a constant step size parameter, provided it is sufficiently small, and with probability **1** if the step size parameter decreases according to some specific (but easy to comply with) stochastic approximation conditions. These proofs mainly apply to the tabular version of the algorithm, which are the versions used for RL theory introduction and understanding. These deal with problems in which states and actions spaces are of limited dimensions so that they can be exhaustively represented by a finite combination of variables.

However, the majority of these proofs can be easily extended to algorithm versions that rely on approximations when they are composed by general linear functions. These approximated versions are used when states and actions spaces are so large that they cannot be represented by a finite combination of variables (for example, when the state space is the space of RGB images).

So far, we have been dealing with state value functions. In order to approach the problem of temporal difference control, we need to learn a state-action value function rather than a state-value function. In fact, in this way, we will be able to associate a value with state-action pairs, thereby building a value map that can then be used to define our policy. How we implement this specifically depends on the method class. First, let's take a look at the on-policy approach, which is implemented by the so-called SARSA algorithm, and then the off-policy one, which is implemented by the so-called Q-learning algorithm.

SARSA – ON-POLICY CONTROL

For an on-policy method, the goal is to estimate $Q_\pi(s, a)$, that is, the state-action value function for the current behavior policy, π, for all states and all actions. To do so, we simply need to apply the equation we saw for the state-value function to the state-action function. Since the two cases are identical (both being Markov chains with a reward process), the theorems stating the convergence of the state-value function to the one corresponding to the optimal policy (and so, solving the problem of finding the optimal policy) are valid in this new setting, where the value function regards state-action pairs. The update equation takes the following form:

$$Q\left(S_t, A_t\right) \leftarrow Q\left(S_t, A_t\right) + \alpha\left[R_{t+1} + \gamma Q\left(S_{t+1}, A_{t+1}\right) - Q\left(S_t, A_t\right)\right]$$

Figure 7.3: State-action value function at time 't'

This update is supposedly performed after every transition from a non-terminal state, S_t. If (S_{t+1}) is a terminal state, then the $Q(S_{t+1}, A_{t+1})$ value is set equal to **0**. As we can see, the update rule uses every element of the quintuple $\left(S_t, A_t, R_{t+1}, S_{t+1}, A_{t+1}\right)$, which explains the transition from one state-action pair to the next, with the reward associated with the transition. This quintuple, written in this form, is the reason why the name **SARSA** was given to this algorithm.

Using these elements, it is straightforward to design an on-policy control algorithm based on them. As we mentioned previously, all on-policy methods estimate q_π for the behavior policy, π, and at the same time, update π based on q_π. A scheme for the SARSA control algorithm can be depicted as follows:

1. Choose the algorithm parameters; that is, the step size, α, which has to be contained in the interval **(0, 1]**, and the **ε** parameter of the ε-greedy policy, which has to be small and greater than 0, since it represents the probability of choosing the non-optimal action to favor exploration. This can be done with the following code:

```
alpha = 0.02
epsilon = 0.05
```

2. Initialize $Q(s, a)$, for all values of $s \in S^+$, $a \in A(s)$, arbitrarily, except that Q(terminal, ·) = 0, as shown by the following code snippet, in the case of an environment with **16** states and **4** actions:

```
q = np.ones((16,4))
```

3. Create a loop for each episode. Initialize S and choose A from S using the policy derived from Q (for example, ε-greedy). This can be done using the following snippet, where the initial state is provided by the environment **reset** function and the action is selected using a dedicated ε-greedy function:

```
for i in range(nb_episodes):
        s = env.reset()
        a = action_epsilon_greedy(q, s, epsilon=epsilon)
```

4. Create a loop for each step of the episode. Take action A and observe R, S'. Choose A' from S' using the policy derived from Q (for example, ε-greedy). Update the state-action value function for the selected state-action pair using the SARSA rule, which defines the new value as the sum of the current one, plus the TD error multiplied by the step size, α, as depicted in the following expression:

$$Q(S, A) \leftarrow Q(S, A) + \alpha[R + \gamma Q(S', A') - Q(S, A)]$$

Figure 7.4: Updating the state-action value function using the SARSA rule

Then, update the state-action pair with the new one using $S \leftarrow S'$; $A \leftarrow A'$ until S is a terminal state. All of this is done using the following code:

```
while not done:
            new_s, reward, done, info = env.step(a)
            new_a = action_epsilon_greedy(q, new_s, epsilon=epsilon)
```

```
q[s, a] = q[s, a] + alpha * (reward + gamma \
            * q[new_s, new_a] - q[s, a])
s = new_s
a = new_a
```

NOTE

The steps and code for this algorithm were originally developed and outlined by *Sutton, Richard S. Introduction to Reinforcement Learning. Cambridge, Mass: MIT Press, 2015.*

The SARSA algorithm can converge to an optimal policy and an optimal action-value function with probability equal to **1** under the following conditions:

- All the state-action pairs need to be visited an infinite number of times.

- The policy converges in the limit to the greedy policy, which can be achieved with ε-greedy policies where **ε** vanishes in time (this can be done by setting **ε = 1/t**).

This algorithm makes use of the ε-greedy algorithm. We will explain this in more detail in the next chapter, so we will only briefly recall what it is here. When learning policies by means of state-action value functions, the value associated with the state-action pairs is used to decide which is the best action to take. At convergence, the best action is chosen among the available ones for a given state, and we opt for the one that has the highest value: this is the greedy approach. This means that for every given state, the same action will always be chosen (if no actions have the same value). This is not a good choice for exploration, especially at the beginning of training. For this reason, the ε-greedy approach is preferred in this phase: the best action is chosen with a probability equal to **1−ε**, while in the other cases, a random action is selected. By making **ε** diminishing, the ε-greedy approach becomes the greedy one in the limit as the number of steps approaches infinity.

In order to consolidate these concepts, let's apply the SARSA control algorithm right away. The following exercise will show you how to implement TD(0) SARSA to solve the FrozenLake-v0 environment, using its deterministic version first.

The goal here is to see how the SARSA algorithm is able to recover the optimal policy, which we humans can estimate in advance, for a given configuration of the problem. Before jumping into it, let's quickly recap what the frozen lake problem is and the optimal policy we aim to make the agent find. The agent sees a grid world whose dimension is 4 x 4.

The grid has a starting position, **S** (upper left-hand side), frozen tiles, **F**, holes, **H**, and a goal, **G** (lower right). The agent is rewarded with +1 when it reaches the terminal goal state, while the episode ends without a reward if it reaches the terminal states constituted by the holes. The following table represents the environment:

S	F	F	F
F	H	F	H
F	F	F	H
H	F	F	G

Figure 7.5: The FrozenLake-v0 environment

As you can see in the preceding diagram, **S** is the starting position, **F** indicates frozen tiles, **H** means holes, and **G** is the goal. For the deterministic environment, the optimal policy is the one that allows the agent to reach the goal in the shortest possible time. To be 100% precise, in this case, since, in this specific environment, no penalty for intermediate steps is applied, there is no need for the optimal path to be the shortest one. Every path that eventually leads to the goal is equally optimal in terms of cumulative expected reward. However, we will see that by appropriately using the discount factor, we will be able to recover the optimal policy, which also accounts for the shortest path. Under this condition, the optimal policy is represented in the following diagram, where each of the four moves (Down, Right, Left, and Up) are represented by their initial letter. There are two tiles for which two actions would result in the same optimal path:

R/D	R	D	L
D	-	D	-
R	R/D	D	-
-	R	R	!

Figure 7.6: Optimal policy

In the preceding diagram, **D** denotes Down, **R** denotes Right, **U** denotes Up, and **L** denotes Left. **!** stands for the goal, and **–** refers to the holes in the environment.

We will use a decreasing **ε** value in order to anneal the exploration from large to small, thereby making it become, in the limit, greedy.

This type of exercise is very useful when learning about classic reinforcement learning algorithms. Being tabular (this is a grid world example, meaning it can be represented by a 4x4 grid) allows us to keep track of everything that's happening in the domain, easily follow state-actions pairs values being updated during algorithm iterations, look at action choices according to the selected policy, and converge to the optimal policy. In this chapter, you will learn how to code a reference algorithm in the RL landscape and get deep hands-on experience with all these fundamental aspects.

Let's now move on to the implementation.

EXERCISE 7.01: USING TD(0) SARSA TO SOLVE FROZENLAKE-V0 DETERMINISTIC TRANSITIONS

In this exercise, we will implement the SARSA algorithm and use it to solve the FrozenLake-v0 environment, where only deterministic transitions are allowed. This means we will look for (and actually find) the optimal policy to retrieve the frisbee in this environment.

The following steps will help you to complete this exercise:

1. Import the required modules:

```
import numpy as np
import matplotlib.pyplot as plt
%matplotlib inline

import gym
```

2. Instantiate the **gym** environment called **FrozenLake-v0**. Set the **is_slippery** flag to **False** to disable its stochasticity:

```
env = gym.make('FrozenLake-v0', is_slippery=False)
```

3. Take a look at the action and the observation spaces:

```
print("Action space = ", env.action_space)
print("Observation space = ", env.observation_space)
```

This will print out the following:

```
Action space =  Discrete(4)
Observation space =  Discrete(16)
```

4. Create two dictionaries to easily translate action numbers into moves:

```
actionsDict = {}
actionsDict[0] = " L "
actionsDict[1] = " D "
actionsDict[2] = " R "
actionsDict[3] = " U "

actionsDictInv = {}
actionsDictInv["L"] = 0
actionsDictInv["D"] = 1
actionsDictInv["R"] = 2
actionsDictInv["U"] = 3
```

5. Reset the environment and render it to be able to take a look at the grid problem:

```
env.reset()
env.render()
```

The output will be as follows:

Figure 7.7: Environment's initial state

6. Visualize the optimal policy for this environment:

```
optimalPolicy = ["R/D"," R "," D "," L ", \
                 " D "," - "," D "," - ", \
                 " R ","R/D"," D "," - ", \
                 " - "," R "," R "," ! ",]
```

```
print("Optimal policy:")
idxs = [0,4,8,12]
for idx in idxs:
    print(optimalPolicy[idx+0], optimalPolicy[idx+1], \
        optimalPolicy[idx+2], optimalPolicy[idx+3])
```

The output will be as follows:

```
Optimal policy:
R/D  R    D  L
 D   -    D  -
 R  R/D   D  -
 -   R    R  !
```

This represents the optimal policy for this environment, showing, for each of the environment states represented in the 4x4 grid, the optimal action among the four available: *move Up*, *move Down*, *move Right*, and *move Left*. Except for two states, all the others have a single optimal action associated with them. In fact, as described previously, optimal actions here are those that bring the agent to the goal using the shortest possible path. Two different possibilities result in the same path length for two states, so they are both equally optimal.

7. Define functions to take ε-greedy actions. The first function implements the ε-greedy policy with a probability of $1 - ε$. The chosen action is the one with the highest value associated with the state-action pair; otherwise, a random action is returned. The second function simply makes the first callable when it's passed as an argument by using a **lambda** function:

```
def action_epsilon_greedy(q, s, epsilon=0.05):
    if np.random.rand() > epsilon:
        return np.argmax(q[s])
    return np.random.randint(4)


def get_action_epsilon_greedy(epsilon):
    return lambda q,s: action_epsilon_greedy\
                        (q, s, epsilon=epsilon)
```

8. Define a function to take greedy actions:

```
def greedy_policy(q, s):
    return np.argmax(q[s])
```

9. Now, define a function that will calculate the mean of the agent's performances. First, we'll define the number of episodes used to calculate the average performance (in this case, **500**), and then execute all these episodes in a loop. We'll reset the environment and start the in-episode loop to do so. We then select an action according to the policy that we want to measure the performance of, step through the environment with the chosen action, and finally add the reward to the accumulated returns. We repeat these environment steps until the episode is complete:

```
def average_performance(policy_fct, q):
    acc_returns = 0.
    n = 500
    for i in range(n):
        done = False
        s = env.reset()
        while not done:
            a = policy_fct(q, s)
            s, reward, done, info = env.step(a)
            acc_returns += reward
    return acc_returns/n
```

10. Set the number of total episodes and number of steps specifying how often the agent's average performance is estimated, as well as the **ε** parameters, which determine its decrease. Use the starting value, minimum value, and range (in terms of the number of episodes) over which the decrease is spread:

```
nb_episodes = 80000
STEPS = 2000
epsilon_param = [[0.2, 0.001, int(nb_episodes/2)]]
```

11. Define the SARSA training algorithm as a function. In this step, the Q-table is initialized. All the values are equal to **1**, but the values at terminal states are set equal to **0**:

```
def sarsa(alpha = 0.02, \
          gamma = 1., \
          epsilon_start = 0.1, \
          epsilon_end = 0.001, \
          epsilon_annealing_stop = int(nb_episodes/2), \
          q = None, \
          progress = None, \
          env=env):
```

```
if q is None:
    q = np.ones((16,4))
    # Set q(terminal,*) equal to 0
    q[5,:] = 0.0
    q[7,:] = 0.0
    q[11,:] = 0.0
    q[12,:] = 0.0
    q[15,:] = 0.0
```

12. Start a **for** loop among all the episodes:

```
for i in range(nb_episodes):
```

13. Inside the loop, first, the epsilon value is defined, depending on the current episode number:

```
inew = min(i,epsilon_annealing_stop)
epsilon = (epsilon_start \
            *(epsilon_annealing_stop - inew)\
            +epsilon_end * inew) / epsilon_annealing_stop
```

14. Next, the environment is reset, and the first action is chosen with an ε-greedy policy:

```
done = False
s = env.reset()
a = action_epsilon_greedy(q, s, epsilon=epsilon)
```

15. Then, we start an in-episode loop:

```
while not done:
```

16. Inside the loop, the environment is stepped throughout using the selected action and the new state and the reward, and the done conditions are retrieved:

```
new_s, reward, done, info = env.step(a)
```

17. Select a new action with the ε-greedy policy, update the Q-table with the SARSA TD(0) rule, and update the state and action with their new values:

```
new_a = action_epsilon_greedy\
            (q, new_s, epsilon=epsilon)
q[s, a] = q[s, a] + alpha * (reward + gamma \
            * q[new_s, new_a] - q[s, a])
s = new_s
a = new_a
```

18. Finally, the agent's average performance is estimated:

```
if progress is not None and i%STEPS == 0:
    progress[i//STEPS] = average_performance\
                        (get_action_epsilon_greedy\
                        (epsilon), q=q)

return q, progress
```

It may be useful to provide a brief description of the **ε** parameter's decrease. This is determined by three parameters: the starting value, minimum value, and decrease range (called **epsilon_annealing_stop**). They are used in the following way: ε starts at the starting value, and then it is decreased linearly across the number of episodes defined by the parameter's "range" until it reaches the minimum value, which is then kept constant.

19. Define an array that will collect all agent performance evaluations during training and the execution of SARSA TD(0) training:

```
sarsa_performance = np.ndarray(nb_episodes//STEPS)
q, sarsa_performance = sarsa(alpha = 0.02, gamma = 0.9, \
                        progress=sarsa_performance, \
                        epsilon_start=epsilon_param[0][0],\
                        epsilon_end=epsilon_param[0][1], \
                        epsilon_annealing_stop = \
                        epsilon_param[0][2])
```

20. Plot the SARSA agent's average reward history during training:

```
plt.plot(STEPS*np.arange(nb_episodes//STEPS), sarsa_performance)
plt.xlabel("Epochs")
plt.title("Learning progress for SARSA")
plt.ylabel("Average reward of an epoch")
```

This generates the following output:

```
Text(0, 0.5, 'Average reward of an epoch')
```

The plot for this can be visualized as follows. This shows the learning progress of the SARSA algorithm:

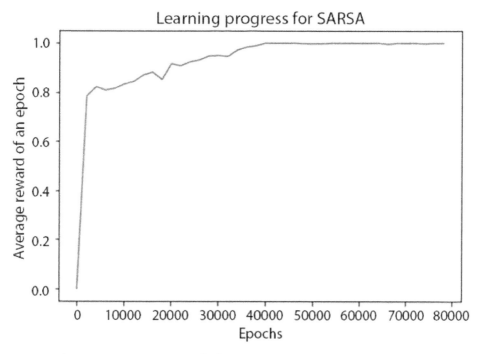

Figure 7.8: Average reward of an epoch trend over training epochs

As we can see, SARSA's performance grows over time as the **ε** parameter is annealed, thus reaching the value of **0** in the limit, thereby obtaining the greedy policy. This also demonstrates that the algorithm is capable of reaching 100% success after learning.

21. Evaluate the greedy policy's performance of the trained agent (Q-table):

```
greedyPolicyAvgPerf = average_performance(greedy_policy, q=q)
print("Greedy policy SARSA performance =", greedyPolicyAvgPerf)
```

The output will be as follows:

```
Greedy policy SARSA performance = 1.0
```

22. Display the Q-table values:

```
q = np.round(q,3)
print("(A,S) Value function =", q.shape)
print("First row")
print(q[0:4,:])
print("Second row")
print(q[4:8,:])
print("Third row")
print(q[8:12,:])
print("Fourth row")
print(q[12:16,:])
```

The output will be as follows:

```
(A,S) Value function = (16, 4)
First row
[[0.505 0.59  0.54  0.506]
 [0.447 0.002 0.619 0.494]
 [0.49  0.706 0.487 0.562]
 [0.57  0.379 0.53  0.532]]
Second row
[[0.564 0.656 0.    0.503]
 [0.    0.    0.    0.   ]
 [0.003 0.803 0.002 0.567]
 [0.    0.    0.    0.   ]]
Third row
[[0.62  0.    0.728 0.555]
 [0.63  0.809 0.787 0.   ]
 [0.707 0.899 0.    0.699]
 [0.    0.    0.    0.   ]]
Fourth row
[[0.    0.    0.    0.   ]
 [0.    0.791 0.9   0.696]
 [0.797 0.895 1.    0.782]
 [0.    0.    0.    0.   ]]
```

This output shows the values of the complete state-action value function for our problem. These values are then used to generate the optimal policy by means of the greedy selection rule.

23. Print out the greedy policy that was found and compare it with the optimal policy. Having calculated the state-action value function, we are able to retrieve the greedy policy from it. In fact, as explained previously, the greedy policy chooses the action that, for a given state, is associated with the maximum value of the Q-table. For this purpose, we are using the **argmax** function. When applied to each of the 16 states (from 0 to 15), it returns the index of the four actions (from 0 to 3) with the highest associated value for that state. Here, we also directly output the label associated with the action index using the pre-built dictionary:

```
policyFound = [actionsDict[np.argmax(q[0,:])],\
               actionsDict[np.argmax(q[1,:])], \
               actionsDict[np.argmax(q[2,:])], \
               actionsDict[np.argmax(q[3,:])], \
               actionsDict[np.argmax(q[4,:])], \
               " - ",\
               actionsDict[np.argmax(q[6,:])], \
               " - ",\
               actionsDict[np.argmax(q[8,:])], \
               actionsDict[np.argmax(q[9,:])], \
               actionsDict[np.argmax(q[10,:])], \
               " - ",\
               " - ",\
               actionsDict[np.argmax(q[13,:])], \
               actionsDict[np.argmax(q[14,:])], \
               " ! "]

print("Greedy policy found:")
idxs = [0,4,8,12]
for idx in idxs:
    print(policyFound[idx+0], policyFound[idx+1], \
          policyFound[idx+2], policyFound[idx+3])

print(" ")

print("Optimal policy:")
idxs = [0,4,8,12]
for idx in idxs:
    print(optimalPolicy[idx+0], optimalPolicy[idx+1], \
          optimalPolicy[idx+2], optimalPolicy[idx+3])
```

The output is as follows:

```
Greedy policy found:
 D    R    D    L
 D    -    D    -
 R    D    D    -
 -    R    R    !

Optimal policy:
R/D   R    D    L
 D    -    D    -
 R   R/D   D    -
 -    R    R    !
```

As the preceding output shows, the TD(0) SARSA algorithm we implemented has been able to successfully learn the optimal policy for this task just by interacting with the environment and collecting experience of it through episodes and then adopting the SARSA state-action pair value function update rule that was defined in the *SARSA – On-Policy Control* section. In fact, as we can see, for every state of the environment, the greedy policy that was obtained with the Q-table calculated by our algorithm prescribes an action that is in accordance with the optimal policy that was defined for analyzing the environment problem. As we already saw, there are two states in which there are two equally optimal actions and the agent correctly implements one of them.

> **NOTE**
>
> To access the source code for this specific section, please refer to https://packt.live/3fJBLBh.
>
> You can also run this example online at https://packt.live/30XeOXj.

THE STOCHASTICITY TEST

Now, let's take a look at what happens if the stochasticity is enabled in the FrozenLake-v0 environment. Enabling the stochasticity for this task means that every transition for a selected action is no longer deterministic. In particular, for a given action, there is a one in three chances that the action is executed as intended and 2 out of 3 equally distributed chances (1/3 and 1/3 each) for the two neighboring actions. Zero probability is assigned to the action in the opposite direction. So, for example, if the Down action is set, the agent will move down 1/3 of the time, move right 1/3 of the time, and move left the remaining 1/3 of the time, never going up, as shown in the following diagram:

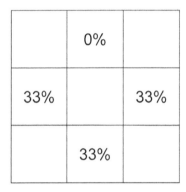

Figure 7.9: Percentages for the resulting states if the Down action is taken from the central tile

The environment setting is the very same as it was for the FrozenLake-v0 deterministic case we saw previously. Again, we want the SARSA algorithm to recover the optimal policy. This can be estimated in advance in this case as well. Just to make the reasoning for this easier, here's the table representing this environment:

S	F	F	F
F	H	F	H
F	F	F	H
H	F	F	G

Figure 7.10: Problem setting

In the preceding diagram, **S** is the starting position, **F** indicates frozen tiles, **H** indicate the holes, and **G** is the goal. For the stochastic environment, the optimal policy is very different with respect to the one that corresponds to the deterministic case, and it may even appear counter-intuitive. The key point is that in order to keep the possibility of obtaining a reward alive, our only chance is to avoid falling into the holes. Since there is no penalty for intermediate steps, we can keep going around for as long as we need to. And the only certain way to do so is as follows:

1. Move in the opposite direction of the hole we find next to us, even if this means moving away from the goal.

2. Avoid, in every possible way, falling into those tiles where there is a chance greater than 0 of falling into a hole:

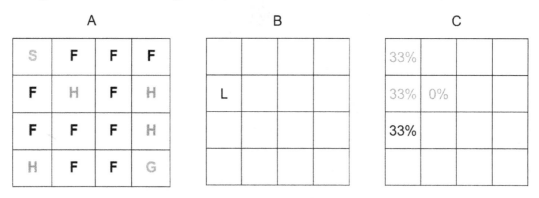

Figure 7.11: Environment setup (A), action executed by the agent (B), and chances of ending near the starting state in each position (C)

For example, let's consider the first tile on the left, in the second row from the top, in our problem setting, as shown in table **B** in the preceding diagram. In the deterministic case, the optimal action was to go down because it would bring us closer to the goal. In this case, instead, the best action to choose is to move left, even if moving left means bouncing into the wall. This is because moving left is the only action that won't make us fall into the hole. In addition, there is a 33% probability that we will end up in the tile on the third row from above, thereby getting closer to the goal.

> **NOTE**
>
> The preceding behavior follows a standard boundary implementation. In that tile, you execute the action "Move Left" (which is a completely legal action) and the environment will understand that this results in "bouncing." The algorithm simply sends a "move left" action to the environment, which, in turn, will take the prescribed action into account.

Similar reasoning can be applied to all the other tiles while keeping the key point mentioned previously in mind. However, it is worth discussing a very peculiar case – one that's the only reason why we cannot achieve 100% success, even with the optimal policy:

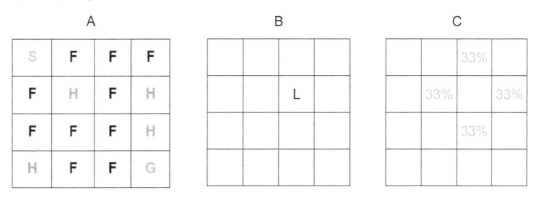

Figure 7.12: Environment setup (A), "Move to the Left" action executed by the agent (B), and chances of ending near the starting state in each position (C)

Now, let's take a look at the third tile from the left in the second row from the top in our problem setting, as shown in table B in the preceding diagram. This tile is between two holes, so there is no way to take an action that is 100% safe. Here, the best action is actually to move toward either the left or the right hole! This is because by moving left or right, we have a 66% chance of moving up or down, and only a 33% chance of falling into the hole. Moving up or down means we would have a 66% chance of moving right or left, falling into the hole, and only a 33% chance of actually moving up or down. And since this tile is the reason why we cannot achieve maximum performance 100% of the time, the best thing is to avoid reaching that tile. In order to do so, optimal actions of the very first row, apart from the starting tile, are all pointing up so that it is not possible to land on the problematic tile.

All other values are constrained by the hole's proximity, except for the tile on the left of the goal: the optimal action choice for this tile is to move down since it maintains the chance of landing in the goal, while at the same time avoiding landing in the tile above it, where, in turn, the agent would be forced to move left to avoid the hole, thus risking landing on the tile between the two holes. The optimal policy is summarized in the following diagram:

L/D/R	U	U	U
L	H	L/R	H
U	D	L	H
H	R	D	G

Figure 7.13: Optimal policy

The preceding diagram displays the optimal policy of the environment explained previously, where **D** denotes a Down move, **R** denotes a Right move, **U** denotes an Up move, and **L** denotes a Left move.

In the following example, we will use the SARSA algorithm to solve this new flavor of the FrozenLake-v0 environment. In order to obtain the optimal policy we just described, we need to adjust our hyperparameters – in particular, the discount factor, γ. In fact, we want to give the agent the freedom to make however many steps they need to. In order to do so, we have to propagate the value of the goal backward so that all the trajectories in the goal will benefit from it, even if those trajectories are not the shortest ones. For this reason, we will use a discount factor equal (or very close) to **1**. In code, this means that instead of using **gamma = 0.9**, we will use **gamma = 1**.

Now, let's see our SARSA algorithm working in this stochastic environment.

EXERCISE 7.02: USING TD(0) SARSA TO SOLVE FROZENLAKE-V0 STOCHASTIC TRANSITIONS

In this exercise, we'll use the TD(0) SARSA algorithm to solve the FrozenLake-v0 environment, with stochastic transitions enabled. As we just saw, the optimal policy looks completely different with respect to the previous exercise since it needs to take care of the stochasticity factor. This imposes a new challenge for the SARSA algorithm, and we will see how it will still be able to solve this task. This exercise will show us how these sound TD methods are able to deal with different challenges, demonstrating a notable robustness.

Follow these steps to complete this exercise:

1. Import the required modules:

    ```
    import numpy as np
    import matplotlib.pyplot as plt
    %matplotlib inline

    import gym
    ```

2. Instantiate the **gym** environment called **FrozenLake-v0** using the **is_slippery** flag set to **True** in order to enable stochasticity:

    ```
    env = gym.make('FrozenLake-v0', is_slippery=True)
    ```

3. Take a look at the action and the observation spaces:

    ```
    print("Action space = ", env.action_space)
    print("Observation space = ", env.observation_space)
    ```

    The output will be as follows:

    ```
    Action space =  Discrete(4)
    Observation space =  Discrete(16)
    ```

4. Create two dictionaries to easily map the **actions** indices (from **0** to **3**) to the labels (Left, Down, Right, and Up):

```
actionsDict = {}
actionsDict[0] = "  L  "
actionsDict[1] = "  D  "
actionsDict[2] = "  R  "
actionsDict[3] = "  U  "

actionsDictInv = {}
actionsDictInv["L"] = 0
actionsDictInv["D"] = 1
actionsDictInv["R"] = 2
actionsDictInv["U"] = 3
```

5. Reset the environment and render it to take a look at the grid problem:

```
env.reset()
env.render()
```

The output will be as follows:

SFFF
FHFH
FFFH
HFFG

Figure 7.14: Environment's initial state

6. Visualize the optimal policy for this environment:

```
optimalPolicy = ["L/R/D","  U  ","  U  ","  U  ",\
                 "  L  ","  -  "," L/R ","  -  ",\
                 "  U  ","  D  ","  L  ","  -  ",\
                 "  -  ","  R  ","  D  ","  !  ",]
```

```
print("Optimal policy:")
idxs = [0,4,8,12]
for idx in idxs:
    print(optimalPolicy[idx+0], optimalPolicy[idx+1], \
          optimalPolicy[idx+2], optimalPolicy[idx+3])
```

The output will be as follows:

```
Optimal policy:
  L/R/D  U    U    U
    L    -   L/R   -
    U    D    L    -
    -    R    D    !
```

This represents the optimal policy for this environment. Except from two states, all the other ones have a single optimal action associated with them. In fact, as described previously, optimal actions here are those that bring the agent away from the holes or from tiles that have a chance greater than zero of leading the agent to tiles placed near holes. Two states have multiple optimal actions associated with them that are all equally optimal, as intended for this task.

7. Define functions that will take ε-greedy actions:

```
def action_epsilon_greedy(q, s, epsilon=0.05):
    if np.random.rand() > epsilon:
        return np.argmax(q[s])
    return np.random.randint(4)


def get_action_epsilon_greedy(epsilon):
    return lambda q,s: action_epsilon_greedy\
                       (q, s, epsilon=epsilon)
```

The first function implements the ε-greedy policy: with a probability of $1 - \varepsilon$, the chosen action is the one with the highest value associated with the state-action pair; otherwise, a random action is returned. The second function simply makes the first callable when passed as an argument using a **lambda** function.

8. Define a function that will take greedy actions:

```
def greedy_policy(q, s):
    return np.argmax(q[s])
```

9. Define a function that will calculate average agent performance:

```
def average_performance(policy_fct, q):
    acc_returns = 0.
    n = 100
    for i in range(n):
        done = False
        s = env.reset()
        while not done:
            a = policy_fct(q, s)
            s, reward, done, info = env.step(a)
            acc_returns += reward
    return acc_returns/n
```

10. Set the number of total episodes, the number of steps representing the interval by which the agent's average performance is evaluated and the ε parameters, ruling its decrease, that is, the starting value, minimum value, and range (in terms of the number of episodes) over which the decrease is spread:

```
nb_episodes = 80000
STEPS = 2000
epsilon_param = [[0.2, 0.001, int(nb_episodes/2)]]
```

11. Define the SARSA training algorithm as a function. Initialize the Q-table with all the values equal to **1**, but with the values at terminal states set equal to **0**:

```
def sarsa(alpha = 0.02, \
          gamma = 1., \
          epsilon_start = 0.1,\
          epsilon_end = 0.001,\
          epsilon_annealing_stop = int(nb_episodes/2),\
          q = None, \
          progress = None, \
          env=env):

    if q is None:
        q = np.ones((16,4))
        # Set q(terminal,*) equal to 0
        q[5,:] = 0.0
        q[7,:] = 0.0
```

```
q[11,:] = 0.0
q[12,:] = 0.0
q[15,:] = 0.0
```

12. Start a loop among all episodes:

```
for i in range(nb_episodes):
```

13. Inside the loop, first, define the epsilon value, depending on the current episode number. Reset the environment and make sure that the first action is chosen with an ε-greedy policy:

```
inew = min(i,epsilon_annealing_stop)
epsilon = (epsilon_start \
            * (epsilon_annealing_stop - inew)\
            + epsilon_end * inew) \
            / epsilon_annealing_stop

done = False
s = env.reset()
a = action_epsilon_greedy(q, s, epsilon=epsilon)
```

14. Then, start an in-episode loop:

```
while not done:
```

15. Inside the loop, step throughout the environment using the selected action and ensure that the new state, the reward, and the done conditions are retrieved:

```
new_s, reward, done, info = env.step(a)
```

16. Select a new action with the ε-greedy policy, update the Q-table with the SARSA TD(0) rule, and ensure that the state and action are updated with their new values:

```
new_a = action_epsilon_greedy\
            (q, new_s, epsilon=epsilon)
q[s, a] = q[s, a] + alpha \
            * (reward + gamma \
            * q[new_s, new_a] - q[s, a])
s = new_s
a = new_a
```

17. Finally, estimate the agent's average performance:

```
        if progress is not None and i%STEPS == 0:
            progress[i//STEPS] = average_performance\
                                (get_action_epsilon_greedy\
                                (epsilon), q=q)
    return q, progress
```

It may be useful to provide a brief description of the ε parameter's decrease. It is ruled by three parameters: starting value, minimum value, and decrease range. They are used in the following way: ε starts at the starting value, and then it is decreased linearly across the number of episodes defined by the parameter's "range" until it reaches the minimum value, which is then kept constant.

18. Define an array that will collect all agent performance evaluations during training and the execution of SARSA TD(0) training:

```
sarsa_performance = np.ndarray(nb_episodes//STEPS)
q, sarsa_performance = sarsa(alpha = 0.02, gamma = 1,\
                            progress=sarsa_performance, \
                            epsilon_start=epsilon_param[0][0],\
                            epsilon_end=epsilon_param[0][1], \
                            epsilon_annealing_stop = \
                            epsilon_param[0][2])
```

19. Plot the SARSA agent's mean reward history during training:

```
plt.plot(STEPS*np.arange(nb_episodes//STEPS), sarsa_performance)
plt.xlabel("Epochs")
plt.title("Learning progress for SARSA")
plt.ylabel("Average reward of an epoch")
```

This generates the following output, showing the learning progress for the SARSA algorithm:

```
Text(0, 0.5, 'Average reward of an epoch')
```

The plot will be as follows:

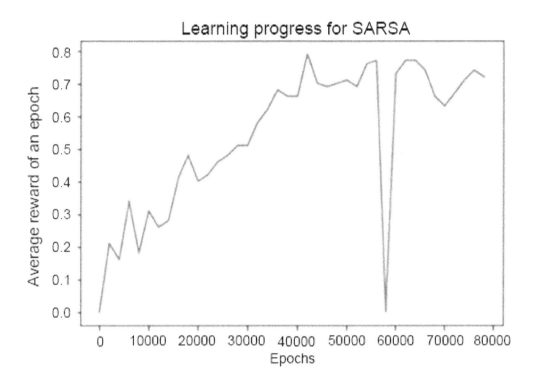

Figure 7.15: Average reward of an epoch trend over training epochs

This plot clearly shows us how the performance of the SARSA algorithm improves over epochs, even when stochastic dynamics are considered. The sudden performance drop around 60k epochs is completely normal when dealing with methods in which random exploration plays a major role, and especially when random transition dynamics are part of the environment, as in this case.

20. Evaluate the greedy policy's performance regarding the trained agent (Q-table):

```
greedyPolicyAvgPerf = average_performance(greedy_policy, q=q)
print("Greedy policy SARSA performance =", greedyPolicyAvgPerf)
```

The output will be as follows:

```
Greedy policy SARSA performance = 0.75
```

21. Display the Q-table values:

```
q = np.round(q,3)
print("(A,S) Value function =", q.shape)
print("First row")
print(q[0:4,:])
print("Second row")
print(q[4:8,:])
print("Third row")
print(q[8:12,:])
print("Fourth row")
print(q[12:16,:])
```

The following output will be generated:

```
(A,S) Value function = (16, 4)

First row
[[0.829 0.781 0.785 0.785]
 [0.416 0.394 0.347 0.816]
 [0.522 0.521 0.511 0.813]
 [0.376 0.327 0.378 0.811]]

Second row
[[0.83  0.552 0.568 0.549]
 [0.    0.    0.    0.   ]
 [0.32  0.195 0.535 0.142]
 [0.    0.    0.    0.   ]]

Third row
[[0.55  0.59  0.546 0.831]
 [0.557 0.83  0.441 0.506]
 [0.776 0.56  0.397 0.342]
 [0.    0.    0.    0.   ]]

Fourth row
[[0.    0.    0.    0.   ]
 [0.528 0.619 0.886 0.506]
 [0.814 0.943 0.877 0.844]
 [0.    0.    0.    0.   ]]
```

This output shows the values of the complete state-action value function for our problem. These values are then used to generate the optimal policy by means of the greedy selection rule.

22. Print out the greedy policy that was found and compare it with the optimal policy. Having calculated the state-action value function, we are able to retrieve the greedy policy from it. In fact, as explained previously, the greedy policy chooses the action that, for a given state, is associated with the maximum value of the Q-table. For this purpose, we are using the **argmax** function. When applied to each of the 16 states (from 0 to 15), it returns the index of the action that, among the four available (from 0 to 3), has the highest associated value for that state. Here, we also directly output the label associated with the action index using the pre-built dictionary:

```
policyFound = [actionsDict[np.argmax(q[0,:])],\
               actionsDict[np.argmax(q[1,:])],\
               actionsDict[np.argmax(q[2,:])],\
               actionsDict[np.argmax(q[3,:])],\
               actionsDict[np.argmax(q[4,:])],\
               "  -  ",\
               actionsDict[np.argmax(q[6,:])],\
               "  -  ",\
               actionsDict[np.argmax(q[8,:])],\
               actionsDict[np.argmax(q[9,:])],\
               actionsDict[np.argmax(q[10,:])],\
               "  -  ",\
               "  -  ",\
               actionsDict[np.argmax(q[13,:])],\
               actionsDict[np.argmax(q[14,:])],\
               "  !  "]

print("Greedy policy found:")
idxs = [0,4,8,12]
for idx in idxs:
    print(policyFound[idx+0], policyFound[idx+1], \
          policyFound[idx+2], policyFound[idx+3])

print(" ")

print("Optimal policy:")
idxs = [0,4,8,12]
```

```
for idx in idxs:
    print(optimalPolicy[idx+0], optimalPolicy[idx+1], \
        optimalPolicy[idx+2], optimalPolicy[idx+3])
```

The output will be as follows:

```
Greedy policy found:
    L    U    U    U
    L    -    R    -
    U    D    L    -
    -    R    D    !


Optimal policy:
  L/R/D  U    U    U
    L    -   L/R   -
    U    D    L    -
    -    R    D    !
```

As you can see, as in the previous exercise, our algorithm has been able to find the optimal policy by simply exploring the environment, and even in the context of stochastic environment transitions. As anticipated, for this setting, it is not possible to achieve the maximum reward 100% of the time. In fact, as we can see, for every state of the environment the greedy policy obtained with the Q-table that is calculated by our algorithm, it prescribes an action that is in accordance with the optimal policy that was defined by analyzing the environment problem. As we already saw, there are two states in which there are many different actions that are equally optimal, and the agent correctly implements one of them.

> **NOTE**
>
> To access the source code for this specific section, please refer to https://packt.live/3eicsGr.
>
> You can also run this example online at https://packt.live/2Z4L1JV.

Now that we've become familiar with on-policy control, it is time for us to change track and look at off-policy control, an early breakthrough in reinforcement learning dating back to 1989 known as Q-learning.

> **NOTE**
>
> The Q-learning algorithm was first formulated by *Watkins in Mach Learn 8, 279–292 (1992)*. Here, we are presenting only an intuitive understanding, along with a brief mathematical description of it. For a much more detailed mathematical discussion, please refer to the original paper at https://link.springer.com/article/10.1007/BF00992698.

Q-LEARNING – OFF-POLICY CONTROL

Q-learning is a name that identifies the family of off-policy control temporal difference algorithms. From a mathematical/implementation point of view, the only difference compared with on-policy algorithms is in the rule used to update the Q-table (or function for approximated methods), which is defined as follows:

$$Q(S_t, A_t) \leftarrow Q(S_t, A_t) + \alpha \left[R_{t+1} + \gamma \max_a Q(S_{t+1}, a) - Q(S_t, A_t) \right]$$

Figure 7.16: Function for approximated methods

The key point regards how the action for the next state, (S_{t+1}), is chosen. In fact, choosing the action with the maximum state-action value directly approximates what happens when the optimal Q value is found and the optimal policy is followed. Moreover, it is independent of the policy used to collect experience while interacting with the environment. The exploration policy can be entirely different to the optimal one; for example, it can be an ε-greedy policy to encourage exploration, and, under some easy-to-satisfy assumptions, it has been proven that Q converges to the optimal values.

In *Chapter 9, What Is Deep Q-Learning?*, you will look at the extension of this approach to non-tabular methods where we use deep neural networks as function approximators. This method is called deep Q-learning. A scheme for the Q-learning control algorithm can be depicted as follows:

1. Choose the algorithm parameters: the step size, α, which has to be contained in the interval (0, 1], and the **ε** parameter of the ε-greedy policy, which has to be small and greater than **0** since it represents the probability of choosing the non-optimal action in order to favor exploration:

```
alpha = 0.02
epsilon_expl = 0.2
```

2. Initialize $Q(s, a)$, for all $s \in S^+$, $a \in A(s)$, arbitrarily, except that Q(terminal, *) = 0:

```
q = np.ones((16, 4))
# Set q(terminal,*) equal to 0
q[5,:] = 0.0
q[7,:] = 0.0
q[11,:] = 0.0
q[12,:] = 0.0
q[15,:] = 0.0
```

3. Create a loop among all episodes. In the loop, initialize **s**:

```
for i in range(nb_episodes):

    done = False
    s = env.reset()
```

4. Create a loop for each step of the episode. Within that loop, choose A from S using the policy derived from Q (for example, ε-greedy):

```
while not done:
    # behavior policy
    a = action_epsilon_greedy(q, s, epsilon=epsilon_expl)
```

5. Taking action, A, observe R, S'. Update the state-action value function for the selected state-action pair using the Q-learning rule, which defines the new value as the sum of the current one, plus the off-policy-specific TD error multiplied by the step size, α. This can be expressed as follows:

$$Q(S, A) \leftarrow Q(S, A) + \alpha \left[R + \gamma \max_a Q(S', a) - Q(S, A) \right]; S \leftarrow S'; \text{ until } S \text{ is terminal}$$

Figure 7.17: Expression for the updated state-action value function

The preceding explanation translates into code as follows:

```
new_s, reward, done, info = env.step(a)
a_max = np.argmax(q[new_s]) # estimation policy
q[s, a] = q[s, a] + alpha \
        * (reward + gamma \
            * q[new_s, a_max] -q[s, a])
s = new_s
```

As we can see, we just substituted the random choice of the action to be taken on the new state with the action associated with the maximum q-value. This (apparently) minor change, which can be easily implemented by adapting the SARSA algorithm, has a relevant impact on the nature of the method. We'll see it at work in the following exercise.

EXERCISE 7.03: USING TD(0) Q-LEARNING TO SOLVE FROZENLAKE-V0 DETERMINISTIC TRANSITIONS

In this exercise, we'll implement the TD(0) Q-learning algorithm to solve the FrozenLake-v0 environment, where only deterministic transitions are allowed. In this exercise, we will consider the same task of retrieving the frisbee with the optimal policy we addressed in *Exercise 7.01, Using TD(0) SARSA to Solve FrozenLake-v0 Deterministic Transitions*, but this time, instead of using the SARSA algorithm (on-policy), we will implement Q-learning (off-policy). We will see how this algorithm behaves and train ourselves in implementing a new approach to estimate a q-value table by means of recovering an optimal policy for our agent.

Follow these steps to complete this exercise:

1. Import the required modules, as follows:

```
import numpy as np
import matplotlib.pyplot as plt
%matplotlib inline

import gym
```

2. Instantiate the **gym** environment called **FrozenLake-v0** using the **is_slippery** flag set to **False** in order to disable stochasticity:

```
env = gym.make('FrozenLake-v0', is_slippery=False)
```

3. Take a look at the action and the observation spaces:

```
print("Action space = ", env.action_space)
print("Observation space = ", env.observation_space)
```

The output will be as follows:

```
Action space =  Discrete(4)
Observation space =  Discrete(16)
```

4. Create two dictionaries to easily translate the **actions** numbers into moves:

```
actionsDict = {}
actionsDict[0] = " L "
actionsDict[1] = " D "
actionsDict[2] = " R "
actionsDict[3] = " U "

actionsDictInv = {}
actionsDictInv["L"] = 0
actionsDictInv["D"] = 1
actionsDictInv["R"] = 2
actionsDictInv["U"] = 3
```

5. Reset the environment and render it to take a look at the grid problem:

```
env.reset()
env.render()
```

The output will be as follows:

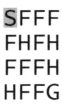

Figure 7.18: Environment's initial state

6. Visualize the optimal policy for this environment:

```
optimalPolicy = ["R/D"," R "," D "," L ",\
                 " D "," - "," D "," - ",\
                 " R ","R/D"," D "," - ",\
                 " - "," R "," R "," ! ",]

print("Optimal policy:")
idxs = [0,4,8,12]
for idx in idxs:
    print(optimalPolicy[idx+0], optimalPolicy[idx+1], \
          optimalPolicy[idx+2], optimalPolicy[idx+3])
```

The output will be as follows:

```
Optimal policy:
R/D  R   D  L
 D   -   D  -
 R  R/D  D  -
 -   R   R  !
```

This represents the optimal policy for this environment and shows, for each of the environment states represented in the 4x4 grid, the optimal action among the four available: move Up, move Down, move Right, and move Left. Except for two states, all the others have a single optimal action associated with them. In fact, as described previously, optimal actions here are those that bring the agent to the goal in the shortest possible path. Two different possibilities result in the same path length for two states, so they are both equally optimal.

7. Next, define functions that will take ε-greedy actions:

```
def action_epsilon_greedy(q, s, epsilon=0.05):
    if np.random.rand() > epsilon:
        return np.argmax(q[s])
    return np.random.randint(4)
```

8. Define a function that will take greedy actions:

```
def greedy_policy(q, s):
    return np.argmax(q[s])
```

9. Define a function that will calculate the mean of the agent's performance:

```
def average_performance(policy_fct, q):
    acc_returns = 0.
    n = 500
    for i in range(n):
        done = False
        s = env.reset()
        while not done:
            a = policy_fct(q, s)
            s, reward, done, info = env.step(a)
            acc_returns += reward
    return acc_returns/n
```

10. Initialize the Q-table so that all the values equal **1**, except for the values at terminal states:

```
q = np.ones((16, 4))
# Set q(terminal,*) equal to 0
q[5,:] = 0.0
q[7,:] = 0.0
q[11,:] = 0.0
q[12,:] = 0.0
q[15,:] = 0.0
```

11. Set the number of total episodes, the number of steps representing the interval by which we evaluate the agent's average performance, the learning rate, the discounting factor, and the **ε** value for the exploration policy and define an array to collect all agent performance evaluations during training:

```
nb_episodes = 40000
STEPS = 2000
alpha = 0.02
gamma = 0.9
epsilon_expl = 0.2


q_performance = np.ndarray(nb_episodes//STEPS)
```

12. Train the agent using the Q-learning algorithm: the external loop takes care of generating the desired number of episodes. Then, the in-episode loop completes the following steps: first, it selects an exploration action with an ε-greedy policy, then the environment is stepped with the selected exploration action, and the **new_s**, **reward**, and **done** condition are retrieved. The new action for the new state is selected with the greedy policy, the Q-table is updated with the Q-learning TD(0) rule, and the state is updated with the new value. Every predefined number of steps, the agent's average performance is estimated:

```
for i in range(nb_episodes):

    done = False
    s = env.reset()
    while not done:
        # behavior policy
        a = action_epsilon_greedy(q, s, epsilon=epsilon_expl)
        new_s, reward, done, info = env.step(a)
        a_max = np.argmax(q[new_s]) # estimation policy
        q[s, a] = q[s, a] + alpha \
                   * (reward + gamma \
                       * q[new_s, a_max] - q[s, a])
        s = new_s

    # for plotting the performance
    if i%STEPS == 0:
        q_performance[i//STEPS] = average_performance\
                                    (greedy_policy, q)
```

13. Plot the Q-learning agent's mean reward history during training:

```
plt.plot(STEPS * np.arange(nb_episodes//STEPS), q_performance)
plt.xlabel("Epochs")
plt.ylabel("Average reward of an epoch")
plt.title("Learning progress for Q-Learning")
```

This generates the following output, showing the learning progress of the Q-learning algorithm:

```
Text(0.5, 1.0, 'Learning progress for Q-Learning')
```

The plot will be as follows:

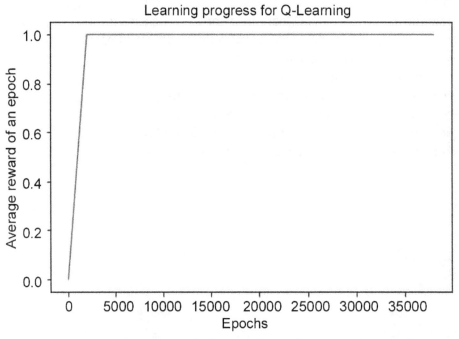

Figure 7.19: Average reward of an epoch trend over training epochs

As we can see, the plot shows how quickly Q-learning performance grows over epochs as the agent collects more and more experience. It also demonstrates that the algorithm is capable of reaching 100% success after learning. It's also evident how, in this case, compared to the SARSA method, the measured algorithm performance increases steadily and much faster.

14. Evaluate the greedy policy's performance of the trained agent (Q-table):

```
greedyPolicyAvgPerf = average_performance(greedy_policy, q=q)
print("Greedy policy Q-learning performance =", \
      greedyPolicyAvgPerf)
```

The output will be as follows:

```
Greedy policy Q-learning performance = 1.0
```

15. Display the Q-table values:

```
q = np.round(q,3)
print("(A,S) Value function =", q.shape)
print("First row")
print(q[0:4,:])
print("Second row")
print(q[4:8,:])
print("Third row")
print(q[8:12,:])
print("Fourth row")
print(q[12:16,:])
```

The following output will be generated:

```
(A,S) Value function = (16, 4)

First row
[[0.531 0.59  0.59  0.531]
 [0.617 0.372 0.656 0.628]
 [0.672 0.729 0.694 0.697]
 [0.703 0.695 0.703 0.703]]

Second row
[[0.59  0.656 0.    0.531]
 [0.    0.    0.    0.   ]
 [0.455 0.81  0.474 0.754]
 [0.    0.    0.    0.   ]]

Third row
[[0.656 0.    0.729 0.59 ]
 [0.656 0.81  0.81  0.   ]
 [0.778 0.9   0.286 0.777]
```

```
   [0.     0.     0.     0.    ]]

Fourth row
[[0.     0.     0.     0.    ]
 [0.     0.81   0.9    0.729]
 [0.81   0.9    1.     0.81  ]
 [0.     0.     0.     0.    ]]
```

This output shows the values of the complete state-action value function for our problem. These values are then used to generate the optimal policy by means of the greedy selection rule.

16. Print out the greedy policy that was found and compare it with the optimal policy:

```
policyFound = [actionsDict[np.argmax(q[0,:])],\
               actionsDict[np.argmax(q[1,:])],\
               actionsDict[np.argmax(q[2,:])],\
               actionsDict[np.argmax(q[3,:])],\
               actionsDict[np.argmax(q[4,:])],\
               " - ",\
               actionsDict[np.argmax(q[6,:])],\
               " - ",\
               actionsDict[np.argmax(q[8,:])],\
               actionsDict[np.argmax(q[9,:])],\
               actionsDict[np.argmax(q[10,:])],\
               " - ",\
               " - ",\
               actionsDict[np.argmax(q[13,:])],\
               actionsDict[np.argmax(q[14,:])],\
               " ! "]

print("Greedy policy found:")
idxs = [0,4,8,12]
for idx in idxs:
    print(policyFound[idx+0], policyFound[idx+1], \
          policyFound[idx+2], policyFound[idx+3])
```

```
print(" ")

print("Optimal policy:")
idxs = [0,4,8,12]
for idx in idxs:
    print(optimalPolicy[idx+0], optimalPolicy[idx+1], \
          optimalPolicy[idx+2], optimalPolicy[idx+3])
```

The output will be as follows:

```
Greedy policy found:
   D    R    D    L
   D    -    D    -
   R    D    D    -
   -    R    R    !


Optimal policy:
 R/D    R    D    L
   D    -    D    -
   R  R/D    D    -
   -    R    R    !
```

As these outputs demonstrate, the Q-learning algorithm has been able to retrieve the optimal policy too, just like SARSA did in *Exercise 07.01, Using TD(0) SARSA to solve FrozenLake-v0 Deterministic Transitions*, only by means of experience and interaction with the environment.

As we can see, for every state of the grid world the greedy policy obtained with the Q-table calculated by our algorithm, this prescribes an action that is in accordance with the optimal policy that was defined by analyzing the environment problem. As we already saw, there are two states in which there are many different actions that are equally optimal, and the agent correctly implements one of them.

> **NOTE**
>
> To access the source code for this specific section, please refer to https://packt.live/2AUIzym.
>
> You can also run this example online at https://packt.live/3fJCnH5.

As for SARSA, it would be interesting to see how Q-learning behaves if we turn on stochastic transitions. This will be the goal of the activity at the end of this chapter. The procedure that the two algorithms follow is the very same one we adopted with SARSA: the same Q-learning algorithm used for the deterministic transition case is applied, and you are expected to adapt hyperparameters (especially the discount factor and the number of episodes) until you obtain convergence to the optimal policy under the stochastic transition dynamics.

To complete the landscape of TD(0) algorithms, we will introduce another specific approach that's obtained by applying very simple modifications of the previous ones: Expected SARSA.

EXPECTED SARSA

Now, let's consider a learning algorithm that is quite similar to Q-learning, with the only difference being the substitution of the maximum over next state-action pairs with the expected value. This is computed by taking into account the probability of each action under the current policy. This modified algorithm can be represented by the following update rule:

$$Q(S_t, A_t) \leftarrow Q(S_t, A_t) + \alpha \left[R_{t+1} + \gamma \sum_a \pi(a|S_{t+1}) Q(S_{t+1}, a) - Q(S_t, A_t) \right]$$

Figure 7.20: State-action value function update rule

The additional computational complexity with respect to SARSA provides the advantage of eliminating variance due to the random selection of A_{t+1}, which is a very powerful trick for improving learning and robustness considerably. It can be used both in an on-policy and off-policy fashion, thus becoming an abstraction of both SARSA and Q-learning with, in general, a performance that dominates both of them. An example of the update rule's implementation is provided in the following snippet:

```
q[s, a] = q[s, a] + alpha * (reward + gamma *
    (np.dot(pi[new_s, :],q[new_s, :]) - q[s, a])
```

In the preceding code, the **pi** variable contains all the probabilities for each action in each state. The dot product involving **pi** and **q** is the operation needed to compute the expected value for the new state, taking into account all the actions for that state with their respective probabilities.

Now that we've studied the TD(0) methods, let's start learning about the N-step TD and TD(λ) algorithms.

N-STEP TD AND TD(λ) ALGORITHMS

In the previous chapter, we looked at Monte Carlo methods, while in the previous sections of this chapter, we learned about TD(0) ones, which, as we will discover soon, are also known as one-step temporal difference methods. In this section, we'll unify them: in fact, they are at the extreme of a spectrum of algorithms (TD(0) on one side, with MC methods at the other end), and often, the best performing methods are somewhere in the middle of this spectrum.

N-step temporal difference algorithms extend one-step TD methods. More specifically, they generalize Monte Carlo and TD approaches, making it possible to smoothly transition between the two. As we already saw, MC methods must wait until the episode finishes to back the reward up into the previous states. One-step TD methods, on the other hand, make direct use of the first available future step to bootstrap and start updating the value function of states or state-action pairs. These extremes are rarely the optimal choices. The optimal choices generally fall in the middle of this broad range. Using N-step methods allows us to adjust the number of steps to consider when updating the value function, thereby distributing the bootstrapping approach to multiple steps.

A similar notion can be recalled in the context of eligibility traces, but they are more general, allowing us to distribute and spread bootstrapping over multiple time intervals at the same time. These two topics will be treated separately for clarity and, so as to enable you to build your knowledge incrementally, we will start with N-step methods first, before moving on to eligibility traces.

N-STEP TD

As we have already seen for one-step TD methods, the first step to approaching the N-step method is to focus on the estimation of the state-value function for sample episodes generated using the policy, π. We already recalled that the Monte Carlo algorithm must wait until the end of an episode before performing an update by using the entire sequence of rewards from a given state. On the other hand, one-step methods just need the next reward. N-step methods use an intermediate rule: instead of relying on just the next reward, or on all future rewards until the episode ends, they use a value in between these two. For example, a three-steps update would use the first three rewards and the estimated state value reached three steps ahead. This can be formalized for a generic number of steps.

This approach gives birth to a family of methods that are still temporal difference ones since they use the N-steps that were encountered after the target state to update its value. It is clear that the methods that we encountered at the beginning of this chapter are a special case of N-step methods. For this reason, they are called "one-step TD methods."

In order to define them more formally, we can consider the estimated value of the state, S_T as a result of the state-reward sequence, $S_t, R_{t+1}, S_{t+1}, R_{t+2}, ..., R_T, S_T$ (except the actions). In MC methods, this estimate is updated only once the episode is complete, and in one-step methods, right after the next step. In N-step methods, on the other hand, the state-value estimate is updated after N-steps using a quantity that discounts n future rewards and the value of the state encountered after N-steps in the future. This quantity, called N-step return, can be defined in an expression, as follows:

$$G_{t:t+n} \overset{def}{=} R_{t+1} + \gamma R_{t+2} + \cdots + \gamma^{n-1} R_{t+n} + \gamma^n V_{t+n-1}\left(S_{t+n}\right)$$

Figure 7.21: N-step return equation (with the state-value function)

A key point to note here is that in order to calculate this N-step return, we have to wait to reach the time **t+1** so that all the terms in the equation are available. By using the N-step return, it is straightforward to formalize the state-value function update rule, as follows:

$$V_{t+n}\left(S_t\right) \overset{def}{=} V_{t+n-1}\left(S_t\right) + \alpha\left[G_{t:t+n} - V_{t+n-1}\left(S_t\right)\right], \qquad 0 \le t < T$$

Figure 7.22: Expression for the natural state-value learning algorithm
for using N-step returns

Note that the values of all the other states remain unchanged, as shown in the following expression:

$$V_{t+n}(S) = V_{t+n-1}(S), \text{ for all } s \ne S_t$$

Figure 7.23: Expression specifying that all the other values are kept constant

This is the equation that formalizes the N-step TD algorithm. It is worth noting again that no changes are made during the first **n−1** steps before we can estimate the N-step return. This needs to be compensated for at the end of the episode, when the remaining **n−1** updates are performed all at once after reaching the terminal state.

Similar to what we already saw for TD(0) methods, and without talking about this in too much data, the state-value function estimation of the N-step TD methods converges to the optimal value under appropriate technical conditions.

N-STEP SARSA

It is quite straightforward to extend the SARSA algorithm, which we looked at when we introduced one-step methods, to its N-step version. As we did previously, the only thing we need to do is substitute the state-action pairs for states in the value function's N-step return and in the update formulations just seen, coupling them with an ε-greedy policy. The definition of the N-step return (update targets) can be described by the following equation:

$$G_{t:t+n} \overset{def}{=} R_{t+1} + \gamma R_{t+2} + \cdots + \gamma^{n-1} R_{t+n} + \gamma^{n} Q_{t+n-1}(S_{t+n}, A_{t+n}), \quad n \geq 1, \quad 0 \leq t < T$$

Figure 7.24: N-step return equation (with the state-action value function)

Here, $G_{t:t+1} = G_t$ if $t + n \geq T$. The update rule for the state-action value function is expressed as follows:

$$Q(S_t, A_t) \overset{def}{=} Q_{t+n-1}(S_t, A_t) + \alpha \left[G_{t:t+n} - Q_{t+n-1}(S_t, A_t) \right], \quad 0 \leq t < T$$

Figure 7.25: Update rule for the state-action value function

Note that the values of all the other state-action pairs remain unchanged: $Q_{t+n}(s, a) = Q_{t+n-1}(s, a)$, for all values of **s**, so that $s \neq S_t$ or $a \neq A_t$. A scheme for the N-step SARSA control algorithm can be depicted as follows:

1. Choose the algorithm's parameters: the step size, α, which has to be contained in the interval (0, 1], and the **ε** parameter of the ε-greedy policy, which has to be small and greater than **0** since it represents the probability of choosing the non-optimal action to favor exploration. A value for the number of steps, **n**, has to be chosen. This can be done, for example, with the following code:

```
alpha = 0.02
n = 4
epsilon = 0.05
```

2. Initialize $Q(s, a)$, for all $s \in S^+$, $a \in A(s)$, arbitrarily, except that Q(terminal, ·) = 0:

```
q = np.ones((16,4))
```

3. Create a loop for each episode. Initialize and store the S0 ≠ terminal. Select and store an action using the ε-greedy policy and initialize time, **T**, as a very high value:

```
for i in range(nb_episodes):
    s = env.reset()
    a = action_epsilon_greedy(q, s, epsilon=epsilon)
    T = 1e6
```

4. Create a loop for t = 0, 1, 2, ... If t < T, then perform action A_t. Observe and store the next reward as R_{t+1} and the next state as S_{t+1} If S_{t+1} is terminal, then set **T** equal to **t+1**:

```
while True:
        new_s, reward, done, info = env.step(a)
        if done:
            T = t+1
```

5. If S_{t+1} is not terminal, select and store a new action for the new state:

```
        new_a = action_epsilon_greedy(q, new_s, epsilon=epsilon)
```

6. Define the time for which the estimate is being updated, **tau**, equal to **t−n+1**:

```
tau = t-n+1
```

7. If **tau** is greater than 0, calculate the N-step return by summing the discounted returns of the previous n steps and adding the discounted value of the next step-next action pair and update the state-action value function:

```
G = sum_n(q, tau, T, t, gamma, R, new_s, new_a)
q[s, a] = q[s, a] + alpha * (G- q[s, a])
```

With a few minor changes, this can easily be extended to accommodate Expected SARSA as well. As seen previously in this chapter, it only requires us to substitute the expected approximate value of the state using the estimated action values at time, t, under the target policy at the last step of the N-steps. When the state in question is terminal, its expected approximate value is defined as 0.

N-STEP OFF-POLICY LEARNING

To define off-policy learning for N-step methods, we will be taking very similar steps as the ones we did for one-step methods. The key point is that, as in all off-policy methods, we are learning the value function for a policy, π, while following a different exploration policy; say, **b**. Typically, π is the greedy policy for the current state-action value function estimate, and b has more randomness so that it effectively explores the environment; for example, ε-greedy. The main difference with respect to what we already saw for one-step off-policy methods is that now, we need to take into account the fact that we are selecting actions using a different policy than the one we want to learn, and we are doing it for more than one step. So, we need to properly weigh the selected actions measuring the relative probability under the two policies of taking those actions.

By means of this correction, it is possible to define the rule for a simple off-policy version of N-step TD: the update for time **t** (actually made at time **t + n**) can simply be weighted by $\rho_{t:t+n-1}$:

$$V_{t+n}\left(S_t\right) \overset{def}{=} V_{t+n-1}\left(S_t\right) + \alpha\rho_{t:t+n-1}\left[G_{t:t+n} - V_{t+n-1}\left(S_t\right)\right], \quad 0 \le t < T$$

Figure 7.26: N-step TD off-policy update rule at time 't'

Here, **V** is the value function, α is the step size, **G** is the N-step return, and $\rho_{t:t+n-1}$ is called the importance sampling ratio. The importance sampling ratio is the relative probability under the two policies of taking **n** actions from A_t to A_{t+n-1}, which can be expressed as follows:

$$\rho_{t:h} \overset{def}{=} \prod_{k=t}^{min(h,T-1)} \frac{\pi\left(A_k|S_k\right)}{b\left(A_k|S_k\right)}$$

Figure 7.27: Sampling ratio equation

Here, π is the agent policy, b is the exploration policy, A is the action, and S is the state.

By this definition, it is evident that actions that would never be selected under the policy we want to learn (that is, their probability is **0**) would be ignored (weight equal to 0). If, on the other hand, an action under the policy we are learning has more probability with respect to the exploratory policy, the weight assigned to it should be higher than **1** since it will be encountered more often. It is also evident that for the on-policy case, the sampling ratio is always equal to **1**, given the fact that π and b are the same policy. For this reason, the N-step SARSA on-policy update can be seen as a special case of the off-policy update. The general form of the update, from which it is possible to derive both on-policy and off-policy methods, is as follows:

$$Q_{t+n}(S_t, A_t) \overset{def}{=} Q_{t+n-1}(S_t, A_t) + \alpha \rho_{t+1:t+n}\left[G_{t:t+n} - Q_{t+n-1}(S_t, A_t)\right]; \text{for } 0 \le t < T$$

Figure 7.28: State-action value function for the off-policy N-step TD algorithm

As you can see, Q is the state-action value function, α is the step size, G is the N-step return, and $\rho_{t:t+n-1}$ is the importance sampling ratio. The scheme for the full algorithm is as follows:

1. Select an arbitrary behavior policy, b, so that the probability for each action of each state is greater than 0 for all states and actions. Choose the algorithm parameters: the step size, α, which has to be contained in the interval (0, 1], and a value for the number of steps, **n**. This can be done, for example, with the following code:

```
alpha = 0.02
n = 4
```

2. Initialize $Q(s, a)$, for all $s \in S^+$, $a \in A(s)$, arbitrarily, except that Q(terminal, ·) = 0:

```
q = np.ones((16,4))
```

3. Initialize the policy, π, to be greedy with respect to Q, or to a fixed given policy. Create a loop for each episode. Initialize and store the $S_0 \neq$ terminal. Select and store an action using the b policy and initialize time, **T**, as a very high value:

```
for i in range(nb_episodes):
        s = env.reset()
        a = action_b_policy(q, s)
        T = 1e6
```

4. Create a loop for t = 0, 1, 2, ... If t < T, then perform action A_t. Observe and store the next reward as R_{t+1} and the next state as S_{t+1}. If S_{t+1} is terminal, then set **T** equal to **t+1**:

```
while True:
            new_s, reward, done, info = env.step(a)
            if done:
                T = t+1
```

5. If S_{t+1} is not terminal, select and store a new action for the new state:

```
            new_a = action_b_policy(q, new_s)
```

6. Define the time for which the estimate is being updated, **tau**, equal to **t−n+1**:

```
tau = t-n+1
```

7. If **tau** is greater than or equal to **0**, calculate the sampling ratio. Calculate the N-step return by summing the discounted returns of the previous n steps and adding the discounted value of the next step-next action pair and update the state-action value function:

```
rho = product_n(q, tau, T, t, R, new_s, new_a)
G = sum_n(q, tau, T, t, gamma, R, new_s, new_a)
q[s, a] = q[s, a] + alpha * rho * (G- q[s, a])
```

Now that we've studied the N-step methods, it is time to proceed to the most general and most performant declination of temporal difference methods, TD(λ).

TD(λ)

The popular TD(λ) algorithm is a temporal difference algorithm that makes use of the eligibility trace concept, which, as we will soon see, is a procedure that allows us to appropriately weight contributions to the state's (or state-action pair's) value function using any possible number of steps. The **lambda** term introduced in the name is a parameter that defines and parameterizes this family of algorithms. As we will see shortly, it is a weighting factor that will allow us to appropriately weight different contributing terms involved in the estimation of the algorithm's return.

It is possible to combine any temporal difference method, such as those we already saw (Q-learning and SARSA), with the eligibility traces concept, which we will implement shortly. This allows us to obtain a more general method, which is also more efficient. This approach, as we already anticipated previously, realizes the final unification and generalization of the TD and Monte Carlo methods. Similarly, regarding what we observed for N-step TD methods, in this case also, we have one-step TD methods on one extreme ($\lambda = 0$) and Monte Carlo methods on the other ($\lambda = 1$). The space between these two boundaries contains intermediate methods (as is the case for N-step methods with finite **n > 1**). In addition to that, eligibility traces allow us to use extended Monte Carlo methods for the so-called online implementation, meaning they become applicable to non-episodic problems.

With respect to what we already saw for N-step TD methods, eligibility traces have an additional advantage, allowing us to generalize these families with significant computational improvement. As we mentioned earlier, choosing the correct value of n for N-step methods can be anything but a straightforward task. Eligibility traces, on the other hand, allow us to "fuse" together the updates corresponding to different timesteps.

To achieve this goal, we need to define a method to weigh the N-step return, G_t^n, using a weight that decays exponentially with time. This is done by introducing a factor, $\lambda \in [0,1]$, and weighting the nth return with λ^{n-1}.

The goal is to define a weighted average so that all these weights must total to **1**. The normalization constant is the limit value of the convergent geometric series: $1/(1-\lambda)$. With this, we can define the so-called λ-return as follows:

$$G_t^\lambda = (1 - \lambda) \sum_{n=1}^{\infty} \lambda^{n-1} G_t^{(n)}$$

Figure 7.29: Expression for the lambda return

This equation defines how our choice of λ influences the speed at which a given return drops exponentially as a function of the number of steps.

We can now use this new return as a target for the state (or state-action pair) value function, thus creating a new value function update rule. It may seem that, at this point, in order to consider all contributes, we should wait until the end of the episode, thus collecting all future returns. This problem is solved by means of the second fundamental novelty introduced by eligibility traces: instead of looking forward in time, the point of view is reversed, and the agent updates all states (state-action pairs) visited in the past according to the eligibility traces rule and using current return and values information.

The eligibility trace is initialized equal to 0 for every state (or state-action pair), is incremented on each time step with a value equal to 1 for the state (or state-action pair), visited so that it gives it the highest weight in contributing to the value function update, and fades away by the $\lambda\gamma$ factor. This factor is the combination of decay in time that's typical of eligibility traces, as explained previously (λ), and the familiar reward discount (γ) we've encountered many times in this chapter. With this new concept, we can now build the new value function update. First, we have the equation that regulates the eligibility trace evolution:

$$E_0(s) \leftarrow 0 \quad \forall s \in S$$

$$E_t(s) \leftarrow \lambda\gamma E_{t-1}(s) + I(S_t = s) \quad \forall s \in S$$

Figure 7.30: Eligibility traces initialization and update rule at time 't' (for states)

Then, we have the new definition of the TD error (or δ). The state-value function update will be as follows:

$$\delta_t = R_{t+1} + \gamma V(S_{t+1}) - V(S_t)$$

$$V(s) \leftarrow V(s) + \alpha\delta_t E_t(s) \quad \forall s \in S$$

Figure 7.31: State-value function update rule using eligibility traces

Now, let's see how this idea is implemented in the SARSA algorithm to obtain an on-policy TD control algorithm with eligibility traces.

SARSA(λ)

Directly translating a state-value update into a state-action-value update allows us to add the eligibility traces feature to our previously seen SARSA algorithm. The eligibility trace equation can be modified as follows:

$$E_0(s) \leftarrow 0 \qquad \forall s \in S, \forall a \in A$$

$$E_t(s, a) \leftarrow \lambda \gamma E_{t-1}(s, a) + I(S_t = s, A_t = a) \qquad \forall s \in S, \forall a \in A$$

Figure 7.32: Eligibility trace initialization and update rule at time 't' (for state-actions pairs)

The TD error and the state-action value function updates are written as follows:

$$\delta_t = R_{t+1} + \gamma V(S_{t+1}, A_{t+1}) - V(S_t, A_t)$$

$$V(s, a) \leftarrow V(s, a) + \alpha \delta_t E_t(s, a) \qquad \forall s \in S \forall a \in A$$

Figure 7.33: State-action pair's value function update rule using eligibility traces

A schema that perfectly summarizes all these steps and presents the complete algorithm is as follows:

1. Choose the algorithm's parameters: the step size α, which has to be contained in the interval (0, 1], and the **ε** parameter of the ε-greedy policy, which has to be small and greater than 0, since it represents the probability of choosing the non-optimal action, to favor exploration. A value for the **lambda** parameter has to be chosen. This can be done, for example, with the following code:

```
alpha = 0.02
lambda = 0.3
epsilon = 0.05
```

2. Initialize $Q(s, a)$, for all $s \in S^+$, $a \in A(s)$, arbitrarily, except that Q(terminal, ·) = 0:

```
q = np.ones((16,4))
```

3. Create a loop for each episode. Initialize the eligibility traces table to **0**:

```
E = np.zeros((16, 4))
```

4. Initialize the state as it is not terminal and select an action using the ε-greedy policy. Then, initiate the in-episode loop:

```
state = env.reset()
action = action_epsilon_greedy(q, state, epsilon)

while True:
```

5. Create a loop for each step of the episode, update the eligibility traces, and assign a value equal to **1** to the last visited state:

```
E = eligibility_decay * gamma * E
E[state, action] += 1
```

6. Step through the environment and choose the next action using the ε-greedy policy:

```
new_state, reward, done, info = env.step(action)
new_action = action_epsilon_greedy\
        (q, new_state, epsilon)
```

7. Calculate the δ update and update the Q-table using the SARSA TD(λ) rule:

```
delta = reward + gamma \
        * q[new_state, new_action] - q[state, action]
q = q + alpha * delta * E
```

8. Update the state and action with new state and action values:

```
state, action = new_state, new_action

if done:
    break
```

We are now ready to test this new algorithm on the environment we already solved with one-step SARSA and Q-learning.

EXERCISE 7.04: USING TD(λ) SARSA TO SOLVE FROZENLAKE-V0 DETERMINISTIC TRANSITIONS

In this exercise, we will implement our SARSA(λ) algorithm to solve the FrozenLake-v0 environment under the deterministic environment dynamics. In this exercise, we will consider the same task we addressed in *Exercise 7.01, Using TD(0) SARSA to Solve FrozenLake-v0 Deterministic Transitions*, and *Exercise 7.03, Using TD(0) Q-Learning to Solve FrozenLake-v0 Deterministic Transitions*, but this time, instead of using one-step TD methods such as SARSA (on-policy) and Q-learning (off-policy), we will implement TD(λ), a temporal difference method coupled with the power of eligibility traces. We will see how this algorithm behaves and train ourselves in implementing a new approach to estimate a Q-value table by means of which we'll recover an optimal policy for our agent.

Follow these steps to complete this exercise:

1. Import the required modules:

```
import numpy as np
from numpy.random import random, choice
import matplotlib.pyplot as plt
%matplotlib inline

import gym
```

2. Instantiate the **gym** environment called **FrozenLake-v0** using the **is_slippery** flag set to **False** in order to disable stochasticity:

```
env = gym.make('FrozenLake-v0', is_slippery=False)
```

3. Take a look at the action and the observation spaces:

```
print("Action space = ", env.action_space)
print("Observation space = ", env.observation_space)
```

The output will be as follows:

```
Action space =  Discrete(4)
Observation space =  Discrete(16)
```

4. Create two dictionaries to easily translate the **actions** numbers into moves:

```
actionsDict = {}
actionsDict[0] = " L "
actionsDict[1] = " D "
actionsDict[2] = " R "
actionsDict[3] = " U "

actionsDictInv = {}
actionsDictInv["L"] = 0
actionsDictInv["D"] = 1
actionsDictInv["R"] = 2
actionsDictInv["U"] = 3
```

5. Reset the environment and render it to take a look at the grid:

```
env.reset()
env.render()
```

The output will be as follows:

Figure 7.34: Environment's initial state

6. Visualize the optimal policy for this environment:

```
optimalPolicy = ["R/D"," R "," D "," L ",\
                 " D "," - "," D "," - ",\
                 " R ","R/D"," D "," - ",\
                 " - "," R "," R "," ! ",]

print("Optimal policy:")
idxs = [0,4,8,12]
for idx in idxs:
    print(optimalPolicy[idx+0], optimalPolicy[idx+1], \
          optimalPolicy[idx+2], optimalPolicy[idx+3])
```

The output will be printed as follows:

```
Optimal policy:
R/D  R    D   L
 D   -    D   -
 R  R/D   D   -
 -   R    R   !
```

This is the optimal policy for the deterministic case we already encountered when dealing with one-step TD methods. It shows the optimal actions we hope our agent will learn within this environment.

7. Define the functions that will take ε-greedy actions:

```python
def action_epsilon_greedy(q, s, epsilon=0.05):
    if np.random.rand() > epsilon:
        return np.argmax(q[s])
    return np.random.randint(4)


def get_action_epsilon_greedy(epsilon):
    return lambda q,s: action_epsilon_greedy\
                    (q, s, epsilon=epsilon)
```

8. Define a function that will take greedy actions:

```python
def greedy_policy(q, s):
    return np.argmax(q[s])
```

9. Define a function that will calculate the agent's average performance:

```python
def average_performance(policy_fct, q):
    acc_returns = 0.
    n = 500
    for i in range(n):
        done = False
        s = env.reset()
        while not done:
            a = policy_fct(q, s)
            s, reward, done, info = env.step(a)
            acc_returns += reward
    return acc_returns/n
```

10. Set the number of total episodes, the number of steps representing the interval by which we evaluate the agent's average performance, the discount factor, the learning rate, and the **ε** parameters ruling its decrease – the starting value, minimum value, and range (in terms of the number of episodes) – over which the decrease is spread, as well as the eligibility trace's decay parameter:

```
# parameters for sarsa(lambda)
episodes = 30000
STEPS = 500
gamma = 0.9
alpha = 0.05
epsilon_start = 0.2
epsilon_end = 0.001
epsilon_annealing_stop = int(episodes/2)
eligibility_decay = 0.3
```

11. Initialize the Q-table, set all values equal to **1** except for terminal states, and set an array that will collect all the agent's performance evaluations during training:

```
q = np.zeros((16, 4))
# Set q(terminal,*) equal to 0
q[5,:] = 0.0
q[7,:] = 0.0
q[11,:] = 0.0
q[12,:] = 0.0
q[15,:] = 0.0
performance = np.ndarray(episodes//STEPS)
```

12. Start the SARSA training loop by looping among all episodes:

```
for episode in range(episodes):
```

13. Define an epsilon value based on the current episode's run:

```
    inew = min(episode,epsilon_annealing_stop)
    epsilon = (epsilon_start * (epsilon_annealing_stop - inew) \
              + epsilon_end * inew) / epsilon_annealing_stop
```

14. Initialize the eligibility traces table to **0**:

```
    E = np.zeros((16, 4))
```

15. Reset the environment, choose the first action with an ε-greedy policy, and start the in-episode loop:

```
state = env.reset()
action = action_epsilon_greedy(q, state, epsilon)


while True:
```

16. Update the eligibility traces and assign a weight of **1** to the last visited state:

```
E = eligibility_decay * gamma * E
E[state, action] += 1
```

17. Step through the environment with the selected action and retrieve the new state, reward, and done conditions:

```
new_state, reward, done, info = env.step(action)
```

18. Select the new action with the ε-greedy policy:

```
new_action = action_epsilon_greedy\
              (q, new_state, epsilon)
```

19. Calculate the δ update and update the Q-table using the SARSA TD(δ) rule:

```
delta = reward + gamma \
          * q[new_state, new_action] - q[state, action]
q = q + alpha * delta * E
```

20. Update the state and action with new state and action values:

```
state, action = new_state, new_action


if done:
    break
```

21. Evaluate the agent's average performance:

```
if episode%STEPS == 0:
    performance[episode//STEPS] = average_performance\
                                  (get_action_epsilon_greedy\
                                  (epsilon), q=q)
```

22. Plot the SARSA agent's mean reward history during training:

```
plt.plot(STEPS*np.arange(episodes//STEPS), performance)
plt.xlabel("Epochs")
plt.title("Learning progress for SARSA")
plt.ylabel("Average reward of an epoch")
```

This generates the following output:

```
Text(0, 0.5, 'Average reward of an epoch')
```

The plot for this can be visualized as follows:

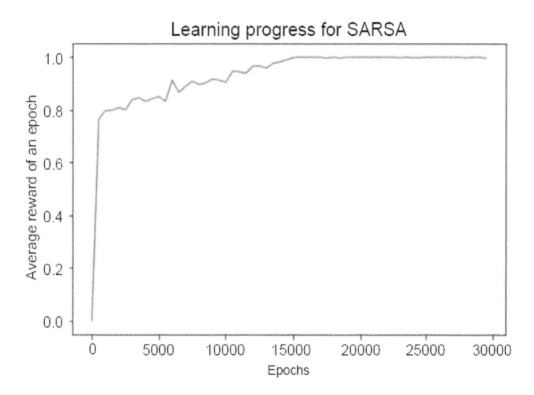

Figure 7.35: Average reward of an epoch trend over training epochs

As we can see, SARSA's TD(λ) performance grows over time as the **ε** parameter is annealed, thus reaching the value of 0 in the limit, and thereby obtaining the greedy policy. It also demonstrates that the algorithm is capable of reaching 100% success after learning. With respect to the one-step SARSA model, as seen in *Figure 7.8*, here, we can see that it reaches maximum performance faster, showing a notable improvement.

23. Evaluate the greedy policy's performance for the trained agent (Q-table):

```
greedyPolicyAvgPerf = average_performance(greedy_policy, q=q)
print("Greedy policy SARSA performance =", greedyPolicyAvgPerf)
```

The output will be as follows:

```
Greedy policy SARSA performance = 1.0
```

24. Display the Q-table values:

```
q = np.round(q,3)
print("(A,S) Value function =", q.shape)
print("First row")
print(q[0:4,:])
print("Second row")
print(q[4:8,:])
print("Third row")
print(q[8:12,:])
print("Fourth row")
print(q[12:16,:])
```

This generates the following output:

```
(A,S) Value function = (16, 4)

First row
[[0.499 0.59  0.519 0.501]
 [0.474 0.    0.615 0.518]
 [0.529 0.699 0.528 0.589]
 [0.608 0.397 0.519 0.517]]

Second row
[[0.553 0.656 0.    0.489]
 [0.    0.    0.    0.    ]
 [0.    0.806 0.    0.593]
 [0.    0.    0.    0.    ]]

Third row
[[0.619 0.    0.729 0.563]
 [0.613 0.77  0.81  0.    ]
 [0.712 0.9   0.    0.678]
```

```
   [0.    0.    0.    0.    ]]

Fourth row
[[0.    0.    0.    0.    ]
 [0.003 0.8   0.9   0.683]
 [0.76  0.892 1.    0.787]
 [0.    0.    0.    0.    ]]
```

This output shows the values of the complete state-action value function for our problem. These values are then used to generate the optimal policy by means of the greedy selection rule.

25. Print out the greedy policy that was found and compare it with the optimal policy:

```
policyFound = [actionsDict[np.argmax(q[0,:])],\
               actionsDict[np.argmax(q[1,:])],\
               actionsDict[np.argmax(q[2,:])],\
               actionsDict[np.argmax(q[3,:])],\
               actionsDict[np.argmax(q[4,:])],\
               " - ",\
               actionsDict[np.argmax(q[6,:])],\
               " - ",\
               actionsDict[np.argmax(q[8,:])],\
               actionsDict[np.argmax(q[9,:])],\
               actionsDict[np.argmax(q[10,:])],\
               " - ",\
               " - ",\
               actionsDict[np.argmax(q[13,:])],\
               actionsDict[np.argmax(q[14,:])],\
               " ! "]

print("Greedy policy found:")
idxs = [0,4,8,12]
for idx in idxs:
    print(policyFound[idx+0], policyFound[idx+1], \
            policyFound[idx+2], policyFound[idx+3])

print(" ")
```

```
print("Optimal policy:")
idxs = [0,4,8,12]
for idx in idxs:
    print(optimalPolicy[idx+0], optimalPolicy[idx+1], \
          optimalPolicy[idx+2], optimalPolicy[idx+3])
```

This produces the following output:

```
Greedy policy found:
  R    R    D    L
  D    -    D    -
  R    D    D    -
  -    R    R    !

Optimal policy:
R/D   R    D    L
  D    -    D    -
  R  R/D   D    -
  -    R    R    !
```

As you can see, our SARSA algorithm has been able to correctly solve the FrozenLake-v0 environment by being able to learn the optimal policy under the deterministic transition dynamics. In fact, as we can see, for every state of the grid world, the greedy policy that was obtained with the Q-table that was calculated by our algorithm prescribes an action that is in accordance with the optimal policy that was defined by analyzing the environment problem. As we already saw, there are two states in which there are two equally optimal actions, and the agent correctly implements one of them.

> **NOTE**
>
> To access the source code for this specific section, please refer to https://packt.live/2YdePoa.
>
> You can also run this example online at https://packt.live/3ek4ZXa.

We can now proceed and test how it behaves when exposed to stochastic dynamics. We'll do this in the next exercise. Just like when using one-step SARSA, in this case, we want to give the agent the freedom to take advantage of the 0 penalty for intermediate steps to minimize risk of falling into the holes, so in this case, we have to set the discount factor's gamma equal to 1. This means that instead of using **gamma = 0.9**, we will use **gamma = 1.0**.

EXERCISE 7.05: USING TD(λ) SARSA TO SOLVE FROZENLAKE-V0 STOCHASTIC TRANSITIONS

In this exercise, we will implement our SARSA(λ) algorithm to solve the FrozenLake-v0 environment under the deterministic environment dynamics. As we saw earlier in this chapter, when talking about one-step TD methods, the optimal policy looks completely different with respect to the previous exercise since it needs to take care of the stochasticity factor. This imposes a new challenge for the SARSA(λ) algorithm. We will see how it will still be able to solve this task in this exercise.

Follow these steps to complete this exercise:

1. Import the required modules:

```
import numpy as np
from numpy.random import random, choice
import matplotlib.pyplot as plt
%matplotlib inline

import gym
```

2. Instantiate the **gym** environment called **FrozenLake-v0** using the **is_slippery** flag set to **True** in order to enable stochasticity:

```
env = gym.make('FrozenLake-v0', is_slippery=True)
```

3. Take a look at the action and observation spaces:

```
print("Action space = ", env.action_space)
print("Observation space = ", env.observation_space)
```

This will print out the following:

```
Action space =  Discrete(4)
Observation space =  Discrete(16)
```

4. Create two dictionaries to easily translate the **actions** numbers into moves:

```
actionsDict = {}
actionsDict[0] = "  L  "
actionsDict[1] = "  D  "
actionsDict[2] = "  R  "
actionsDict[3] = "  U  "
```

```
actionsDictInv = {}
actionsDictInv["L"] = 0
actionsDictInv["D"] = 1
actionsDictInv["R"] = 2
actionsDictInv["U"] = 3
```

5. Reset the environment and render it to take a look at the grid problem:

```
env.reset()
env.render()
```

The output will be as follows:

SFFF
FHFH
FFFH
HFFG

Figure 7.36: Environment's initial state

6. Visualize the optimal policy for this environment:

```
optimalPolicy = ["L/R/D"," U "," U "," U ",\
                 " L "," - "," L/R "," - ",\
                 " U "," D "," L "," - ",\
                 " - "," R "," D "," ! ",]

print("Optimal policy:")
idxs = [0,4,8,12]
for idx in idxs:
    print(optimalPolicy[idx+0], optimalPolicy[idx+1], \
          optimalPolicy[idx+2], optimalPolicy[idx+3])
```

This prints out the following output:

```
Optimal policy:
 L/R/D  U    U    U
   L    -   L/R   -
   U    D    L    -
   -    R    D    !
```

This represents the optimal policy for this environment. Except for two states, all the others have a single optimal action associated with them. In fact, as described earlier in this chapter, optimal actions here are those that bring the agent away from the holes, or from tiles that have a chance greater than zero to lead the agent into tiles placed near holes. Two states have multiple optimal actions associated with them that are all equally optimal, as intended for this task.

7. Define the functions that will take ε-greedy actions:

```
def action_epsilon_greedy(q, s, epsilon=0.05):
    if np.random.rand() > epsilon:
        return np.argmax(q[s])
    return np.random.randint(4)


def get_action_epsilon_greedy(epsilon):
    return lambda q,s: action_epsilon_greedy\
                    (q, s, epsilon=epsilon)
```

8. Define a function that will take greedy actions:

```
def greedy_policy(q, s):
    return np.argmax(q[s])
```

9. Define a function that will calculate the agent's average performance:

```
def average_performance(policy_fct, q):
    acc_returns = 0.
    n = 500
    for i in range(n):
        done = False
        s = env.reset()
        while not done:
            a = policy_fct(q, s)
            s, reward, done, info = env.step(a)
            acc_returns += reward
    return acc_returns/n
```

10. Set the number of total episodes, the number of steps representing the interval by which we will evaluate the agent's average performance, the discount factor, the learning rate, and the **ε** parameters ruling its decrease – the starting value, minimum value, and range (in terms of the number of episodes) – over which the decrease is spread, as well as the eligibility trace's decay parameter:

```
# parameters for sarsa(lambda)
episodes = 80000
STEPS = 2000
gamma = 1
alpha = 0.02
epsilon_start = 0.2
epsilon_end = 0.001
epsilon_annealing_stop = int(episodes/2)
eligibility_decay = 0.3
```

11. Initialize the Q-table, set all the values equal to one except for terminal states, and set an array so that it collects all agent performance evaluations during training:

```
q = np.zeros((16, 4))
# Set q(terminal,*) equal to 0
q[5,:] = 0.0
q[7,:] = 0.0
q[11,:] = 0.0
q[12,:] = 0.0
q[15,:] = 0.0
performance = np.ndarray(episodes//STEPS)
```

12. Start the SARSA training loop by looping among all episodes:

```
for episode in range(episodes):
```

13. Define the epsilon value based on the current episode run:

```
inew = min(episode,epsilon_annealing_stop)
epsilon = (epsilon_start * (epsilon_annealing_stop - inew) \
            + epsilon_end * inew) / epsilon_annealing_stop
```

14. Initialize the eligibility traces table to 0:

```
E = np.zeros((16, 4))
```

15. Reset the environment and state your choice for the first action with an ε-greedy policy. Then, start the in-episode loop:

```
state = env.reset()
action = action_epsilon_greedy(q, state, epsilon)


while True:
```

16. Update the eligibility traces by applying decay and making the last state-action pair the most important one:

```
E = eligibility_decay * gamma * E
E[state, action] += 1
```

17. Define the environment step with the selected action and retrieval of the new state, reward, and done conditions:

```
new_state, reward, done, info = env.step(action)
```

18. Select a new action with the ε-greedy policy:

```
new_action = action_epsilon_greedy(q, new_state, epsilon)
```

19. Calculate the δ update and update the Q-table with the SARSA TD(λ) rule:

```
delta = reward + gamma \
        * q[new_state, new_action] - q[state, action]
q = q + alpha * delta * E
```

20. Update the state and action with new values:

```
state, action = new_state, new_action


if done:
    break
```

21. Evaluate the average agent performance:

```
if episode%STEPS == 0:
    performance[episode//STEPS] = average_performance\
                                (get_action_epsilon_greedy\
                                (epsilon), q=q)
```

22. Plot the SARSA agent's mean reward history during training:

```
plt.plot(STEPS*np.arange(episodes//STEPS), performance)
plt.xlabel("Epochs")
plt.title("Learning progress for SARSA")
plt.ylabel("Average reward of an epoch")
```

This generates the following output:

```
Text(0, 0.5, 'Average reward of an epoch')
```

The plot for this can be visualized as follows:

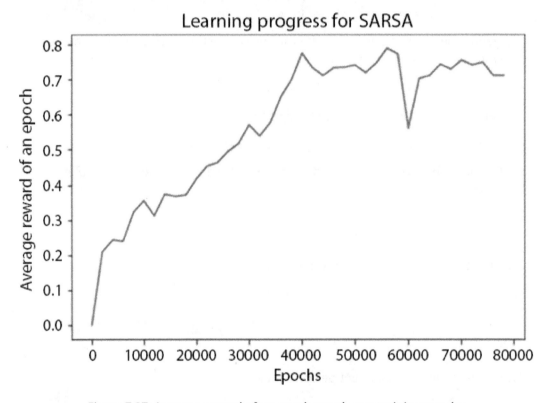

Figure 7.37: Average reward of an epoch trend over training epochs

Again, in comparison to the previous TD(0) SARSA case seen in *Figure 7.15*, this plot clearly shows us how the algorithm's performance improves over epochs, even when stochastic dynamics are considered. The behavior is very similar, and it also shows that, in the case of stochastic dynamics, it is not possible to obtain a perfect performance, in other words, reaching the goal 100% of the time.

23. Evaluate the greedy policy's performance of the trained agent (Q-table):

```
greedyPolicyAvgPerf = average_performance(greedy_policy, q=q)
print("Greedy policy SARSA performance =", greedyPolicyAvgPerf)
```

This prints out the following output:

```
Greedy policy SARSA performance = 0.734
```

24. Display the Q-table values:

```
q = np.round(q,3)
print("(A,S) Value function =", q.shape)
print("First row")
print(q[0:4,:])
print("Second row")
print(q[4:8,:])
print("Third row")
print(q[8:12,:])
print("Fourth row")
print(q[12:16,:])
```

This generates the following output:

```
(A,S) Value function = (16, 4)

First row
[[0.795 0.781 0.79  0.786]
 [0.426 0.386 0.319 0.793]
 [0.511 0.535 0.541 0.795]
 [0.341 0.416 0.393 0.796]]

Second row
[[0.794 0.515 0.541 0.519]
 [0.    0.    0.    0.    ]
 [0.321 0.211 0.469 0.125]
 [0.    0.    0.    0.    ]]

Third row
[[0.5   0.514 0.595 0.788]
 [0.584 0.778 0.525 0.46 ]
 [0.703 0.54  0.462 0.365]
```

```
    [0.    0.    0.    0.    ]]

Fourth row
[[0.    0.    0.    0.    ]
 [0.563 0.557 0.862 0.508]
 [0.823 0.94  0.878 0.863]
 [0.    0.    0.    0.    ]]
```

This output shows the values of the complete state-action value function for our problem. These values are then used to generate the optimal policy by means of the greedy selection rule.

25. Print out the greedy policy that was found and compare it with the optimal policy:

```
policyFound = [actionsDict[np.argmax(q[0,:])],\
               actionsDict[np.argmax(q[1,:])],\
               actionsDict[np.argmax(q[2,:])],\
               actionsDict[np.argmax(q[3,:])],\
               actionsDict[np.argmax(q[4,:])],\
               " - ",\
               actionsDict[np.argmax(q[6,:])],\
               " - ",\
               actionsDict[np.argmax(q[8,:])],\
               actionsDict[np.argmax(q[9,:])],\
               actionsDict[np.argmax(q[10,:])],\
               " - ",\
               " - ",\
               actionsDict[np.argmax(q[13,:])],\
               actionsDict[np.argmax(q[14,:])],\
               " ! "]

print("Greedy policy found:")
idxs = [0,4,8,12]
for idx in idxs:
    print(policyFound[idx+0], policyFound[idx+1], \
          policyFound[idx+2], policyFound[idx+3])

print(" ")
```

```
print("Optimal policy:")
idxs = [0,4,8,12]
for idx in idxs:
    print(optimalPolicy[idx+0], optimalPolicy[idx+1], \
        optimalPolicy[idx+2], optimalPolicy[idx+3])
```

This produces the following output:

```
Greedy policy found:
    L     U     U     U
    L     -     R     -
    U     D     L     -
    -     R     D     !


Optimal policy:
  L/R/D   U     U     U
    L     -    L/R    -
    U     D     L     -
    -     R     D     !
```

Also, as in the case of stochastic environment dynamics, the SARSA algorithm with eligibility traces has been able to correctly learn the optimal policy.

> **NOTE**
>
> To access the source code for this specific section, please refer to https://packt.live/2CiyZVf.
>
> You can also run this example online at https://packt.live/2Np7zQ9.

With this exercise, we've completed our study of temporal difference methods and covered many of the aspects, from their most simple one-step formulation to the most advanced ones. We are now able to combine multi-step methods without the restriction of having to wait until the end of the episode to update the state-value (or state-action pair) function. To complete our journey, we'll conclude with a quick comparison of the methods we explained in this chapter with those explained in *Chapter 5, Dynamic Programming*, and *Chapter 6, Monte Carlo Methods*.

THE RELATIONSHIP BETWEEN DP, MONTE-CARLO, AND TD LEARNING

From what we've learned in this chapter, and as we've stated multiple times, it is clear how temporal difference learning has characteristics in common with both Monte Carlo methods and dynamic programming ones. Like the former, it learns directly from experience, without leveraging a model of the environment representing transition dynamics or knowledge of the reward function involved in the task. Like the latter, it bootstraps, meaning that it updates the value function estimate partially based on other estimates, thereby circumventing the need to wait until the end of the episode. This point is particularly important since, in practice, very long episodes (or even infinite ones) can be encountered, making MC methods impractical and too slow. This strict relation plays a central role in reinforcement learning theory.

We have also learned about N-step methods and eligibility traces, two different but related topics that allow us to frame TD method's theory as a general picture capable of fusing together MC and TD methods. In particular, the eligibility traces concept allowed us to formally represent both of them, with the additional advantage of implementing a perspective change from a forward view to a more efficient incremental backward view, which allows us to extend MC methods even to non-episodic problems.

When bringing TD and MC methods under the same theory umbrella, eligibility traces demonstrate their value in making TD methods more robust to non-Markovian tasks, a typical problem in which MC algorithms behave better than TD ones. Thus, eligibility traces, even if typically coupled with an increased computational overhead, offer a better learning capability in general since they are both faster and more robust.

It is now time for us to tackle the final activity of this chapter, where we will apply what we have learned from the theory and exercises we've covered on TD methods.

ACTIVITY 7.01: USING TD(0) Q-LEARNING TO SOLVE FROZENLAKE-V0 STOCHASTIC TRANSITIONS

The goal of this activity is for you to adapt the TD(0) Q-learning algorithm to solve the FrozenLake-v0 environment under the stochastic transition dynamics. We have already seen that the optimal policy appears as follows:

L/D/R	U	U	U
L	H	L/R	H
U	D	L	H
H	R	D	G

Figure 7.38: Optimal policy – D = Down move, R = Right move, U = Up move, and L = Left move

Making Q-learning converge on this environment is not a simple task, but it is possible. In order to make this a little bit easier, we can use a value for the discount factor gamma that's equal to **0.99**. The following steps will help you to complete this exercise:

1. Import all the required modules.

2. Instantiate the gym environment and print out the observation and action spaces.

3. Reset the environment and render the starting state.

4. Define and print out the optimal policy for reference.

5. Define the functions for implementing the greedy and ε-greedy policies.

6. Define a function that will evaluate the agent's average performance and initialize the Q-table.

7. Define the learning method hyperparameters (ε, discount factor, total number of episodes, and so on).

8. Implement the Q-learning algorithm.

9. Train the agent and plot the average performance as a function of training epochs.

10. Display the Q-values found and print out the greedy policy while comparing it with the optimal one.

The final output of this activity is very similar to the ones you've encountered for all of the exercises in this chapter. We want to compare the policy found by our agent that was trained using the prescribed method with the optimal one to make sure we succeeded in making it learn the optimal policy correctly.

The optimal policy should be as follows:

```
Greedy policy found:
    L    U    U    U
    L    -    R    -
    U    D    L    -
    -    R    D    !


Optimal policy:
  L/R/D  U    U    U
    L    -   L/R   -
    U    D    L    -
    -    R    D    !
```

NOTE

The solution to this activity can be found on page 726.

By completing this activity, we've learned how to correctly implement and set up a one-step Q-learning algorithm by appropriately tuning its hyperparameters to solve an environment with stochastic transition dynamics. We monitored the agent's performance during training, and we confronted ourselves with the role of the reward discount factor. We selected a value for it, allowing us to make our agent learn the optimal policy for this specific task, even if the maximum reward for this environment is bound and there is no possibility of completing the episode 100% of the time.

SUMMARY

This chapter dealt with temporal difference learning. We started by studying one-step methods in both their on-policy and off-policy implementations, leading to us learning about the SARSA and Q-learning algorithms, respectively. We tested these algorithms on the FrozenLake-v0 problem and covered both deterministic and stochastic transition dynamics. Then, we moved on to the N-step temporal difference methods, the first step toward the unification of TD and MC methods. We saw how on-policy and off-policy methods are extended to this case. Finally, we studied TD methods with eligibility traces, which constitute the most relevant step toward the formalization of a unique theory describing both TD and MC algorithms. We extended SARSA to eligibility tracing, too, and learned about this through implementing two exercises where it has been implemented and applied to the FrozenLake-v0 environment under both deterministic and stochastic transition dynamics. With this, we have been able to successfully learn about the optimal policy in all cases, thereby demonstrating how these methods are sound and robust.

Now, it is time to move on to the next chapter, in which we will address the multi-armed bandit problem, a classic setting that's often encountered when studying reinforcement learning theory and the application of RL algorithms.

8

THE MULTI-ARMED BANDIT PROBLEM

OVERVIEW

In this chapter, we will introduce the popular Multi-Armed Bandit problem and some common algorithms used to solve it. We will learn how to implement some of these algorithms, such as Epsilon Greedy, Upper Confidence Bound, and Thompson Sampling, in Python via an interactive example. We will also learn about contextual bandits as an extension of the general Multi-Armed Bandit problem. By the end of this chapter, you will have a deep understanding of the general Multi-Armed Bandit problem and the skill to apply some common ways to solve it.

INTRODUCTION

In the previous chapter, we discussed the technique of temporal difference learning, a popular model-free reinforcement learning algorithm that predicts a quantity via the future values of a signal. In this chapter, we will focus on another common topic, not only in reinforcement learning but also in artificial intelligence and probability theory – the **Multi-Armed Bandit (MAB)** problem.

Framed as a sequential decision-making problem to maximize the reward while playing at the slot machines in a casino, the MAB problem is highly applicable for any situation where sequential learning under uncertainty is needed, such as A/B testing or designing recommender systems. In this chapter, we will be introduced to the formalization of the problem, learn about the different common algorithms as solutions to the problem (namely Epsilon Greedy, Upper Confidence Bound, and Thompson Sampling), and finally implement them in Python.

Overall, this chapter will offer you a deep understanding of the MAB problem in different contexts of sequential decision-making and offer you the opportunity to apply that knowledge to solve a variation of the problem called the queueing bandit.

First, let's begin by discussing the background and the theoretical formulation of the problem.

FORMULATION OF THE MAB PROBLEM

In its most simple form, the MAB problem consists of multiple slot machines (casino gambling machines), each of which can return a stochastic reward to the player each time it is played (specifically, when its arm is pulled). The player, who would like to maximize their total reward at the end of a fixed number of rounds, does not know the probability distribution or the average reward that they will obtain from each slot machine. The problem, therefore, boils down to the design of a learning strategy where the player needs to explore what possible reward values each slot machine can return and from there, quickly identify the one that is most likely to return the greatest expected reward.

In this section, we will briefly explore the background of the problem and establish the notation and terminology that we will be using throughout this chapter.

APPLICATIONS OF THE MAB PROBLEM

The slot machines we mentioned earlier are just a simplification of our settings. In the general case of a MAB problem, we are faced with a set of multiple decisions that we can choose from at each step, and we need to sufficiently explore each of the decisions so that we become more informed about the environment we are in, all while making sure we converge on the optimal decision soon so that our total reward is maximized at the end of the process. This is the classic trade-off between exploration and exploitation we are faced with in common reinforcement learning problems.

Popular applications of the MAB problem include recommender systems, clinical trials, network routing, and as we will see at the end of this chapter, queueing theory. Each of these applications contains the quintessential characteristics that define the MAB problem: at each step of a sequential process, a decision-maker needs to select from a predetermined set of possible choices and, depending on the past observations, the decision-maker needs to find a balance between exploring different choices and exploiting the one that they believe is the most beneficial.

As an example, one of the goals of recommender systems is to display the products that their customers are most likely to consider/buy. When a new customer logs into a system such as a shopping website or an online streaming service, the recommender system can observe the customer's past behaviors and choices and make a decision regarding what kind of product advertisement should be shown to the customer. It does this so that the probability that they click on the advertisement is maximized.

As another example, which we will see in more detail later on, in a queueing system consisting of multiple customer classes, each is characterized by an unknown service rate. The queue coordinator needs to figure out how to best order these customers so that a given objective, such as the cumulative waiting time of the whole queue, is optimized.

Overall, MAB is an increasingly ubiquitous problem in artificial intelligence and, specifically, reinforcement learning that has many interesting applications. In the next section, we will officially formalize the problem and the terminologies that we will be using throughout this chapter.

BACKGROUND AND TERMINOLOGY

A MAB problem is characterized by the following elements:

- A set of "K" actions to choose from. Each of these actions is called an arm, following the colloquial terminology with respect to slot machines.

- A central decision-maker who needs to choose between this set of actions at each step of the process. We call the act of choosing an action pulling an arm, and the decision-maker the player.

- When one of the "K" available arms is pulled, the player receives a stochastic, or random, reward drawn from a probability distribution that is specific to that arm. It is important that the rewards are randomly chosen from their respective distributions; if they were otherwise fixed, the player could identify the arm that will return the highest reward quickly and the problem would become less interesting.

- The goal of the player is, again, to choose one of the "K" arms during each step of a running process so that their reward is maximized at the end. The number of steps in the process is called the horizon, which may or may not be known by the player beforehand.

- In most cases, each arm can be pulled infinitely. When the player is certain that a specific arm is the optimal one, they can keep choosing that arm for the rest of the process without deviating. However, in various settings, the number of times an arm can be pulled is finite, thus increasing the complexity of the problem.

The following diagram visualizes an iterative step in the environment that we are working with, where we have four arms whose success rates are estimated as 70%, 30%, 55%, and 40%.

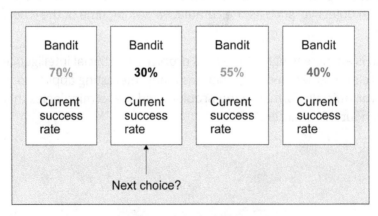

Figure 8.1: A typical MAB iteration

At each step, we need to make a decision about which arm we should choose to pull next:

- Opposite to the language of reward and the corresponding maximization objective, a MAB problem can also be framed in the context of a cost minimization objective. The queueing example can, once again, be used: the cumulative waiting time of the whole queue is a negative quantity, or in other words, the cost that needs to be minimized.

- It is common to compare the performance of a strategy to the optimal strategy or the genie strategy, which knows ahead of time what the optimal arm to pull is and always pulls that arm at every step of the process. Of course, it is highly improbable that any real learning strategy can simulate the performance of the genie strategy, but it does provide us with a fixed metric to compare our approaches against. The difference in performance of a given strategy and the genie strategy is known as the regret, which is to be minimized.

The central question of a MAB problem is how to identify the arm with the greatest *expected* reward (or lowest *expected* cost) with minimal exploration (pulling the suboptimal arms). This is because the more the player explores, the less frequent their choice of the optimal arm becomes, and the more their final reward decreases. However, if the player does not sufficiently explore all the arms, chances are they will misidentify the optimal arm and their total reward will be negatively affected in the end.

These situations arise when the stochastic rewards of the true optimal arm appear to be lower than those from other arms in the first few examples (due to randomness), causing the player to misidentify the optimal arm. Depending on the actual reward distribution that each arm has, this event can be quite likely to happen.

So, that is the general problem we set out to solve in this chapter. We now need to briefly consider the concept of probability distributions of reward in the context of MAB in order to fully understand the problem we are trying to solve.

MAB REWARD DISTRIBUTIONS

In the traditional MAB problem, the reward from each arm of the bandit is associated with a Bernoulli distribution. Each Bernoulli distribution is, in turn, parameterized by a non-negative number, p, that is, at most, 1. When a number is drawn from a Bernoulli distribution, it can take on two possible values: 1, which has a probability of p, and 0, which consequently has a probability of $1 - p$. A high value of p therefore corresponds to a good arm for the player to pull. This is because the player is more likely to receive 1 as their reward. Of course, a high value of p does not guarantee that the reward obtained from a specific arm is always 1, and chances are, out of many pulls from even the arm with the highest value of p (in other words, the optimal arm), some of the rewards will be 0.

The following diagram is an example of a Bernoulli bandit setting:

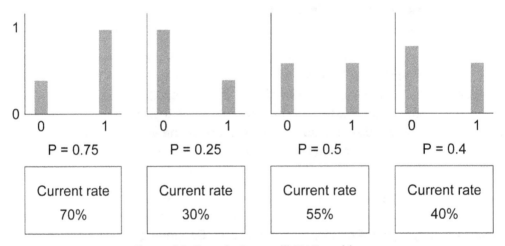

Figure 8.2: Sample Bernoulli MAB problem

Each arm has its own reward distribution: the first has a true probability of 75% of returning 1 and 25% of returning 0, the second has 25% for 1 and 75% for 0, and so on. Note that the rates that we empirically observe do not always match the true rates.

From here, we can generalize a MAB problem where the reward of an arm follows any probability distribution. While the inner workings of these distributions are different, the goal of a MAB algorithm remains constant: identifying the arm associated with the distribution with the highest expectation in order to maximize the final cumulative reward.

Throughout this chapter, we will be working with Bernoulli-distributed rewards, as they are among the most natural and intuitive reward distributions and will provide us with the context in which we can study various MAB algorithms. Finally, before we consider the different algorithms that will be covered in this chapter, let's take a moment to familiarize ourselves with the programming interface that we will be working with.

THE PYTHON INTERFACE

The Python environment that will help facilitate our discussions of MAB algorithms is included in the **utils.py** file of this chapter's code repository on GitHub: https://packt.live/3cWiZ8j.

From this file, we can import the **Bandit** class into a separate script or a Jupyter script. This class is the interface we will use to create, interact, and solve various MAB problems. If the code we are working with is in the same directory as this file, we can import the **Bandit** class by simply using the following code:

```
from utils import Bandit
```

Then, we can declare an MAB problem as an instance of a **Bandit** object:

```
my_bandit = Bandit()
```

Since we are not passing any arguments to this declaration, this **Bandit** instance takes on its default value: an MAB problem with two Bernoulli arms with probabilities of 0.7 and 0.3 (although our algorithms technically cannot know this).

The most integral method of the **Bandit** class that we need to be aware of is **pull()**. This method takes in an integer as an argument, denoting the index of the arm we would like to pull at a given step, and returns a number representing the stochastic reward drawn from the distribution associated with that same arm.

For example, in the following code snippet, we call this **pull()** method with the **0** parameter to pull the first arm and record the returned reward, like so:

```
reward = my_bandit.pull(0)
reward
```

Here, you might see the number **0** or the number **1** printed out, which denotes the reward that you receive by pulling arm 0. Say we'd like to pull arm 1 once; the same API can be used:

```
reward = my_bandit.pull(1)
reward
```

Again, the output might be **0** or **1** since we are drawing from a Bernoulli distribution.

Say we'd like to inspect what the reward distribution of each arm might look like, or more specifically, which out of the two arms is the one more likely to return more reward. To do this, we pull from each arm 10 times and record the returned reward at each step:

```
running_rewards = [[], []]

for _ in range(10):
    running_rewards[0].append(my_bandit.pull(0))
    running_rewards[1].append(my_bandit.pull(1))

running_rewards
```

This code produces the following output:

```
[[1, 1, 1, 0, 0, 1, 1, 1, 0, 0], [0, 0, 1, 0, 0, 1, 0, 1, 1, 1]]
```

Again, due to randomness, you might get a different output. Considering the preceding output, we can see that arm 0 returned a positive reward 6 out of 10 pulls, while arm 1 returned a positive reward 5 times.

We'd also like to plot the cumulative reward throughout the process of 20 steps from each arm. Here, we can use the **np.cumsum()** function from the NumPy library to compute that quantity and plot it using the Matplotlib library, like so:

```
rounds = [i for i in range(1, 11)]

plt.plot(rounds, np.cumsum(running_rewards[0]),\
        label='Cumulative reward from arm 0')

plt.plot(rounds, np.cumsum(running_rewards[1]), \
        label='Cumulative reward from arm 1')

plt.legend()

plt.show()
```

The following graph will then be produced:

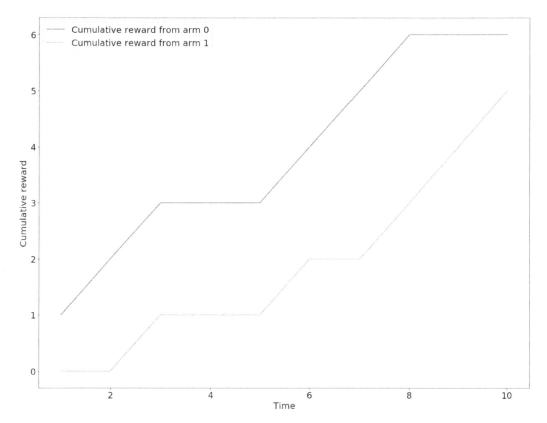

Figure 8.3: Sample graph of cumulative reward

This graph allows us to visually inspect how fast the cumulative reward we receive from each arm grew throughout the process of 10 pulls. We can also see that the cumulative reward from arm 0 is always greater than that from arm 1, indicating that out of the two arms, arm 0 is the optimal one. This is consistent with the fact that arm 0 was initialized with a Bernoulli reward distribution where $p = 0.7$, and arm 1 has one where $p = 0.3$.

The **pull()** method is the lower-level API that facilitates processing at each step. However, when we design various MAB algorithms, we will be allowing the algorithms to interact with the bandit problem automatically, on their own, without any human interference. This leads us to the second method from the **Bandit** class, which we will be using to test out our algorithms: **automate()**.

As we will see in the next section, this method takes in an algorithm object implementation and streamlines the testing process for us. Specifically, this method will call the algorithm object, record its decisions, and return the corresponding rewards in an automatic manner. Aside from the algorithm object, it also takes in two other optimal parameters: **n_rounds**, which is used to specify the number of times we can interact with the bandit, and **visualize_regret**, which is a Boolean flag indicating whether we would like to plot the regret of the algorithm we are considering against the genie algorithm.

This whole process is called an experiment, where an algorithm that does not have any prior knowledge is tested against an MAB problem. To fully analyze the performance of a given algorithm, we need to put that algorithm through many experiments and study its performance in the general case across all experiments. This is because a specific initialization of the MAB problem might favor one algorithm over another; by comparing the performance of different algorithms across multiple experiments, our resulting insight regarding which algorithms are better will be more robust.

This is where the **repeat()** method of the **Bandit** class comes in. This method takes in an algorithm class' implementation (as opposed to an object implementation) and repeatedly calls the **automate()** method described previously on the instances of the algorithm class. Doing this facilitates multiple experiments on the algorithm we are considering and, again, will give us a more holistic view of its performance.

In order to interact with the methods of this **Bandit** class, we will be implementing our MAB algorithms as Python classes. The **pull()** method, and therefore both the **automate()** and **repeat()** methods as well, require these algorithm class implementations to have two distinct methods: **decide()**, which should return the index of the arm that the algorithm thinks should be pulled next at any given time, and **update()**, which takes in an arm index and a new reward that was just returned from that arm of the bandit. We will be keeping these two methods in mind while writing our algorithms later in this chapter.

As a final note about the bandit API, due to randomness, it is entirely possible that, in your own implementation, you will obtain different results from the results shown in this chapter. For better reproducibility, we have fixed the random seed number of all the scripts in this chapter to **0** so that it will be possible for you to run the code and obtain the same results, which can be done by taking any of the Jupyter Notebooks from this book's GitHub repository and running the code using the option shown in the following screenshot:

Figure 8.4: Reproducing results with Jupyter Notebooks

With that said, even with randomness, we will see that some algorithms are better than others at solving the MAB problem. This is also why we will be analyzing the performance of our algorithms via many repeated experiments, ensuring that any performance superiority is robust to randomness.

And that is all the background information we need in order to understand the MAB problem. We are now ready to begin discussing the approaches that are commonly employed on this problem, starting with the Greedy algorithm.

THE GREEDY ALGORITHM

Recall the brief interaction we had with the **Bandit** instance in the previous section, in which we pulled the first arm 10 times and the second 10 times. This might not be the best strategy to maximize our cumulative reward as we are spending 10 rounds pulling a sub-optimal arm, whichever it is among the two. The naïve approach is, therefore, to simply pull both (or all) of the arms once and greedily commit to the one that returns a positive reward.

A generalization of this strategy is the Greedy algorithm, in which we maintain the list of reward averages across all available arms and at each step, we choose to pull the arm with the highest average. While the intuition is simple, it follows the probabilistic rationale that after a large number of samples, the empirical mean (the average of the samples) is a good approximation of the actual expectation of the distribution. If the reward average of an arm is larger than that of any other arm, the probability that that given arm is indeed the optimal arm should not be low.

IMPLEMENTING THE GREEDY ALGORITHM

Now, let's try implementing this algorithm. As explained in the previous section, we will be writing our MAB algorithms as Python classes to interact with the bandit API that is provided in this book. Here, we will require this algorithm class to have two attributes: the number of available arms to pull and the list of rewards that the algorithm has observed from each arm:

```
class Greedy:
    def __init__(self, n_arms=2):
        self.n_arms = n_arms
        self.reward_history = [[] for _ in range(n_arms)]
```

Here, **reward_history** is a list of lists, where each sub-list contains the past rewards returned from a given arm. The data stored in this attribute will be used to drive the decision of our MAB algorithms.

Recall that an algorithm class implementation needs two specific methods to interact with the bandit API, **decide()** and **update()**, the latter of which is simpler and is implemented here:

```
class Greedy:
    ...

    def update(self, arm_id, reward):
        self.reward_history[arm_id].append(reward)
```

Again, this **update()** method needs to take in two arguments: an arm index (corresponding to the **arm_id** variable) and a number representing the most recent reward we obtain by pulling that arm (the **reward** variable). In this method, we simply need to store this information in the **reward_history** attribute by appending the number to the corresponding sub-list of rewards.

As for the **decide()** method, we need to implement the greedy logic that we described previously: the reward averages across all the arms are to be computed, and the arm with the highest average should be returned. However, before that, we need to handle the first few rounds where the algorithm has not observed any reward from any arm. The convention here is to simply force the algorithm to pull each arm at least once, which is implemented by the conditional given at the beginning of the code:

```
def decide(self):
    for arm_id in range(self.n_arms):
        if len(self.reward_history[arm_id]) == 0:
            return arm_id

    mean_rewards = [np.mean(history) for history in self.reward_
history]

    return int(np.random.choice\
            (np.argwhere(mean_rewards == np.max(mean_rewards))\
            .flatten()))
```

As you can see, we first find out whether any reward sub-list has a length of 0, indicating that the corresponding arm has not been pulled by the algorithm. If this is the case, we simply return the index of that arm.

Otherwise, we compute the reward averages with the **mean_rewards** variable: the **np.mean()** method computes the mean of each sub-list that is stored in the **reward_history** attribute, which we iterate through using list comprehension.

Finally, we find the arm index with the highest average, which is computed using **np.max(mean_rewards)**. A subtle point about the algorithm we're implemented here is the **np.random.choice()** function: there will be scenarios where multiple arms have the same highest value of reward average, in which case the algorithm should randomly choose among these arms without biasing any of them. The hope here is that if a suboptimal arm is chosen, future rewards will reveal that the arm is indeed less likely to yield a positive reward, and we will be converging to the optimal arm anyway.

And that is all there is to it. As noted earlier, the Greedy algorithm is fairly straightforward but also makes intuitive sense. Now, we want to see the algorithm in action by having it interact with our bandit API. First, we need to create a new instance of a MAB problem:

```
N_ARMS = 3

bandit = Bandit(optimal_arm_id=0,\
                n_arms=3,\
                reward_dists=[np.random.binomial \
                              for _ in range(N_ARMS)],\
                reward_dists_params=[(1, 0.9), (1, 0.8), (1, 0.7)])
```

Here, our MAB problem has three arms whose rewards all follow Bernoulli distributions (implemented by the **np.random.binomial** random function from NumPy). The first arm has a reward probability of $p = 0.9$, while it has $p = 0.8$ in the second arm and $p = 0.7$ in the third; the first arm is, therefore, the optimal arm that our algorithms have to identify.

(As a side note, to draw from a Bernoulli distribution with the p parameter, we call **np.random.binomial(1, p)**, so that is why we are pairing each value of p with the number **1** in the preceding code snippet.)

Now, we declare an instance of our Greedy algorithm with the appropriate number of arms and call the **automate()** method of the bandit problem to have the algorithm interact with the bandit for 500 rounds, as follows:

```
greedy_policy = Greedy(n_arms=N_ARMS)

history, rewards, optimal_rewards = bandit.automate\
                                    (greedy_policy, n_rounds=500,\
                                    visualize_regret=True)
```

As we can see, the **automate()** method returns three objects in a tuple: **history**, which is the sequential list of arms chosen by the algorithm throughout the process; **rewards**, the corresponding reward obtained by pulling those arms; and **optimal_rewards**, which is a list of what our rewards would be, had we chosen the optimal arm at every step throughout the process (in other words, this is the reward list of the genie algorithm). The tuple is visualized by the following plot, which is the actual output for the preceding code.

From within the **automate()** method, we also have the option to visualize the difference in cumulative sum between the two lists, **rewards** and **optimal_rewards**, specified by the **visualize_regret** parameter. Essentially, the option will plot out the cumulative regret of our algorithm as a function of a round number. Since we are enabling this option in our call, the following plot will be generated:

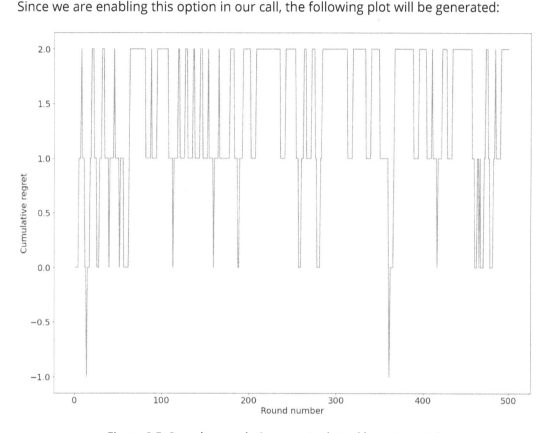

Figure 8.5: Sample cumulative regret, plotted by automate()

While we don't have any other algorithm to compare it with, from this graph, we can see that our Greedy algorithm did significantly well as it was able to keep the cumulative regret no higher than 2 at all times throughout the 500 rounds. Another way to inspect the performance of our algorithm is to consider the **history** list, which, again, contains the arms that the algorithm chose to pull:

```
print(*history)
```

This will print out the list in the following format:

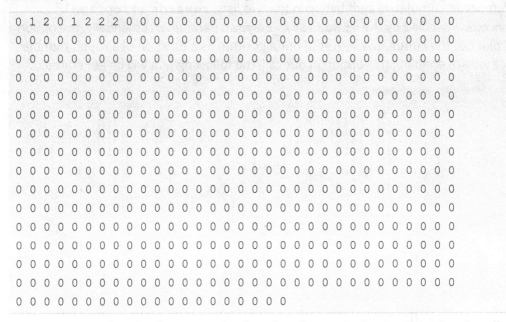

As we can see, after the three rounds of exploration at the beginning, when it pulled each arm once, the algorithm vacillated a bit between the arms but then quickly converged to choosing arm 0, the actual optimal arm, for the rest of the rounds. This is why the resulting cumulative regret of the algorithm is so low.

With that said, this is simply one single experiment. As mentioned previously, to fully benchmark the performance of our algorithms, we need to repeat this experiments many times, making sure that the single experiment we are considering is not an outlier where the algorithm does especially well or badly due to randomness.

To facilitate repeated experiments, we utilize the **repeat()** method of the bandit API, as follows:

```
regrets = bandit.repeat(Greedy, [N_ARMS], n_experiments=100, \
                    n_rounds=300, visualize_regret_dist=True)
```

Remember that the **repeat()** method takes in the class implementation of a given algorithm, as opposed to simply an instance of the algorithm as **automate()** does. This is why we are passing the whole **Greedy** class to the method. Additionally, with the second argument of the method, we can specify whatever arguments the class implementation of our algorithm takes in. In this case, it is simply the number of arms available to be pulled, but we will have different parameters with different algorithms in later sections.

Here, we are putting our Greedy algorithm through 100 experiments with the same bandit problem of the three Bernoulli arms we declared previously, specified by the **n_experiments** parameter. To save time, we only require that each experiment lasts for 300 rounds with the **n_rounds** parameter. Finally, we specify **visualize_regret_dist** to be **True**, which will help us plot the distribution of the cumulative regret obtained by the algorithm at the end of each experiment.

Indeed, when this code finishes running, the following plot will be produced:

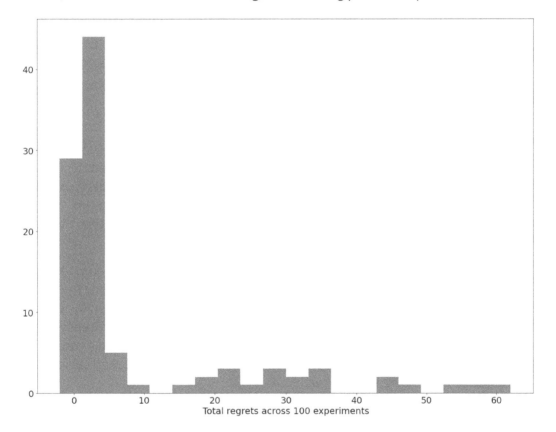

Figure 8.6: Distribution of cumulative regret by the Greedy algorithm

Here, we can see that in most cases, the Greedy algorithm does sufficiently well, keeping the cumulative regret below **10**. However, there are instances where the cumulative regret gets as high as **60**. We speculate that these are the situations where the algorithm misestimates the true expected reward from each arm and commits too early.

As the final way to gauge how well an algorithm performs, we consider the mean and the max cumulative regret across these experiments, as follows:

```
np.mean(regrets), np.max(regrets)
```

In our current experiment, the following numbers will be printed out:

```
(8.66, 62)
```

This is consistent with the distribution that we have here: most of the regrets are low enough, causing the mean to be relatively low (**8.66**), but the maximum regret can get as high as **62**.

And that is the end of our discussion on the Greedy algorithm. For the rest of this section, we will discuss two popular variations of the algorithm, namely Explore-then-commit and ε-Greedy.

THE EXPLORE-THEN-COMMIT ALGORITHM

We mentioned that a potential reason for poor performance of the Greedy algorithm in some cases is committing too early when a sufficient number of sample rewards from each arm have not been observed. The Explore-then-commit algorithm attempts to address this problem by formalizing the number of rounds that should be spent exploring each arm at the beginning of the process.

Specifically, each Explore-then-commit algorithm is parameterized by a number, *T*. In each bandit problem, an Explore-then-commit algorithm will spend exactly *T* rounds pulling each of the available arms. Only after these forced exploration rounds does the algorithm start choosing the arm with the greatest reward average. Greedy is a special case of the Explore-then-commit algorithm where *T* is set to 1. This general algorithm, therefore, gives us the option to customize this parameter and set it appropriately, depending on the situation.

The implementation of this algorithm is mostly similar to what we have for Greedy, so we will not consider it here. In short, instead of the conditional used to ensure the Greedy algorithm pulls each arm at least once, we can modify the conditional like so in its **decide()** method, given that a value for the **T** variable has been set:

```
def decide(self):
        for arm_id in range(self.n_arms):
            if len(self.reward_history[arm_id]) < T:
                return arm_id
```

```
mean_rewards = [np.mean(history) \
                for history in self.reward_history]

return int(np.random.choice\
       (np.argwhere(mean_rewards == np.max(mean_rewards))\
       .flatten()))
```

While Explore-then-commit is a more flexible version of Greedy, it does leave open the question of how to choose the value for T. Indeed, it is not obvious how we should set T for a specific bandit problem without any prior knowledge about the problem. Most of the time, T is set with respect to the horizon if it is known beforehand; common values for T could range from 3, 5, 10, or even 20.

THE ε-GREEDY ALGORITHM

Another variation of the Greedy algorithm is the ε-Greedy algorithm. For Explore-then-commit, the amount of forced exploration depends on the settable parameter, T, which again gives rise to the question of how to best set it. For ε-Greedy, we do not explicitly require the algorithm to explore more than one round for each arm. Instead, we leave it to chance to determine when the algorithm should carry on exploitation, and when it should explore a seemingly suboptimal arm.

Formally, an ε-Greedy algorithm is parameterized by a number, ε, between zero and one, denoting the exploration probability of the algorithm. After the first exploration rounds, the algorithm will choose to pull the arm with the greatest running reward average with probability 1 - ε. Otherwise, it will uniformly choose one out of all the available arms (with probability ε). Unlike Explore-then-commit, where we know for sure the algorithm will be forced to explore for the first few rounds, an ε-Greedy algorithm might explore arms with suboptimal reward averages during later rounds too. However, when exploration happens, this is entirely due to chance, and the choice of the parameter, ε, controls how often these exploration rounds are expected to happen.

For example, a common choice for ε is 0.01. In a typical bandit problem, an ε-Greedy algorithm will pull each arm once at the start of the process and begin choosing the arm with the best reward history. However, at each step, with probability 0.01 (one percent), the algorithm might choose to explore this, in which case it will randomly choose one of all the arms without any bias. ε, like T in the Explore-then-commit algorithm, is used to control how much an MAB algorithm should explore. A high value of ε will cause the algorithm to explore more often, although, again, when it does explore, this is completely random.

The intuition behind ε-Greedy is clear: we still want to preserve the greedy nature of the Greedy algorithm, but to avoid incorrect committing to a suboptimal arm due to nonrepresentative reward samples, we also want exploration to happen every now and then throughout the entire process. Hopefully, ε-Greedy will kill two birds with one stone, being able to greedily exploit the temporarily good arms while leaving the possibility that other seemingly suboptimal arms are better open.

Implementation-wise, the **decide()** method of the algorithm should have an additional conditional where we check whether the algorithm should explore:

```
def decide(self):
    ...

    if np.random.rand() < self.e:
        return np.random.randint(0, self.n_arms)

    ...
```

And with that, let's move on and complete this chapter's first exercise, where we will implement the ε-Greedy algorithm.

EXERCISE 8.01 IMPLEMENTING THE ε-GREEDY ALGORITHM

Similar to what we did to implement the Greedy algorithm, in this exercise, we will learn how to implement the ε-Greedy algorithm. This exercise will consist of three main sections: implementing the logic of ε-Greedy, testing it in a sample bandit problem, and finally putting it through multiple experiments to benchmark its performance.

We will follow these steps to achieve this:

1. Create a new Jupyter Notebook and import NumPy, Matplotlib, and the **Bandit** class from the **utils.py** file included in the code repository for this chapter:

```
import numpy as np
np.random.seed(0)
import matplotlib.pyplot as plt

from utils import Bandit
```

Note that we are now fixing the random seed number of NumPy to ensure the reproducibility of our code.

2. Now, let's begin implementing the logic of the ε-Greedy algorithm. First, its initialization method should take in two parameters: the number of arms for the bandit problem it is to solve and ε, the exploration probability:

```
class eGreedy:
    def __init__(self, n_arms=2, e=0.01):
        self.n_arms = n_arms
        self.e = e
        self.reward_history = [[] for _ in range(n_arms)]
```

Similar to what we had with Greedy, here, we are also keeping track of the reward history, which is stored in the **reward_history** attribute of the class object.

3. In the same code cell, implement the **decide()** method for the **eGreedy** class.

This method should be mostly similar to its counterpart in the **Greedy** class. However, before computing the reward averages of the arms, it should draw a random number between 0 and 1 and check to see if it is less than its parameter, ε. If this is the case, it should randomly return the index of one of the arms:

```
def decide(self):
    for arm_id in range(self.n_arms):
        if len(self.reward_history[arm_id]) == 0:
            return arm_id

    if np.random.rand() < self.e:
        return np.random.randint(0, self.n_arms)

    mean_rewards = [np.mean(history) \
                    for history in self.reward_history]

    return int(np.random.choice(np.argwhere\
              (mean_rewards == np.max(mean_rewards))\
              .flatten()))
```

4. In the same code cell, implement the **update()** method for the **eGreedy** class, which should be identical to the corresponding method in the **Greedy** class:

```
def update(self, arm_id, reward):
    self.reward_history[arm_id].append(reward)
```

Again, this method only needs to append the most recent reward from an arm to the reward history of that arm.

And that is the complete implementation of our ε-Greedy algorithm.

5. In the next code cell, create a single experiment with the bandit problem with three Bernoulli arms with respective probabilities of **0.9**, **0.8**, and **0.7** and run it with an instance of the **eGreedy** class (with **ε = 0.01**, which is the default value that does not need to be specified) using the **automate()** method.

Make sure to specify the **visualize_regret=True** parameter to plot out the cumulative regret of the algorithm throughout the process:

```
N_ARMS = 3

bandit = Bandit(optimal_arm_id=0, \
                n_arms=3, \
                reward_dists=[np.random.binomial \
                        for _ in range(N_ARMS)], \
                        reward_dists_params=[(1, 0.9), \
                                             (1, 0.8), \
                                             (1, 0.7)])


egreedy_policy = eGreedy(n_arms=N_ARMS)

history, rewards, optimal_rewards = bandit.automate\
                                (egreedy_policy, \
                                 n_rounds=500, \
                                 visualize_regret=True)
```

This should produce the following graph:

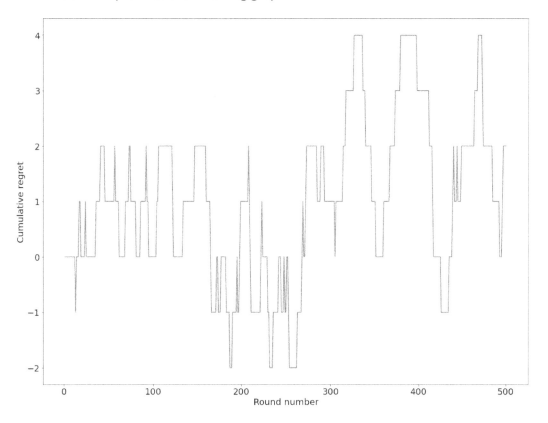

Figure 8.7: Sample cumulative regret of the ε-Greedy algorithm

Compared to the corresponding plot we saw with Greedy, our cumulative regret here has more variation, sometimes growing to **4** and sometimes dropping to **-2**. This is an effect of the increase in exploration of the algorithm.

6. In the next code cell, we print out the **history** variable and see how it compares to the Greedy algorithm:

```
print(*history)
```

This will produce the following output:

```
0 1 2 1 2 1 0 0 1 2 1 0 0 2 0 1 1 0 0 0 0 0 0 0 0 0 0 0 0 0 0 0
0 0 0 0 0 0 0 0 0 0 0 0 0 0 0 0 0 0 0 0 0 0 0 0 0 0 0 0 0 0 0 0
0 0 0 0 0 0 0 0 0 0 0 0 0 0 0 0 0 0 0 0 0 0 0 0 0 0 0 0 0 0 0 0
0 0 0 0 0 0 0 0 0 0 0 0 0 0 0 0 0 0 0 0 0 0 0 0 0 0 0 0 0 0 0 0
0 0 0 0 0 0 0 0 0 1 0 0 0 0 0 0 0 0 0 0 0 0 0 0 0 0 0 0 0 0 0 0
0 0 0 0 0 0 0 0 0 0 0 0 0 0 0 0 0 0 0 0 0 0 0 0 0 0 0 0 0 0 0 0
0 0 0 0 0 0 0 0 0 0 0 0 0 0 0 0 0 0 0 0 0 0 0 0 0 0 0 0 0 0 0 0
0 0 0 0 0 0 0 0 0 0 0 0 0 0 0 0 0 0 0 0 0 0 0 0 0 0 0 0 0 0 0 0
0 0 1 1 1 1 1 1 0 0 0 0 0 0 0 0 0 0 0 0 0 0 0 2 0 0 0 0 0 0 0
0 0 0 0 0 0 0 0 0 0 0 0 0 0 0 0 0 0 0 0 0 0 0 0 0 1 1 0 0 0 0 0
0 0 0 0 0 0 0 0 0 0 0 0 0 0 0 0 0 0 0 0 0 0 0 0 0 0 0 0 0 0 0 0
0 0 0 0 0 0 0 0 0 0 0 0 0 0 0 0 0 0 0 0 0 0 0 0 0 0 0 0 0 0 0 0
0 0 0 0 0 0 1 0 0 0 0 0 0 0 0 0 0 0 0 0 0 0 0 0 0 0 0 0 0 0 0 0
0 0 0 0 0 0 1 0 0 0 0 0 0 0 0 0 0 0 0 0 0 0 0 0 0 0 0 0 0 0 0 0
0 0 0 0 0 0 0 0 0 0 0 0 0 0 0 0 0 0 0 0 0 0 0 0 0 0 0 0 0 0 0 0
0 0 0 0 0 0 0 0 0 0 0 0 0 0 0 0 0 0 0 0 0 0 0 0 0 0 0 0 0 0 0 0
0 0 0 0 0 0 0 0 0 0 0 0 0 0 0 0 0 0 0 0 0 0
```

Here, we can see that after the first few rounds, most of the choices made by the algorithm were all arm 0. But from time to time, arm 1 or arm 2 would be chosen, presumably from the random exploration probability.

7. In the next code cell, we will conduct the same experiment, but this time, we will set $\varepsilon = 0.1$:

```
egreedy_policy_v2 = eGreedy(n_arms=N_ARMS, e=0.1)

history, rewards, optimal_rewards = bandit.automate\
                                    (egreedy_policy_v2, \
                                     n_rounds=500, \
                                     visualize_regret=True)
```

This will produce the following graph:

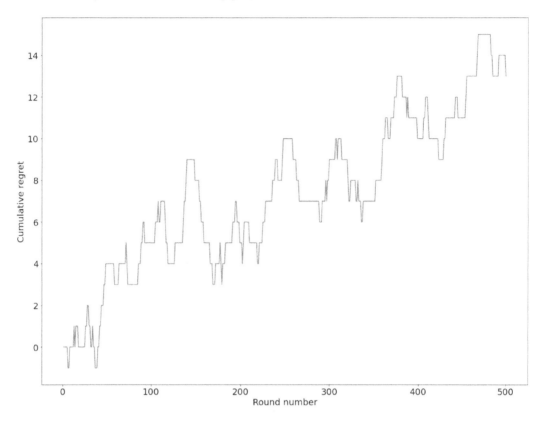

Figure 8.8: Sample cumulative regret with increased exploration probability

Here, our cumulative regret is a lot higher than what we got with ε = 0.01 in *step 5*. This is presumably due to the increased exploration probability, which is too high.

8. To analyze this experiment further, we can print out the action history once more:

```
print(*history)
```

This will produce the following output:

```
0 1 2 2 0 1 0 1 2 2 0 2 2 0 2 0 0 0 0 0 0 0 0 0 0 0 0 0 0 0 0 0
0 0 0 0 0 1 0 0 0 2 0 0 0 0 0 0 0 0 0 0 0 0 0 0 0 0 0 0 0 0 0 0
0 0 0 2 0 0 0 0 0 0 0 0 0 0 0 0 0 0 2 0 0 0 0 0 1 2 0 1 0
0 0 0 0 0 0 0 2 0 0 0 0 0 0 0 0 0 0 0 0 0 0 0 0 0 0 0 0 0
0 0 0 0 0 0 0 0 0 0 1 2 0 0 0 0 0 1 0 0 0 1 0 0 0 0 0 0 0
0 0 0 0 0 0 0 0 0 0 0 0 0 1 0 0 0 0 0 0 0 0 0 0 0 0 0 2 0
0 0 0 1 0 0 0 0 0 0 0 0 0 0 0 0 0 0 0 0 0 0 0 0 0 0 0 0 0
0 0 0 0 0 0 0 0 0 0 0 1 0 0 0 0 0 0 0 0 0 0 0 0 0 0 0 1 1
0 0 0 0 0 0 2 0 0 0 1 0 0 0 0 0 0 0 0 0 0 0 0 0 0 0 2 0
0 0 0 0 0 0 0 0 0 0 0 0 0 0 1 0 1 0 0 0 0 0 0 0 0 2 0 0
0 0 0 0 0 0 0 0 2 0 0 0 0 0 0 0 0 0 0 0 0 0 0 0 0 0 0 0
0 0 0 0 0 0 0 2 2 1 0 0 0 0 0 0 0 0 0 0 0 0 0 0 0 0 0 0
0 0 2 0 0 0 0 0 0 0 1 0 0 0 0 0 0 0 0 1 0 0 0 0 0 0 0
0 0 0 0 0 0 0 0 0 0 0 0 0 0 1 0 0 0 0 0 0 0 0 0 0 0 0 0
0 0 0 0 0 0 0 0 0 0 0 0 0 0 0 1 0 0 0 0 0 0 0 0 0 0 0 0
0 0 0 0 0 0 1 0 0 0 0 0 0 0 0 0 2 0 0 2 0 0 0 0 0 0
0 0 0 0 1 0 0 0 0 0 0 0 0 0 0 0 0 0 0 0
```

Comparing this with the same history of the previous algorithm, we can see that this algorithm did indeed explore significantly more during the late rounds. All of this indicates to us that $\varepsilon = 0.1$ might not be an appropriate exploration probability.

9. As the last component of our analysis of the ε-Greedy algorithm, let's utilize the repeated-experiment option. This time, we will choose $\varepsilon = 0.03$, like so:

```
regrets = bandit.repeat(eGreedy, [N_ARMS, 0.03], \
                         n_experiments=100, n_rounds=300, \
                         visualize_regret_dist=True)
```

The following graph will be produced, which visualizes the distribution of cumulative regret resulting from these repeated experiments:

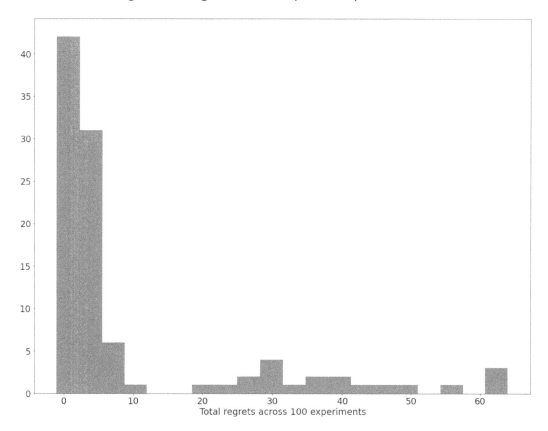

Figure 8.9: Distribution of cumulative regret by the ε-Greedy algorithm

This distribution is quite similar to what we obtained with the Greedy algorithm. Next, we will compare the two algorithms further.

10. Calculate the mean and max of these cumulative regret values with the following code:

```
np.mean(regrets), np.max(regrets)
```

The output will be as follows:

```
(9.95, 64)
```

Comparing this with what we had with the Greedy algorithm (**8 . 66** and **62**), this result indicates that the ε-Greedy algorithm might be inferior in this specific bandit problem. However, it has managed to formalize the choice between exploration and exploitation using its exploration rate, which was lacking in the Greedy algorithm. This is a valuable characteristic of a MAB algorithm, and will be the focus of other algorithms that we will be discussing in the rest of this chapter.

> **NOTE**
>
> To access the source code for this specific section, please refer to https://packt.live/3fiE3Y5.
>
> You can also run this example online at https://packt.live/3cYT4fY.

Before we move on to the next section, let's briefly discuss yet another so-called variant of the Greedy algorithm, the Softmax algorithm.

THE SOFTMAX ALGORITHM

The Softmax algorithm attempts to quantify the trade-off between exploration and exploitation by choosing each of the available arms with a probability that is proportional to its average reward. Formally, the probability that arm *i* is chosen by the algorithm at each time step, *t*, is as follows:

$$p_i(t) = \frac{e^{m_i(t-1)}}{\sum_j^k e^{m_j(t-1)}}$$

Figure 8.10: Expression for the probability that the arm is chosen
by the algorithm at each time step

Each term in the exponent $m_i(t-1)$ is the average reward observed from arm i in the first *(t - 1)* time steps. Given the way the probabilities are defined, the larger this average reward is, the more likely the corresponding arm will be chosen. In its most general form, this average term is divided by a tunable parameter, τ, which controls the exploration rate of the algorithm. Specifically, when τ tends to infinity, the probability of the largest arm will approach one while the other probabilities approach zero, making the algorithm purely greedy (which is why we consider it to be a generalization of the Greedy algorithm). The smaller τ is, the more likely it is that a temporarily sub-optimal arm is chosen. As it tends to 0, the algorithm uniformly explores all the available arms indefinitely.

Similar to the problem we encounter while designing the ε-Greedy algorithm, it is not entirely clear how we should set the value of this parameter, τ, for each specific bandit problem, even though the performance of the algorithm can be highly dependent on this parameter. For that reason, the Softmax algorithm is not as popular as the algorithms we will be discussing in this chapter.

And with that, we conclude our discussion of the Greedy algorithm, our first approach to solving the MAB problem, and three of its variations: Explore-then-commit, ε-Greedy, and Softmax. Overall, these algorithms focus on exploiting the arm with the greatest reward mean while sometimes deviating from that to explore other, seemingly suboptimal, arms.

In the next section, we will move on to another common MAB algorithm called **Upper Confidence Bound** (**UCB**), the intuition of which is slightly different from what we have seen so far.

THE UCB ALGORITHM

The term *upper confidence bound* denotes the fact that instead of considering the average of past rewards returned from each arm like Greedy, the algorithm computes an upper bound for its estimates of the expected reward for each arm.

This concept of a confidence bound is quite common in probability and statistics, where the distribution of a quantity that we care about (in this case, the reward from each arm) cannot be represented well using simply the average of past observations. Instead, a confidence bound is a numerical range that aims to estimate and narrow down where most of the values in the distribution in question will lie. For example, this idea is widely used in Bayesian analyses and Bayesian optimization.

In the following section, we will discuss how UCB establishes its use of a confidence bound.

OPTIMISM IN THE FACE OF UNCERTAINTY

Consider the middle of the process of a bandit with only two arms. We have already pulled the first arm 100 times and observed an average reward of **0.61**; for the second arm, we have only seen five samples, three of which were **1**, so its average reward is **0.6**. Should we commit to exploring the first arm for the rest of the remaining rounds and ignore the second?

Many would say no; we should at least explore the second arm more to get a better estimation of its expected reward. The motivation for this observation is that since we only have very few samples of the reward from the second arm, we should not be *confident* that the mean reward of the second arm is actually lower than that of the first. How, then, should we formalize our intuition? The UCB algorithm, or specifically, its most common variant – the UCB1 algorithm – states that instead of the mean reward, we will use the following sum of the average reward and the confidence bound:

$$Reward\ estimation\ of\ arm\ i = \frac{X_1 + X_2 + \cdots X_{T_i(t-1)}}{T_i(t-1)} + \sqrt{\frac{2log(t)}{T_i(t-1)}}$$

Figure 8.11: Expression for the UCB algorithm

Here, t denotes the time step, or the round number, that we are currently in while interacting with a bandit, and $T_i(t)$ denotes the number of times we have pulled arm i up to round t. The rest of UCB works in the same way as the Greedy algorithm: at each step, we choose to pull the arm that maximizes the preceding sum, observe the returned reward, add it to our reward, and repeat the process.

To implement this logic, we can use the **decide()** method, which contains the following code:

```
def decide(self):
        for arm_id in range(self.n_arms):
            if len(self.reward_history[arm_id]) == 0:
                return arm_id
```

```
conf_bounds = [np.mean(history) \
                + np.sqrt(2 * np.log(self.t) / len(history))\
                for history in self.reward_history]

return int(np.random.choice\
            (np.argwhere(conf_bounds == np.max(conf_bounds))\
            .flatten()))
```

Here, **self.t** should be equal to the current step time. As we can see, the method returns the arm that maximizes the element in **conf_bounds**, which is the list storing the optimistic estimation of each arm.

You might be wondering why using the preceding quantity can capture the idea behind the confidence bound that we'd like to apply to our estimations of the expected reward. Remember the example of a two-arm bandit we sketched out earlier, where we would like to have a formalization that encourages exploration of the rarely explored arm (the second one, in our example). As you can see, at any given round, this quantity is a decreasing function of T_i. In other words, the quantity gets smaller when T_i is large and grows larger when the opposite is true. So, this quantity is maximized by the arm that has the lower number of pulls – the arm that is explored the least. In our example, the estimation of the first arm is as follows:

$$0.61 + \sqrt{\frac{2\log(105)}{100}} = 0.811$$

Figure 8.12: Estimation of the first arm

The estimation of the second arm is as follows:

$$0.6 + \sqrt{\frac{2\log(105)}{5}} = 1.499$$

Figure 8.13: Estimation of the second arm

Using UCB, we choose to pull the second arm next, which is what we argued was the correct choice. By adding what is called an exploration term to the mean reward, we are, in a way, estimating the largest possible value of the expected mean, not just the expected mean itself. This intuition is best summed up with the term *optimism in the face of uncertainty*, and it is the quintessential characteristic of the UCB algorithm.

OTHER PROPERTIES OF UCB

UCB is not unjustifiably optimistic. When an arm is significantly under-explored, the exploration term will make the sum larger, making it more likely to be chosen by UCB, but it is never guaranteed that the arm will surely be chosen. Specifically, when the mean reward of an arm is so low that a large value of the term cannot compensate for it, UCB will choose to exploit the good arms anyway.

We should also discuss its variation in cost-centric MAB problems, known as the **Lower Confidence Bound** (**LCB**). With respect to a reward-centric problem, we are adding the exploration term to the mean reward to compute an optimistic estimation of the true mean. When the MAB problem is the minimization of costs returned by the arms, our optimistic estimation becomes the mean cost *subtracted* by the exploration term, and the arm that minimizes this quantity will be chosen by UCB, or in this case, LCB.

In particular, we are saying that if an arm is under-explored, its true mean cost might be lower than what we have observed so far, so we subtract the average cost from the exploration term to estimate the lowest possible cost of an arm. Aside from this, the implementation of this variation of UCB remains the same.

That is enough theory about UCB. To conclude our discussions of this algorithm, we will implement it for the Bernoulli three-arm bandit problem that we have been using in the next exercise.

EXERCISE 8.02 IMPLEMENTING THE UCB ALGORITHM

In this exercise, we will be implementing the UCB algorithm. This exercise will walk us through the familiar workflow that we have been using to analyze the performance of an MAB algorithm: implement it as a Python class, put it through a single experiment and observe its behavior, and finally repeat the experiment many times to consider the resulting distribution of its regret.

We will follow these steps to do so:

1. Create a new Jupyter Notebook and import **NumPy**, **Matplotlib**, and the **Bandit** class from the **utils.py** file included in the code repository for this chapter:

```
import numpy as np
np.random.seed(0)
import matplotlib.pyplot as plt

from utils import Bandit
```

2. Declare a Python class named **UCB** with the following initialization method:

```
class UCB:
    def __init__(self, n_arms=2):
        self.n_arms = n_arms
        self.reward_history = [[] for _ in range(n_arms)]
        self.t = 0
```

Different from Greedy and its variants, our implementation of **UCB** needs to keep track of an additional piece of information, the current round number, in its attribute, **t**. This information is used when calculating the exploration term of the upper confidence bound.

3. Implement the **decide()** method of the class, as follows:

```
def decide(self):
    for arm_id in range(self.n_arms):
        if len(self.reward_history[arm_id]) == 0:
            return arm_id

    conf_bounds = [np.mean(history) \
                    + np.sqrt(2 * np.log(self.t) \
                        / len(history))\
                    for history in self.reward_history]
    return int(np.random.choice\
            (np.argwhere\
            (conf_bounds == np.max(conf_bounds))\
            .flatten()))
```

The preceding code is self-explanatory: after pulling each arm at least once, we compute the confidence bounds as the sum of the empirical mean reward and the exploration term. Finally, we return the arm with the largest sum, randomly tie-breaking if necessary.

4. In the same code cell, implement the **update()** method of the class, like so:

```
def update(self, arm_id, reward):
    self.reward_history[arm_id].append(reward)
    self.t += 1
```

We are already familiar with most of the logic here from the previous algorithms. Notice here that with each call to **update()**, we also need to increment the attribute, **t**.

5. Declare the Bernoulli three-arm bandit problem that we have been considering and run it on an instance of the UCB algorithm we just implemented:

```
N_ARMS = 3

bandit = Bandit(optimal_arm_id=0,\
                n_arms=3,\
                reward_dists=[np.random.binomial \
                            for _ in range(N_ARMS)],\
                reward_dists_params=[(1, 0.9), (1, 0.8), \
                                    (1, 0.7)])

ucb_policy = UCB(n_arms=N_ARMS)

history, rewards, optimal_rewards = bandit.automate\
                            (ucb_policy, n_rounds=500, \
                            visualize_regret=True)
```

This code will produce the following graph:

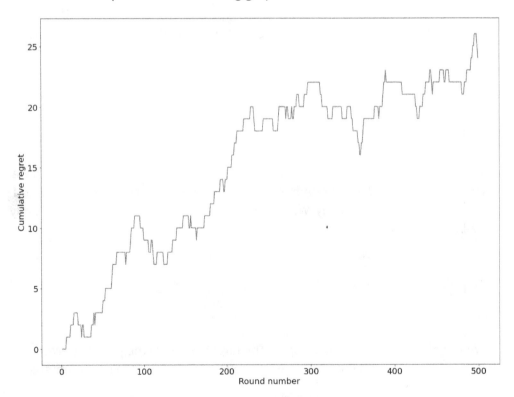

Figure 8.14: Sample cumulative regret from UCB

Here, we can see that this cumulative regret is significantly worse than what we saw with Greedy, which was, at most, 2. We hypothesize that the difference is a direct result of the optimistic nature of the algorithm.

6. To understand this behavior better, we will inspect the pulling history of the algorithm:

```
print(*history)
```

This produces the following output:

```
0 1 2 1 0 2 0 1 1 0 1 0 2 0 2 0 0 1 0 1 0 1 0 1 2 0 1 0 1 0 0
1 0 1 0 1 0 0 0 2 2 1 1 0 1 0 1 0 1 0 1 1 1 2 2 2 2 0 2 0 2 0
1 1 1 1 1 0 0 0 0 0 2 2 0 0 1 0 1 0 0 0 0 0 1 0 2 2 2 0 0 0 0
0 0 0 0 1 0 1 0 1 1 0 1 0 1 0 1 0 0 0 0 2 2 2 0 0 0 0 0 0 0 0
0 1 1 0 1 0 0 0 0 0 2 2 2 2 2 0 1 0 1 1 0 1 0 0 0 0 0 0 2 2
2 2 0 0 1 0 1 1 0 1 0 0 0 0 0 0 0 0 0 1 0 1 1 1 0 2 1 1 0 1 0
1 1 1 1 1 1 1 0 0 0 0 0 0 1 1 1 0 0 2 2 2 0 0 0 1 1 1 0 0 0 0
0 0 2 2 0 0 0 0 0 1 0 0 0 0 0 0 0 0 0 0 0 0 0 1 1 1 0 0 0 0
0 0 0 0 0 0 2 2 2 2 2 2 2 0 0 0 0 0 0 2 2 2 2 1 1 1 1 1 1 0
0 0 0 0 0 1 1 1 1 1 1 1 1 1 2 2 2 2 0 0 0 0 0 0 1 1 0 0 0 0 0
0 0 0 0 0 0 0 0 0 0 0 0 0 0 2 2 0 0 0 0 0 0 0 0 0 0 0 0 0 1
1 1 1 0 0 0 0 0 0 0 0 0 0 0 0 0 0 0 0 0 0 0 2 2 1 0 0 0 0 0 0 0
0 0 0 0 0 0 0 0 0 0 0 1 1 1 1 0 2 2 0 0 0 0 0 0 0 0 0 0 0 0 1
1 1 1 0 0 0 0 0 0 0 0 0 0 0 0 0 0 0 0 0 0 0 0 0 0 0 0 0 2 1 1
1 0 0 0 0 0 0 0 0 1 1 1 1 0 0 0 0 0 0 0 0 2 2 2 2 2 2 2 0 0 0
0 0 0 0 0 0 0 0 0 1 1 1 1 1 1 1 1 1 1 0 0 0 2 2 2 2 0 0 1 1 0 0 0 0
```

Here, we can observe that instead of exploiting the true optimal arm (arm 0), UCB frequently chose to deviate. This is a direct effect of its tendency to optimistically explore seemingly suboptimal arms.

7. At face value, we might conclude that for this bandit problem, UCB is, in fact, not superior than the Greedy algorithm, but to truly confirm whether that is true or not, we need to inspect how the algorithm does across multiple experiments. Use the **repeat()** method from the bandit API to confirm this:

```
regrets = bandit.repeat(UCB, [N_ARMS], n_experiments=100, \
                        n_rounds=300, visualize_regret_dist=True)
```

This code snippet will generate the following plot:

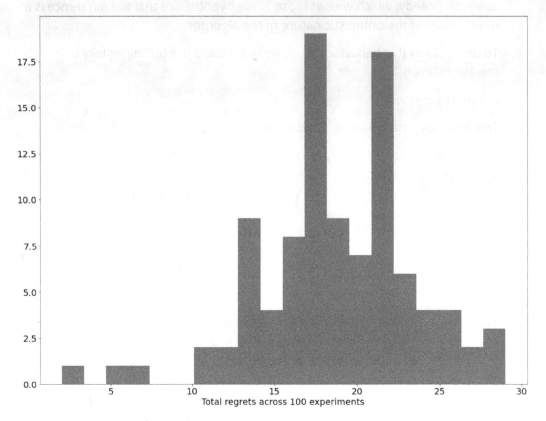

Figure 8.15: Distribution of regret of UCB

To our surprise, the regret values in this distribution are significantly lower than those resulting from the Greedy algorithm.

8. In addition to visualizing the distribution, we also need to consider the average and max regret across all experiments:

```
np.mean(regrets), np.max(regrets)
```

The output will be as follows:

```
(18.78, 29)
```

As you can see, the values are significantly lower than the corresponding statistics we saw in Greedy, which were **8.66** and **62**. Here, we can say that we have evidence supporting the claim that UCB is better than Greedy in terms of minimizing the cumulative regret of a bandit problem.

> **NOTE**
>
> To access the source code for this specific section, please refer to https://packt.live/3fhxSmX.
>
> You can also run this example online at https://packt.live/2XXuJmK.

This example also illustrates the importance of repeating experiments when analyzing the performance of a MAB algorithm. As we saw earlier, using just a single experiment, we could have arrived at the wrong conclusion that UCB is inferior to the Greedy algorithm in the specific bandit problem we are considering. However, across many repeated experiments, we can see that the opposite is true.

Throughout this exercise, we have implemented UCB, as well as learned about the need for comprehensive analysis with multiple experiments while working with MAB algorithms. This also marks the end of the topic surrounding the UCB algorithm. In the next section, we will begin talking about the last MAB algorithm in this chapter: Thompson Sampling.

THOMPSON SAMPLING

The algorithms we have seen so far make up a set of diverse insights: Greedy and its variants mostly focus on exploitation and might need to be explicitly forced to employ exploration; UCB, on the other hand, tends to be optimistic about the true expected reward of under-explored arms and therefore naturally, but also justifiably, focuses on exploration.

Thompson Sampling also uses a completely different intuition. However, before we can understand the idea behind the algorithm, we need to discuss one of its principal building blocks: the concept of Bayesian probability.

INTRODUCTION TO BAYESIAN PROBABILITY

Generally speaking, the workflow of using Bayesian probability to describe a quantity consists of the following elements:

- A prior probability representing whatever prior knowledge or belief we have about the quantity.

- A likelihood probability that denotes, as the name of the term suggests, how likely the data that we have observed so far is.

- And finally, a posterior probability, which is the combination of the preceding two elements.

One fundamental component of Bayesian probability is Bayes' theorem:

$$P(A|B) = \frac{P(B|A).P(A)}{P(B)}$$

Figure 8.16: Bayes' theorem

Here, *P(X)* denotes the probability of a given event, *X*, while *P(X | Y)* is the probability of a given event, *X*, provided that event *Y* has already happened. The latter is an example of conditional probabilities, which is a common object in machine learning, especially when different events/quantities are conditionally dependent on each other.

This specific formula outlines the general idea of Bayesian probability that we have here: say we are given a prior probability for an event, *A*, and we also know how likely event *B* happens *given* event *A*. Here, the posterior probability of the same event, *A*, given event *B*, is proportional to the product of the two aforementioned probabilities. Event A is typically what we care about, while event B is the data that we have observed. To put this into perspective, let's consider the application of this formula in the context of a Bernoulli distribution.

We'd like to estimate the unknown parameter, *p*, that characterizes a Bernoulli distribution, from which we have observed five samples. Due to how a Bernoulli distribution is defined, the probability that the sum of these five samples is equal to *x*, an integer between 0 and 5, is $\binom{5}{x}p^x(1-p)^{5-x}$ (don't worry if you aren't familiar with this expression).

But what if the samples are what we can observe, and we are unsure what the actual value of p is? How can we "flip" the direction of the preceding probabilistic quantity so that we can draw some conclusions about the value of p from the samples? This is where Bayes' theorem comes into play. In the Bernoulli example, from the likelihood of the sum of the observed samples given any value of p, we can calculate the probability that p is indeed that value, given the observations that we have.

This is directly connected to the MAB problem. We, of course, always start out not knowing what the actual value, p, that parameterizes the reward distribution of a given arm is, but we can observe the reward samples drawn from it by pulling that arm. So, from a number of samples, we can calculate what the probability that p is equal to, say, 0.5 is, and whether that probability is larger than the probability that p is equal to 0.6.

A question remains about how to choose the prior distribution for p. In our case, when we start out without any prior information about p, we might say that p is equally likely to be any number between 0 and 1. So, we model p using a uniform distribution between 0 and 1. The Beta distribution is a generalization of the uniform distribution where its parameters are α = 1 and β = 1, so let's say p, for now, follows Beta(1, 1).

Bayes' theorem allows us to *update* this Beta distribution to another Beta distribution with different parameters after seeing some observations. Following our running example, say, after five separate observations from this Bernoulli distribution that we are modeling, we have three instances of 1 and two instances of 0. According to the Bayesian updating rule (the math of which is out of scope for this book), a Beta distribution with α and β parameters will be updated to α + 3 and β + 2.

> **NOTE**
>
> In general, out of n observations, **x** of which are **1** and the others are **0**, a `Beta(α, β)` distribution will be updated to `Beta(α + x, β + n - x)`. Roughly speaking, in an update, α should be incremented by the number of samples observed, while β should be incremented by the number of zero samples.

From this newly updated distribution, which reflects that data that we can observe, the new estimation of p, which is the mean of the distribution, can be computed as α / ($\alpha + \beta$). We said that we typically start out by modeling p using a uniform, or Beta(1, 1), distribution; the expectation of p, in this case, is 1 / (1 + 1) = 0.5. As we see more and more samples from the Bernoulli distribution with the true value of p, we will update this Beta distribution that we are using to better model p to reflect which values of p are now likely, given those samples.

Let's consider a visual illustration to tie all of this together. Consider a Bernoulli distribution with $p = 0.9$, which we consider unknown to us. We can, again, only draw samples from this distribution and we'd like to use the Bayesian update rule described previously to model our belief about p. Say that at each timestep out of 1,000 timesteps, we can draw one sample from the distribution. Our observations are as follows:

- At timestep 0, we don't have any observations yet.

- At timestep 5, we have all observations being ones, and none being zero.

- At timestep 10, we have 9 positive observations and 1 zero.

- At timestep 20, we have 18 positive observations and 2 zeros.

- At timestep 100, we have 91 positive observations and 9 zeros.

- At timestep 1,000, we have 892 positive observations and 108 zeros.

First of all, we can see that the fraction of positive observations is roughly equal to the true value of $p = 0.9$, which is unknown to us. Additionally, we don't have any prior knowledge on this value of p, so we choose to model it using Beta(1, 1). This corresponds to the horizontal probability density function that we have in the upper-left panel of the following plot.

For the rest of the panels, we use the Bayesian update rule to compute a Beta distribution with new parameters to fit the data we observe better. The blue lines are the probability density function of p, indicating how likely it is that p is equal to one specific value between 0 and 1, given the observations we have.

At timestep 5, all of our observations are one, so our belief gets update to reflect that the probability that p is some value close to 1 is very large. This is indicated by an increase in probability mass on the right-hand side of the plot. At timestep 10, one zero observation occurs, so the probability that p is exactly 1 decreases, giving more mass to the values close to but below 1. In latter timesteps, the curve grows tighter and tighter, indicating that the model is becoming more and more confident about what values p can take on. Finally, at timestep 1,000, the function peaks around the point 0.9 and nowhere else, indicating that it is extremely confident p is roughly 0.9:

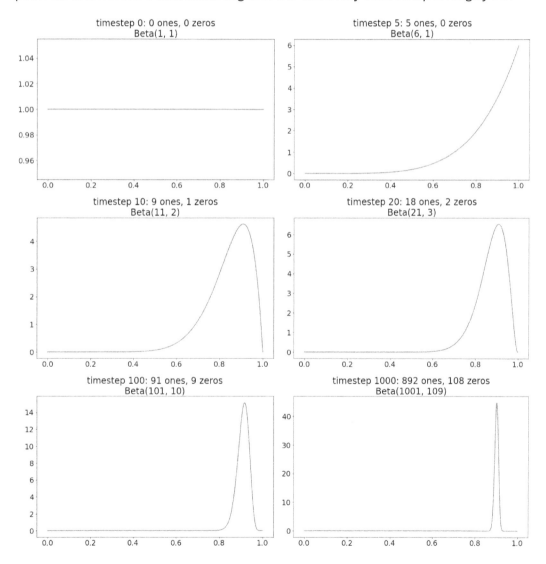

Figure 8.17: A visual illustration of the Bayesian updating process

In our example, Beta distributions are used to model the unknown parameter of a Bernoulli distribution; it is important that Beta distributions are used because when Bayes' theorem is applied, the prior probability associated with a Beta distribution, combined with the likelihood probability of a Bernoulli distribution, simplifies the math significantly, allowing the posterior to become a different Beta distribution with newly updated parameters. If another distribution aside from Beta were to be used, the formula would not be simplified in such a way. The Beta distribution is therefore called the *conjugate prior* of the Bernoulli distribution. In Bayesian probability, when we'd like to model the unknown parameters of a given distribution, the conjugate prior of that distribution should be used so that the math will work out.

If this process is still confusing to you, don't worry, as most of the theory behind Bayesian updating and conjugate priors has already been worked out for common probability distributions. For our purposes, we simply need to remember the update rule for the Bernoulli/Beta distribution that we just discussed.

> **NOTE**
>
> For those of you who are interested, please feel free to consult the following material from MIT, which further introduces conjugate priors of various probability distributions: https://ocw.mit.edu/courses/mathematics/18-05-introduction-to-probability-and-statistics-spring-2014/readings/MIT18_05S14_Reading15a.pdf.

So far, we have learned how to model the unknown parameter, p, of a Bernoulli distribution in a Bayesian fashion when given data that we can observe. In the next section, we will finally connect this topic back to our original point of discussion: the Thompson Sampling algorithm.

THE THOMPSON SAMPLING ALGORITHM

Consider the Bayesian technique of modeling p that we just learned about in the context of an MAB problem with Bernoulli reward distributions. We now have a way to quantify, probabilistically, our belief about the value of p, given the reward samples we have observed from the corresponding arm. From here, we can simply employ a greedy strategy again and choose the arm with the greatest expectation of p, which is, again, computed as $\alpha / (\alpha + \beta)$, where α and β are the running parameters of the current Beta distribution modeling p.

Instead, to implement Thompson Sampling, we draw a sample from each of the Beta distributions that model the p parameter of each of the Bernoulli distributions and select the maximal one. In other words, each arm in the bandit problem has a Bernoulli reward distribution whose parameter, p, is being modeled by some Beta distribution. We sample from each of these Beta distributions and pick the arm with the highest-valued sample.

Let's say that, in the class object syntax that we have been using to implement MAB algorithms, we store the running values of alpha and beta used by a Beta distribution in the **temp_beliefs** attribute to model the parameter, p, of each arm. The logic of Thompson Sampling can be applied as follows:

```
def decide(self):
    for arm_id in range(self.n_arms):
        if len(self.reward_history[arm_id]) == 0:
            return arm_id

    draws = [np.random.beta(alpha, beta, size=1)\
            for alpha, beta in self.temp_beliefs]

    return int(np.random.choice\
            (np.argwhere(draws == np.max(draws)).flatten()))
```

Different from Greedy or UCB, to estimate the true value of p for each arm, we draw a random sample from the corresponding Beta distribution whose parameters have been updated by the Bayesian updating rule throughout the process (as can be seen in the **draws** variable). To choose an arm to pull, we simply identify the arm that has the best sample.

Two immediate questions come to mind: first, why is this sampling process a good way to estimate the reward expectation of each arm, and second, how does the technique address the trade-off between exploration and exploitation?

When we sample from each of the Beta distributions, the more likely p is equal to a given value, the more likely that value will be chosen as our sample – this is simply the nature of a probability distribution. So, in a way, a sample from a distribution is an approximation of the quantity that the distribution models. This is why samples from the Beta distributions can be justifiably used as estimations of the true value of p of each of the Bernoulli distributions.

With that said, when the current distribution representing our belief about parameter p of a given Bernoulli is flat and does not have a sharp peak (as opposed to the one in the final panel of the preceding visualization), it indicates that we still have a lot of uncertainty about what value p might be, which is why many numbers are given more probability mass than in a distribution with a single sharp peak. When a distribution is relatively flat, the samples drawn from it are likely to be dispersed across the range of the distribution, as opposed to surrounding one single region, again indicating our uncertainty about the true value. All of this is to say that even though samples can be used as approximations of a given quantity, the accuracy of those approximations depends on how flat the modeling distribution is (and therefore, ultimately, how certain our belief is about the true value).

This fact directly helps us address the exploration-exploitation dilemma. When samples for p's are drawn from distributions with single sharp peaks, they are more likely to be very close to the true values of the corresponding p's, so choosing the arm with the highest sample is equivalent to choosing the arm with the highest p (or expected reward). When a distribution is still flat, the values of the samples drawn from it are likely to be volatile and might, therefore, take on large values. If, somehow, an arm is chosen because of this reason, this means that we are not certain enough about the value of p for this arm, and it's therefore worth exploring.

Thompson Sampling, by sampling from the modeling distributions, offers an elegant method of balancing exploitation and exploration: if we are certain with our beliefs about each arm, picking the best sample is likely to be equivalent to picking the actual optimal arm; if we are not certain about an arm enough that its corresponding sample has the best value, exploring it will be beneficial.

As we will see in the upcoming exercise, the actual implementation of Thompson Sampling is quite straightforward, and we won't need to include much of the theoretical Bayesian probability that we have discussed in the implementation.

EXERCISE 8.03: IMPLEMENTING THE THOMPSON SAMPLING ALGORITHM

In this exercise, we will be implementing the Thompson Sampling algorithm. As always, we will be implementing the algorithm as a Python class and subsequently applying it to the Bernoulli three-arm bandit problem. Specifically, we will walk through the following steps:

1. Create a new Jupyter Notebook and import **NumPy**, **Matplotlib**, and the **Bandit** class from the **utils.py** file included in the code repository for this chapter:

```
import numpy as np
np.random.seed(0)
import matplotlib.pyplot as plt

from utils import Bandit
```

2. Declare a Python class named **BernoulliThompsonSampling** (indicating that the class will be implementing the Bayesian update rule for Bernoulli/Beta distributions) with the following initialization method:

```
class BernoulliThompsonSampling:
    def __init__(self, n_arms=2):
        self.n_arms = n_arms
        self.reward_history = [[] for _ in range(n_arms)]
        self.temp_beliefs = [(1, 1) for _ in range(n_arms)]
```

Remember that in Thompson Sampling, we maintain a running belief about p of each Bernoulli arm using a Beta distribution whose two parameters are updated according to the update rule. Therefore, we only need to keep track of the running values of these parameters; the **temp_beliefs** attribute contains this information for each of the arms, whose default value is (1, 1).

3. Implement the **decide()** method, using the **np.random.beta** function from NumPy to draw a sample from a Beta distribution, like so:

```
def decide(self):
    for arm_id in range(self.n_arms):
        if len(self.reward_history[arm_id]) == 0:
            return arm_id
```

```
        draws = [np.random.beta(alpha, beta, size=1)\
                    for alpha, beta in self.temp_beliefs]

        return int(np.random.choice\
                    (np.argwhere(draws == np.max(draws)).flatten()))
```

Here, we can see that instead of computing the mean reward or its upper confidence bound, we simply draw a sample from each of the Beta distributions defined by the parameters stored in the **temp_beliefs** attribute.

Finally, we pick the arm that corresponds to the maximum sample.

4. In the same code cell, implement the **update()** method for the class. In addition to appending the most recent reward to the history of the appropriate arm, we need to implement the logic of the update rule:

```
    def update(self, arm_id, reward):
        self.reward_history[arm_id].append(int(reward))

        # Update parameters according to Bayes rule
        alpha, beta = self.temp_beliefs[arm_id]
        alpha += reward
        beta += 1 - reward
        self.temp_beliefs[arm_id] = alpha, beta
```

Remember that the first parameter, α, should be incremented once for every sample we observe, while β should be incremented if the sample is zero. The preceding code implements this logic.

5. Next, set up the familiar Bernoulli three-arm bandit problem and apply an instance of the Thompson Sampling class implementation to it to plot out the cumulative regret in that single experiment:

```
N_ARMS = 3

bandit = Bandit(optimal_arm_id=0,\
                n_arms=3,\
                reward_dists=[np.random.binomial \
                                for _ in range(N_ARMS)],\
                reward_dists_params=[(1, 0.9), (1, 0.8), \
                                        (1, 0.7)])
```

```
ths_policy = BernoulliThompsonSampling(n_arms=N_ARMS)

history, rewards, optimal_rewards = bandit.automate\
                                    (ths_policy, n_rounds=500, \
                                     visualize_regret=True)
```

The following graph will be produced:

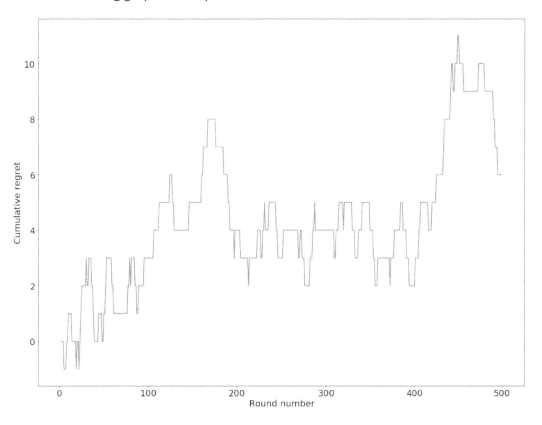

Figure 8.18: Sample cumulative regret from Thompson Sampling

This regret plot is better than the one we obtained with UCB but worse than the one from Greedy. The plot will be used in conjunction with the pulling history in the next step for further analysis. Let's analyze the pulling history further.

6. Print out the pulling history:

```
print(*history)
```

The output will be as follows:

```
0 1 2 0 0 2 0 0 0 1 0 2 2 0 0 0 2 0 2 2 0 0 0 2 2 0 0 0 0 0 0 0
0 0 0 0 0 2 2 0 0 0 0 0 0 0 0 0 0 1 0 0 0 0 0 0 0 0 0 0 0 0 0 1
0 1 2 0 0 0 0 0 2 0 0 0 0 0 0 0 0 0 1 0 1 0 1 0 0 2 1 1 0 2
0 1 0 0 0 0 0 0 0 0 0 0 0 0 0 0 2 0 0 0 0 0 0 0 0 1 1 0 0 0 1
0 0 0 0 0 0 0 0 0 0 2 0 0 0 1 0 0 0 0 0 0 0 0 0 0 1 0 0 0 0
0 0 0 0 0 0 0 0 0 0 1 0 0 0 0 0 0 2 0 0 0 0 0 0 0 0 0 0 0 0
0 0 0 0 0 0 0 0 0 0 1 2 0 0 0 0 0 0 0 0 0 0 2 0 0 0 0 0 0 0
0 0 0 0 0 0 1 0 0 0 0 0 0 0 0 0 0 0 0 0 0 0 0 0 0 0 0 0 0 0
0 0 0 0 0 0 0 0 0 0 0 0 0 0 0 0 0 0 0 0 0 0 0 0 0 0 0 0 0 0
0 0 0 2 1 0 0 0 0 0 0 2 0 0 0 0 0 0 2 0 0 0 0 0 0 0 0 2 0 2
0 0 0 0 0 0 0 2 2 0 0 0 0 0 0 2 0 0 0 0 0 0 0 0 0 0 0 0 2 0
0 0 0 0 0 0 0 0 0 0 0 0 0 0 0 0 0 0 0 0 0 0 0 0 0 0 0 1 0
0 0 1 0 0 0 0 1 0 0 0 0 0 0 0 0 0 0 0 0 0 0 0 0 0 0 0 0 0 0
0 0 0 0 0 0 0 0 0 0 0 0 0 0 0 0 0 0 0 1 0 0 0 0 0 0 0 0 0 0
0 0 0 0 2 0 0 2 0 0 0 0 0 0 0 1 0 0 0 0 0 0 0 0 0 0 0 0 0 0
1 0 0 0 0 0 0 0 0 0 0 0 0 0 0 0 0 0 0 0 2 0 0 0 0 0 0 0 0 0 0 0
```

As you can see, the algorithm was able to identify the optimal arm but deviated to arms 1 and 2 from time to time. However, the exploration frequency decreased as time went on, indicating that the algorithm was growing more and more certain about its beliefs (in other words, each running Beta distribution was consolidating around a single peak). This is the typical behavior of Thompson Sampling.

7. As we have learned, just considering one single experiment might not be sufficient for the analysis of an algorithm. To facilitate a fuller analysis on the performance of Thompson Sampling, let's set up the usual repeated experiments:

```
regrets = bandit.repeat(BernoulliThompsonSampling, [N_ARMS], \
                        n_experiments=100, n_rounds=300,\
                        visualize_regret_dist=True)
```

This will generate the following distribution of regret:

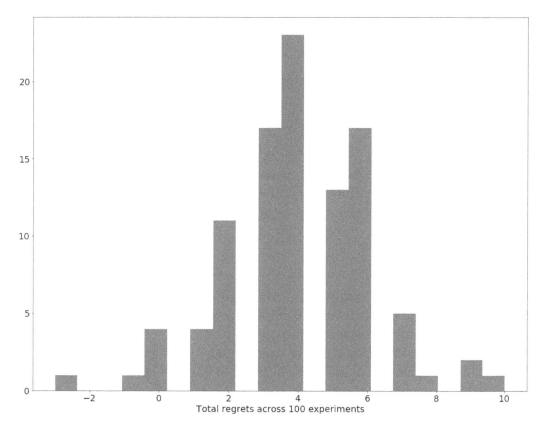

Figure 8.19: Distribution of cumulative regret from Thompson Sampling

Here, we can see that Thompson Sampling is able to minimize the cumulative regret across all experiments by a large margin compared to other algorithms (the maximum value in the distribution is only **10**).

8. To quantify this claim, let's print out the mean and max regret from these experiments:

```
np.mean(regrets), np.max(regrets)
```

The output will be as follows:

```
(4.03, 10)
```

This is significantly better than the counterpart statistics from other algorithms: Greedy had **8.66** and **62** while UCB had **18.78** had **29**.

> **NOTE**
>
> To access the source code for this specific section, please refer to https://packt.live/2UCbZXw.
>
> You can also run this example online at https://packt.live/37oQrTz.

Thompson Sampling is also the last of the common MAB algorithms that will be discussed in this book. In the next and final section of this chapter, we will briefly consider a common variant of the classical MAB problem, namely the contextual bandit problem.

CONTEXTUAL BANDITS

In the classical bandit problem, the reward from pulling an arm solely depends on the reward distribution associated with that arm, and our goal is to identify the optimal arm as soon as possible and keep pulling it until the end of the process. A contextual bandit problem, on the other hand, includes an additional element to the problem: the environment, or the context. Similar to its definition in the context of reinforcement learning, an environment contains all of the information about the problem settings, the state of the world at any given time, as well as other agents that might be participating in the same environment as our player.

CONTEXT THAT DEFINES A BANDIT PROBLEM

In the traditional MAB problem, we only care about what potential reward each arm will return if we pull it at any time. In contextual bandits, we are provided with the contextual information about the environment that we are operating in, and depending on the setting, the reward distribution of an arm might vary. In other words, the choice of which arm to pull that we make at every step should be dependent on the state of the environment, or the context.

This setting complicates the model that we have been working with, as now, we will need to consider the quantities that we are interested in as conditional probabilities: given the context we are seeing, what is the probability that arm 0 is the optimal arm, for example? In a contextual bandit problem, the context might have a minor role in the decision process of our algorithms, or it could be the main factor that drives the algorithm's decisions.

A real-world example is in order. We mentioned at the beginning of this chapter that recommender systems are a common application of the MAB problem, where, for each user who has just arrived at a website, the system needs to decide which kind of ads/recommendations would maximize the probability that the user would be interested in it. Every user has their own preferences and liking, and those factors might very well play an important role in helping the system decide whether they will be interested in a specific ad or not.

For example, dog owners will be significantly more likely to click on dog toy advertisements than the average user and potentially less likely to click on cat food advertisements. This information about the users is a part of the context of the MAB problem, which is the current recommender system we are considering. Other factors might include their profiles, buying/viewing history, and so on.

Overall, in a contextual bandit problem, we need to consider the expectation of reward of each arm/decision and do so while keeping the current context that we are in in mind. Now, let's start talking about the contextual bandit problem that we will be solving in the upcoming activity; it is also a problem that we have mentioned a couple of times throughout this book: queueing bandits.

QUEUEING BANDITS

Our bandit problem has the following elements:

- We start out with a queue of customers, each of whom belongs to one of a predetermined set of customer classes. For example, let's say our queueing contains 300 customers in total. Among these customers, 100 customers belong to class 0, another 100 belong to class 1, and the other 100 belong to class 2.

- We also have a single server that is to serve all of these customers in a specific order. Only one customer can be served at any given time, and when a customer is being served, the remaining ones in the queue will have to wait until it is their turn to be served. Once a customer has been served, they leave the queue completely.

- Each customer has a specific job length, that is, the amount of time that it will take for the server to begin and end the customer's service. The job length of a customer belonging to class i (i is 0, 1, or 2) is a random sample drawn from an exponential distribution with parameter λ_i, called the rate parameter. The larger the parameter is, the more likely it is that a sample drawn from the distribution is small. In other words, the expected value of a sample is inversely proportional to the rate parameter. (In fact, the mean of an exponential distribution is 1, divided by its rate parameter.)

> **NOTE**
>
> If you are interested in learning more about the exponential distribution, you can find more information here: https://mathworld.wolfram.com/ExponentialDistribution.html. For our purposes, we only need to know that the expected value of an exponentially distribution random variable is inversely proportional to the rate parameter.

- When a customer is being served, all the customers remaining in the queue will contribute to the total cumulative waiting time that we will incur at the end of the process. Our goal, as the queue coordinator, is to come up with a way of ordering these customers so that the total cumulative waiting time of all of the customers at the end of the process is minimized. It is known that the optimal ordering to minimize this total cumulative waiting time is shortest job first, where out of the remaining customers at any given time, the one with the shortest job length should be chosen.

With this, we can see the parallel between this queueing problem and the classical MAB problem. If the true rate parameter that characterizes the job length distribution of customers belonging to a given class is not known, we need to find a way to estimate that quantity by observing the job length of a few sample customers from each class. The sooner we can identify and converge on processing the customers with the highest rate parameter, the lower our total cumulative waiting time at the end will be. Here, pulling an arm is equivalent to picking a customer of a given class to serve next, and the negative reward (or cost) that we need to minimize at the end is the cumulative waiting time of the whole queue.

As a contextual bandit problem, a queueing problem also contains some extra context at each step that needs to be considered in the decision-making process. For example, we mentioned that in each experiment, we start out with a queue of finitely many customers (specifically, 100 customers for each of three different classes), and once a customer is processed, they will leave the queue forever. This means each of the three "arms" of our bandit problem has to be pulled exactly 100 times, and an algorithm needs to find a way to arrange the order of these pulls optimally.

In the next section, the API for the queueing bandit problem we have provided for you will be discussed.

WORKING WITH THE QUEUEING API

To have the problem defined via an API, make sure to download the two following files from the code repository for this chapter, **utils.py**, which contains the API for traditional bandit problems that we have been using, as well as queueing bandit problems, and **data.csv**, which includes the input data that will be used for our queueing experiments.

Now, different from the API that we have been using, we need to do the following to interact with a queueing bandit. First, from the **utils.py** file, the **QueueBandit** class needs to be imported. An instance of this class is declared like so:

```
queue_bandit = QueueBandit(filename='../data.csv')
```

The **filename** argument takes in the relative location of your code and the **data.csv** file, so that might change, depending on where your own notebook is. Unlike the **Bandit** class, because the **data.csv** file contains data generated from multiple experiments with different randomly chosen parameters, we don't need to declare those specific details ourselves. In fact, what we mentioned previously applies to all experiments that we will be using: in each experiment, we have the input of a 300-customer queue belonging to three different customer classes with varying unknown rate parameters.

This API also offers us the **repeat()** method so that we have an algorithm interact with the queueing problem, which similarly takes in a class implementation of that algorithm and any potential parameters as its two main arguments. The method will run the input algorithm through many different starting queues (which, again, were generated with different rate parameters for the three classes) and return the cumulative waiting time for each of those queues. The method also has an argument named **visualize_cumulative_times**, which, if set to **True**, will visualize the distribution of that cumulative waiting time in a histogram.

A call to this method should look as follows:

```
cumulative_times = queue_bandit.repeat\
                   ([ALG NAME], [ANY ALG ARGUMENTS], \
                   visualize_cumulative_times=True)
```

Finally, the last difference that we need to keep in mind is the requirement for algorithm implementations. The class implementation of an algorithm should have an **update()** method that acts the same way as we have become familiar with (it should take in the index of an arm (or of a customer class) and the most recent corresponding cost (or job length) and update whatever appropriate information that the algorithm keeps track of).

More importantly, the **decide()** method should now take in an argument that indicates how many customers of each class we have left in the queue at any given time, stored in a three-item Python list. Remember that we always start out with a queue consisting of 100 customers for each class, so the list at the beginning will be **[100, 100, 100]**. As customers are chosen by our algorithms and served, this list of customer numbers will be updated accordingly. This is the context that our algorithm needs to keep in mind while making its decisions; obviously, it cannot choose to serve a customer from class 1 next, for example, if there is no class-1 customer left in the queue. Finally, the **decide()** method should return the index of the class that should be chosen to serve, similar to what we had with the traditional MAB problem.

And that is what we need to know about this queueing bandit problem. While marking the end of the materials covered in this chapter, this section also prepares us for the upcoming activity: implementing various algorithms to solve the queueing bandit problem.

ACTIVITY 8.01: QUEUEING BANDITS

As mentioned previously, a queueing problem where the true rate parameters of the customer job lengths are unknown can be framed as a MAB problem. In this activity, we will be reimplementing the algorithms that we have learned about in this chapter in the context of queueing and comparing their performance. This activity will, therefore, reinforce the concepts that we have discussed throughout this chapter, while giving us the opportunity to tackle a contextual bandit problem.

With that, let's start the activity by following these steps:

1. Create a new Jupyter Notebook and in its first code cell, import **NumPy** and the QueueBandit class from **utils.py**. Be sure to set the random seed of NumPy to **0**.

2. Declare an instance of this class using the code included in the preceding text.

3. In a new code cell, implement the Greedy algorithm for this queueing problem and apply it to the bandit object using the code included in the preceding text. Along with the histogram of the cumulative waiting time distribution, print out the mean and max items among them.

   Again, the Greedy algorithm should choose the customer class that has the lower average job length out of the remaining classes at each iteration of a queue.

   The output will be as follows:

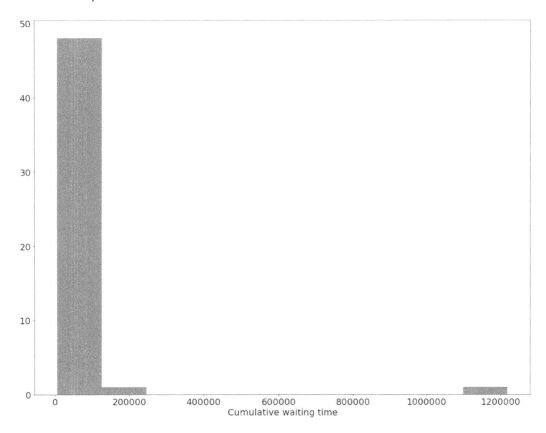

Figure 8.20: Distribution of cumulative waiting time from Greedy

The max and mean cumulative waiting times will be as follows:

```
(1218887.7924350922, 45155.236786598274)
```

4. In a new code cell, implement the Explore-then-commit algorithm for the problem. The class implementation of the algorithm should take in a parameter named **T** that specifies how may exploration rounds the algorithm should take in at the beginning of an experiment.

5. Similar to the Greedy algorithm, apply Explore-then-commit with **T=2** to the bandit object. Compare the distribution of the cumulative waiting times, as well as the mean and max resulting from this algorithm, with what we have for Greedy.

This will produce the following graph:

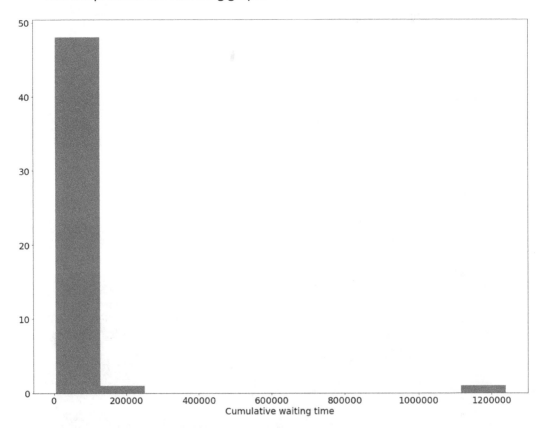

Figure 8.21: Distribution of cumulative waiting time from Explore-then-commit

The max and mean cumulative waiting times will be as follows:

```
(1238591.3208636027, 45909.77140562623)
```

6. In a new code cell, implement the Thompson Sampling algorithm for the problem.

To model an unknown rate parameter of an exponential distribution, a Gamma distribution should be used as the conjugate prior. A Gamma distribution is also parameterized by two numbers, α and β; their update rule with respect to a sample job length, x, is $\alpha = \alpha + 1$ and $\beta = \beta + x$. At the beginning, both parameters should be initialized to **0**.

To draw a sample from a Gamma distribution, the **np.random.gamma()** function could be used, which takes in α and $1 / \beta$. Similar to our logic for Greedy and Explore-then-commit, the class with the highest sampled rate should be chosen at each iteration.

7. Apply the algorithm to the bandit object and analyze its performance via the cumulative waiting times. Compare it to Greedy and Explore-then-commit.

The following plot will be produced:

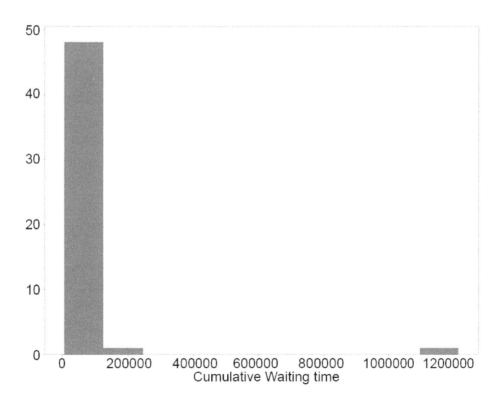

Figure 8.22: Distribution of cumulative waiting time from Thompson Sampling

The max and mean cumulative waiting times will be as follows:

```
(1218887.7924350922, 45129.343871806814)
```

8. In contextual bandit problems, specialized algorithms are commonly developed. These algorithms are variants of common MAB algorithms, specifically designed to use the contextual information.

   In a new code cell, implement an exploitative variant of Thompson Sampling where its logic is similar to Thompson Sampling at the beginning of each experiment, and solely exploits (like Greedy) by choosing the class with the lowest average job length when at least half of the customers have been served.

9. Apply the algorithm to the bandit object. Compare its performance with traditional Thompson Sampling, as well as the other algorithms we have implemented.

   The plot will be as follows:

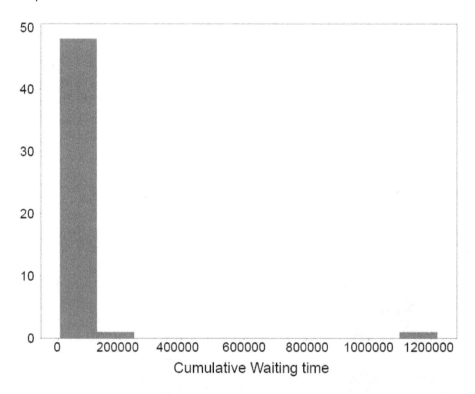

Figure 8.23: Distribution of cumulative waiting time from modified Thompson Sampling

The max and mean cumulative waiting times will be as follows:

```
(1218887.7924350922, 45093.244027644556)
```

> **NOTE**
>
> The solution to this activity can be found on page 734.

SUMMARY

In this chapter, the MAB problem and its motivation as a reinforcement learning and artificial intelligence problem were introduced. We explored a plethora of algorithms that are commonly used to solve the MAB problem, including the Greedy algorithm and its variants, UCB, and Thompson Sampling. Via these algorithms, we were exposed to unique insights and heuristics on how to balance exploration and exploitation (which is one of the most fundamental components of reinforcement learning) such as random exploration, optimism under uncertainty, or sampling from Bayesian posterior distributions.

This knowledge was put into practice as we learned how to implement these algorithms from scratch in Python. During this process, we also examined the importance of analyzing MAB algorithms over many repeated experiments to obtain robust results. This procedure is integral for any analysis framework that involves randomness. Finally, in this chapter's activity, we applied our knowledge to a queueing bandit problem and learned how to modify MAB algorithms so that they fit a given contextual bandit.

This chapter also marks the end of the topic of Markov decision problems, which spanned the last four chapters. From the next chapter onward, we will start looking at the exciting field of Deep Q Learning as a reinforcement learning framework.

9

WHAT IS DEEP Q-LEARNING?

OVERVIEW

In this chapter, we will be learning about deep Q learning in detail along with all other possible variations. You will learn how to implement the Q function and use the Q learning algorithm along with deep learning to solve complex **Reinforcement Learning** (**RL**) problems. By the end of this chapter, you will be able to describe and implement the deep Q learning algorithm in PyTorch, and we will also do a hands-on implementation of some of the advanced variants of deep Q learning, such as double deep Q learning with PyTorch.

INTRODUCTION

In the previous chapter, we learned about the **Multi-Armed Bandit** (**MAB**) problem – a popular sequential decision-making problem that aims to maximize your reward when playing on the slot machines in a casino. In this chapter, we will combine deep learning techniques with a popular **Reinforcement Learning** (**RL**) technique called Q learning. Put simply, Q learning is an RL algorithm that decides the best action to be taken by an agent for maximum rewards. The "Q" in Q learning represents the quality of the action that is used to gain future rewards. In many RL environments, we may not have state transition dynamics (that is, the probability of going from one state to another), or it is too complex to gather state transition dynamics. In these complex RL environments, we can use the Q learning approach to implement RL.

In this chapter, we will start by understanding the very basics of deep learning, such as what a perceptron and a gradient descent are and what steps need to be followed to build a deep learning model. Next, we will learn about PyTorch and how to build deep learning models using PyTorch. Once you have been introduced to Q learning, we will learn and implement a **Deep Q Network** (**DQN**) with the help of PyTorch. Then, we will improve the performance of DQNs with the help of experience replay and target networks. Finally, you will implement another variant of DQN called a **Double Deep Q Network** (**DDQN**).

BASICS OF DEEP LEARNING

We have already implemented deep learning algorithms in *Chapter 03, Deep Learning in Practice using TensorFlow 2*. Before we begin with deep Q learning, which is the focus of this chapter, it is essential that we quickly revise the basics of deep learning.

Let us first understand what a perceptron is before we look into neural networks. The following figure represents a general perceptron:

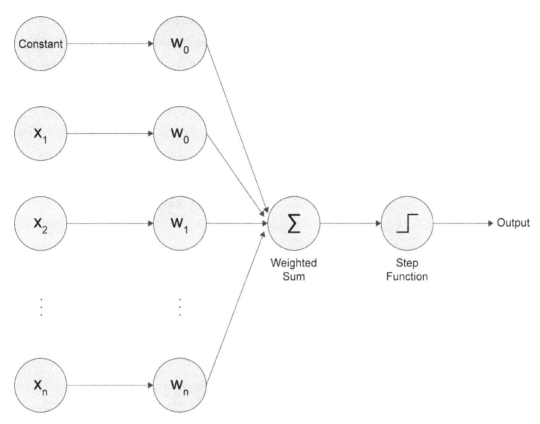

Figure 9.1: Perceptron

A perceptron is a binary linear classifier, where the inputs are first multiplied by the weights, and then we take a weighted sum of all these multiplied values. Then, we pass this weighted sum through an activation function or step function. The activation function is used to convert the input values into certain values, such as (0,1), as output for binary classification. This whole process can be visualized in the preceding figure.

Deep feedforward networks, which we also refer to as **Multilayer Perceptrons (MLPs)**, have multiple perceptrons at multiple layers, as shown in *Figure 9.2*. The goal of MLPs is to approximate any function. For example, for a classifier, the function $y = f(x, W)$, maps an input, x, to a category of y (for binary classification, either to 0 or 1) by learning the value of parameter W or the weights. The following figure shows a general deep neural network:

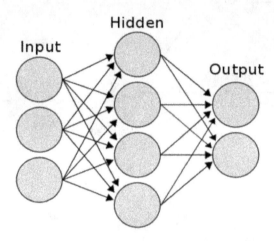

Figure 9.2: A deep neural network

A basic building block of MLPs consists of artificial neurons (also called nodes). They automatically learn the optimal weight coefficients that are then multiplied with the input's features in order to decide whether a neuron fires or not. The network consists of multiple layers where the first layer is called the input layer and the last layer is called the output layer. The intermediate layers are called hidden layers. The number of hidden layers can be of size "one" or more depending on how deep you want to make the network.

The following figure represents the general steps that are needed to train a deep learning model:

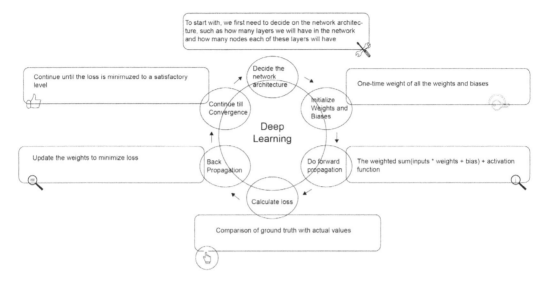

Figure 9.3: Deep learning model training flow

The training process of a typical deep learning model can be explained as follows:

1. **Decide on the network architecture**:

   To begin, we first need to decide on the network architecture, such as how many layers we will have in the network and how many nodes each of these layers will have.

2. **Initialize the weights and biases**:

   In the network, each neuron in a layer will be connected to all of the neurons in the previous layer. These connections between the neurons have a corresponding weight associated with them. During the training of the whole neural network, we first initialize the values of these weights. Each of the neurons will also have an associated bias component attached to it. This initialization is a one-time process.

3. **Perform forward propagation:**

   During forward propagation, the previous layer's input values are multiplied by the weights and summed up with the bias units in order to get a linear output for each of the neurons. These linear outputs are then passed through a non-linear activation function (for example, **sigmoid**, **relu**, or **tanh**) to produce a non-linear output. These values get propagated through each of the hidden layers and finally produce the output at the output layer.

4. **Calculate the loss:**

   The output of the network is then compared with the true/actual values or labels of the training dataset to calculate the loss of the network. The loss of the network is a measure of how well the network is performing. The lower the loss, the better the performance of the network.

5. **Update the weights (backpropagation):**

   Once we have calculated the loss of the network, the aim is to then minimize the loss of the network. This is done by using the gradient descent algorithm to adjust the weights associated with each node. Gradient descent is an optimization algorithm used for minimizing the loss in various machine learning algorithms:

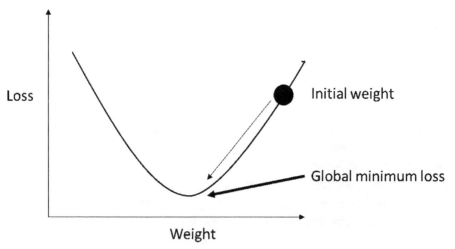

Figure 9.4: Gradient descent

Loss minimization, in turn, pushes the predicted values closer to the actual values during the training process. The learning rate plays a crucial part in deciding the rate at which these weights are updated. Other examples of optimizers are Adam, RMSProp, and Momentum.

6. Continue the iteration:

The preceding steps (*steps 3 to 5*) will be continued until the loss is minimized to a certain threshold or we have completed a certain number of iterations to complete the training process.

The following lists a few of the hyperparameters that should be tuned during the training process:

- The number of layers in the network

- The number of neurons or nodes in each of the layers

- The choice of activation functions in each of the layers

- The choice of learning rate

- The choice of variants of the gradient algorithm

- The batch size if we are using the mini-batch gradient descent algorithm

- The number of iterations to be done for weight optimization

Now that we have a good recollection of the basic concepts of deep learning, let's move toward understanding PyTorch.

BASICS OF PYTORCH

In this chapter, we will use PyTorch to build deep learning solutions. The obvious question that comes to mind is, why PyTorch? The following describes a number of reasons as to why we should use PyTorch to build deep learning models:

- **Pythonic deep integration**:

  The learning curve of PyTorch is smooth due to the Pythonic approach of the coding style and the adoption of object-oriented methods. One example of this is deep integration with the NumPy Python library, where you can easily convert a NumPy array into a torch tensor and vice versa. Also, Python debuggers work smoothly with PyTorch, which makes code debugging easier when using PyTorch.

- **Dynamic graph computation**:

  Many other deep learning frameworks come with a static computation graph; however, in PyTorch, dynamic graph computation is supported, which gives the developer a far more in-depth understanding of what is going on in each algorithm and allows them to change the network behavior programmatically at runtime.

- **OpenAI adoption of PyTorch**:

  In general academics for RL, PyTorch gained a huge momentum due to its speed and ease of use. As you will have already noted, nowadays, OpenAI Gym is often the default environment for solving RL problems. Recently, OpenAI announced that it is adopting PyTorch as its primary framework for research and development work.

The following are a few steps that should be followed in order to build a deep neural network in PyTorch:

1. Import the required libraries, prepare the data, and define the source and target data. Please note that you will need to convert your data into torch tensors when working with any PyTorch models.

2. Build the model architecture using a class.

3. Define the loss function and optimizer to be used.

4. Train the model.

5. Make predictions using the model.

Let's do an exercise to build a simple PyTorch deep learning model.

EXERCISE 9.01: BUILDING A SIMPLE DEEP LEARNING MODEL IN PYTORCH

The aim of this exercise is to build a working end-to-end deep learning model in PyTorch. This exercise will take you through the steps of how to create a neural network model in PyTorch and how to train the same model in PyTorch using sample data. This will demonstrate the backbone process in PyTorch, which we will later use in the *Deep Q Learning* section:

1. Open a new Jupyter notebook. We will import the required libraries:

```
# Importing the required libraries
import numpy as np
```

```
import torch
from torch import nn, optim
```

2. Then, using a NumPy array, we will convert the source and target data into torch tensors. Please remember that for the PyTorch model to work, you should always convert the source and target data into torch tensors, as shown in the following code snippet:

```
#input data and converting to torch tensors
inputs = np.array([[73, 67, 43],\
                   [91, 88, 64],\
                   [87, 134, 58],\
                   [102, 43, 37],\
                   [69, 96, 70]], dtype = 'float32')
inputs = torch.from_numpy(inputs)

#target data and converting to torch tensors
targets = np.array([[366], [486], [558],\
                    [219], [470]], dtype = 'float32')
targets = torch.from_numpy(targets)
#Checking the shapes
inputs.shape , targets.shape
```

The output will be as follows:

```
(torch.Size([5, 3]), torch.Size([5, 1]))
```

It is very important to be aware of the input and target dataset shapes. This is because the deep learning model should be compatible with the shapes of the input and target data for the matrix multiplication operations.

3. Define the network architecture as follows:

```
class Model(nn.Module):
    def __init__(self):
        super().__init__()
        self.fc1 = nn.Linear(3, 10)
        self.fc2 = nn.Linear(10, 1)

    def forward(self, x):
        x = torch.relu(self.fc1(x))
        x = self.fc2(x)
```

```
        return x

# Instantiating the model
model = Model()
```

Once we have the source and target data in tensor format, we should create a class for the neural network model architecture. This class inherits the properties from the **nn** base class using a package called **Module**. This new class, called **Model**, will have a forward function along with a regular constructor, called (**__init__**).

The **__init__** method, at first, will call the **super** method to gain access to the base class. Then, all the layer definitions will be written within this constructor method. The role of the forward method is to provide the steps that are required to do the forward propagation steps of the neural network.

nn.Linear() has the syntax of (input size, output size) to define the linear layers of the model. We can use a non-linear function such as **relu** or **tanh** in combination with the linear layers in the forward function.

The neural network architecture represents the 3 nodes in the input layer, 10 in the hidden layer, and 1 in the output layer. Inside the forward function, we will use the **relu** activation function in the hidden layer. Once we define the model class, then we must instantiate the model.

Now you should have successfully created and initiated a model.

4. Now define the loss function and optimizer. The exercise we are working on is a regression problem; we generally use the mean squared error as the loss function in regression problems. In PyTorch, we use the **MSELoss()** function for a regression problem. Generally, the loss is assigned to **criterion**.

The **Model** parameter and learning rate must be passed as mandatory arguments to the optimizers for backpropagation. Model parameters can be accessed using the **model.parameters()** function. Now define the loss function and optimizer using the Adam optimizer. While creating the Adam optimizer, pass **0.01** as the learning rate along with the model parameter:

```
# Loss function and optimizer
criterion = nn.MSELoss()
optimizer = torch.optim.Adam(model.parameters(), lr=0.01)
```

At this point, you should have successfully defined the loss and optimization functions.

> **NOTE**
>
> **optim.Adam**, **optim.SGD**, **optim.RMSprop**, and **optim.Adagrad** are the available optimizers in PyTorch. All of them are slightly different variants of gradient descent and are present in the **torch.optim** package.

5. Train the model for 20 epochs and monitor the loss. To form a training loop, create a variable called **n_epochs** and initialize the value as 20. Create a **for** loop to run the loop for **n_epoch** times. Inside the loop, complete these steps: zero out the parameter gradients using **optimizer.zero_grad()**. Pass the input through the model to get the output. Obtain the loss using **criterion** by passing the outputs and targets. Use **loss.backward()** and **optimizer.step()** to do the backpropagation step. Print the loss after every epoch:

```
# Train the model
n_epochs = 20
for it in range(n_epochs):
    # zero the parameter gradients
    optimizer.zero_grad()

    # Forward pass
    outputs = model(inputs)
    loss = criterion(outputs, targets)


    # Backward and optimize
    loss.backward()
    optimizer.step()

    print(f'Epoch {it+1}/{n_epochs}, Loss: {loss.item():.4f}')
```

PyTorch, by default, accumulates the gradients calculated at each step. We need to handle this during the training process so that the weights are updated with their proper gradients. `optimizer.zero_grad()` will zero out the gradients from the previous training step to stop the gradient accumulations. This step should be done before calculating the gradients at each epoch. To calculate the loss, we should pass the predicted and actual values to the loss function. The `criterion(outputs, targets)` step is used to calculate the loss. The `loss.backward()` step is used to calculate the weight gradients, and we will use these gradients to update the weights to get our optimum weights. Weight updates are done using the `optimizer.step()` function.

The output will be as follows:

```
Epoch 1/20, Loss: 185159.9688
Epoch 2/20, Loss: 181442.8125
Epoch 3/20, Loss: 177829.2188
Epoch 4/20, Loss: 174210.5938
Epoch 5/20, Loss: 170534.4375
Epoch 6/20, Loss: 166843.9531
Epoch 7/20, Loss: 163183.2500
Epoch 8/20, Loss: 159532.0625
Epoch 9/20, Loss: 155861.8438
Epoch 10/20, Loss: 152173.0000
Epoch 11/20, Loss: 148414.5781
Epoch 12/20, Loss: 144569.6875
Epoch 13/20, Loss: 140625.1094
Epoch 14/20, Loss: 136583.0625
Epoch 15/20, Loss: 132446.6719
Epoch 16/20, Loss: 128219.9688
Epoch 17/20, Loss: 123907.7422
Epoch 18/20, Loss: 119515.7266
Epoch 19/20, Loss: 115050.4375
Epoch 20/20, Loss: 110519.2969
```

As you can see, the output prints the loss after every epoch. You should closely monitor the training loss. From the preceding output, we can see that training loss decreases.

6. Once the model is trained, we can use the trained model to make predictions. Pass the input data through the model to get the predictions and observe the output:

```
#Prediction using the trained model
preds = model(inputs)
print(preds)
```

The output is as follows:

```
tensor([[ 85.6779],
        [115.3034],
        [146.7106],
        [ 69.4034],
        [120.5457]], grad_fn=<AddmmBackward>)
```

The preceding output shows the model prediction of the corresponding input data.

> **NOTE**
>
> To access the source code for this specific section, please refer to https://packt.live/3e2DscY.
>
> You can also run this example online at https://packt.live/37q0J68.

We now know how a PyTorch model works. This example will be useful for you when training your deep Q neural network. However, in addition to this, there are few other important PyTorch utilities that you should be aware of, which you will study in the next section. Understanding these utilities is essential for the implementation of deep Q learning.

PYTORCH UTILITIES

To work with the utilities, first, we will create a torch tensor of size 10 with numbers starting from 1 to 9 using the **arange** function of PyTorch. A torch tensor is essentially a matrix of elements that belongs to a single data type, which can have multiple dimensions. Please note that, like Python, PyTorch also excludes the number given in the **arange** function:

```
import torch
t = torch.arange(10)
print(t)
print(t.shape)
```

The output will be as follows:

```
tensor([0, 1, 2, 3, 4, 5, 6, 7, 8, 9])
torch.Size([10])
```

Let's now begin exploring the various functions, one by one.

THE VIEW FUNCTION

Use the **view** function to reshape your tensor as follows:

```
t.view(2,5) # reshape the tensor to of size - (2,5)
```

The output will be as follows:

```
tensor([[0, 1, 2, 3, 4],
        [5, 6, 7, 8, 9]])
```

Let's try a new shape now:

```
t.view(-1,5)
# -1 will by default infer the first dimension
# use when you are not sure about any dimension size
```

The output will be as follows:

```
tensor([[0, 1, 2, 3, 4],
        [5, 6, 7, 8, 9]])
```

THE SQUEEZE FUNCTION

The **squeeze** function is used to remove any dimensions with a value of 1. The following is an example of a tensor with a shape of (5,1):

```
x = torch.zeros(5, 1)
print(x)
print(x.shape)
```

The output will be as follows:

```
tensor([[0.],
        [0.],
        [0.],
        [0.],
        [0.]])
torch.Size([5, 1])
```

Apply the **squeeze** function to the tensor:

```
# squeeze will remove any dimension with a value of 1
y = x.squeeze(1)
# turns a tensor of shape [5, 1] to [5]
y.shape
```

The output will be as follows:

```
torch.Size([5])
```

As you can see, after using **squeeze**, the dimension of 1 is removed.

THE UNSQUEEZE FUNCTION

As the name suggests, the **unsqueeze** function does the reverse of **squeeze**. It adds a dimension of 1 to the input data.

Consider the following example. First, we create a tensor of shape **5**:

```
x = torch.zeros(5)
print(x)
print(x.shape)
```

The output will be as follows:

```
tensor([0., 0., 0., 0., 0.])
torch.Size([5])
```

Apply the **unsqueeze** function to the tensor:

```
y = x.unsqueeze(1) # unsqueeze will add a dimension of 1
print(y.shape) # turns a tensor of shape [5] to [5,1]
```

The output will be as follows:

```
torch.Size([5, 1])
```

As you can see, a dimension of 1 has been added to the tensor.

THE MAX FUNCTION

If a multidimensional tensor is passed to the **max** function, the function returns the max values and corresponding index in the specified axis. Please refer to the code comments for more details.

First, we create a tensor of dimensions **(4, 4)**:

```
a = torch.randn(4, 4)
a
```

The output will be as follows:

```
tensor([[-0.5462,  1.3808,  1.4759,  0.1665],
        [-1.6576, -1.2805,  0.5480, -1.7803],
        [ 0.0969, -1.7333,  1.0639, -0.4660],
        [ 0.3135, -0.4781,  0.3603, -0.6883]])
```

Now, let's apply the **max** function to the tensor:

```
"""
returns max values in the specified dimension along with index
specifying 1 as dimension means we want to do the operation row-wise
"""
torch.max(a , 1)
```

The output will be as follows:

```
torch.return_types.max(values=tensor([1.4759, 0.5480, \
                                       1.0639, 0.3603]),\
                   indices=tensor([2, 2, 2, 2]))
```

Let's now try to find the max values from the tensor:

```
torch.max(a , 1)[0] # to fetch the max values
```

The output will be as follows:

```
tensor([1.4759, 0.5480, 1.0639, 0.3603])
```

To find the index of the max values, use the following code:

```
# to fetch the index of the corresponding max values
torch.max(a , 1)[1]
```

The output will be as follows:

```
tensor([2, 2, 2, 2])
```

As you can see, the indices of the max values have been displayed.

THE GATHER FUNCTION

The **gather** function works by collecting values along an axis specified by **dim**. The general syntax of the **gather** function is as follows:

```
torch.gather(input, dim, index)
```

The syntax can be explained as follows:

- **input** (tensor): Specify the source tensor here.

- **dim** (python:int): Specify the axis along which to index.

- **index** (**LongTensor**): Specify the indices of elements to gather.

In the following example, we have **q_values**, which is a torch tensor of shape (**4 , 4**), and the action is a **LongTensor** that has indexes that we want to extract from the **q_values** tensor:

```
q_values = torch.randn(4, 4)
print(q_values)
```

The output will be as follows:

```
q_values = torch.randn(4, 4)
print(q_values)
tensor([[-0.2644, -0.2460, -1.7992, -1.8586],
        [ 0.3272, -0.9674, -0.2881,  0.0738],
        [ 0.0544,  0.5494, -1.7240, -0.8058],
        [ 1.6687,  0.0767,  0.6696, -1.3802]])
```

Next, we will apply **LongTensor** to specify the indices of the elements of the tensor that are to be gathered:

```
# index must be defined as LongTensor
action =torch.LongTensor([0 , 1, 2, 3])
```

Then, find the shape of the **q_values** tensor:

```
q_values.shape , action.shape

# q_values -> 2-dimensional tensor
# action -> 1-dimension tensor
```

The output will be as follows:

```
(torch.Size([4, 4]), torch.Size([4]))
```

Let's now apply the **gather** function:

```
"""
unsqueeze is used to take care of the error - Index tensor
must have same dimensions as input tensor
returns the values from q_values using the action as indexes
"""
torch.gather(q_values , 1, action.unsqueeze(1))
```

The output will be as follows:

```
tensor([[-0.2644],
        [-0.9674],
        [-1.7240],
        [-1.3802]])
```

Now you have a basic understanding of neural networks and how to implement a simple neural network in PyTorch. Aside from a vanilla neural network (the combination of linear layers with a non-linear activation function), there are two other variants called **Convolutional Neural Networks** (**CNNs**) and **Recurrent Neural Networks** (**RNNs**). A CNN is mainly used for image classification and image segmentation tasks, and an RNN is used for data with a sequential pattern, such as time series data, or for language translation tasks.

Now, as we have gained some knowledge of deep learning and how to build deep learning models in PyTorch, we will shift our focus to Q learning and how to use deep learning in RL with the help of PyTorch. First, we will start with the state-value function and the Bellman equation, and then we will move on to Q learning.

THE STATE-VALUE FUNCTION AND THE BELLMAN EQUATION

As we are slowly moving toward the Q function and Q learning process, let's revisit the Bellman equation, which is the backbone of the Q learning process. In the following section, we will first revise our definition of an "expected value" and how it is used in a Bellman equation.

EXPECTED VALUE

The following figure depicts the expected value in a state space:

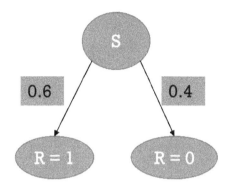

Figure 9.5: Expected value

Suppose that an agent is in state **S**, and it has two paths on which it can travel. The first path has a transition probability of 0.6 and an associated reward of 1, and the second path has a transition probability of 0.4 and an associated reward of 0.

Now, the expected value or reward of state **S** would be as follows:

```
(0.6 * 1) + (0.4 * 1) = 0.6
```

Mathematically, it can be expressed as follows:

$$E(X) = \sum_{X} p(X)X$$

Figure 9.6: Expression for the expected value

THE VALUE FUNCTION

When an agent is in an environment, the value function provides the required information about the states. The value function provides a methodology through which an agent can know how good any given state is for the agent. So, if an agent has the option of going two states from the current state, the agent will always choose the state with the larger value function.

The value function can be expressed recursively using the value function of future states. When we are working in a stochastic environment, we will use the concept of expected values, as discussed in the previous section.

THE VALUE FUNCTION FOR A DETERMINISTIC ENVIRONMENT

For a deterministic world, the value of a state is just the sum of all future rewards.

The value function of state 1 can be expressed as follows:

$$V(S_1) = \gamma_2 + \gamma_3 + \gamma_4 + \ldots + \gamma_n$$

Figure 9.7: Value function of state 1

The value function of state 1 can be expressed in terms of state 2, as follows:

$$V(S_2) = \gamma_3 + \gamma_4 + \ldots + \gamma_n$$

Figure 9.8: Value function of state 2

The simplified value function of state 1, using the value function of state 2, can be expressed as follows:

$$V(S_1) = \gamma_2 + V(S_2)$$

Figure 9.9: Simplified value function of state 1 using the value function of state 2

The simplified value function of state 1 with the discount factor can be expressed as follows:

$$V(S_1) = \gamma_2 + \gamma * V(S_2)$$

Figure 9.10: Simplified value function of state 1 with the discount factor

In general, we can rewrite the value function as follows:

$$V(s) = \left[r + \gamma * V(s') \right]$$

Figure 9.11: Value function for a deterministic environment

THE VALUE FUNCTION FOR A STOCHASTIC ENVIRONMENT:

For stochastic behavior, due to the randomness or uncertainty that is present in the environment, instead of taking the raw future rewards, we take the expected total reward from a state to come up with the value function. The new addition to the preceding equation is the expectation part. The equation is as follows:

$$V(s) = E[r + \gamma * V(s')]$$

Figure 9.12: Value function for a stochastic environment

Here, **s** is the current state, s' is the next state, and **r** is the reward of going from **s** to s'.

THE ACTION-VALUE FUNCTION (Q VALUE FUNCTION)

In the previous sections, we learned about the state-value function, which tells us how rewarding it is to be in a particular state for an agent. Now we will learn about another function where we can combine the state with actions. The action-value function will tell us how good it is for the agent to take any given action from a given state. We also call the action value the **Q value**. The equation can be written as follows:

$$Q(s, a) = E\left[G_t \big| S_t = s, A_t = a\right]$$

Figure 9.13: Expression for the Q value function

The preceding equation can be written in an iterative fashion, as follows:

$$Q(s, a) = r + \gamma \max_{a'} Q(s', a')$$

Figure 9.14: Expression for the Q value function with iterations

This equation is also known as the **bellman equation**. From the equation, we can express $Q(s, a)$ recursively in terms of the Q value of the next state, s'. A Bellman equation can be described as follows:

"The total expected reward being in state s and taking action a is the sum of two components: the reward (which is r) that we can get from state 's' by taking action a, plus the maximum expected discounted return $(Q(s', a'))$ that we can get from any possible next state-action pair (s', a'). a' is the next best possible action."

IMPLEMENTING Q LEARNING TO FIND OPTIMAL ACTIONS

The process of finding the optimal action from any state using the Q function is called Q learning. Q learning is also a tabular method, where state and action combinations are stored in a tabular format. In the following section, we will learn how to find the optimal action using the Q learning method in a stepwise fashion. Consider the following table:

Action

States	↑	↓	→	←
001	.69	.708	.98	.03
002	.91	.56	.45	.33
......	......	......	......	......
100	.55	.34	.86	.10

Figure 9.15: Sample table for Q learning

As you can see in the preceding table, the Q values are stored in terms of a table, where the rows represent the states that are present in the environment, and the columns represent all of the possible actions for the agent. You can see that all of the states are represented as rows, and all the actions, such as going up, down, right, and left, are stored as columns.

The values present in the intersection of any row and column is the Q value for that particular state-action pair.

Initially, all values of state-action pairs are initialized with zero. The agent, being in a state, will choose the action with the highest Q value. For example, as shown in the preceding figure, being in state 001, the agent will choose to go right, which has the highest Q value (0.98).

During the initial phase, when most of the state-action pairs will have zero values, we will make use of the previously discussed epsilon-greedy strategy to tackle the exploration-exploitation dilemma as follows:

- Set the value of ε (a high value such as 0.90).

- Choose a random number between 0 and 1:

```
if random_number > ε :
    choose the best action(exploitation)
else:
    choose the random action (exploration)
decay ε
```

The state with a higher value of ε decays gradually. The idea then would be to initially explore and then exploit.

Using the **Temporal Difference (TD)** method described in the previous chapter, we iteratively update the Q values as follows:

$$Q_t(s, a) = Q_{t-1}(s, a) + \alpha * TD$$

Figure 9.16: Updating the Q values using iterations

The timestamp, **t**, is the current iteration, and timestamp **(t-1)** is the previous iteration. In this way, we are updating the previous timestamp's Q value with the **TD** method and pushing the Q value as close as we can to the optimal Q value, which is also called $Q*(s, a)$.

We can rewrite the preceding equation as follows:

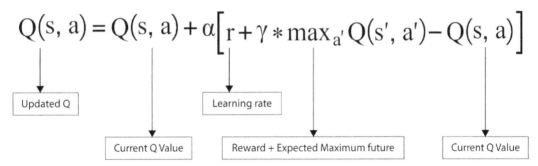

Figure 9.17: Updated expression for updating Q values

Using simple math, we can further simplify the equation as follows:

$$Q(s, a) = (1 - \alpha)Q(s, a) + \alpha\left[r + \gamma * \max_{a'} Q(s', a')\right]$$

Figure 9.18: Updated equation for Q values

The learning rate decides how big or small the steps we should take should be in order to update the Q values. This iteration and updating of Q values will continue until the Q values come closer to Q^* or if we reach a certain predefined number of iterations. The iteration can be visualized as follows:

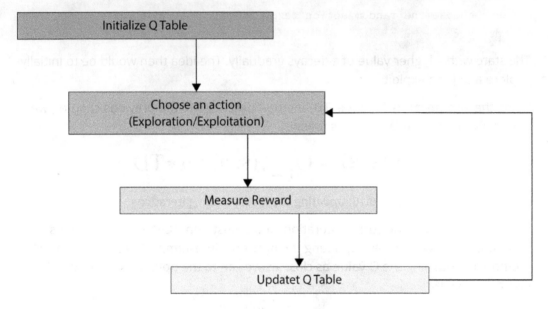

Figure 9.19: The Q learning process

As you can see, after multiple iterations, the Q table is finally ready.

ADVANTAGES OF Q LEARNING

The following describes the advantages of using Q learning in the RL domain:

- We don't need to know the full transition dynamic; this means that we don't have to know all of the state transition probabilities that may not be available for some environments.

- As we store the state-action combination in a tabular format, it is easy to understand and implement the Q learning algorithm by fetching the details from the tables.

- We don't have to wait for the entire episode to finish to update the Q value for any state due to the continuous online update of the learning process, unlike in the Monte Carlo method where we have to wait for an episode to finish in order to update the action-value function.

- It works well when the combination of states and action spaces is low.

As we have now learned about the basics of Q learning, we can implement Q learning using an OpenAI Gym environment. So, before going ahead with the exercise, let's review the concept of OpenAI Gym.

OPENAI GYM REVIEW

Before we implement the Q learning tabular method, let's quickly review and revisit the Gym environment. OpenAI Gym is a toolkit for developing RL algorithms. It supports teaching agents everything from walking to playing games such as CartPole or FrozenLake-v0. Gym provides an environment, and it is up to the developer to write and implement any RL algorithms such as tabular methods or deep Q learning. We can also write algorithms using existing deep learning frameworks, such as PyTorch or TensorFlow. The following is a sample code example to work with an existing Gym environment:

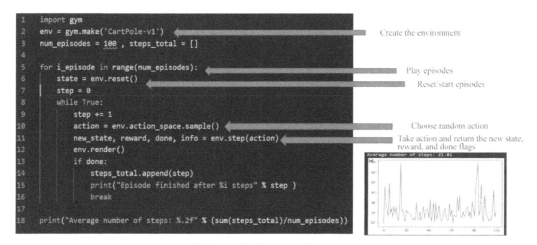

Figure 9.20: Gym environment

Let's understand a few parts of the code as follows:

- **gym.make("CartPole-v1")**

  This creates an existing Gym environment (**CartPole-v1**).

- **env.reset()**

  This resets the environment, so the environment will be at the starting state.

- **env.action_space.sample()**

  This selects a random action from the action space (a collection of available actions).

- **`env.step(action)`**

  This performs the action selected from the previous step. Once you take the actions, the environment will return the **new_state**, **reward**, and **done** flags (to indicate whether the game is over), and some extra information.

- **`env.render()`**

  This renders to see the agent performing the actions or playing the game.

We now have a theoretical understanding of the Q learning process, and we have also reviewed the Gym environment. Now it's your turn to implement Q learning using a Gym environment.

EXERCISE 9.02: IMPLEMENTING THE Q LEARNING TABULAR METHOD

In this exercise, we will implement the tabular Q Learning method using the OpenAI Gym environment. We will use the **FrozenLake-v0** Gym environment to implement the tabular Q learning method. The goal is to play and collect maximum rewards with the help of the Q learning process. You should already be familiar with the FrozenLake-v0 environment from *Chapter 5, Dynamic Programming*. The following steps will help you to complete the exercise:

1. Open a new Jupyter notebook file.

2. Import the required libraries:

```
# Importing the required libraries
import gym
import numpy as np
import matplotlib.pyplot as plt
```

3. Create the Gym environment with **'FrozenLake-v0'** for a stochastic environment:

```
env = gym.make('FrozenLake-v0')
```

4. Fetch the number of states and actions:

```
number_of_states = env.observation_space.n
number_of_actions = env.action_space.n
# checking the total number of states and action
print('Total number of States : {}'.format(number_of_states))
print('Total number of Actions : {}'.format(number_of_actions))
```

The output will be as follows:

```
Total number of States : 16
Total number of Actions : 4
```

5. Create the Q table with details fetched from the previous step:

```
# Creation of Q table
Q_TABLE = np.zeros([number_of_states, number_of_actions])
# Looking at the initial values Q table
print(Q_TABLE)
print('shape of Q table : {}'.format(Q_TABLE.shape)
```

The output is as follows:

```
[[0. 0. 0. 0.]
 [0. 0. 0. 0.]
 [0. 0. 0. 0.]
 [0. 0. 0. 0.]
 [0. 0. 0. 0.]
 [0. 0. 0. 0.]
 [0. 0. 0. 0.]
 [0. 0. 0. 0.]
 [0. 0. 0. 0.]
 [0. 0. 0. 0.]
 [0. 0. 0. 0.]
 [0. 0. 0. 0.]
 [0. 0. 0. 0.]
 [0. 0. 0. 0.]
 [0. 0. 0. 0.]
 [0. 0. 0. 0.]]
shape of Q table : (16, 4)
```

Now we know the shape of the Q table and that the initial values are all zero for every state-action pair.

6. Set all of the required hyperparameter values to be used for Q learning:

```
# Setting the Hyper parameter Values for Q Learning

NUMBER_OF_EPISODES = 10000
MAX_STEPS = 100
LEARNING_RATE = 0.1
```

```
DISCOUNT_FACTOR = 0.99

EGREEDY = 1
MAX_EGREEDY = 1
MIN_EGREEDY = 0.01
EGREEDY_DECAY_RATE = 0.001
```

7. Create empty lists to store the values of the rewards and the decayed egreedy values for visualization:

```
# Creating empty lists to store rewards of all episodes
rewards_all_episodes = []
# Creating empty lists to store egreedy_values of all episodes
egreedy_values = []
```

8. Implement the Q learning training process to play the episode for a fixed number of episodes. Use the previously learned Q learning process (from the *Implementing Q Learning to Find Optimal Actions* section) in order to find the optimal actions from any given state.

   Create a **for** loop to iterate for **NUMBER_OF_EPISODES**. Reset the environment and set the **done** flag equal to **False** and **current_episode_rewards** as **zero**. Create another **for** loop to run a single episode for **MAX_STEPS**. Inside the **for** loop, choose the best action using the epsilon-greedy strategy. Perform the action and update the Q values using the equation shown in *Figure 9.18*. Collect the reward and assign **new_state** as the current state. If the episode is over, break out from the loop, else continue taking the steps. Decay the epsilon value to be able to continue for the next episode:

```
# Training Process
for episode in range(NUMBER_OF_EPISODES):

    state = env.reset()
    done = False

    current_episode_rewards = 0

    for step in range(MAX_STEPS):

        random_for_egreedy = np.random.rand()

        if random_for_egreedy > EGREEDY:
            action = np.argmax(Q_TABLE[state, :])
```

```
        else:
            action = env.action_space.sample()

        new_state, reward, done, info = env.step(action)

        Q_TABLE[state, action] = (1 - LEARNING_RATE) \
                                * Q_TABLE[state, action] \
                                + LEARNING_RATE \
                                * (reward + DISCOUNT_FACTOR \
                                   * np.max(Q_TABLE[new_state,:]))

        state = new_state
        current_episode_rewards += reward

        if done:
            break
    egreedy_values.append(EGREEDY)

    EGREEDY = MIN_EGREEDY + (MAX_EGREEDY - MIN_EGREEDY) \
            * np.exp(-EGREEDY_DECAY_RATE*episode)

    rewards_all_episodes.append(current_episode_rewards)
```

9. Implement a function called **rewards_split** that will split the 10,000 rewards into 1,000 individual lists of rewards, and calculate the average rewards for each of these 1,000 lists of rewards:

```
def rewards_split(rewards_all_episodes , total_episodes , split):
    """
    Objective:
    To split and calculate average reward or percentage of
    completed rewards per splits
    inputs:
    rewards_all_episodes - all the per episode rewards
    total_episodes - total of episodes
    split - number of splits on which we will check the reward
    returns:
    average reward of percentage of completed rewards per splits
    """
    splitted = np.split(np.array(rewards_all_episodes),\
                        total_episodes/split)
```

```
        avg_reward_per_splits = []
        for rewards in splitted:
            avg_reward_per_splits.append(sum(rewards)/split)
        return avg_reward_per_splits
    avg_reward_per_splits = rewards_split\
                            (rewards_all_episodes , \
                            NUMBER_OF_EPISODES , 1000)
```

10. Visualize the average rewards or percentage of completed episodes:

```
plt.figure(figsize=(12,5))
plt.title("% of Episodes completed")
plt.plot(np.arange(len(avg_reward_per_splits)), \
        avg_reward_per_splits, 'o-')
plt.show()
```

The output is as follows:

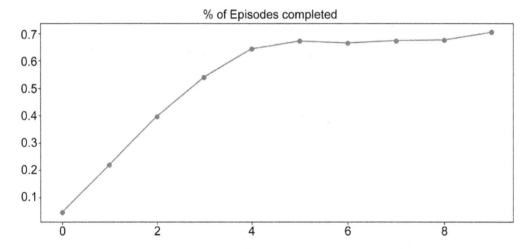

Figure 9.21: Visualizing the percentage of episodes completed

As you can see from the preceding figure, the episodes are completed, and the percentage rises exponentially until it reaches a point where it becomes constant.

11. Now we will visualize the **Egreedy** value decay:

```
plt.figure(figsize=(12,5))
plt.title("Egreedy value")
plt.bar(np.arange(len(egreedy_values)), egreedy_values, \
        alpha=0.6, color='blue', width=5)
plt.show()
```

The plot will be produced as follows:

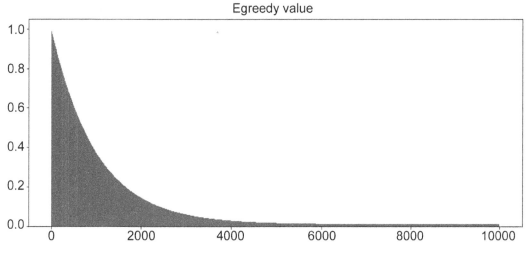

Figure 9.22: Egreedy value decay

In *Figure 9.22*, we can see the **Egreedy** value has been gradually decayed with the increasing number of steps. This means that, as the value drops toward zero, the algorithm becomes more and more greedy, taking the action with maximum reward without exploring the less rewarding actions, which, with enough exploration, may turn out to be more rewarding in the long term, but we do not know enough about the model in the initial stages.

This highlights the need for a higher exploration of the environment when we are in the early stages of learning. This is achieved with higher epsilon values. The epsilon value is reduced as training progresses. This results in less exploration and more exploitation of knowledge gained from past runs.

Thus, we have successfully implemented the tabular method of Q learning.

> **NOTE**
>
> To access the source code for this specific section, please refer to https://packt.live/2B3NziM.
>
> You can also run this example online at https://packt.live/2AjbACJ.

Now that we have a good understanding of the required entities, we will study another important concept of RL: deep Q learning.

DEEP Q LEARNING

Before diving into the details of the deep Q learning process, let's first discuss the disadvantages of the traditional tabular Q learning process, and then we will look at how combining deep learning with Q learning can help us to resolve these disadvantages of tabular methods.

The following describes several disadvantages of the tabular Q learning approach:

- Performance issues: When the state spaces are very large, the tabular iterative lookup operations will be much slower and more costly.

- Storage issues: Along with the performance issues, storage will also be costly when it comes to storing the tabular data for large combinations of state and action spaces.

- The tabular method will work well only when an agent comes across seen discrete states that are present in the Q table. For the unseen states that are not present in the Q table, the agent's performance may be the optimal performance.

- For continuous state spaces for the previously mentioned issues, the tabular Q learning method won't be able to approximate the Q values in an efficient or proper manner.

Keeping all of these issues in mind, we can consider using a function approximator that will work as a mapping between states and Q values. In machine learning terms, we can think of this problem as using a non-linear function approximator to solve regression problems. Since we are thinking of a function approximator, a neural network works best as a function approximator through which we can approximate the Q values for each state-action pair. This act of combining Q learning with a neural network is called deep Q learning or DQN.

Let's break down and explain each part of this puzzle:

- **Inputs to the DQN**:

  The neural network accepts the states of the environment as input. For example, in the case of the FrozenLake-v0 environment, a state can be a simple coordinate of the grid at any given point in time. For more complex games such as Atari, the input can be a few consecutive snapshots of the screen in the form of an image as state representation. The number of nodes in the input layer will be the same as the number of states present in the environment.

- **Outputs from the DQN**:

  The output would be the Q values for each action. For example, for any given environment, if there are four possible actions, then the output would have four Q values for each action. To choose the optimal action, we will select the action with the maximum Q value.

- **The loss function and learning process**:

  The DQN will accept states from the environment and, for each given input or state, the network will output an estimated Q value for each action. The objective of this is that it approximates the optimal Q value, which will satisfy the right-hand side of the Bellman equation, as shown in the following expression:

$$Q(s, a) = r + \gamma \max_{a'} Q(s', a')$$

Figure 9.23: Bellman equation

To calculate the loss, we need the target Q value and the Q values coming from the network. From the preceding Bellman equation, the target Q values are calculated on the right-hand side of the equation. The loss from the DQN is calculated by comparing the output Q values from the DQN to the target Q values. Once we calculate the loss, we then update the weights of the DQN via backpropagation to minimize the loss and to push the DQN-output Q values closer to the optimal Q values. In this way, with the help of DQN, we treat the RL problem as a supervised learning problem with a source and a target.

The DQN implementation can be visualized as follows:

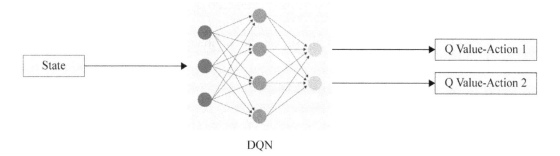

DQN

Figure 9.24: DQN

We can write the steps of the deep Q learning process as follows:

1. Initialize the weights to get an initial approximation of $Q(s,a)$:

```python
class DQN(nn.Module):
    def __init__(self , hidden_layer_size):
        super().__init__()
        self.hidden_layer_size = hidden_layer_size
        self.fc1 = nn.Linear\
                    (number_of_states,self.hidden_layer_size)
        self.fc2 = nn.Linear\
                    (self.hidden_layer_size,number_of_actions)

    def forward(self, x):
        output = torch.tanh(self.fc1(x))
        output = self.fc2(output)
        return output
```

    As you can see, we have initialized the DQN class with the weights. The two code lines in the **__init__** functions, where we create the neural network, are responsible for giving random weights to the network connections. We can also explicitly initialize the weights. A common practice nowadays is to let PyTorch or TensorFlow use their internal default initialization logic to create initial weight vectors, as you can see in the following sample code:

```python
self.fc1 = nn.Linear(number_of_states,self.hidden_layer_size)
self.fc2 = nn.Linear(self.hidden_layer_size,number_of_actions)
```

2. Doing one forward pass through the network, obtain the flags (**state**, **action**, **reward**, and **new_state**). The action is selected by taking the argmax of the Q values (selecting the index of the max Q value) or by taking random actions during the exploration phase. We can achieve this using the following code sample:

```python
def select_action(self,state,EGREEDY):
        random_for_egreedy = torch.rand(1)[0]

        if random_for_egreedy > EGREEDY:
            with torch.no_grad():
                state = torch.Tensor(state).to(device)
                q_values = self.dqn(state)
                action = torch.max(q_values,0)[1]
```

```
                action = action.item()
        else:
            action = env.action_space.sample()

        return action
```

As you can see in the preceding code snippet, the egreedy algorithm is being used to select the action. The **select_action** function passes the state through the DQN to obtain the Q values and selects the action with the highest Q value during exploitation. The **if** statement decides whether the exploration should be carried out or not.

3. If the episode is ended, the target Q value will be the reward obtained; otherwise, use the Bellman equation to estimate the target Q value. You can realize this in the following code sample:

```
def optimize(self, state, action, new_state, reward, done):
        state = torch.Tensor(state).to(device)
        new_state = torch.Tensor(new_state).to(device)

        reward = torch.Tensor([reward]).to(device)

        if done:
            target_value = reward
        else:
            new_state_values = self.dqn(new_state).detach()
            max_new_state_values = torch.max(new_state_values)
            target_value = reward + DISCOUNT_FACTOR \
                          * max_new_state_values
```

4. The loss obtained is as follows.

    If the episode ended, then the loss will be $[Q(s, a; \theta) - r]^2$.

    Otherwise, the loss will be termed as $\left[Q(s, a; \theta) - \left(r(s, a) + \gamma \max_a Q(s', a; \theta)\right)\right]^2$.

    The following is sample code for **loss**:

```
        loss = self.criterion(predicted_value, target_value)
```

5. Using backpropagation, we update the network weights (θ). This iteration will run for each state until we sufficiently minimize the loss and get an approximate optimal Q function. The following is the sample code:

```
self.optimizer.zero_grad()
loss.backward()
self.optimizer.step()
```

Now that we have a fair understanding of the implementation of deep Q learning, let's test our understanding with an exercise.

EXERCISE 9.03: IMPLEMENTING A WORKING DQN NETWORK WITH PYTORCH IN A CARTPOLE-V0 ENVIRONMENT

In this exercise, we will implement the deep Q learning algorithm with the OpenAI Gym CartPole environment. The aim of this exercise is to build a PyTorch-based DQN model that will learn to balance the cart in the CartPole environment. Please refer to the PyTorch example for building neural networks, which was explained at the start of this chapter.

Our main aims are to apply a Q learning algorithm, keep the pole steady during each step, and collect the maximum reward during each episode. A reward of +1 is given for every step when the pole remains straight. The episode will end when the pole is more than 15 degrees away from the vertical position or when the cart moves more than 2.4 units away from the center position in the CartPole environment:

1. Open a new Jupyter notebook and import the required libraries:

```
import gym
import matplotlib.pyplot as plt
import torch
import torch.nn as nn
from torch import optim
import numpy as np
import math
```

2. Create a device based on the availability of a **Graphics Processing Unit (GPU)** environment:

```
# selecting the available device (cpu/gpu)

use_cuda = torch.cuda.is_available()

device = torch.device("cuda:0" if use_cuda else "cpu")

print(device)
```

3. Create a Gym environment using the **'CartPole-v0'** environment:

```
env = gym.make('CartPole-v0')
```

4. Set the **seed** for torch and the environment to guarantee reproducible results:

```
seed = 100
env.seed(seed)
torch.manual_seed(seed)
```

5. Set all of the hyperparameter values required for the DQN process:

```
NUMBER_OF_EPISODES = 700
MAX_STEPS = 1000
LEARNING_RATE = 0.01
DISCOUNT_FACTOR = 0.99
HIDDEN_LAYER_SIZE = 64


EGREEDY = 0.9
EGREEDY_FINAL = 0.02
EGREEDY_DECAY = 500
```

6. Implement a function for decaying the epsilon values after every step. We will decay with the epsilon value exponentially. The epsilon value will start with the value of **EGREEDY** and will be decayed until it reaches the value of **EGREEDY_FINAL**. Use the following formula:

```
EGREEDY_FINAL + (EGREEDY - EGREEDY_FINAL) \
* math.exp(-1. * steps_done / EGREEDY_DECAY )
```

The code will be as follows:

```
def calculate_epsilon(steps_done):
    """
    Decays epsilon with increasing steps
    Parameter:
    steps_done (int) : number of steps completed
    Returns:
    int - decayed epsilon
    """
    epsilon = EGREEDY_FINAL + (EGREEDY - EGREEDY_FINAL) \
                * math.exp(-1. * steps_done / EGREEDY_DECAY )
    return epsilon
```

7. Fetch the number of states and actions from the environment:

```
number_of_states = env.observation_space.shape[0]
number_of_actions = env.action_space.n
print('Total number of States : {}'.format(number_of_states))
print('Total number of Actions : {}'.format(number_of_actions))
```

The output will be as follows:

```
Total number of States : 4
Total number of Actions : 2
```

8. Create a class, called **DQN**, that accepts the number of states as inputs and outputs Q values for the number of actions present in the environment, and has a network with a hidden layer of size **64**:

```
class DQN(nn.Module):
    def __init__(self , hidden_layer_size):
        super().__init__()
        self.hidden_layer_size = hidden_layer_size
        self.fc1 = nn.Linear\
                    (number_of_states,self.hidden_layer_size)
        self.fc2 = nn.Linear\
                    (self.hidden_layer_size,number_of_actions)

    def forward(self, x):
        output = torch.tanh(self.fc1(x))
        output = self.fc2(output)
        return output
```

9. Create a **DQN_Agent** class and implement the constructor's **_init_** function.
    This function will create an instance of the DQN class within which the hidden
    layer size is passed. It will also define the **MSE** as a loss criterion. Next, define
    Adam as the optimizer with model parameters and a predefined learning rate:

```
class DQN_Agent(object):
    def __init__(self):
        self.dqn = DQN(HIDDEN_LAYER_SIZE).to(device)

        self.criterion = torch.nn.MSELoss()

        self.optimizer = optim.Adam\
                    (params=self.dqn.parameters() , \
                     lr=LEARNING_RATE)
```

10. Next, define the **select_action** function that will accept **state** and egreedy
    values as input parameters. Use the **egreedy** algorithm to select the action.
    This function will pass the **state** through the DQN to get the Q value, and then
    select the action with the highest Q value using the **torch.max** operation
    during the exploitation phase. During this process, gradient computation is not
    required; that's why we use the **torch.no_grad()** function to turn off the
    gradient calculation:

```
    def select_action(self,state,EGREEDY):
        random_for_egreedy = torch.rand(1)[0]

        if random_for_egreedy > EGREEDY:
            with torch.no_grad():
                state = torch.Tensor(state).to(device)
                q_values = self.dqn(state)
                action = torch.max(q_values,0)[1]
                action = action.item()
        else:
            action = env.action_space.sample()

        return action
```

11. Define the **optimize** function that will accept **state**, **action**, **new_state**, **reward**, and **done** as inputs and convert them into tensors, keeping their compatibility with the device used. If the episode is over, then we make the reward the target value; otherwise, the new state is passed through the DQN (which is used to detach and turn off the gradient calculation) to calculate the max part present in the right-hand side of the Bellman equation. Using the reward obtained and the discount factor, we can calculate the target value:

$$Q(s, a) = r + \gamma \max_{a'} Q(s', a')$$

Figure 9.25: Target value equation

Calculate the predicted value by passing the current state through the network. Calculate the loss, followed by clearing the previous gradients using the **optimizer.zero_grad()** function. Calculate the gradients and weight updates:

```
def optimize(self, state, action, new_state, reward, done):
    state = torch.Tensor(state).to(device)
    new_state = torch.Tensor(new_state).to(device)

    reward = torch.Tensor([reward]).to(device)

    if done:
        target_value = reward
    else:
        new_state_values = self.dqn(new_state).detach()
        max_new_state_values = torch.max(new_state_values)
        target_value = reward + DISCOUNT_FACTOR \
                        * max_new_state_values

    predicted_value = self.dqn(state)[action].view(-1)

    loss = self.criterion(predicted_value, target_value)

    self.optimizer.zero_grad()
    loss.backward()
    self.optimizer.step()
```

Deep Q Learning | 523

12. Write a training process using a **for** loop. At first, instantiate the DQN agent using the class created earlier. Create a **steps_total** empty list to collect the total number of steps for each episode. Initialize **steps_counter** with zero and use it to calculate the decayed epsilon value for each step. Use two loops during the training process. The first one is to play the game for a certain number of steps. The second loop ensures that each episode goes on for a fixed number of steps. Inside the second **for** loop, the first step is to calculate the epsilon value for the current step. Using the present state and epsilon value, you select the action to perform. The next step is to take the action. Once you take the action, the environment returns the **new_state**, **reward**, and **done** flags. Using the **optimize** function, perform one step of gradient descent to optimize the DQN. Now make the new state the present state for the next iteration. Finally, check whether the episode is over or not. If the episode is over, you can collect and record the reward for the current episode:

```
# Instantiating the DQN Agent
dqn_agent = DQN_Agent()

steps_total = []
steps_counter = 0
for episode in range(NUMBER_OF_EPISODES):

    state = env.reset()
    done = False
    step = 0

    for I in range(MAX_STEPS):
        step += 1
        steps_counter += 1

        EGREEDY = calculate_epsilon(steps_counter)

        action = dqn_agent.select_action(state, EGREEDY)

        new_state, reward, done, info = env.step(action)
```

```
            dqn_agent.optimize(state, action, new_state, reward, done)


        state = new_state


        if done:
            steps_total.append(step)
            break
```

13. Now observe the reward, as the reward is scalar feedback and gives you an indication of how well the agent is performing. You should look at the average reward and the average reward for the last 100 episodes:

```
print("Average reward: %.2f" \
    % (sum(steps_total)/NUMBER_OF_EPISODES))
print("Average reward (last 100 episodes): %.2f" \
    % (sum(steps_total[-100:])/100))
```

The output will be as follows:

```
Average reward: 158.83
Average reward (last 100 episodes): 176.28
```

14. Perform the graphical representation of rewards. Check how the agent is performing while playing more episodes, and check what the reward average is for the last 100 episodes:

```
plt.figure(figsize=(12,5))
plt.title("Rewards Collected")
plt.bar(np.arange(len(steps_total)), steps_total, \
        alpha=0.5, color='green', width=6)
plt.show()
```

The output plot should be as follows:

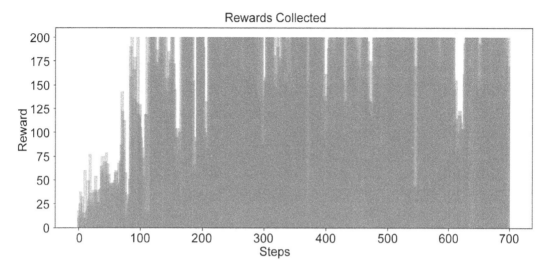

Figure 9.26: Rewards collected

Figure 9.26 shows that the initial number of steps and rewards is low. However, with the increasing numbers of steps, we have collected a stable and higher value of rewards with the help of the DQN algorithm.

> **NOTE**
>
> To access the source code for this specific section, please refer to https://packt.live/3cUE8Q9.
>
> You can also run this example online at https://packt.live/37zeUpz.

Thus, we have successfully implemented a working DQN using PyTorch in a CartPole environment. Let's now look at a few challenging aspects of DQN.

CHALLENGES IN DQN

Everything that was explained in the preceding sections looks good; however, there are a few challenges with DQNs. Here are a couple of the challenges of a DQN:

- The correlation between the steps causes a convergence issue during the training process

- The challenge of having a non-stationary target.

These challenges and their corresponding solutions are explained in the following sections.

CORRELATION BETWEEN STEPS AND THE CONVERGENCE ISSUE

From the previous exercise, we have seen that, during Q learning, we treat the RL problem as a supervised machine learning problem, where we have predictions and target values, and, using gradient descent optimization, we try to reduce the loss to find the optimal Q function.

The gradient descent algorithm assumes that the training data points are independent and identically distributed (that is, **i.i.d**), which is generally true in the case of traditional machine learning data. However, in the case of RL, each data point is highly correlated and dependent on the other. Put simply, the next state depends on the action taken from the previous state. Due to the correlation present in the RL data, we have a convergence issue in the case of the gradient descent algorithm.

To solve this issue of convergence, we will now look at a possible solution, known as **Experience Replay**, in the following section.

EXPERIENCE REPLAY

To break the correlation between the data points in the case of RL, we can use a technique called experience replay. Here, at each timestep during training, we store the agent's experience in a **Replay Buffer** (which is just a Python list).

For example, during the training at time t, the following agent experience is stored as a tuple in the replay buffer $(s_t, a_t, s_{t+1}, r_t, done)$, where:

- s_t - current state

- a_t - action taken

- s_{t+1} - new state

- r_t - reward

- done - indicates whether the episode is complete or not

We set a maximum size for the replay buffer; we will keep on adding new tuples of experience as we encounter them. So, when we reach the maximum size, we will throw out the oldest value. At any given point in time, the replay buffer will always store the latest experience of maximum size.

During training, to break the correlation, we will randomly sample these experiences from the replay buffer to train the DQN. This process of gaining experience and sampling from the replay buffer that stores these experiences is called experience replay.

During the Python implementation, we will use a push function to store the experiences in the replay buffer. An example function will be implemented to sample the experience from the buffer, the pointer, and the length method, which will help us to keep track of the replay buffer size.

The following is a detailed code implementation example of experience replay.

We will implement an **ExperienceReplay** class with all of the functionality explained earlier. In the class, the constructor will contain the following variables: **capacity**, which indicates the maximum size of the replay buffer; **buffer**, which is an empty Python list that acts as the memory buffer; and **pointer**, which points to the current location of the memory buffer while pushing the memory to the buffer.

The class will contain the **push** function, which checks whether there is any space in the buffer using the **pointer** variable. If there is an empty space, **push** adds an experience tuple at the end of the buffer, else the function will replace the memory from the starting point of the buffer. It also contains the **sample** function, which will return the experience tuple of the batch size, and the **__len__** function, which will return the length of the current buffer, as part of the implementation.

The following is an example of how the pointer, capacity, and modular division will work in experience replay.

We initialize the pointer with zero and capacity with three. After every operation, we increase the pointer value and, using modular division, we get the current value of the pointer. When the pointer exceeds the maximum capacity, the value will reset to zero:

```
pointer = 0 ; capacity = 3
print(pointer)
pointer = (pointer + 1) % capacity
print(pointer)

0
1

pointer = (pointer + 1) % capacity
print(pointer)

2

pointer = (pointer + 1) % capacity
print(pointer)

0
```

Figure 9.27: Pointer, capacity, and modular division in the experience replay class

Adding all of the previously mentioned functionality, we can implement the **ExperienceReplay** class as shown in the following code snippet:

```
class ExperienceReplay(object):
    def __init__(self , capacity):

        self.capacity = capacity
        self.buffer = []
        self.pointer = 0


    def push(self , state, action, new_state, reward, done):
        experience = (state, action, new_state, reward, done)

        if self.pointer >= len(self.buffer):
            self.buffer.append(experience)
        else:
            self.buffer[self.pointer] = experience
```

```
        self.pointer = (self.pointer + 1) % self.capacity

    def sample(self , batch_size):
        return zip(*random.sample(self.buffer , batch_size))

    def __len__(self):
        return len(self.buffer)
```

As you can see, the experience class has been initiated.

THE CHALLENGE OF A NON-STATIONARY TARGET

Consider the following code snippet. If you look closely at the following **optimize** function, you will see that we have two passes through the DQN network: one pass to calculate the target Q value (using the Bellman equation) and the other pass to calculate the predicted Q value. After that, we have calculated the loss:

```
def optimize(self, state, action, new_state, reward, done):
        state = torch.Tensor(state).to(device)
        new_state = torch.Tensor(new_state).to(device)

        reward = torch.Tensor([reward]).to(device)

        if done:
            target_value = reward
        else:
            # first pass
            new_state_values = self.dqn(new_state).detach()
            max_new_state_values = torch.max(new_state_values)
            target_value = reward + DISCOUNT_FACTOR \
                            * max_new_state_values
        # second pass
        predicted_value = self.dqn(state)[action].view(-1)

        loss = self.criterion(predicted_value, target_value)

        self.optimizer.zero_grad()
        loss.backward()
        self.optimizer.step() # weight optimization
```

The first pass is just an approximation of the optimal Q value using the Bellman equation; however, to calculate the target and predicted Q values, we use the same weights from the network. This process makes the whole deep Q learning process unstable. Consider the following equation during the loss calculation:

$$\left[Q(s,\ a;\ \theta) - \left(r(s,\ a) + \gamma \max_a Q(s',\ a;\ \theta)\right)\right]^2$$

Figure 9.28: Expression for the loss calculation

After the loss is calculated, we perform one step of gradient descent and optimize the weights to minimize the loss. Once the weights are updated, the predicted Q value will change. However, our target Q values will also change because, to calculate the target Q values, we are using the same weights. Due to the unavailability of the fixed target Q value, this whole process is unstable in the current architecture.

One solution to this problem would be to have a fixed target Q value during the whole training process.

THE CONCEPT OF A TARGET NETWORK

To resolve the issue of the non-stationary target, we can fix the issue by introducing a target neural network architecture in the pipeline. We call this network the **Target Network**. The target network will have the same network architecture as the base neural network in the architecture. We can call this base neural network the predicted DQN.

As discussed previously, to calculate the loss, we must do two passes through the DQN: one pass is to calculate the target Q values, and the second one is to calculate the predicted Q values.

Due to the architectural change, target Q values will be calculated using the target network and the predicted Q values process will remain the same, as shown in the following figure:

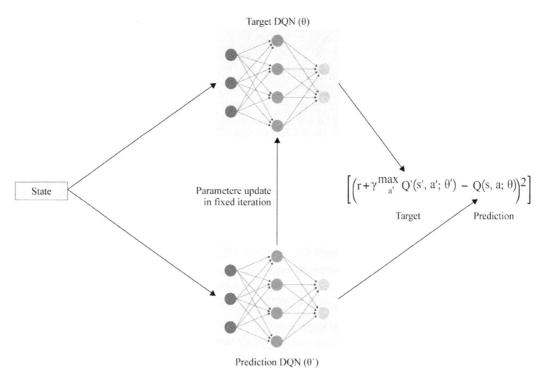

Figure 9.29: Target network

As inferred from the preceding figure, the loss function can be written as follows:

$$\left[\left(r+\gamma \max_{a'} Q'(s', a'; \theta') - Q(s, a; \theta)\right)^2\right]$$

Figure 9.30: Expression for the loss function

The entire purpose of the target network is to calculate the max part in the Bellman equation using the new state-action pair.

At this point in time, the obvious question you may ask is, what about the weights or parameters of this target, and how can we get the target values in an optimum way from this target network? To ensure a balance between fixing the target value and the optimal target approximation using the target network, we will update the weights of the target network from the predicted values after every fixed iteration. But after how many iterations should we update the weights of the target network from the prediction network? Well, that's a hyperparameter that should be tuned during the training process of the DQN. This whole process makes the training process stable as the target Q values are fixed for a while.

We can summarize the steps of training a DQN with experience replay and a target network as follows:

1. Initialize the replay buffer.

2. Create and initialize the prediction network.

3. Create a copy of the prediction network as a target network.

4. Run through the fixed number of episodes.

Within every episode, perform the following steps:

1. Use the egreedy algorithm to choose an action.

2. Perform the action and collect the reward and new state.

3. Store the whole experience in the replay buffer.

4. Select a random batch of experience from the replay buffer.

5. Pass the batch of states through the prediction network to get the predicted Q values.

6. Using a new state, pass through the target network to calculate the target Q value.

7. Perform gradient descent to optimize the weights of the prediction network.

8. After a fixed iteration, clone the weights of the prediction network to the target network.

Now we understand the concept of DQN, the disadvantages of DQN, and how we can overcome these disadvantages of DQN using experience replay and a target network; we can combine all of these to build a robust DQN algorithm. Let's implement our learning in the following exercise.

EXERCISE 9.04: IMPLEMENTING A WORKING DQN NETWORK WITH EXPERIENCE REPLAY AND A TARGET NETWORK IN PYTORCH

In the previous exercise, you implemented a working DQN to work with the CartPole environment. Then, we looked at the disadvantages of a DQN. Now, in this exercise, let's implement the DQN network with experience replay and a target network using the same CartPole environment in PyTorch to build a more stable DQN learning process:

1. Open a new Jupyter notebook and import the required libraries:

```
import gym
import matplotlib.pyplot as plt
import torch
import torch.nn as nn
from torch import optim
import numpy as np
import random
import math
```

2. Write code that will create a device based on the availability of a GPU environment:

```
use_cuda = torch.cuda.is_available()
device = torch.device("cuda:0" if use_cuda else "cpu")

print(device)
```

3. Create a **gym** environment using the **'CartPole-v0'** environment:

```
env = gym.make('CartPole-v0')
```

4. Set the seed for torch and the environment for reproducibility:

```
seed = 100
env.seed(seed)
torch.manual_seed(seed)
random.seed(seed)
```

5. Fetch the number of states and actions from the environment:

```
number_of_states = env.observation_space.shape[0]
number_of_actions = env.action_space.n
print('Total number of States : {}'.format(number_of_states))
print('Total number of Actions : {}'.format(number_of_actions))
```

The output is as follows:

```
Total number of States : 4
Total number of Actions : 2
```

6. Set all of the hyperparameter values required for the DQN process. Please add several new hyperparameters, as stated here, along with the usual parameters:

REPLAY_BUFFER_SIZE – This sets the replay buffer's maximum length size.

BATCH_SIZE – This indicates how many sets of experiences $(s_t, a_t, s_{t+1}, r_t, \text{done})$ are to be drawn to train the DQN.

UPDATE_TARGET_FREQUENCY – This is the periodic frequency at which target network weights will be refreshed from the prediction network:

```
NUMBER_OF_EPISODES = 500
MAX_STEPS = 1000
LEARNING_RATE = 0.01
DISCOUNT_FACTOR = 0.99
HIDDEN_LAYER_SIZE = 64

EGREEDY = 0.9
EGREEDY_FINAL = 0.02
EGREEDY_DECAY = 500

REPLAY_BUFFER_SIZE = 6000
BATCH_SIZE = 32

UPDATE_TARGET_FREQUENCY = 200
```

7. Use the previously implemented **calculate_epsilon** function to decay the epsilon value with increasing values of steps:

```
def calculate_epsilon(steps_done):
    """
    Decays epsilon with increasing steps
    Parameter:
    steps_done (int) : number of steps completed
    Returns:
    int - decayed epsilon
    """
    epsilon = EGREEDY_FINAL + (EGREEDY - EGREEDY_FINAL) \
            * math.exp(-1. * steps_done / EGREEDY_DECAY )
    return epsilon
```

8. Create a class, called **DQN**, that accepts the number of states as inputs and outputs Q values for the number of actions present in the environment, with the network that has a hidden layer of size **64**:

```
class DQN(nn.Module):
    def __init__(self , hidden_layer_size):
        super().__init__()
        self.hidden_layer_size = hidden_layer_size
        self.fc1 = nn.Linear\
                   (number_of_states,self.hidden_layer_size)
        self.fc2 = nn.Linear\
                   (self.hidden_layer_size,number_of_actions)

    def forward(self, x):
        output = torch.tanh(self.fc1(x))
        output = self.fc2(output)
        return output
```

9. Implement the **ExperienceReplay** class:

```
class ExperienceReplay(object):
    def __init__(self , capacity):

        self.capacity = capacity
        self.buffer = []
        self.pointer = 0

    def push(self , state, action, new_state, reward, done):
        experience = (state, action, new_state, reward, done)
            if self.pointer >= len(self.buffer):
            self.buffer.append(experience)
        else:
            self.buffer[self.pointer] = experience

        self.pointer = (self.pointer + 1) % self.capacity

    def sample(self , batch_size):
        return zip(*random.sample(self.buffer , batch_size))

    def __len__(self):
        return len(self.buffer)
```

10. Now instantiate the **ExperienceReplay** class by passing the buffer size as input:

```
memory = ExperienceReplay(REPLAY_BUFFER_SIZE)
```

11. Implement the **DQN_Agent** class.

Please note, here are several changes in the **DQN_Agent** class (which we used in *Exercise 9.03, Implementing a Working DQN Network with PyTorch in a CartPole-v0 Environment*) that need to be incorporated with the previously implemented **DQN_Agent** class.

Create a replica of the normal DQN network and name it **target_dqn**. Use **target_dqn_update_counter** to periodically update the weights of the target DQN from the DQN network. Add the following steps. **memory.sample(BATCH_SIZE)** will randomly pull the experiences from the replay buffer for training. Pass **new_state** in the target network to get the target Q values from the target network. Finally, update the weights of the target network from the normal or predicted DQN after a certain iteration is specified in **UPDATE_TARGET_FREQUENCY**.

Note that we have used the **gather**, **squeeze**, and **unsqueeze** functions, which we studied in the dedicated *PyTorch Utilities* section:

```
class DQN_Agent(object):
    def __init__(self):

        self.dqn = DQN(HIDDEN_LAYER_SIZE).to(device)
        self.target_dqn = DQN(HIDDEN_LAYER_SIZE).to(device)

        self.criterion = torch.nn.MSELoss()

        self.optimizer = optim.Adam(params=self.dqn.parameters(),\
                                    lr=LEARNING_RATE)

        self.target_dqn_update_counter = 0

    def select_action(self,state,EGREEDY):

        random_for_egreedy = torch.rand(1)[0]

        if random_for_egreedy > EGREEDY:

            with torch.no_grad():

                state = torch.Tensor(state).to(device)
                q_values = self.dqn(state)
                action = torch.max(q_values,0)[1]
                action = action.item()
        else:
            action = env.action_space.sample()

        return action
```

```python
    def optimize(self):

        if (BATCH_SIZE > len(memory)):
            return

        state, action, new_state, reward, done = memory.sample\
                                        (BATCH_SIZE)

        state = torch.Tensor(state).to(device)
        new_state = torch.Tensor(new_state).to(device)
        reward = torch.Tensor(reward).to(device)
        # to be used as index
        action = torch.LongTensor(action).to(device)
        done = torch.Tensor(done).to(device)

        new_state_values = self.target_dqn(new_state).detach()
        max_new_state_values = torch.max(new_state_values , 1)[0]
        # when done = 1 then target = reward
        target_value = reward + (1 - done) * DISCOUNT_FACTOR \
                    * max_new_state_values

        predicted_value = self.dqn(state)\
                        .gather(1, action.unsqueeze(1))\
                        .squeeze(1)
        loss = self.criterion(predicted_value, target_value)

        self.optimizer.zero_grad()
        loss.backward()
        self.optimizer.step()

        if self.target_dqn_update_counter \
           % UPDATE_TARGET_FREQUENCY == 0:
            self.target_dqn.load_state_dict(self.dqn.state_dict())

        self.target_dqn_update_counter += 1
```

12. Write the training process of the DQN network. The training process with experience relay and target DQN simplifies the process with less code.

First, instantiate the DQN agent using the class created earlier. Create a **steps_total** empty list to collect the total number of steps for each episode. Initialize **steps_counter** with zero and use it to calculate the decayed epsilon value for each step. Use two loops during the training process: the first one to play the game for a certain number of episodes; the second loop ensures that each episode goes on for a fixed number of steps.

Inside the second **for** loop, the first step is to calculate the epsilon value for the current step. Using the present state and epsilon value, you select the action to perform.

The next step is to take the action. Once you take the action, the environment returns the **new_state**, **reward**, and **done** flags. Push **new_state**, **reward**, **done**, and **info** in the experience replay buffer. Using the **optimize** function, perform one step of gradient descent to optimize the DQN.

Now make the new state the present state for the next iteration. Finally, check whether the episode is over or not. If the episode is over, then you can collect and record the reward for the current episode:

```
dqn_agent = DQN_Agent()

steps_total = []
steps_counter = 0
for episode in range(NUMBER_OF_EPISODES):

    state = env.reset()
    done = False

    step = 0
    for i in range(MAX_STEPS):
        step += 1
        steps_counter += 1

        EGREEDY = calculate_epsilon(steps_counter)

        action = dqn_agent.select_action(state, EGREEDY)
```

```
        new_state, reward, done, info = env.step(action)

        memory.push(state, action, new_state, reward, done)

        dqn_agent.optimize()

        state = new_state

        if done:
            steps_total.append(step)
            break
```

13. Now observe the reward. As the reward is scalar feedback and gives you an indication of how well the agent is performing, you should look at the average reward and the average reward for the last 100 episodes. Also, perform the graphical representation of rewards. Check how the agent is performing while playing more episodes and what the reward average is for the last 100 episodes:

```
print("Average reward: %.2f" \
      % (sum(steps_total)/NUMBER_OF_EPISODES))
print("Average reward (last 100 episodes): %.2f" \
      % (sum(steps_total[-100:])/100))
```

The output will be as follows:

```
Average reward: 154.41
Average reward (last 100 episodes): 183.28
```

Now we can see that the average reward for the last 100 episodes is higher for the DQN with experience replay, and the fixed target is higher than the vanilla DQN implemented in the previous exercise. This is because we have achieved stability during the DQN training process and because we have incorporated experience replay and a target network.

14. Plot the rewards in the y axis along with the number of steps in the x axis to see how the rewards have been collected with the increasing number of steps:

```
plt.figure(figsize=(12,5))
plt.title("Rewards Collected")
plt.xlabel('Steps')
plt.ylabel('Reward')
plt.bar(np.arange(len(steps_total)), steps_total, alpha=0.5, \
        color='green', width=6)
plt.show()
```

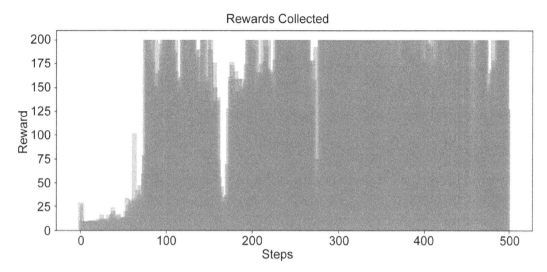

Figure 9.31: Rewards collected

As you can see in the preceding plot, using experience replay with the target network, the rewards are initially a bit low compared to the previous version (refer to *Figure 9.26*); however, after certain episodes, the rewards are relatively stable and the average rewards are high in the last 100 episodes.

> **NOTE**
>
> To access the source code for this specific section, please refer to https://packt.live/2C1KikL.
>
> You can also run this example online at https://packt.live/3dVwiqB.

In this exercise, we have added experience replay and a target network in the vanilla DQN network (which was explained in *Exercise 9.03, Implementing a Working DQN Network with PyTorch in a CartPole-v0 Environment*) to overcome the drawbacks of a vanilla DQN. This results in much better performance in terms of rewards, as we have seen a more stable performance in terms of the average reward for the last 100 episodes. A comparison of the outputs is shown here:

Vanilla DQN Outputs:

```
Average reward: 158.83
Average reward (last 100 episodes): 176.28
```

DQN with Experience Replay and Target Network Outputs:

```
Average reward: 154.41
Average reward (last 100 episodes): 183.28
```

Still, we have another issue with the DQN process, that is, overestimation in the DQN. We will learn more about this and how to tackle it in the next section.

THE CHALLENGE OF OVERESTIMATION IN A DQN

In the previous section, we introduced a target network as a solution to fix the non-stationary target problem. Using this target network, we calculated the target Q value and calculated the loss. This whole process of introducing a new target network to calculate a fixed target value has somehow made the training process a bit more stable. However, in 2015, *Hado van Hasselt*, in his paper called *Deep Reinforcement Learning with Double Q-learning*, showed through multiple experiments that this process overestimates the target Q values and makes the whole training process unstable:

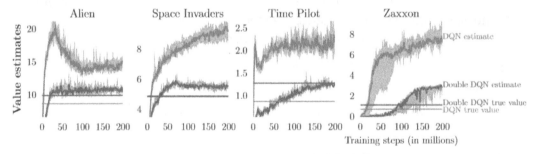

Figure 9.32: Q value estimation in DQN and DDQN

> **NOTE**
>
> The preceding diagram has been sourced from the paper *Deep Reinforcement Learning with Double Q-learning* by *Hasselt et al., 2015*. Please refer to the following link for more in-depth reading on DDQN: https://arxiv.org/pdf/1509.06461.pdf.

After performing experiments on multiple Atari games, the authors of the paper showed that using a DQN network can lead to a high estimation of Q values (shown in orange), which indicates a high deviation from the true DQN values. In the paper, the authors proposed a new algorithm called **Double DQN**. We can see that, by using Double DQN, Q value estimations are much closer to true values and any overestimations are much lower. Now, let's discuss what Double DQN is and how it is different from a DQN with a target network.

DOUBLE DEEP Q NETWORK (DDQN)

In comparison to a DQN with a target network, the minor differences of a DDQN are given as follows:

- A DDQN uses our prediction network to select the best action to take for the next state, by selecting the action with the highest Q values.

- A DDQN uses the action from the prediction network to calculate the corresponding estimate of the target Q value (using the target network) at the next state.

As described in the *Deep Q Learning* section, the loss function for a DQN is as follows:

$$\left[\left(r + \gamma \max_{a'} Q'(s', a'; \theta') - Q(s, a; \theta)\right)^2\right]$$

Figure 9.33: Loss function for a DQN

The updated loss function for a DDQN will be as follows:

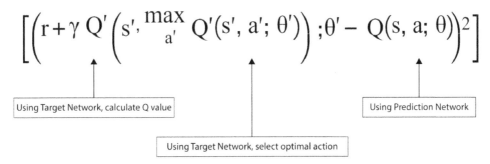

$$\left[\left(r + \gamma\, Q'\left(s', \max_{a'} Q'(s', a'; \theta')\right); \theta' - Q(s, a; \theta)\right)^2\right]$$

Using Target Network, calculate Q value

Using Prediction Network

Using Target Network, select optimal action

Figure 9.34: Updated loss function for a DDQN

The following figure depicts the functioning of a typical DDQN:

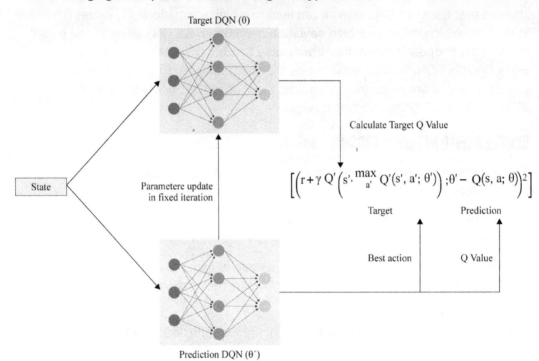

$$\left[\left(r + \gamma\, Q'\left(s',\ \underset{a'}{\max}\ Q'(s', a';\ \theta')\right);\theta' - Q(s, a;\ \theta)\right)^2\right]$$

Figure 9.35: DDQN

The following outlines the required changes in the optimize function for a DDQN implementation:

1. Select an action using the prediction network.

   We will pass the **new_state** through the prediction network to get the Q values for the **new_state**, as shown in the following code:

   ```
   new_state_indxs = self.dqn(new_state).detach()
   ```

   To select the action, we will select the max index value from the output Q values, as shown here:

   ```
   max_new_state_indxs = torch.max(new_state_indxs, 1)[1]
   ```

2. Select the Q value for the best action using the target network.

   We will pass the **new_state** through the target network to get the Q values for the **new_state**, as shown in the following code:

   ```
   new_state_values = self.target_dqn(new_state).detach()
   ```

For the Q values associated with the best action in the **new_state**, we use the target network, as shown in the following code:

```
max_new_state_values = new_state_values.gather\
                            (1, max_new_state_indxs\
                                .unsqueeze(1))\
                        .squeeze(1)
```

The **gather** function is used to select the Q values using the indexes fetched from the prediction network.

The following is a complete implementation of the DDQN with the required changes:

```
def optimize(self):

    if (BATCH_SIZE > len(memory)):
        return

    state, action, new_state, reward, done = memory.sample\
                                        (BATCH_SIZE)

    state = torch.Tensor(state).to(device)
    new_state = torch.Tensor(new_state).to(device)
    reward = torch.Tensor(reward).to(device)
    action = torch.LongTensor(action).to(device)
    done = torch.Tensor(done).to(device)

    """
    select action : get the index associated with max q value
    from prediction network
    """
    new_state_indxs = self.dqn(new_state).detach()
    # to get the max new state indexes
    max_new_state_indxs = torch.max(new_state_indxs, 1)[1]
    """
    Using the best action from the prediction nn get the max new
state
    value in target dqn
    """
    new_state_values = self.target_dqn(new_state).detach()
    max_new_state_values = new_state_values.gather\
                            (1, max_new_state_indxs\
```

```
                            .unsqueeze(1))\
                        .squeeze(1)
    #when done = 1 then target = reward
    target_value = reward + (1 - done) * DISCOUNT_FACTOR \
                    * max_new_state_values

    predicted_value = self.dqn(state).gather\
                        (1, action.unsqueeze(1)).squeeze(1)
    loss = self.criterion(predicted_value, target_value)

    self.optimizer.zero_grad()
    loss.backward()
    self.optimizer.step()

    if self.target_dqn_update_counter \
        % UPDATE_TARGET_FREQUENCY == 0:
        self.target_dqn.load_state_dict(self.dqn.state_dict())

    self.target_dqn_update_counter += 1
```

Now that we have studied the various concepts of DQN and DDQN, let's now concretize our understanding with an activity.

ACTIVITY 9.01: IMPLEMENTING A DOUBLE DEEP Q NETWORK IN PYTORCH FOR THE CARTPOLE ENVIRONMENT

In this activity, you are tasked with implementing a DDQN in PyTorch for the CartPole environment to tackle the issue of overestimation in a DQN. We can summarize the steps of training a DQN with experience replay and the target network.

The following steps will help you to complete the activity:

1. Open a new Jupyter notebook and import the required libraries:

```
import gym
import matplotlib.pyplot as plt
import torch
import torch.nn as nn
from torch import optim
import numpy as np
import random
import math
```

2. Write code that will create a device based on the availability of a GPU environment.

3. Create a **gym** environment using the **CartPole-v0** environment.

4. Set the seed for torch and the environment for reproducibility.

5. Fetch the number of states and actions from the environment.

6. Set all of the hyperparameter values required for the DQN process.

7. Implement the **calculate_epsilon** function.

8. Create a class, called **DQN**, that accepts the number of states as inputs and outputs Q values for the number of actions present in the environment, with the network that has a hidden layer of size 64.

9. Initialize the replay buffer.

10. Create and initialize the prediction network in the **DQN_Agent** class, as shown in *Exercise 9.03*, *Implementing a Working DQN Network with Experience Replay and a Target Network in PyTorch*. Create a copy of the prediction network as the target network.

11. Make changes to the **optimize** function of the **DQN_Agent** class according to the code example shown in the *Double Deep Q Network (DDQN)* section.

12. Run through a fixed number of episodes. Inside the episode, use the egreedy algorithm to choose an action.

13. Perform the action and collect the reward and new state. Store the whole experience in the replay buffer.

14. Select a random batch of experience from the replay buffer. Pass the batch of states through the prediction network to get the predicted Q values.

15. Use our prediction network to select the best action to take for the next state by selecting the action with the highest Q values. Use the action from the prediction network to calculate the corresponding estimate of the target Q value at the next state.

16. Perform gradient descent to optimize the weights of the prediction network. After a fixed iteration, clone the weights of the prediction network to the target network.

17. Check the average reward and the average reward for the last 100 episodes once you have trained the DDQN agent.

18. Plot the rewards collected in the y axis and the number of episodes in the x axis to visualize how the rewards have been collected with the increasing number of episodes.

The output for the average rewards should be similar to the following:

```
Average reward: 174.09
Average reward (last 100 episodes): 186.06
```

The plot for the rewards should be similar to the following:

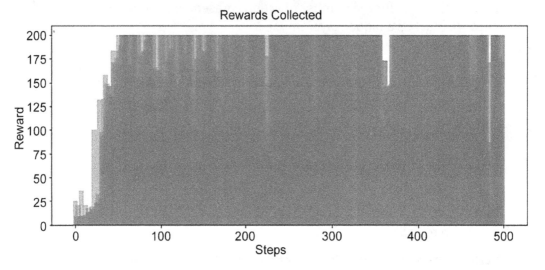

Figure 9.36: Plot for the rewards collected

> **NOTE**
>
> The solution to this activity can be found on page 743.

Before we end the chapter, we present the following comparison of the average rewards for different DQN techniques and DDQN:

Vanilla DQN Outputs:

```
Average reward: 158.83
Average reward (last 100 episodes): 176.28
```

DQN with Experience Replay and Target Network Outputs:

```
Average reward: 154.41
Average reward (last 100 episodes): 183.28
```

DDQN Outputs:

```
Average reward: 174.09
Average reward (last 100 episodes): 186.06
```

As you can see from the preceding figure, along with the comparison of the results shown earlier, DDQN has the highest average reward, compared to other DQN implementations, and the average reward for the last 100 episodes is also higher. We can say that DDQN improves performance significantly in comparison to the other two DQN techniques. After completing this whole activity, we have learned how to combine a DDQN network with experience replay to overcome the issues of a vanilla DQN and achieve more stable rewards.

SUMMARY

In this chapter, we started with an introduction to deep learning, and we looked at the different components of the deep learning process. Then, we learned how to build deep learning models using PyTorch.

Next, we slowly shifted our focus to RL, where we learned about value functions and Q learning. We demonstrated how Q learning can help us to build RL solutions without knowing the transition dynamics of the environment. We also investigated the problems associated with tabular Q learning and how to solve those performance and memory-related issues with deep Q learning.

Then, we looked into the issues related to a vanilla DQN implementation and how we can use a target network and experience replay mechanism to overcome issues such as correlated data and non-stationary targets during the training of a DQN. Finally, we learned how double deep Q learning helps us to overcome the issue of overestimation in a DQN. In the next chapter, you will learn how to use CNNs and RNNs in combination with a DQN to play the very popular Atari game Breakout.

10

PLAYING AN ATARI GAME WITH DEEP RECURRENT Q-NETWORKS

INTRODUCTION

In this chapter, we will be introduced to **Deep Recurrent Q Networks (DRQNs)** and their variants. You will train **Deep Q Network (DQN)** models with **Convolutional Neural Networks (CNNs)** and **Recurrent Neural Networks (RNNs)**. You will acquire hands-on experience of using the OpenAI Gym package to train reinforcement learning agents to play an Atari game. You will also learn how to analyze long sequences of input and output data using attention mechanisms. By the end of this chapter, you will have a good understanding of what DRQNs are and how to implement them with TensorFlow.

INTRODUCTION

In the previous chapter, we learned that DQNs achieved higher performance compared to traditional reinforcement learning techniques. Video games are a perfect example of where DQN models excel. Training an agent to play video games can be quite difficult for traditional reinforcement learning agents as there is a huge number of possible combinations of states, actions, and Q-values to be processed and analyzed during the training.

Deep learning algorithms are renowned for handling high-dimensional tensors. Some researchers combined Q-learning techniques with deep learning models to overcome this limitation and came up with DQNs. A DQN model comprises a deep learning model that is used as a function approximation of Q-values. This technique constituted a major breakthrough in the reinforcement learning field as it helped to handle much larger state and action spaces than traditional models.

Since then, further research has been undertaken and different types of DQN models have been designed, such as DRQNs or **Deep Attention Recurrent Q Networks (DARQNs)**. In this chapter, we will see how DQN models can benefit from CNN and RNN models, which have achieved amazing results in computer vision and natural language processing. We will look at how to train such models to play the famous Atari game Breakout in the next section.

UNDERSTANDING THE BREAKOUT ENVIRONMENT

We will be training different deep reinforcement learning agents to play the game Breakout in this chapter. Before diving in, let's learn some more about the game.

Breakout is an arcade game designed and released in 1976 by Atari. Steve Wozniak, co-founder of Apple, was part of the design and development team. The game was extremely popular at that time and multiple versions were developed over the years.

The goal of the game is to break all the bricks located at the top of the screen with a ball (since the game was developed in 1974 with low screen definition, the ball is represented by pixels and so its shape can be seen as a rectangle in the following screenshot) without dropping it. The player can move a paddle horizontally at the bottom of the screen to hit the ball before it drops and bounce it back toward the bricks. Also, the ball will bounce back after hitting the side walls or the ceiling. The game ends when either the ball drops (in this case, the player loses) or when all the bricks have been broken and the player wins and can proceed to the next stage:

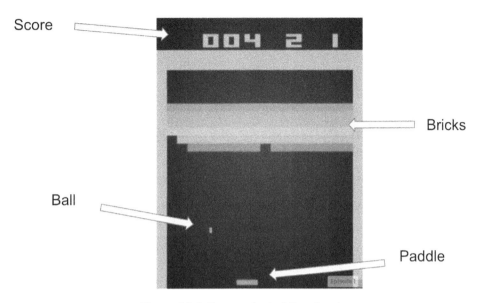

Figure 10.1: Screenshot of Breakout

The **gym** package from OpenAI provides an environment that emulates this game and allows deep reinforcement learning agents to train and play on it. The name of the environment that we will be using is **BreakoutDeterministic-v4**. Given below are some basic code implementations of this environment.

You will need to load the Breakout environment from the **gym** package before being able to train an agent to play this game. To do so, we will use the following code snippet:

```
import gym

env = gym.make('BreakoutDeterministic-v4')
```

This is a deterministic game where the actions chosen by the agent will happen every time as intended and with a frame skipping rate of **4**. Frame skipping corresponds to the number of frames an action is repeated until a new action is performed.

The game comprises four deterministic actions, as shown by the following code:

```
env.action_space
```

The following is the output of the code:

```
Discrete(4)
```

The observation space is a color image (a box of **3** channels) of size **210** by **160**:

```
env.observation_space
```

The following is the output of the code:

```
Box(210, 160, 3)
```

To initialize the game and get the first initial state, we need to call the `.reset()` method, as shown in the following code:

```
state = env.reset()
```

To sample an action (that is, taking a random action from all the possible actions) from the action space, we can use the `.sample()` method:

```
action = env.action_space.sample()
```

Finally, to perform a single action and get its results from the environment, we need to call the `.step()` method:

```
new_state, reward, is_done, info = env.step(action)
```

The following screenshot is a **new_state** result of the environment state after performing an action:

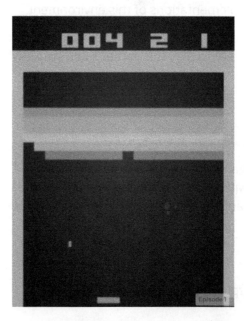

Figure 10.2: Result of the new state after performing an action

The `.step()` method returns four different objects:

- The new environment state resulting from the previous action.

- The reward related to the previous action.

- A flag indicating whether the game has ended after the previous action (either a win or the game is over).

- Some additional information from the environment. This information cannot be used to train the agent, as stated in the OpenAI instructions.

Having gone through some basic code implementation of Breakout in OpenAI, let's perform our first exercise where we will have our agent play this game.

EXERCISE 10.01: PLAYING BREAKOUT WITH A RANDOM AGENT

In this exercise, we will be implementing some functions for playing the game Breakout that will be useful for the remainder of the chapter. We will also create an agent that takes random actions:

1. Open a new Jupyter Notebook file and import the **gym** library:

```
import gym
```

2. Create a class called **RandomAgent** that takes a single input parameter named **env**, the game environment. This class will have a method called **get_action()** that will return a random action from the environment:

```
class RandomAgent():
    def __init__(self, env):
        self.env = env
    def get_action(self, state):
        return self.env.action_space.sample()
```

3. Create a function called **initialize_env()** that will return the initial state of the given input environment, a **False** value that corresponds to the initial value of a done flag, and **0** as the initial value of the reward:

```
def initialize_env(env):
    initial_state = env.reset()
    initial_done_flag = False
    initial_rewards = 0
    return initial_state, initial_done_flag, initial_rewards
```

4. Create a function called **play_game()** that takes an agent, a state, a done flag, and a list of rewards as inputs. This will return the total reward received. This **play_game()** function will iterate until the done flag equals **True**. At each iteration, it will perform the following actions: get an action from the agent, perform the action on the environment, accumulate the reward received, and prepare for the next state:

```
def play_game(agent, state, done, rewards):
    while not done:
        action = agent.get_action(state)
        next_state, reward, done, _ = env.step(action)
        state = next_state
        rewards += reward
    return rewards
```

5. Create a function called **train_agent()** that takes as inputs an environment, a number of episodes, and an agent. This function will create a **deque** object from the collections package and iterate through the number of episodes provided. At each iteration, it will perform the following actions: initialize the environment with **initialize_env()**, play a game with **play_game()**, and append the received rewards to the **deque** object. Finally, it will print the average score of the games played:

```
def train_agent(env, episodes, agent):
    from collections import deque
    import numpy as np

    scores = deque(maxlen=100)

    for episode in range(episodes)
        state, done, rewards = initialize_env(env)
        rewards = play_game(agent, state, done, rewards)
        scores.append(rewards)
    print(f"Average Score: {np.mean(scores)}")
```

6. Instantiate a Breakout environment called **env** using the **gym.make()** function:

```
env = gym.make('BreakoutDeterministic-v4')
```

7. Instantiate a **RandomAgent** object called **agent**:

```
agent = RandomAgent(env)
```

8. Create a variable called **episodes** that will take the value **10**:

```
episodes = 10
```

9. Call the **train_agent** function by providing **env**, episodes, and the agent:

```
train_agent(env, episodes, agent)
```

After training the agent, you will expect to achieve something approaching the following score (your score may be slightly different due to the randomness of the game):

```
Average Score: 0.6
```

The random agent is achieving a low score after 10 episodes, that is, 0.6. We will consider that the agent will have learned to play this game if it achieves a score above 10. However, since we have use a low number of episodes, we have not yet reached a stage where we achieve a score above 10. At this stage, however, we have created some functions for playing the game Breakout that we will reuse and update for the coming sections.

> **NOTE**
>
> To access the source code for this specific section, please refer to https://packt.live/30CfVeH.
>
> You can also run this example online at https://packt.live/3hi12nU.

In the next section, we will look at CNN models and how to build them in TensorFlow.

CNNS IN TENSORFLOW

CNNs are a type of deep learning architecture that achieved amazing results in computer vision tasks such as image classification, object detection, and image segmentation. Self-driving cars are an example of a real-life application of such technology.

The main element of CNNs is the convolutional operation, where a filter is applied to different parts of an image to detect specific patterns and generate a feature map. A feature map can be thought of as an image with the detected patterns highlighted, as shown in the following example:

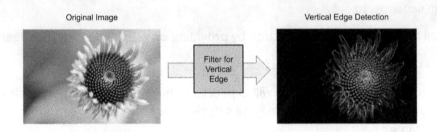

Figure 10.3: Example of a vertical edge feature map

A CNN is composed of several convolutional layers that apply the convolutional operation with different filters. The final layers of a CNN are usually one or several fully connected layers that are responsible for making the right predictions for a given dataset. For example, the final layer of a CNN trained to predict images of digits will be a fully connected layer of 10 neurons. Each neuron will be responsible for predicting the probability of occurrence of each digit (0 to 9):

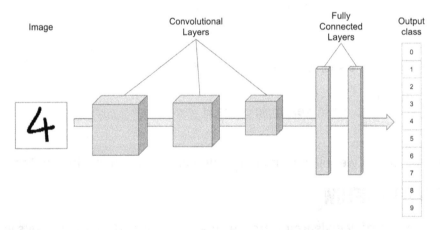

Figure 10.4: Example of a CNN architecture for classifying images of digits

Building CNN models is extremely easy with TensorFlow, thanks to the Keras API. To define a convolutional layer, we just need to use the **Conv2D ()** class, as shown in the following code:

```
from tensorflow.keras.layers import Conv2D

Conv2D(128, kernel_size=(3, 3), activation="relu")
```

In the preceding example, we have created a convolutional layer with **128** filters (or kernels) of size **3** by **3**, and **relu** as the **activation** function.

> **NOTE**
>
> Throughout the course of this chapter, we'll be using the ReLU activation function for CNN models, as it is one of the most performant activation functions.

To define a fully connected layer, we will use the **Dense()** class:

```
from tensorflow.keras.layers import Dense

Dense(units=10, activation='softmax')
```

In Keras, we can use the **Sequential()** class to create a multi-layer CNN:

```
import tensorflow as tf
from tensorflow.keras.layers import Conv2D, Dense

model = tf.keras.Sequential()
model.add(Conv2D(128, kernel_size=(3, 3), activation="relu"), \
          input_shape=(100, 100, 3))
model.add(Conv2D(128, kernel_size=(3, 3), activation="relu"))
model.add(Dense(units=100, activation="relu"))
model.add(Dense(units=10, activation="softmax"))
```

Please note that you need to provide the dimensions of the input images for the first convolutional layer only. After defining the layers of your model, you will need to compile it by providing the loss function, the optimizer, and the metrics to be displayed:

```
model.compile(loss='sparse_categorical_crossentropy', \
              optimizer="adam", metrics=['accuracy'])
```

Finally, the last step is to train the CNN with the training set on a specified number of **epochs**:

```
model.fit(features_train, label_train, epochs=5)
```

Another useful method from TensorFlow is **tf.image.rgb_to_grayscale()**, which is used to convert a color image to grayscale:

```
img = tf.image.rgb_to_grayscale(img)
```

To resize an input image, we will use the **tf.image.resize()** method:

```
img = tf.image.resize(img, [50, 50])
```

Now that we know how to build a CNN model, let's put this into practice in the following exercise.

EXERCISE 10.02: DESIGNING A CNN MODEL WITH TENSORFLOW

In this exercise, we will be designing a CNN model with TensorFlow. This model will be used for our DQN agent in *Activity 10.01, Training a DQN with CNNs to Play Breakout*, where we will train this model to play the game Breakout. Perform the following steps to implement the exercise:

1. Open a new Jupyter Notebook file and import the **tensorflow** package:

```
import tensorflow as tf
```

2. Import the **Sequential** class from **tensorflow.keras.models**:

```
from tensorflow.keras.models import Sequential
```

3. Instantiate a sequential model and save it to a variable called **model**:

```
model = Sequential()
```

4. Import the **Conv2D** class from **tensorflow.keras.layers**:

```
from tensorflow.keras.layers import Conv2D
```

5. Instantiate a convolutional layer with **Conv2D** with **32** filters of size **8**, a stride of 4 by 4, relu as the activation function, and an input shape of (**84**, **84**, **1**). These dimensions are related to the size of the screen for the game Breakout. Save it to a variable called **conv1**:

```
conv1 = Conv2D(32, 8, (4,4), activation='relu', \
               padding='valid', input_shape=(84, 84, 1))
```

6. Instantiate a second convolutional layer with **Conv2D** with **64** filters of size **4**, a stride of **2** by **2**, and **relu** as the activation function. Save it to a variable called **conv2**:

```
conv2 = Conv2D(64, 4, (2,2), activation='relu', \
               padding='valid')
```

7. Instantiate a third convolutional layer with **Conv2D** with **64** filters of size **3**, a stride of **1** by **1**, and **relu** as the activation function. Save it to a variable called **conv3**:

```
conv3 = Conv2D(64, 3, (1,1), activation='relu', padding='valid')
```

8. Add the three convolutional layers to the model by means of the **add()** method:

```
model.add(conv1)
model.add(conv2)
model.add(conv3)
```

9. Import the **Flatten** class from **tensorflow.keras.layers**. This class will resize the output of the convolutional layers to a one-dimension vector:

```
from tensorflow.keras.layers import Flatten
```

10. Add an instantiated **Flatten** layer to the model by means of the **add()** method:

```
model.add(Flatten())
```

11. Import the **Dense** class from **tensorflow.keras.layers**:

```
from tensorflow.keras.layers import Dense
```

12. Instantiate a fully connected layer with **256** units and **relu** as the activation function:

```
fc1 = Dense(256, activation='relu')
```

13. Instantiate a fully connected layer with **4** units, which corresponds to the number of possible actions from the game Breakout:

```
fc2 = Dense(4)
```

14. Add the two fully connected layers to the model by means of the **add()** method:

```
model.add(fc1)
model.add(fc2)
```

15. Import the **RMSprop** class from **tensorflow.keras.optimizers**:

```
from tensorflow.keras.optimizers import RMSprop
```

16. Instantiate an **RMSprop** optimizer with **0.00025** as the learning rate:

```
optimizer=RMSprop(lr=0.00025)
```

17. Compile the model by specifying **mse** as the loss function, **RMSprop** as **optimizer**, and **accuracy** as the metric to be displayed during training to the **compile** method:

```
model.compile(loss='mse', optimizer=optimizer, \
              metrics=['accuracy'])
```

18. Print a summary of the model using the **summary** method:

```
model.summary()
```

Following is the output of the code:

```
Model: "sequential"

Layer (type)              Output Shape              Param #
=================================================================
conv2d (Conv2D)           (None, 20, 20, 32)        2080

conv2d_1 (Conv2D)         (None, 9, 9, 64)          32832

conv2d_2 (Conv2D)         (None, 7, 7, 64)          36928

flatten (Flatten)         (None, 3136)              0

dense (Dense)             (None, 256)               803072

dense_1 (Dense)           (None, 4)                 1028
=================================================================
Total params: 875,940
Trainable params: 875,940
Non-trainable params: 0
```

Figure 10.5: Summary of the CNN model

The output shows the architecture of the model we just built, together with the different layers and the number of parameters that will be used during the training of the model.

> **NOTE**
>
> To access the source code for this specific section, please refer to https://packt.live/2YrqiiZ.
>
> You can also run this example online at https://packt.live/3fiNMxE.

We have designed a CNN model with three convolutional layers. In the next section, we will see how we can use this model in relation to a DQN agent.

COMBINING A DQN WITH A CNN

Humans play video games using their sight. They look at the screen, analyze the situation, and decide what the best action to be performed is. In video games, there can be a lot of things happening on the screen, so being able to see all these patterns can give a significant advantage in playing the game. Combining a DQN with a CNN can help a reinforcement learning agent to learn the right action to take given a particular situation.

Instead of just using fully connected layers, a DQN model can be extended with convolutional layers as inputs. The model will then be able to analyze the input image, find the relevant patterns, and feed them to the fully connected layers responsible for predicting the Q-values, as shown in the following:

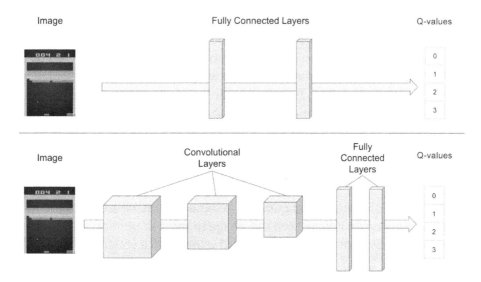

Figure 10.6: Difference between a normal DQN and a DQN combined
with convolutional layers

Adding convolutional layers helps the agent to better understand the environment. The DQN agent that we will build in the coming activity will use the CNN model from *Exercise 10.02, Designing a CNN Model with TensorFlow*, to output the Q-values for a given state. But rather than using a single model, we will use two models instead. The models will share the exact same architecture.

The first model will be responsible for predicting the Q-values for playing the game, while the second one (referred to as the target model) will be responsible for learning what should be the optimal Q-values. This technique helps the target model to converge faster on the optimal solution.

ACTIVITY 10.01: TRAINING A DQN WITH CNNS TO PLAY BREAKOUT

In this activity, we will build a DQN with additional convolutional layers and train it to play the game Breakout with CNNs. We will add experience replay to the agent. We will need to preprocess the images in order to create a sequence of four images for our Breakout game.

The following instructions will help you to complete this activity:

1. Import the relevant packages (**gym**, **tensorflow**, **numpy**).

2. Reshape the training and test sets.

3. Create a DQN class with the **build_model()** method, which will instantiate a CNN model composed of the **get_action()** method, which will apply the epsilon-greedy algorithm to choose the action to be played, the **add_experience()** method to store in memory the experience acquired by playing the game, the **replay()** method, which will perform experience replay by sampling experiences from the memory and train the DQN model, and the **update_epsilon()** method to gradually decrease the epsilon value for epsilon-greedy.

4. Use the **initialize_env()** function to initialize the environment by returning the initial state, **False** for the done flag, and **0** as the initial reward.

5. Create a function called **preprocess_state()** that will perform the following preprocessing on an image: crop the image to remove unnecessary parts, convert to a grayscale image, and resize the image to a square shape.

6. Create a function called **play_game()** that will play a game until it is over, and then store the experience and the accumulated reward.

7. Create a function called **train_agent()** that will iterate through a number of episodes where the agent will play a game and perform experience replay.

8. Instantiate a Breakout environment and train a DQN agent to play this game for **50** episodes. Please note that it might take longer for this step to execute as we are training large models.

The expected output will be close to the one shown here. You may have slightly different values on account of the randomness of the game and the randomness of the epsilon-greedy algorithm in choosing the action to be played:

```
[Episode 0] - Average Score: 3.0
Average Score: 0.59
```

> **NOTE**
>
> The solution to this activity can be found on page 752.

In the next section, we will see how we can extend this model with another type of deep learning architecture: the RNN.

RNNS IN TENSORFLOW

In the previous section, we saw how to integrate a CNN into a DQN model to improve the performance of a reinforcement learning agent. We added a few convolutional layers as inputs to the fully connected layers of the DQN model. These convolutional layers helped the model to analyze visual patterns from the game environment and make better decisions.

There is a limitation, however, to using a traditional CNN approach. CNNs can only analyze a single image. While playing video games such as Breakout, analyzing a sequence of images is a much more powerful tool when it comes to understanding the movements of the ball. This is where RNNs come to the fore:

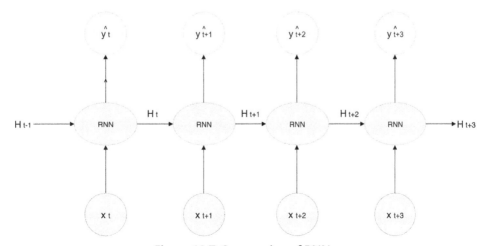

Figure 10.7: Sequencing of RNNs

RNNs are a specific architecture of neural networks that take a sequence of inputs. They are very popular in natural language processing for treating corpora of texts for speech recognition, chatbots, or text translation. Texts can be defined as sequences of words that are correlated with one another. It is hard to determine the topic of a sentence or a paragraph just by looking at a single word. You have to look at a sequence of multiple words before being able to make a guess.

There are different types of RNN models. The most popular ones are **Gated Recurrent Unit** (**GRU**) and **Long Short-Term Memory** (**LSTM**). Both of these models have a memory that keeps a record of the different inputs the model has already processed (for instance, the first five words of a sentence) and combines them with new inputs (such as the sixth word of a sentence).

In TensorFlow, we can build an **LSTM** layer of **10** units as follows:

```
from tensorflow.keras.layers import LSTM

LSTM(10, activation='tanh', recurrent_activation='sigmoid')
```

The sigmoid activation function is the most popular one used for RNN models.

The syntax will be very similar for defining a **GRU** layer:

```
from tensorflow.keras.layers import GRU

GRU(10, activation='tanh', recurrent_activation='sigmoid')
```

In Keras, we can use the **Sequential()** class to create a multi-layer LSTM:

```
import tensorflow as tf
from tensorflow.keras.layers import LSTM, Dense

model = tf.keras.Sequential()
model.add(LSTM(128, activation='tanh', \
               recurrent_activation='sigmoid'))
model.add(Dense(units=100, activation="relu")))
model.add(Dense(units=10, activation="softmax"))
```

Before fitting the model, you will need to compile it by providing the loss function, the optimizer, and the metrics to be displayed:

```
model.compile(loss='sparse_categorical_crossentropy', \
              optimizer="adam", metrics=['accuracy'])
```

We already saw how to define LSTM layers previously, but in order to combine them with a CNN model, we need to use a wrapper in TensorFlow called **TimeDistributed()**. This class is used to apply the same specified layer to each timestep of an input tensor, such as the following:

```
TimeDistributed(Dense(10))
```

In the preceding example, the same fully connected layer is applied to each of the timesteps received. In our case, we want to apply a convolutional layer to each image of a sequence before feeding an LSTM model. To build such a sequence, we will need to stack multiple images together to create a sequence that the RNN model will take as input. Let's now perform an exercise to design a combination of CNN and RNN models.

EXERCISE 10.03: DESIGNING A COMBINATION OF CNN AND RNN MODELS WITH TENSORFLOW

In this exercise, we will be designing a combination of CNN and RNN models with TensorFlow. This model will be used by our DRQN agent in *Activity 10.02, Training a DRQN to Play Breakout*, to play the game Breakout:

1. Open a new Jupyter Notebook and import the **tensorflow** package:

```
import tensorflow as tf
```

2. Import the **Sequential** class from **tensorflow.keras.models**:

```
from tensorflow.keras.models import Sequential
```

3. Instantiate a **sequential** model and save it to a variable called **model**:

```
model = Sequential()
```

4. Import the **Conv2D** class from **tensorflow.keras.layers**:

```
from tensorflow.keras.layers import Conv2D
```

5. Instantiate a convolutional layer with **Conv2D** with **32** filters of size **8**, a stride of **4** by **4**, and **relu** as the activation function. Save it to a variable called **conv1**:

```
conv1 = Conv2D(32, 8, (4,4), activation='relu', \
               padding='valid', input_shape=(84, 84, 1))
```

6. Instantiate a second convolutional layer with **Conv2D** with **64** filters of size **4**, a stride of **2** by **2**, and **relu** as the activation function. Save it to a variable called **conv2**:

```
conv2 = Conv2D(64, 4, (2,2), activation='relu', \
               padding='valid')
```

7. Instantiate a third convolutional layer with **Conv2D** with **64** filters of size **3**, a stride of **1** by **1**, and **relu** as the activation function. Save it to a variable called **conv3**:

```
conv3 = Conv2D(64, 3, (1,1), activation='relu', \
               padding='valid')
```

8. Import the **TimeDistributed** class from **tensorflow.keras.layers**:

```
from tensorflow.keras.layers import TimeDistributed
```

9. Instantiate a time-distributed layer that will take **conv1** as the input and (**4, 84, 84, 1**) as the input shape. Save it to a variable called **time_conv1**:

```
time_conv1 = TimeDistributed(conv1, input_shape=(4, 84, 84, 1))
```

10. Instantiate a second time-distributed layer that will take **conv2** as the input. Save it to a variable called **time_conv2**:

```
time_conv2 = TimeDistributed(conv2)
```

11. Instantiate a third time-distributed layer that will take **conv3** as the input. Save it to a variable called **time_conv3**:

```
time_conv3 = TimeDistributed(conv3)
```

12. Add the three time-distributed layers to the model using the **add()** method:

```
model.add(time_conv1)
model.add(time_conv2)
model.add(time_conv3)
```

13. Import the **Flatten** class from **tensorflow.keras.layers**:

```
from tensorflow.keras.layers import Flatten
```

14. Instantiate a time-distributed layer that will take a **Flatten()** layer as input. Save it to a variable called **time_flatten**:

```
time_flatten = TimeDistributed(Flatten())
```

15. Add the **time_flatten** layer to the model with the **add()** method:

```
model.add(time_flatten)
```

16. Import the **LSTM** class from **tensorflow.keras.layers**:

```
from tensorflow.keras.layers import LSTM
```

17. Instantiate an LSTM layer with **512** units. Save it to a variable called **lstm**:

```
lstm = LSTM(512)
```

18. Add the LSTM layer to the model with the **add()** method:

```
model.add(lstm)
```

19. Import the **Dense** class from **tensorflow.keras.layers**:

```
from tensorflow.keras.layers import Dense
```

20. Instantiate a fully connected layer with **128** units and **relu** as the activation function:

```
fc1 = Dense(128, activation='relu')
```

21. Instantiate a fully connected layer with **4** units:

```
fc2 = Dense(4)
```

22. Add the two fully connected layers to the model with the **add()** method:

```
model.add(fc1)
model.add(fc2)
```

23. Import the **RMSprop** class from **tensorflow.keras.optimizers**:

```
from tensorflow.keras.optimizers import RMSprop
```

24. Instantiate **RMSprop** with **0.00025** as the learning rate:

```
optimizer=RMSprop(lr=0.00025)
```

25. Compile the model by specifying **mse** as the loss function, **RMSprop** as the optimizer, and **accuracy** as the metric to be displayed during training to the **compile** method:

```
model.compile(loss='mse', optimizer=optimizer, \
              metrics=['accuracy'])
```

26. Print a summary of the model using the **summary** method:

```
model.summary()
```

Following is the output of the code:

```
Model: "sequential"

Layer (type)                    Output Shape              Param #
=================================================================
time_distributed (TimeDistri (None, 4, 20, 20, 32)        2080

time_distributed_1 (TimeDist (None, 4, 9, 9, 64)          32832

time_distributed_2 (TimeDist (None, 4, 7, 7, 64)          36928

time_distributed_3 (TimeDist (None, 4, 3136)              0

lstm (LSTM)                     (None, 512)               7473152

dense (Dense)                   (None, 128)               65664

dense_1 (Dense)                 (None, 4)                 516
=================================================================
Total params: 7,611,172
Trainable params: 7,611,172
Non-trainable params: 0
```

Figure 10.8: Summary of the CNN+RNN model

We have successfully combined a CNN model with an RNN model. The preceding output shows the architecture of the model we just built with the different layers and the number of parameters that will be used during training. This model takes as input a sequence of four images and passes it to the RNN, which will analyze their relationship before feeding the results to the fully connected layers, which will be responsible for predicting the Q-values.

> **NOTE**
>
> To access the source code for this specific section, please refer to https://packt.live/2UDB3h4.
>
> You can also run this example online at https://packt.live/3dVrf9T.

Now that we know how to build an RNN, we can combine this technique with a DQN model. This kind of model is called a DRQN, and this is what we are going to look at in the next section.

BUILDING A DRQN

A DQN can benefit greatly from RNN models facilitating the processing of sequential images. Such an architecture is known as **Deep Recurrent Q Network** (**DRQN**). Combining a GRU or LSTM model with a CNN model will allow the reinforcement learning agent to understand the movement of the ball. To do so, we just need to add an LSTM (or GRU) layer between the convolutional and fully connected layers, as shown in the following figure:

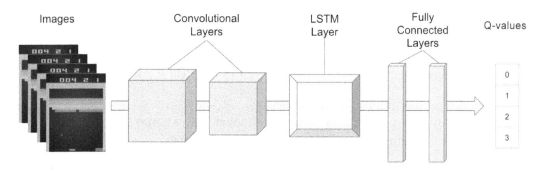

Figure 10.9: DRQN architecture

To feed the RNN model with a sequence of images, we need to stack several images together. For the Breakout game, after initializing the environment, we will need to take the first image and duplicate it several times in order to have the first initial sequence of images. Having done this, after each action, we can append the latest image to the sequence and remove the oldest one in order to maintain the exact same size of sequence (for instance, a sequence of a maximum of four images).

ACTIVITY 10.02: TRAINING A DRQN TO PLAY BREAKOUT

In this activity, we will build a DRQN model by replacing the DQN model from *Activity 10.01, Training a DQN with CNNs to Play Breakout*. We will then train the DRQN model to play the Breakout game and analyze the performance of the agent. The following instructions will help you to complete this activity:

1. Import the relevant packages (**gym**, **tensorflow**, **numpy**).

2. Reshape the training and test sets.

3. Create the **DRQN** class with the following methods: the **build_model()** method to instantiate a CNN combined with an RNN model, the **get_action()** method to apply the epsilon-greedy algorithm to choose the action to be played, the **add_experience()** method to store in memory the experience acquired by playing the game, the **replay()** method, which will perform experience replay by sampling experiences from the memory and train the DRQN model with a callback to save the model every two episodes, and the **update_epsilon()** method to gradually decrease the epsilon value for epsilon-greedy.

4. Use the **initialize_env()** function to train the agent, which will initialize the environment by returning the initial state, **False** for the done flag, and **0** as the initial reward.

5. Create a function called **preprocess_state()** that will perform the following preprocessing on an image: crop the image to remove unnecessary parts, convert to a grayscale image, and then resize the image to a square shape.

6. Create a function called **combine_images()** that will stack a sequence of images.

7. Create a function called **play_game()** that will play a game until it is over, and then store the experience and the accumulated reward.

8. Create a function called **train_agent()** that will iterate through a number of episodes where the agent will play a game and perform experience replay.

9. Instantiate a Breakout environment and train a **DRQN** agent to play this game for **200** episodes.

> **NOTE**
>
> We recommend training for 200 (or 400) episodes in order to train the models properly and achieve good performance, but this may take a few hours depending on the system configuration. Alternatively, you can reduce the number of episodes, which will reduce the training time but will impact the performance of the agent.

The expected output will be close to the one shown here. You may have slightly different values on account of the randomness of the game and the randomness of the epsilon-greedy algorithm in choosing the action to be played:

```
[Episode 0] - Average Score: 0.0
[Episode 50] - Average Score: 0.43137254901960786
[Episode 100] - Average Score: 0.4
[Episode 150] - Average: 0.54
Average Score: 0.53
```

> **NOTE**
>
> The solution to this activity can be found on page 756.

In the next section, we will see how we can improve the performance of our model by adding an attention mechanism to DRQN and building a DARQN model.

INTRODUCTION TO THE ATTENTION MECHANISM AND DARQN

In the previous section, we saw how adding an RNN model to a DQN helped to increase its performance. RNNs are known for handling sequential data such as temporal information. In our case, we used a combination of CNNs and RNNs to help our reinforcement learning agent to better understand sequences of images from the game.

However, RNN models do have some limitations when it comes to analyzing long sequences of input or output data. To overcome this situation, researchers have come up with a technique called attention, which is the principal technique behind a **Deep Attention Recurrent Q-Network** (**DARQN**). The DARQN model is the same as the DRQN model, with just an attention mechanism added to it. To better understand this concept, we will go through an example of its application: neural translation. Neural translation is the field of translating text from one language to another, such as translating Shakespeare's plays, which were written in English, into French.

Sequence-to-sequence models are the best fit for such a task. They comprise two components: an encoder and a decoder. Both of them are RNN models, such as an LSTM or GRU model. The encoder is responsible for processing a sequence of words from the input data (in our previous example, this would be a sentence of English words) and generates an encoded version called the context vector. The decoder will take this context vector as input and will predict the relevant output sequence (a sentence of French words, in our example):

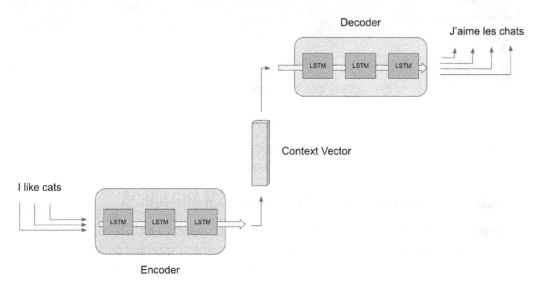

Figure 10.10: Sequence-to-sequence model

The size of the context vector is fixed. It is an encoded version of the input sequence with only the relevant information. You can think of it as a summary of the input data. However, the set size of this vector limits the model in terms of retaining sufficient relevant information from long sequences. It will tend to "forget" the earlier elements of a sequence. But in the case of translation, the beginning of a sentence usually contains very important information, such as its subject.

The attention mechanism not only provides the decoder with the context vector, but also the previous states of the encoder. This enables the decoder to find relevant relationships between previous states, the context vector, and the desired output. This will help in our example to understand the relationship between two elements that are far away from one another in the input sequence:

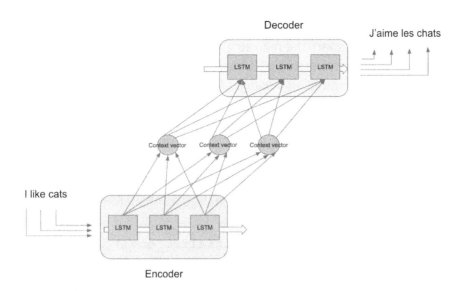

Figure 10.11: Sequence-to-sequence model with an attention mechanism

TensorFlow provides an **Attention** class. It takes as input a tensor of shape **[output, states]**. It is better to use it by using the functional API, where each layer acts as a function that takes inputs and provides outputs as results. In this case, we can simply extract the output and states from a GRU layer and provide them as inputs for the attention layer:

```
from tensorflow.keras.layers import GRU, Attention

out, states = GRU(512, return_sequences=True, \
                  return_state=True)(input)
att = Attention()([out, states])
```

To build a DARQN model, we just need to add this attention mechanism to a DRQN model.

Let's add this attention mechanism to our previous DRQN agent (in *Activity 10.02, Training a DRQN to Play Breakout*) and build a DARQN model in the next activity.

ACTIVITY 10.03: TRAINING A DARQN TO PLAY BREAKOUT

In this activity, we will build a DARQN model by adding an attention mechanism to our previous DRQN from *Activity 10.02, Training a DRQN to Play Breakout*. We will then train the model to play the Breakout game and then analyze the performance of the agent. The following instructions will help you to complete this activity:

1. Import the relevant packages (**gym**, **tensorflow**, and **numpy**).

2. Reshape the training and test sets.

3. Create a **DARQN** class with the following methods: the **build_model()** method, which will instantiate a CNN combined with an RNN model (similar to *Exercise 10.03, Designing a Combination of CNN and RNN Models with TensorFlow*); the **get_action()** method, which will apply the epsilon-greedy algorithm to choose the action to be played; the **add_experience()** method to store in memory the experience acquired by playing the game; the **replay()** method, which will perform experience replay by sampling experiences from the memory and train the DARQN model with a callback to save the model every two episodes; and the **update_epsilon()** method to gradually decrease the epsilon value for epsilon-greedy.

4. Initialize the environment using the **initialize_env()** function by returning the initial state, **False** for the done flag, and **0** as the initial reward.

5. Use the **preprocess_state()** function to perform the following preprocessing on an image: crop the image to remove unnecessary parts, convert to a grayscale image, and resize the image to a square shape.

6. Create a function called **combine_images()** that will stack a sequence of images.

7. Use the **play_game()** function to play a game until it is over, and then store the experience and the accumulated reward.

8. Iterate through a number of episodes where the agent will play a game and perform experience replay using the **train_agent()** function.

9. Instantiate a Breakout environment and train a **DARQN** agent to play this game for **400** episodes.

> **NOTE**
>
> We recommend training for 400 episodes in order to properly train the model and achieve good performance, but this may take a few hours depending on the system configuration. Alternatively, you can reduce the number of episodes, which will reduce the training time but will impact the performance of the agent.

The output will be close to what you see here. You may have slightly different values on account of the randomness of the game and the randomness of the epsilon-greedy algorithm in choosing the action to be played:

```
[Episode 0] - Average Score: 1.0
[Episode 50] - Average Score: 2.4901960784313726
[Episode 100] - Average Score: 3.92
[Episode 150] - Average Score: 7.37
[Episode 200] - Average Score: 7.76
[Episode 250] - Average Score: 7.91
[Episode 300] - Average Score: 10.33
[Episode 350] - Average Score: 10.94
Average Score: 10.83
```

> **NOTE**
>
> The solution to this activity can be found on page 761.

SUMMARY

In this chapter, we learned how to combine deep learning techniques to a DQN model and train it to play the Atari game Breakout. We first looked at adding convolutional layers to the agent for processing screenshots from the game. This helped the agent to better understand the game environment.

We then took things a step further and added an RNN to the outputs of the CNN model. We created a sequence of images and fed it to an LSTM layer. This sequential model provided the DQN agent with the ability to "visualize" the direction of the ball. This kind of model is called a DRQN.

Finally, we used an attention mechanism and trained a DARQN model to play the Breakout game. This mechanism helped the model to better understand previous relevant states and improved its performance drastically. This field is still evolving as new deep learning techniques and models are designed, outperforming previous generations in the process.

In the next chapter, you will be introduced to policy-based methods and the actor-critic model, which consists of multiple models responsible for computing an action based on a state and calculating the Q-values.

11

POLICY-BASED METHODS FOR REINFORCEMENT LEARNING

OVERVIEW

In this chapter, we will implement different policy-based methods of **Reinforcement Learning** (**RL**), such as policy gradients, **Deep Deterministic Policy Gradients** (**DDPGs**), **Trust Region Policy Optimization** (**TRPO**), and **Proximal Policy Optimization** (**PPO**). You will be introduced to the math behind some of the algorithms and you'll also learn how to code policies for RL agents within the OpenAI Gym environment. By the end of this chapter, you will not only have a base-level understanding of policy-based RL methods but you'll also be able to create complete working prototypes using the previously mentioned policy-based RL methods.

INTRODUCTION

The focus of this chapter is policy-based methods for RL. However, before diving into a formal introduction to policy-based methods for RL, let's spend some time understanding the motivation behind them. Let's go back a few hundred years when the globe was still mostly undiscovered and maps were incomplete. Brave sailors at that time sailed the great oceans with only indomitable courage and unyielding curiosity on their side. But they weren't completely blind in the vastness of the oceans. They looked up to the night sky for direction. The stars and planets in the night sky guided them to their destination. The night sky is viewed differently at different times of the year from different parts of the globe. This information, along with highly accurate maps of the night sky, guided these brave explorers to their destinations and sometimes to unknown, uncharted lands.

Now, you might question what this story has to do with RL at all. A map of the night sky wasn't always available to those sailors. They were created by globetrotters, sailors, skywatchers, and astronomers over centuries. Sailors actually voyaged blindly at one time. They looked at the stars during night time, and every time they took a turn, they marked their position relative to the position of the stars in the night sky. Upon reaching their destination, they evaluated each turn they took and worked out which was more effective during their voyage. Every other ship that sailed to the same destination could do the same. With time, they had a good assessment of the turns that are the most effective for reaching a certain destination with respect to a ship's position in the sea, as assessed by looking at the position of the stars in the night sky. You can think of this as computing the value function where you know the immediate best move. But once sailors had a complete map of the night sky, they could simply derive a policy that would lead them to their destination.

You can consider the sea and the night sky as the environment and the sailors as agents within it. Over the span of a few centuries, our agents (sailors) built a model of their environment and so were able to come up with a value function (calculating a ship's relative position) that would lead them to the immediate best possible step (immediate navigational step) and also helped them build the optimal policy (a complete navigation route).

In the last chapter, you learned about **Deep Recurrent Q Networks (DRQNs)** and their advantage over simple deep Q networks. You also modeled a DRQN network for playing the very popular Atari video game *Breakout*. In this chapter, you'll learn about policy-based approaches to RL.

We will also learn about the policy gradient, which will help you learn about the model in real time. We will then learn about a policy gradient technique called DDPG to understand the continuous action space. Here, we will also learn how to code the Lunar Lander simulation to understand DDPGs using classes such as the **OUActionNoise** class, the **ReplayBuffer** class, the **ActorNetwork** class, and the **CriticNetwork** class. We will learn about these classes in detail later in this chapter. Finally, we will learn how we can improve the policy gradient technique by using the TRPO, PPO, and **Advantage Actor-Critic** (**A2C**) techniques. These techniques will help us reduce the operating cost of training the model and so will improve the policy-gradient technique.

Let's begin by learning about some basic concepts, such as value-based RL, model-based RL, actor-critic, and action space, in the following sub-sections.

INTRODUCTION TO VALUE-BASED AND MODEL-BASED RL

While it is useful to have a good model of the environment to be able to predict whether a particular move is better with regard to other possible moves, you still need to assess all the possible moves from every possible state in order to come up with an optimal policy. This is a non-trivial problem and is also computationally expensive if, say, our environment is a simulation and our agent is **Artificial Intelligence** (**AI**). This approach of model-based learning, when applied within a simulation, can look like the following scenario.

Consider the game of *Pong* (*Figure 11.1*). (*Pong*—released in 1972—was one of the first arcade video games manufactured by Atari.) Now, let's see how the model-based learning approach could be beneficial for an optimal policy playing *Pong* and what could be its drawbacks. So, suppose our agent has learned how to play *Pong* by looking at the game environment—that is, by looking at the black and white pixels of each frame. We can then ask our agent to predict the next possible state given a certain frame of black and white pixels from the game environment. But if there is any background noise in the environment (for example, a random, unrelated video playing in the background), our agent would also take that into consideration.

Now, in most cases, those background noises would not help us in our planning—that is, determining an optimal policy—but would still eat up our computational resources. Following is a screenshot of *Pong* game:

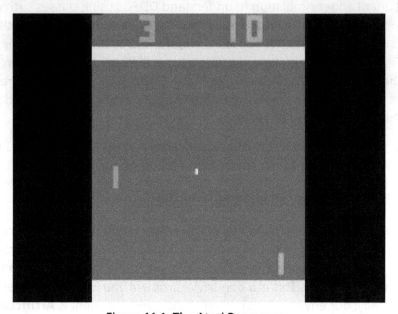

Figure 11.1: The Atari Pong game

A value-based approach is better than the model-based approach because while performing the transition from one state to another, a value-based approach would only care about the value of the action in terms of the cumulative reward that we are predicting for each action. It would deem any background noise as mostly irrelevant. A value-based approach is well-suited for deriving an optimal policy. Imagine you have learned an action-value function—a Q function. Then, you can simply look at the highest values in each state and that gives you the optimal policy. However, value-based functions could still be inefficient. Let me try to explain why with an example. In order to travel from Europe to North America, or from South Africa to the southern coasts of India, the optimal policy for our explorer ship might just be to go straight. However, the ship might encounter icebergs, small islands, or ocean currents that might set it off course temporarily. It might still be the optimal policy for the ship to head straight, but the value function might change arbitrarily. So, a value-based method, in this case, would try to approximate all the arbitrary values, while a policy can be blind and, therefore, be more efficient in terms of computational cost. So, in many cases, it might be less efficient to compute an optimal policy based on the value function.

INTRODUCTION TO ACTOR-CRITIC MODEL

So, we have briefly explained the trade-offs between the value-based and model-based approaches. Now, can we somehow take the best of both the worlds and create a hybrid of them? The actor-critic model will help us to do that. If we draw a Venn diagram (*Figure 11.2*), we will see that the actor-critic model lies at the intersection of the value-based and policy-based RL approaches. They can basically learn a value function as well as a policy. We will discuss actor-critic model further in the following sections.

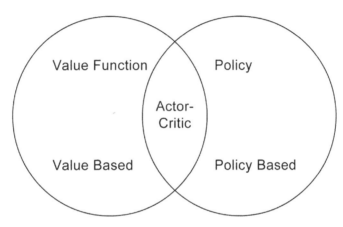

Figure 11.2: The relation between different RL approaches

In practice, most of the time, we try to learn a policy based on the values yielded by the value function, but we actually learn the policy and the values simultaneously. To end this introduction to actor-critic, let me share a quote by Bertrand Russell. Russell, in his book *The Problems of Philosophy*, said: "*We can know the general proposition without inferring it from instances, although some instances are usually necessary to make clear to us what the general proposition means.*" Treat that as food for thought. The code on how to implement the actor-critic model is shown later in this chapter. Next, we will learn about action spaces, the basics of which we already covered in *Chapter 1, Introduction to Reinforcement Learning*.

In the previous chapters, we already covered the basic definition and the types of action spaces. Here, we will quickly revise the concept of action spaces. Action spaces define the properties of the game environment. Let's look at the following diagram to understand the types:

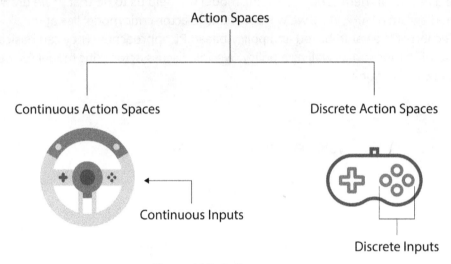

Figure 11.3: Action spaces

There are two types of action spaces—discrete and continuous. Discrete action spaces allow discrete inputs—for example, the buttons on a gamepad. These discrete actions can move in either the left or right direction, going either up or down, moving forward or backward, and so on.

On the other hand, continuous action spaces allow continuous inputs—for example, inputs from a steering wheel or a joystick. In the following section, we will learn how to apply a policy gradient to a continuous action space.

POLICY GRADIENTS

Now that we have established the motivation behind favoring policy-based methods over value-based ones with the navigation example in the previous section, let's begin our formal introduction to policy gradients. Unlike Q-learning, which uses a storage buffer to store past experiences, policy-gradient methods learn in real time (that is, they learn from the most recent experience or action). A policy gradient's learning is driven by whatever the agent encounters in the environment. After each gradient update, the experience is discarded and the policy moves on. Let's look at a pictorial representation of what we have just learned:

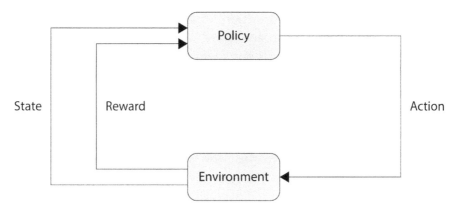

Figure 11.4: The policy gradient method explained pictorially

- One thing that should immediately catch our attention is that the policy gradient method is, in general, less sample-efficient than Q-learning because the experiences are discarded after each gradient update. The mathematical representation of the gradient estimator is given as follows:

$$\hat{g} = \widehat{\mathbb{E}}_t \left[\nabla_{\theta l} \log \pi_\theta(a_t | s_t) \widehat{A}_t \right]$$

Figure 11.5: Mathematical representation of policy gradient estimator

In this equation, π_θ is the stochastic policy and $\hat{A}_t$ is our advantage estimation function at time t —the estimate of the relative value of the selected action. The expectation, $\widehat{\mathbb{E}}_t$, indicates the average over a finite batch of samples in our algorithm, where we perform sampling and optimization, alternatively. Here, $\hat{g}$ is the gradient estimator. The a_t and s_t variables define the action and state at the time interval, t.

Finally, the policy gradient loss is defined as follows:

$$L^{PG}(\theta) = \widehat{\mathbb{E}}_t \left[\log \pi_\theta(a_t | s_t) \widehat{A}_t \right]$$

Figure 11.6: Policy gradient loss defined

In order to calculate the advantage function, $\hat{A}_t$, we need the **discounted reward** and the **baseline estimate**. The discounted reward is also known as the **return**, which is the weighted sum of all the rewards our agent got during the current episode. It is called the discounted reward as there is a discount factor associated with it that prioritizes the immediate rewards over the long-term ones. $\hat{A}_t$ is basically the difference between the discounted reward and the baseline estimate.

Note that if you still have any problems with wrapping your head around the concept, then it's not a big problem. Just try to grasp the overall idea and you'll be able to grasp the full concept of it eventually. Having said that, let me also introduce you to a stripped-down version of the vanilla policy gradient algorithm.

We start by initializing the policy parameter, θ, and the baseline, b:

```
for iteration=1, 2, 3, … do
    Execute the current policy and collect a set of trajectories
    At each timestep in each trajectory, compute
        the return R_t and the advantage estimate  A_t = R_t - b(s_t) .
    Refit the baseline minimizing ‖b(s_t)-R_t‖² ,
        summed over all trajectories and timesteps.
    Update the policy using the policy gradient estimate ĝ
end for
```

One suggestion would be to go through the algorithm multiple times, along with the initial explanation, in order to properly understand the concept of policy gradients. But again, an overall understanding of things should be your first priority.

Before implementing the practical elements, please install OpenAI Gym and the Box2D environment (which includes environments such as Lunar Lander) using PyPI. To carry out the installation, type the following commands into Terminal/Command Prompt:

```
pip install torch==0.4.1
pip install pillow
pip install gym "gym[box2d]"
```

Now, let's implement an exercise using the policy gradient method.

EXERCISE 11.01: LANDING A SPACECRAFT ON THE LUNAR SURFACE USING POLICY GRADIENTS AND THE ACTOR-CRITIC METHOD

In this exercise, we will work on a toy problem (OpenAI Lunar Lander) and help land the Lunar Lander inside the OpenAI Gym Lunar Lander environment using vanilla policy gradients and actor-critic. The following are the steps to implement this exercise:

1. Open a new Jupyter Notebook, import all the necessary libraries (**gym**, **torch**, and **numpy**):

```
import gym
import torch as T
import numpy as np
```

2. Define the **ActorCritic** class:

```
class ActorCritic(T.nn.Module):
    def __init__(self):
        super(ActorCritic, self).__init__()
        self.transform = T.nn.Linear(8, 128)
        self.act_layer = T.nn.Linear(128, 4) # Action layer
        self.val_layer = T.nn.Linear(128, 1) # Value layer
        self.log_probs = []
        self.state_vals = []
        self.rewards = []
```

So, during the initialization of the **ActorCritic** class in the preceding code, we are creating our action and value networks. We are also creating blank arrays for storing the log probabilities, state values, and rewards.

3. Next, create a function to pass our state through the layers and name it **forward**:

```
def forward(self, state):
    state = T.from_numpy(state).float()
    state = T.nn.functional.relu(self.transform(state))
    state_value = self.val_layer(state)

    act_probs = T.nn.functional.softmax\
                (self.act_layer(state))
    act_dist = T.distributions.Categorical(act_probs)
```

```
action = act_dist.sample()

self.log_probs.append(act_dist.log_prob(action))
self.state_vals.append(state_value)

return action.item()
```

Here, we are taking the state and passing it through the value layer after a ReLU transformation to get the state value. Similarly, we are passing the state through the action layer, followed by a softmax function, to get the action probabilities. Then, we are transforming the probabilities to discrete values for the purpose of sampling. Finally, we are adding our log probabilities and state values to their respective arrays and returning an action item.

4. Create the **computeLoss** function to calculate a discounted reward first. This will help give greater priority to the immediate reward. Then, we will calculate the loss as described in the policy gradient loss equation:

```
def computeLoss(self, gamma=0.99):
    rewards = []
    discounted_reward = 0
    for reward in self.rewards[::-1]:
        discounted_reward = reward + gamma \
                                * discounted_reward
        rewards.insert(0, discounted_reward)

    rewards = T.tensor(rewards)
    rewards = (rewards - rewards.mean()) / (rewards.std())

    loss = 0
    for log_probability, value, reward in zip\
    (self.log_probs, self.state_vals, rewards):
        advantage = reward - value.item()
        act_loss = -log_probability * advantage
        val_loss = T.nn.functional.smooth_l1_loss\
                    (value, reward)
        loss += (act_loss + val_loss)

    return loss
```

5. Next, create a **clear** method to clear the arrays that store the log probabilities, state values, and rewards after each episode:

```
def clear(self):
    del self.log_probs[:]
    del self.state_vals[:]
    del self.rewards[:]
```

6. Now, let's start with the main code, which will help us to call the classes that we defined previously in the exercise. We start by assigning a random seed:

```
np.random.seed(0)
```

7. Then, we need to set up our environment and initialize our policy:

```
env = gym.make(""LunarLander-v2"")

policy = ActorCritic()
optimizer = T.optim.Adam(policy.parameters(), \
                         lr=0.02, betas=(0.9, 0.999))
```

8. Finally, we iterate for at least **10000** iterations for proper convergence. In each iteration, we sample an action and get the state and reward for that action. Then, we update our policy based on that action and clear our observations:

```
render = True
np.random.seed(0)
running_reward = 0
for i in np.arange(0, 10000):
    state = env.reset()
    for t in range(10000):
        action = policy(state)
        state, reward, done, _ = env.step(action)
        policy.rewards.append(reward)
        running_reward += reward
        if render and i > 1000:
            env.render()
        if done:
            break
    print("Episode {}\tReward: {}".format(i, running_reward))

    # Updating the policy
    optimizer.zero_grad()
```

```
loss = policy.computeLoss(0.99)
loss.backward()
optimizer.step()
policy.clear()

if i % 20 == 0:
    running_reward = running_reward / 20
    running_reward = 0
```

Now, when you run the code, you'll see the running reward for each episode. The following is the reward for the first 20 episodes out of the total 10,000 episodes:

```
Episode 0Reward: -320.65657506841114
Episode 1Reward: -425.64874914703705
Episode 2Reward: -671.2867424162646
Episode 3Reward: -1032.281198268248
Episode 4Reward: -1224.3354097571892
Episode 5Reward: -1543.1792365484055
Episode 6Reward: -1927.4910808775028
Episode 7Reward: -2023.4599189797761
Episode 8Reward: -2361.9002491621986
Episode 9Reward: -2677.470775357419
Episode 10      Reward: -2932.068423127369
Episode 11      Reward: -3204.4024449864355
Episode 12      Reward: -3449.3136628102934
Episode 13      Reward: -3465.3763860613317
Episode 14      Reward: -3617.162199366013
Episode 15      Reward: -3736.83983321837
Episode 16      Reward: -3883.140249551331
Episode 17      Reward: -4100.137703945375
Episode 18      Reward: -4303.308164747067
Episode 19      Reward: -4569.71587308837
Episode 20      Reward: -4716.304224574078
```

NOTE

The output for only the first 20 episodes is shown here for ease of presentation.

To access the source code for this specific section, please refer to https://packt.live/3hDibst.

This section does not currently have an online interactive example and will need to be run locally.

This output indicates that our agent, the Lunar Lander, has started taking actions. The negative reward indicates that in the beginning, the agent is not smart enough to take the right actions and so it takes random actions, for which it is rewarded negatively. A negative reward is a penalty. With time, the agent will start getting positive rewards as it starts learning. Soon, you'll see the game window popping up on your screen showing the real-time progress of your Lunar Lander, as in the following screenshot:

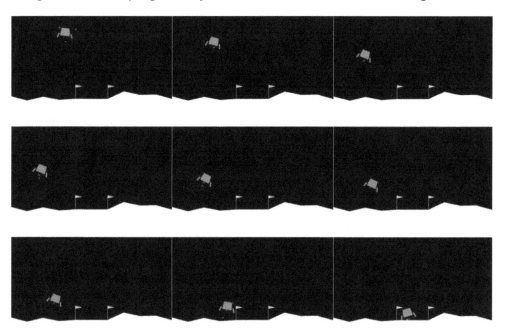

Figure 11.7: The real-time progress of the Lunar Lander

In the next section, we will look into DDPGs, which extend the idea of policy gradients.

DEEP DETERMINISTIC POLICY GRADIENTS

In this section, we will apply the DDPG technique to understand the continuous action space. Moreover, we will learn how to code a moon lander simulation to understand DDPGs.

> **NOTE**
>
> We suggest that you type all the code given in this section into your Jupyter notebook as we will be using it later, in *Exercise 11.02, Creating a Learning Agent.*

We are going to use the OpenAI Gym Lunar Lander environment for continuous action spaces here. Let's start by importing the essentials:

```
import os
import gym
import torch as T
import numpy as np
```

Now, we will learn how to define some classes, such as the **OUActionNoise** class, the **ReplayBuffer** class, the **ActorNetwork** class, and the **CriticNetwork** class, which will help us to implement the DDGP technique. At the end of this section, you'll have the complete code base that applies the DDPG within our OpenAI Gym game environment.

ORNSTEIN-UHLENBECK NOISE

First, we will define a class that will provide us with something known as Ornstein-Uhlenbeck noise. This Ornstein–Uhlenbeck process, in physics, is used to model the velocity of a Brownian particle under the influence of friction. Brownian motion, as you may already know, is the random motion of particles when suspended in a fluid (liquid or gas) resulting from their collisions with other particles in the same fluid. Ornstein–Uhlenbeck noise gives you a type of noise that is temporally correlated and is centered on a mean of 0. Since the agent has zero knowledge of the model, it becomes difficult to train it. Here, Ornstein–Uhlenbeck noise can be used as a sample to generate that knowledge. Let's look at the code implementation of this class:

```
class OUActionNoise(object):
    def __init__(self, mu, sigma=0.15, theta=.2, dt=1e-2, x0=None):
        self.theta = theta
```

```
        self.mu = mu
        self.sigma = sigma
        self.dt = dt
        self.x0 = x0
        self.reset()

    def __call__(self):
        x = self.x_previous
        dx = self.theta * (self.mu -- x) * self.dt + self.sigma \
            * np.sqrt(self.dt) * np.random.normal\
            (size=self.mu.shape)
        self.x_previous = x + dx
        return x

    def reset(self):
        self.x_previous = self.x0 if self.x0 is not None \
                            else np.zeros_like(self.mu)
```

In the preceding code, we defined three different functions—that is, **_init_()**, **_call()_**, and **reset()**. In the next section, we will learn how to implement the **ReplayBuffer** class to store the agent's past learnings.

THE REPLAYBUFFER CLASS

Replay buffer is a concept we have borrowed from Q-learning. This buffer is basically a space to store all of our agent's past learnings, which will help us to train the model better. We will initialize the class by defining the memory size for our state, action, and rewards, respectively. So, the initialization would look something like this:

```
class ReplayBuffer(object):
    def __init__(self, max_size, inp_shape, nb_actions):
        self.memory_size = max_size
        self.memory_counter = 0
        self.memory_state = np.zeros\
                        ((self.memory_size, *inp_shape))
        self.new_memory_state = np.zeros\
                            ((self.memory_size, *inp_shape))
        self.memory_action = np.zeros\
                        ((self.memory_size, nb_actions))
        self.memory_reward = np.zeros(self.memory_size)
    self.memory_terminal = np.zeros(self.memory_size, \
                            dtype=np.float32)
```

Next, we need to define the **store_transition** method. This method takes the state, action, reward, and new state as arguments and stores the transitions from one state to another. There's also a **done** flag to indicate the terminal state of our agent. Note that the index here is just a counter that we initialized previously, and it starts from **0** when its value is equal to the maximum memory size:

```
def store_transition(self, state, action, \
                     reward, state_new, done):
    index = self.memory_counter % self.memory_size
    self.memory_state[index] = state
    self.new_memory_state[index] = state_new
    self.memory_action[index] = action
    self.memory_reward[index] = reward
    self.memory_terminal[index] = 1  - done
    self.memory_counter += 1
```

Finally, we need the **sample_buffer** method, which will be used to randomly sample the buffer:

```
def sample_buffer(self, bs):
    max_memory = min(self.memory_counter, self.memory_size)

    batch = np.random.choice(max_memory, bs)

    states = self.memory_state[batch]
    actions = self.memory_action[batch]
    rewards = self.memory_reward[batch]
    states_ = self.new_memory_state[batch]
    terminal = self.memory_terminal[batch]

    return states, actions, rewards, states_, terminal
```

So, the entire class, at a glance, looks like this:

`DDPG_Example.ipynb`

```
class ReplayBuffer(object):
    def __init__(self, max_size, inp_shape, nb_actions):
        self.memory_size = max_size
        self.memory_counter = 0
        self.memory_state = np.zeros((self.memory_size, *inp_shape))
        self.new_memory_state = np.zeros\
                                ((self.memory_size, *inp_shape))
        self.memory_action = np.zeros\
                             ((self.memory_size, nb_actions))
        self.memory_reward = np.zeros(self.memory_size)
        self.memory_terminal = np.zeros(self.memory_size, \
                                        dtype=np.float32)
```

The complete code for this example can be found at https://packt.live/2YNL2BO.

In this section, we learned how to store the agent's past learnings to train the model better. Next, we will learn in more detail about the actor-critic model, which we briefly explained in this chapter's introduction.

THE ACTOR-CRITIC MODEL

Next, in the DDPG technique, we will define the actor and critic networks. Now, we have already introduced actor-critic, but we haven't talked much about it. Take the actor as the current policy and the critic as the value. You may conceptualize the actor-critic model as a guided policy. We will define our actor-critic model using fully connected neural networks.

The **CriticNetwork** class starts with an initialization. First, we will explain the parameters. The **beta** is our learning rate. Then, we have our input dimensions, followed by the dimensions for the two fully connected layers we will be using. Finally, we have the number of actions. We haven't written any mechanisms for saving the models yet. After initializing the input dimension, the dimensions for the fully connected layers, and the number of actions, we will initialize our first layer. It will just be a **Linear** layer and we will initialize it using our input and output dimensions. Next, is the initialization of the weights and biases of our fully connected layer. This initialization restricts the values of the weights and biases to a very narrow band of the parameter space when we sample between the **-f1** to **f1** range, as seen in the following code. This helps our network to better converge. Our initial layer is followed by a batch normalization, which again helps to better converge our network. We will repeat the same process with our second fully connected layer. The **CriticNetwork** class will also get an action value. Finally, the output is a single scalar value, which we will initialize next with a constant initialization.

We will optimize our **CriticNetwork** class using the **Adam** optimizer with a learning rate beta:

```
class CriticNetwork(T.nn.Module):
    def __init__(self, beta, inp_dimensions,\
                 fc1_dimensions, fc2_dimensions,\
                 nb_actions):
        super(CriticNetwork, self).__init__()
        self.inp_dimensions = inp_dimensions
        self.fc1_dimensions = fc1_dimensions
        self.fc2_dimensions = fc2_dimensions
        self.nb_actions = nb_actions

        self.fc1 = T.nn.Linear(*self.inp_dimensions, \
                          self.fc1_dimensions)
        f1 = 1./np.sqrt(self.fc1.weight.data.size()[0])
        T.nn.init.uniform_(self.fc1.weight.data, -f1, f1)
        T.nn.init.uniform_(self.fc1.bias.data, -f1, f1)

        self.bn1 = T.nn.LayerNorm(self.fc1_dimensions)

        self.fc2 = T.nn.Linear(self.fc1_dimensions, \
                          self.fc2_dimensions)
        f2 = 1./np.sqrt(self.fc2.weight.data.size()[0])

        T.nn.init.uniform_(self.fc2.weight.data, -f2, f2)
        T.nn.init.uniform_(self.fc2.bias.data, -f2, f2)

        self.bn2 = T.nn.LayerNorm(self.fc2_dimensions)

        self.action_value = T.nn.Linear(self.nb_actions, \
                                    self.fc2_dimensions)
        f3 = 0.003
        self.q = T.nn.Linear(self.fc2_dimensions, 1)
        T.nn.init.uniform_(self.q.weight.data, -f3, f3)
        T.nn.init.uniform_(self.q.bias.data, -f3, f3)
```

```
    self.optimizer = T.optim.Adam(self.parameters(), lr=beta)


    self.device = T.device(""gpu"" if T.cuda.is_available() \
                            else ""cpu"")
    self.to(self.device)
```

Now, we have to write the **forward** function for our network. This takes a state and an action as input. We get the state-action value from this method. So, our state goes through the first fully connected layer, followed by the batch normalization and the ReLU activation. The activation is passed through the second fully connected layer, followed by another batch normalization, and before the final activation, we take into account the action value. Notice that we are adding the state and action values together to form the state-action value. The state-action value is then passed through the final layer and there we have our output:

```
def forward(self, state, action):
    state_value = self.fc1(state)
    state_value = self.bn1(state_value)
    state_value = T.nn.functional.relu(state_value)
    state_value = self.fc2(state_value)
    state_value = self.bn2(state_value)

    action_value = T.nn.functional.relu(self.action_value(action))
    state_action_value = T.nn.functional.relu\
                        (T.add(state_value, action_value))
    state_action_value = self.q(state_action_value)

    return state_action_value
```

So, finally, the **CriticNetwork** class would look like this:

DDPG_Example.ipynb

```
class CriticNetwork(T.nn.Module):
    def __init__(self, beta, inp_dimensions,\
                 fc1_dimensions, fc2_dimensions,\
                 nb_actions):
        super(CriticNetwork, self).__init__()
        self.inp_dimensions = inp_dimensions
        self.fc1_dimensions = fc1_dimensions
        self.fc2_dimensions = fc2_dimensions
        self.nb_actions = nb_actions
```

The complete code for this example can be found at https://packt.live/2YNL2BO.

Next, we will define **ActorNetwork**. This would be mostly similar to the **CriticNetwork** class but with some minor yet important changes. Let's code it first and then we will explain it:

```python
class ActorNetwork(T.nn.Module):
    def __init__(self, alpha, inp_dimensions,\
                 fc1_dimensions, fc2_dimensions, nb_actions):
        super(ActorNetwork, self).__init__()
        self.inp_dimensions = inp_dimensions
        self.fc1_dimensions = fc1_dimensions
        self.fc2_dimensions = fc2_dimensions
        self.nb_actions = nb_actions

        self.fc1 = T.nn.Linear(*self.inp_dimensions, \
                               self.fc1_dimensions)
        f1 = 1./np.sqrt(self.fc1.weight.data.size()[0])
        T.nn.init.uniform_(self.fc1.weight.data, -f1, f1)
        T.nn.init.uniform_(self.fc1.bias.data, -f1, f1)

        self.bn1 = T.nn.LayerNorm(self.fc1_dimensions)

        self.fc2 = T.nn.Linear(self.fc1_dimensions, \
                               self.fc2_dimensions)
        f2 = 1./np.sqrt(self.fc2.weight.data.size()[0])

        T.nn.init.uniform_(self.fc2.weight.data, -f2, f2)
        T.nn.init.uniform_(self.fc2.bias.data, -f2, f2)

        self.bn2 = T.nn.LayerNorm(self.fc2_dimensions)

        f3 = 0.003
        self.mu = T.nn.Linear(self.fc2_dimensions, \
                              self.nb_actions)
        T.nn.init.uniform_(self.mu.weight.data, -f3, f3)
        T.nn.init.uniform_(self.mu.bias.data, -f3, f3)

        self.optimizer = T.optim.Adam(self.parameters(), lr=alpha)

        self.device = T.device("gpu" if T.cuda.is_available() \
                               else "cpu")
```

```
        self.to(self.device)

def forward(self, state):
    x = self.fc1(state)
    x = self.bn1(x)
    x = T.nn.functional.relu(x)
    x = self.fc2(x)
    x = self.bn2(x)
    x = T.nn.functional.relu(x)
    x = T.tanh(self.mu(x))


    return x
```

As you can see, this is similar to our **CriticNetwork** class. The main difference here is that we don't have an action value here and that we have written the **forward** function in a slightly different way. Notice that the final output from the **forward** function is a **tanh** function, which will bind our output between **0** and **1**. This is necessary for the environment we are going to play around with. Let's implement an exercise that will help us to create a learning agent.

EXERCISE 11.02: CREATING A LEARNING AGENT

In this exercise, we will write our **Agent** class. We are already familiar with the concept of a learning agent, so let's see how we can implement one. This exercise will conclude the DDPG example that we have been building. Please make sure that you have run all the example code in this section before starting the exercise.

> **NOTE**
>
> We have assumed you have typed the code presented in the preceding section into a new notebook. Specifically, we have assumed you already have the code for importing the necessary libraries and creating the **OUActionNoise**, **ReplayBuffer**, **CriticNetwork**, and **ActorNetwork** classes in your notebook. This exercise begins by creating the **Agent** class.
>
> For convenience, the complete code for this exercise, including the code in the example, can be found at https://packt.live/37Jwhnq.

The following are the steps to implement this exercise:

1. Let's start by using the **__init__** method and passing the alpha and the beta, which are the learning rates for our actor and critic networks, respectively. Then, pass the input dimensions and a parameter called **tau**, which we will explain in a bit. Then, we want to pass the environment, which is our continuous action space, gamma, which is the agent's discount factor, which we talked about earlier. Then, the number of actions, the maximum size of the memory, the size of the two layers, and the batch size are passed. Then, initialize our actor and critic. Finally, we will introduce our noise and the **update_params** function:

```python
class Agent(object):
    def __init__(self, alpha, beta, inp_dimensions, \
                 tau, env, gamma=0.99, nb_actions=2, \
                 max_size=1000000, l1_size=400, \
                 l2_size=300, bs=64):
        self.gamma = gamma
        self.tau = tau
        self.memory = ReplayBuffer(max_size, inp_dimensions, \
                                    nb_actions)
        self.bs = bs

        self.actor = ActorNetwork(alpha, inp_dimensions, \
                                   l1_size, l2_size, \
                                   nb_actions=nb_actions)
        self.critic = CriticNetwork(beta, inp_dimensions, \
                                     l1_size, l2_size, \
                                     nb_actions=nb_actions)

        self.target_actor = ActorNetwork(alpha, inp_dimensions, \
                                          l1_size, l2_size, \
                                          nb_actions=nb_actions)
        self.target_critic = CriticNetwork(beta, inp_dimensions, \
                                            l1_size, l2_size, \
                                            nb_actions=nb_actions)

        self.noise = OUActionNoise(mu=np.zeros(nb_actions))

        self.update_params(tau=1)
```

The **update_params** function updates our parameters, but there's a catch. We basically have a moving target. This means we are using the same network to calculate the action and the value of the action simultaneously as we are updating the estimate in every episode. Because we are using the same parameters for both, it may lead to divergence. To tackle that, we use the target network, which learns the value and the action combinations, and the other network is used to learn the policy. We will periodically update the target network's parameters with the parameters of the evaluation network.

2. Next, we have the **select_action** method. Here, we take the observation from our actor and pass it through the feed-forward network. **mu_prime** here is basically the noise we add to the network. It is also called exploration noise. Finally, we call **actor.train()** and return the **numpy** value for **mu_prime**:

```
def select_action(self, observation):
    self.actor.eval()
    observation = T.tensor(observation, dtype=T.float)\
                   .to(self.actor.device)
    mu = self.actor.forward(observation).to(self.actor.device)
    mu_prime = mu + T.tensor(self.noise(),\
                         dtype=T.float).to(self.actor.device)
    self.actor.train()
    return mu_prime.cpu().detach().numpy()
```

3. Next comes our **remember** function, which is self-explanatory. This takes the **state**, **action**, **reward**, **new_state**, and **done** flags in order to store them in memory:

```
def remember(self, state, action, reward, new_state, done):
    self.memory.store_transition(state, action, reward, \
                              new_state, done)
```

4. Next, we will define the **learn** function:

```
def learn(self):
    if self.memory.memory_counter < self.bs:
        return
    state, action, reward, new_state, done = self.memory\
                                          .sample_buffer\
                                          (self.bs)

    reward = T.tensor(reward, dtype=T.float)\
             .to(self.critic.device)
```

```python
done = T.tensor(done).to(self.critic.device)
new_state = T.tensor(new_state, dtype=T.float)\
            .to(self.critic.device)
action = T.tensor(action, dtype=T.float).to(self.critic.device)
state = T.tensor(state, dtype=T.float).to(self.critic.device)

self.target_actor.eval()
self.target_critic.eval()
self.critic.eval()

target_actions = self.target_actor.forward(new_state)
critic_value_new = self.target_critic.forward\
                    (new_state, target_actions)
critic_value = self.critic.forward(state, action)

target = []
for j in range(self.bs):
    target.append(reward[j] + self.gamma\
                *critic_value_new[j]*done[j])
target = T.tensor(target).to(self.critic.device)
target = target.view(self.bs, 1)

self.critic.train()
self.critic.optimizer.zero_grad()
critic_loss = T.nn.functional.mse_loss(target, critic_value)
critic_loss.backward()
self.critic.optimizer.step()

self.critic.eval()
self.actor.optimizer.zero_grad()
mu = self.actor.forward(state)
self.actor.train()
actor_loss = -self.critic.forward(state, mu)
actor_loss = T.mean(actor_loss)
actor_loss.backward()
self.actor.optimizer.step()

self.update_params()
```

Here, we first check whether we have enough samples in our memory buffer for learning. So, if our memory counter is less than the batch size—meaning we do not have the batch size number of samples in our memory buffer—we simply return the value. Otherwise, we sample from our memory buffer the **state, action, reward, new_state**, and **done** flags. Once sampled, we must convert all of these flags into tensors for implementation. Then, we need to calculate the target actions, followed by the calculation of the new critic value using the target action states and the new state. Next, we calculate the critic value, which is the value we met for the states and actions in the current replay buffer. After that, we calculate the targets. Note that the part where we multiply **gamma** with the new critic value becomes **0** when the **done** flag is **0**. This basically means that when the episode is over, we only take into account the reward from the current state. The target is then converted into a tensor and reshaped for implementation purposes. Now, we can calculate and backpropagate our loss for the critic. Then, we do the same for our actor network. Finally, we update the parameters for our target actor and target critic network.

5. Next, define the **update_params** function:

```
def update_params(self, tau=None):
    if tau is None:
        tau = self.tau # tau is 1

    actor_params = self.actor.named_parameters()
    critic_params = self.critic.named_parameters()
    target_actor_params = self.target_actor.named_parameters()
    target_critic_params = self.target_critic.named_parameters()

    critic_state_dict = dict(critic_params)
    actor_state_dict = dict(actor_params)
    target_critic_dict = dict(target_critic_params)
    target_actor_dict = dict(target_actor_params)

    for name in critic_state_dict:
        critic_state_dict[name] = tau*critic_state_dict[name]\
                                  .clone() + (1-tau)\
                                  *target_critic_dict[name]\
                                  .clone()

    self.target_critic.load_state_dict(critic_state_dict)
```

```
    for name in actor_state_dict:
        actor_state_dict[name] = tau*actor_state_dict[name]\
                                  .clone() + (1-tau)\
                                  *target_actor_dict[name]\
                                  .clone()
    self.target_actor.load_state_dict(actor_state_dict)
```

Here, the **update_params** function takes a **tau** value, which basically allows us to update the target network in very small steps. The value for **tau** is typically very small, much smaller than **1**. One thing to note is that we start with **tau** equal to **1** but later, the value is reduced to a much smaller number. What the function does is that it first gets all the names of the parameters for the critic, actor, target critic, and target actor. It then updates those parameters with the target critic and target actor. Now, we can create the main part of our Python code.

6. If you have created the **Agent** class properly, then, along with the preceding example code, you'll be able to initialize our learning agent with the following bit of code:

```
env = gym.make("LunarLanderContinuous-v2")

agent = Agent(alpha=0.000025, beta=0.00025, \
              inp_dimensions=[8], tau=0.001, \
              env=env, bs=64, l1_size=400, \
              l2_size=300, nb_actions=2)

for i in np.arange(100):
    observation = env.reset()
    action = agent.select_action(observation)
    state_new, reward, _, _ = env.step(action)
    observation = state_new
    env.render()
    print("Episode {}\tReward: {}".format(i, reward))
```

For the output, you'll see the reward for each episode. Here is the output for the first 10 episodes:

```
Episode 0 Reward: -0.2911892911560017
Episode 1 Reward: -0.4945150137594737
Episode 2 Reward: 0.5150667951556557
Episode 3 Reward: -1.33324749569461
```

```
Episode  4Reward:  -0.9969126433110092
Episode  5Reward:  -1.8466220765944854
Episode  6Reward:  -1.6207456680346013
Episode  7Reward:  -0.4027838988393455
Episode  8Reward:  0.42631743995534066
Episode  9Reward:  -1.1961709218053898
Episode 10        Reward:  -1.0679394471159185
```

What you see in the preceding output is that the reward oscillates between negative and positive. That is because until now, our agent was sampling random actions from all the actions it could take.

> **NOTE**
>
> To access the source code for this specific section, please refer to https://packt.live/37Jwhnq.
>
> This section does not currently have an online interactive example and will need to be run locally.

In the next activity, we will make the agent remember its past learnings and learn from them. Here's how the game environment will look:

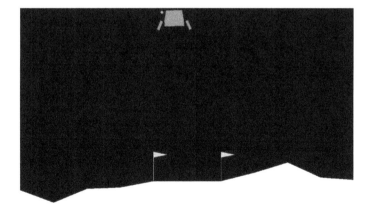

Figure 11.8: The output window showing the Lunar Lander hovering
in the game environment

However, you'll find that the Lander doesn't attempt to land, and rather, it hovers over the lunar surface in our game environment. That's because we haven't enabled the agent to learn yet. We will do that in the following activity.

In the next activity, we will create an agent that will help to learn a model using DDPG.

ACTIVITY 11.01: CREATING AN AGENT THAT LEARNS A MODEL USING DDPG

In this activity, we will implement what we have learned in this section and create an agent that learns through DDPG.

> **NOTE**
>
> We have created a Python file for the actual DDPG implementation to be imported as a module using **from ddpg import ***. The module and the code of the activity can be downloaded from GitHub at https://packt.live/2YksdXX.

The following are the steps to perform for this activity:

1. Import the necessary libraries (**os**, **gym**, and **ddpg**).

2. First, we create our Gym environment (**LunarLanderContinuous-v2**), as we did previously.

3. Initialize the agent with some sensible hyperparameters, as in *Exercise 11.02, Creating a Learning Agent*.

4. Set up a random seed so that our experiments are reproducible.

5. Create a blank array to story the scores; you can name it **history**. Iterate for at least **1000** episodes and in each episode, set a running score variable to **0** and the **done** flag to **False**, then reset the environment. Then, when the **done** flag is not **True**, carry out the following step.

6. Select an action from the observations and get the new **state**, **reward**, and **done** flags. Save the **observation, action, reward, state_new**, and **done** flags. Call the **learn** function of the agent and add the current reward to the running score. Set the new state as the observation and finally, when the **done** flag is **True**, append **score** to **history**.

> **NOTE**
>
> To observe the rewards, we can simply add a **print** statement.
> The rewards will be similar to those in the previous exercise.

The following is the expected simulation output:

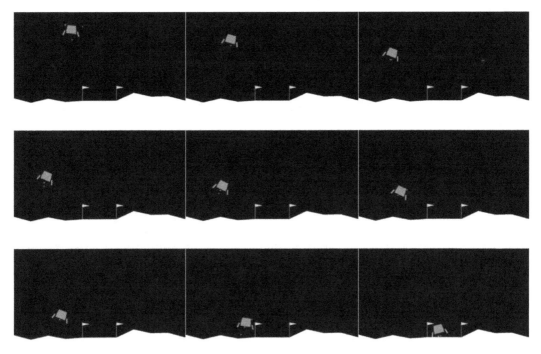

Figure 11.9: Screenshots from the environment after 1,000 rounds of training

> **NOTE**
>
> The solution to this activity can be found on page 766.

In the next section, we will see how we can improve the policy gradient approach that we just implemented.

IMPROVING POLICY GRADIENTS

In this section, we will learn the various approaches that will help us improve the policy gradient approach that we learned about in the previous section. We will learn about techniques such as TRPO and PPO.

We will also learn about the A2C technique in brief. Let's understand the TRPO optimization technique in the next section.

TRUST REGION POLICY OPTIMIZATION

In most cases, RL is very sensitive to the initialization of weights. Take, for instance, the learning rate. If our learning rate is too high, then it may so happen that our policy update takes our policy network to a region of the parameter space where the next batch of data it collects is gathered against a very poor policy. This might cause our network to never recover again. Now, we will talk about newer methods that try to get rid of this problem. But before we do that, let's have a quick recap of what we have already covered.

In the *Policy Gradients* section, we defined the estimator of the advantage function, $\hat{A}_t$, as the difference between the discounted reward and the baseline estimate. Intuitively, the advantage estimator quantifies how good the action taken by our agent in a certain state was compared to what would typically happen in that state. One problem with the advantage function is that if we simply keep on updating our weights based on one batch of samples using gradient descent, then our parameter updates might stray far from the range where the data was sampled from. That could lead to an inaccurate estimate of the advantage function. In short, if we keep running gradient descent on a single batch of experiences, we might corrupt our policy.

One way to make sure this problem doesn't occur is to ensure that the updated policy doesn't differ too much from the old policy. This is basically the main crux of TRPO.

We already understand how the gradient estimator works for vanilla policy gradients:

$$\hat{g} = \widehat{\mathbb{E}}_t \left[\nabla_{\theta l} \, log \, \pi_\theta \big(a_t | s_t\big) \widehat{A}_t \right]$$

Figure 11.10: The vanilla policy gradient method

Here's how it looks for TRPO:

$$\underset{\theta}{maximize} \, \widehat{\mathbb{E}}_t \left[\frac{\pi_\theta \big(a_t | s_t\big)}{\pi_{\theta_{old}} \big(a_t | s_t\big)} \widehat{A}_t \right]$$

Figure 11.11: Mathematical representation of TRPO

The only change here is that the log operator in the preceding equation has been replaced by a division by $\pi_{\theta_{old}}$. This is known as the TRPO objective and optimizing it yields the same result as the vanilla policy gradients. In order to ensure that the new and updated policy doesn't differ much from the old policy, TRPO introduces a constraint known as the KL constraint.

This constraint, in simple words, makes sure that our new policy doesn't stray too far from the old one. Note that the actual TRPO strategy, however, proposes a penalty instead of a constraint.

PROXIMAL POLICY OPTIMIZATION

It might seem like everything is fine and good for TRPO, but the introduction of the KL constraint introduces an additional operating cost to our policy. To address that problem, and to basically solve the problems with vanilla policy gradients once and for all, the researchers at OpenAI have introduced PPO, which we will look into now.

The main motivations behind PPO are as follows:

- Ease of implementation

- Ease of parameter tuning

- Efficient sampling

One thing to note is that the PPO method doesn't use a replay buffer to store past experiences and learns straight from whatever the agent encounters in the environment. This is also known as an **online** method of learning while the former—using a replay buffer to store past experiences—is known as an **offline** method of learning.

The authors of PPO define a probability ratio, $r_t(\theta)$, which is basically the probability ratio between the new and the old policy. So, we have the following:

$$r_t(\theta) = \frac{\pi_\theta(a_t|s_t)}{\pi_{\theta_{old}}(a_t|s_t)}$$

Figure 11.12: The probability ratio between the old and new policy

When provided with a sampled batch of actions and states, this ratio would be greater than **1** if the action is more likely now than in the old policy. Otherwise, it would remain between **0** and **1**. Now, the final objective of PPO when written down looks like this:

$$L^{CLIP}(\theta) = \widehat{\mathbb{E}}_t\left[min\left(r_t(\theta)\widehat{A}_t, clip\left(r_t(\theta), 1-\varepsilon, 1+\varepsilon\right)\widehat{A}_t\right)\right]$$

Figure 11.13: The final objective of PPO

Let's explain. Like the vanilla policy gradient, PPO tries to optimize the expectation and so we compute this expectation operator over batches of trajectories. Now, this is a minimum value of the modified policy gradient objective that we saw in TRPO and the second part is a clipped version of it. The clipping operation keeps the policy gradient objective between $1 - \varepsilon$ and $1 + \varepsilon$. ε here is a hyperparameter, often equal to **0.2**.

Although the function looks simple at a glance, its ingenuity is remarkable. $\hat{\mathbb{E}}$, can be both negative and positive, suggesting a negative advantage estimation and a positive advantage estimation, respectively. This behavior of the advantage estimator determines how the **min** operator works. Here's an illustration of the **clip** parameter from the actual PPO paper:

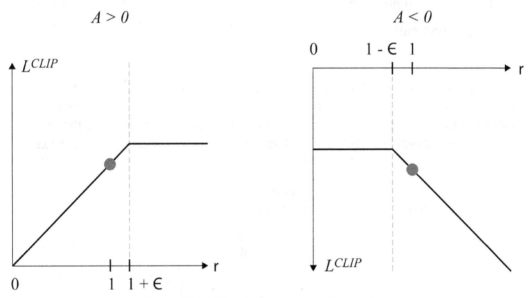

Figure 11.14: The clip parameter illustration

The plot on the left is where the advantage is positive. This means our actions yielded results that were better than expected. On the other plot on the right are the cases where our actions yielded less than the expected return.

Notice how on the plot on the left the loss flattens out when **r** is too high. This might occur when the current action is much more plausible under the current policy than the old one. In this case, the objective function is clipped here, thereby ensuring the gradient update doesn't go beyond a certain limit.

On the other hand, when the objective function is negative, the loss flattens when **r** is approaching **0**. This relates to actions that are more unlikely under the current policy than the old one. It may be clear by now how the clipping operation keeps the updates to the network parameters within a desirable range. It would be a better approach to learn about the PPO technique while implementing it. So, let's start with an exercise to reduce the operating cost of our policy using PPO.

EXERCISE 11.03: IMPROVING THE LUNAR LANDER EXAMPLE USING PPO

In this exercise, we'll implement the Lunar Lander example using PPO. We will follow almost the same structure as before and you'll be able to follow through this exercise easily if you have gone through the previous exercises and examples:

1. Open a new Jupyter Notebook and import the necessary libraries (**gym**, **torch**, and **numpy**):

```
import gym
import torch as T
import numpy as np
```

2. Set our device as we did in the DDPG example:

```
device = T.device("cuda:0" if T.cuda.is_available() else "cpu")
```

3. Next, we will create the **ReplayBuffer** class. Here, we will create arrays to store the actions, states, log probabilities, reward, and terminal states:

```
class ReplayBuffer:
    def __init__(self):
        self.memory_actions = []
        self.memory_states = []
        self.memory_log_probs = []
        self.memory_rewards = []
        self.is_terminals = []

    def clear_memory(self):
        del self.memory_actions[:]
        del self.memory_states[:]
        del self.memory_log_probs[:]
        del self.memory_rewards[:]
        del self.is_terminals[:]
```

4. Now, we will define our **ActorCritic** class. We will define our **action** and **value** layers first:

```
class ActorCritic(T.nn.Module):

    def __init__(self, state_dimension, action_dimension, \
                nb_latent_variables):
        super(ActorCritic, self).__init__()

        self.action_layer = T.nn.Sequential\
                            (T.nn.Linear(state_dimension, \
                                        nb_latent_variables),\
                            T.nn.Tanh(),\
                            T.nn.Linear(nb_latent_variables, \
                                        nb_latent_variables),\
                            T.nn.Tanh(),\
                            T.nn.Linear(nb_latent_variables, \
                                        action_dimension),\
                            T.nn.Softmax(dim=-1))

        self.value_layer = T.nn.Sequential\
                            (T.nn.Linear(state_dimension, \
                                        nb_latent_variables),\
                            T.nn.Tanh(), \
                            T.nn.Linear(nb_latent_variables, \
                                        nb_latent_variables),\
                            T.nn.Tanh(),\
                            T.nn.Linear(nb_latent_variables, 1))
```

5. Now, we will define methods to sample from the action space and evaluate the log probabilities of the actions, state value, and entropy of the distribution:

```
# Sample from the action space
def act(self, state, memory):
    state = T.from_numpy(state).float().to(device)
    action_probs = self.action_layer(state)
    dist = T.distributions.Categorical(action_probs)
    action = dist.sample()

    memory.memory_states.append(state)
    memory.memory_actions.append(action)
```

```
            memory.memory_log_probs.append(dist.log_prob(action))


        return action.item()


    # Evaluate log probabilities
    def evaluate(self, state, action):
        action_probs = self.action_layer(state)
        dist = T.distributions.Categorical(action_probs)


        action_log_probs = dist.log_prob(action)
        dist_entropy = dist.entropy()


        state_value = self.value_layer(state)


        return action_log_probs, \
                T.squeeze(state_value), dist_entropy
```

Finally, the **ActorCritic** class looks like this:

Exercise11_03.ipynb

```
class ActorCritic(T.nn.Module):
    def __init__(self, state_dimension, \
                action_dimension, nb_latent_variables):
        super(ActorCritic, self).__init__()

        self.action_layer = T.nn.Sequential(T.nn.Linear\
                                        (state_dimension, \
                                         nb_latent_variables),\
                            T.nn.Tanh(), \
                            T.nn.Linear(nb_latent_variables, \
                                    nb_latent_variables),\
                            T.nn.Tanh(),\
                            T.nn.Linear(nb_latent_variables, \
                                    action_dimension),\
                            T.nn.Softmax(dim=-1))
```

The complete code for this example can be found at https://packt.live/2zM1Z6Z.

6. Now, we will define our **Agent** class using the **__init__()** and **update()** functions. First let's define **__init__()** function:

```
class Agent:
    def __init__(self, state_dimension, action_dimension, \
                nb_latent_variables, lr, betas, gamma, \
                K_epochs, eps_clip):
        self.lr = lr
        self.betas = betas
        self.gamma = gamma
```

```
            self.eps_clip = eps_clip
            self.K_epochs = K_epochs

            self.policy = ActorCritic(state_dimension,\
                                    action_dimension,\
                                    nb_latent_variables)\
                                    .to(device)
            self.optimizer = T.optim.Adam\
                        (self.policy.parameters(), \
                        lr=lr, betas=betas)
            self.policy_old = ActorCritic(state_dimension,\
                                    action_dimension,\
                                    nb_latent_variables)\
                                    .to(device)
            self.policy_old.load_state_dict(self.policy.state_dict())

            self.MseLoss = T.nn.MSELoss()
```

7. Now let's define the **update** function:

```
    def update(self, memory):
        # Monte Carlo estimate
        rewards = []
        discounted_reward = 0
        for reward, is_terminal in \
            zip(reversed(memory.memory_rewards), \
                        reversed(memory.is_terminals)):
            if is_terminal:
                discounted_reward = 0
            discounted_reward = reward + \
                            (self.gamma * discounted_reward)
            rewards.insert(0, discounted_reward)
```

8. Next, normalize the rewards and convert them to tensors:

```
            rewards = T.tensor(rewards).to(device)
            rewards = (rewards - rewards.mean()) \
                    / (rewards.std() + 1e-5)

            # Convert to Tensor
            old_states = T.stack(memory.memory_states)\
                        .to(device).detach()
```

```
old_actions = T.stack(memory.memory_actions)\
            .to(device).detach()
old_log_probs = T.stack(memory.memory_log_probs)\
            .to(device).detach()
# Policy Optimization
for _ in range(self.K_epochs):
    log_probs, state_values, dist_entropy = \
    self.policy.evaluate(old_states, old_actions)
```

9. Next, find the probability ratio, find the loss and propagate our loss backwards:

```
# Finding ratio: pi_theta / pi_theta__old
ratios = T.exp(log_probs - old_log_probs.detach())

# Surrogate Loss
advantages = rewards - state_values.detach()
surr1 = ratios * advantages
surr2 = T.clamp(ratios, 1-self.eps_clip, \
                1+self.eps_clip) * advantages
loss = -T.min(surr1, surr2) \
       + 0.5*self.MseLoss(state_values, rewards) \
       - 0.01*dist_entropy

# Backpropagation
self.optimizer.zero_grad()
loss.mean().backward()
self.optimizer.step()
```

10. Update the old policy with the new weights:

```
# New weights to old policy
self.policy_old.load_state_dict(self.policy.state_dict())
```

So, here in *steps 6-10* of this exercise we are defining an agent by starting with the initialization of our policy, the optimizer, and the old policy. Then, in the **update** function, we are at first taking the Monte Carlo estimate of the state rewards. After normalizing the rewards, we are converting them into tensors.

Then, we are carrying out the policy optimization for **K_epochs**. Here, we have to find the probability ratio, $r_t(\theta)$, which is the probability ratio between the new and the old policy, as described previously.

After that, we are finding the loss, $L^{CLIP}(\theta)$, and propagating our loss backward. Finally, we are updating our old policy with the new weights.

11. Now, we can run the simulation as we did in the previous exercise and save the policy for future use:

Exercise11_03.ipynb

```
env = gym.make("LunarLander-v2")

render = False
solved_reward = 230
logging_interval = 20
update_timestep = 2000

np.random.seed(0)

memory = ReplayBuffer()
agent = Agent(state_dimension=env.observation_space.shape[0],\
              action_dimension=4, nb_latent_variables=64, \
              lr=0.002, betas=(0.9, 0.999), gamma=0.99,\
              K_epochs=4, eps_clip=0.2)

current_reward = 0
avg_length = 0
timestep = 0

for i_ep in range(50000):
    state = env.reset()
    for t in range(300):
        timestep += 1
```

The complete code for this example can be found at https://packt.live/2zM1Z6Z.

The following is the first 10 lines of the output:

```
Episode 0, reward: -8
Episode 20, reward: -182
Episode 40, reward: -154
Episode 60, reward: -175
Episode 80, reward: -136
Episode 100, reward: -178
Episode 120, reward: -128
Episode 140, reward: -137
Episode 160, reward: -140
Episode 180, reward: -150
```

Note that we are saving our policy at certain intervals. This is useful if you want to load the policy at a later stage and simply run the simulation from there. The simulation output of this exercise would be the same as in *Figure 11.9*, only the operating cost is reduced here.

Here, if you look at the difference between the rewards, the points given in each consequent episode are much less as we have used the PPO technique. That means the learning is not going haywire as it was in *Exercise 11.01, Landing a Spacecraft on the Lunar Surface Using Policy Gradients and the Actor-Critic Method*, where the difference between the rewards was higher.

> **NOTE**
>
> To access the source code for this specific section, please refer to https://packt.live/2zM1Z6Z.
>
> This section does not currently have an online interactive example and will need to be run locally.

We have almost covered all the important topics relating to policy-based RL. So, now we will talk about the last topic, which is the A2C method.

THE ADVANTAGE ACTOR-CRITIC METHOD

We have already learned about the actor-critic method and the reason for using it in the introduction, and we have also seen it used in our coding examples. But, a quick recap—actor-critic methods lie at the intersection of the value-based and policy-based methods, where we simultaneously update our policy and our value, which acts as a judge quantifying how good our policy actually is.

Next, we will learn how A2C works:

1. We start by initializing the policy parameter, θ, with random weights.

2. Next, we play *N* number of steps with the current policy, π_θ, and store the state, action, reward, and transitions.

3. We set our reward to **0** if we reach the final episode of the state; otherwise, we set the reward to the value of the current state.

4. Then, we calculate the discounted reward, policy loss, and value loss by looping backward from the final episode.

5. Finally, we apply **Stochastic Gradient Descent (SGD)** using the mean policy and value loss for each batch.

6. Repeat the steps from *step 2* onward until it reaches convergence.

The first coding example (*Exercise 11.01, Landing a Spacecraft on the Lunar Surface Using Policy Gradients and the Actor-Critic Method*) that we covered follows the basic A2C method. However, there's another technique, called the **Asynchronous Advantage Actor-Critic (A3C)** method. Remember, our policy gradient methods work online. That is, we only train on the data obtained using the current policy and we do not keep track of past experiences. However, to keep our data independent and identically distributed, we need a large buffer of transitions. The solution provided by A3C is to run multiple training environments in parallel to acquire large amounts of training data. With multiprocessing in Python, this actually becomes very fast in practice.

In the next activity, we will write code to run the Lunar Lander simulation that we learned about in *Exercise 11.03, Improving the Lunar Lander Example Using PPO*. We will also render the environment to see the Lunar Lander. To do that, we will have to import the PIL library. The code to render the image is as follows:

```
if render:
    env.render()

    img = env.render(mode = "rgb_array")
    img = Image.fromarray(img)
    image_dir = "./gif"
    if not os.path.exists(image_dir):
        os.makedirs(image_dir)
    img.save(os.path.join(image_dir, "{}.jpg".format(t)))
```

Let's begin with the implementation of our final activity.

ACTIVITY 11.02: LOADING THE SAVED POLICY TO RUN THE LUNAR LANDER SIMULATION

In this activity, we will combine multiple aspects of RL that we have explained in previous sections. We will use what we learned in *Exercise 11.03, Improving the Lunar Lander Example Using PPO*, to write simple code to load the saved policy. This activity combines all the essential components of building a working RL prototype—in our case, the Lunar Lander simulation.

The steps to take are as follows:

1. Open Jupyter and in a new notebook, import the essential Python libraries, including the PIL library to save the image.

2. Set your device using the device parameter.

3. Define the **ReplayBuffer**, **ActorCritic**, and **Agent** classes. We already defined these in the previous exercise.

4. Create the Lunar Lander environment. Initialize the random seed.

5. Create the memory buffer and initialize the agent with hyperparameters, as in the previous exercise.

6. Load the saved policy as an old policy.

7. Finally, loop through your desired number of episodes. In every iteration, start by initializing the episode reward as **0**. Do not forget to reset the state. Run another loop, specifying the **max** timestamp. Get the **state**, **reward**, and **done** flags for each action taken and add the reward to the episode reward.

8. Render the environment to see how your Lunar Lander is doing.

    The following is the expected output:

    ```
    Episode: 0, Reward: 272
    Episode: 1, Reward: 148
    Episode: 2, Reward: 249
    Episode: 3, Reward: 169
    Episode: 4, Reward: 35
    ```

The following screenshot shows the simulation output of some of the stages:

Figure 11.15: The environment showing the simulation of the Lunar Lander

The complete simulation output can be found in the form of images at https://packt.live/3ehPaAj.

> **NOTE**
>
> The solution to this activity can be found on page 769.

SUMMARY

In this chapter, we learned about policy-based methods, principally the drawbacks to value-based methods such as Q-learning, which motivate the use of policy gradients. We discussed the purposes of policy-based methods of RL, along with the trade-offs of other RL approaches.

You learned about the policy gradients that help a model to learn in a real-time environment. Next, we learned how to implement the DDPG using the actor-critic model, the `ReplayBuffer` class, and Ornstein–Uhlenbeck noise to understand the continuous action space. We also learned how you can improve policy gradients by using techniques such as TRPO and PPO. Finally, we talked in brief about the A2C method, which is an advanced version of the actor-critic model.

Also, in this chapter, we played around with the Lunar Lander environment in OpenAI Gym—for both continuous and discrete action spaces—and coded the multiple policy-based RL approaches that we discussed.

In the next chapter, we will learn about a gradient-free method to optimize neural networks and RL-based algorithms. We will then discuss the limitations of gradient-based methods. The chapter presents an alternative optimization solution to gradient methods through genetic algorithms as they ensure global optimum convergence. We will also learn about the hybrid neural networks that use genetic algorithms to solve complex problems.

12

EVOLUTIONARY STRATEGIES FOR RL

OVERVIEW

In this chapter, we will be identifying the limitations of gradient-based methods and the motivation for evolutionary strategies. We will break down the components of genetic algorithms and implement them in **Reinforcement Learning** (**RL**). By the end of this chapter, you will be able to combine evolutionary strategies with traditional machine learning methods, specifically in the selection of neural network hyperparameters, and also identify the limitations of these evolutionary methods.

INTRODUCTION

In the previous chapter, we looked at various policy-based methods and their advantages. In this chapter, we are going to learn about gradient-free methods, namely genetic algorithms; develop these algorithms step by step; and use them to optimize neural networks and RL-based algorithms. This chapter discusses the limitations of gradient-based methods, such as getting stuck at local optima and slower convergence when dealing with noisy input. This chapter presents an alternative optimization solution to gradient methods through genetic algorithms, as they ensure global optimum convergence. You will examine and implement the structure of genetic algorithms and implement them through hyperparameter selection for neural networks and evolving network topologies, as well as using them in combination with RL for a cart-pole balancing activity. Hybrid neural networks that use genetic algorithms are used to solve complex problems, such as modeling plasma chemical reactors, designing fractal frequency selective surfaces, or optimizing production processes. In the following section, you will be examining the problems posed by gradient-based methods.

PROBLEMS WITH GRADIENT-BASED METHODS

In this section, you will learn about the differences between value-based and policy-based methods and the use of gradient-based methods in policy search algorithms. You will then examine the advantages and disadvantages of using gradient-based methods in policy-based approaches and implement stochastic gradient descent using TensorFlow to solve a cubic function with two unknowns.

There are two approaches when doing RL: value-based and policy-based. These approaches are used to solve complex decision problems related to **Markov Decision Processes (MDPs)** and **Partially Observable Markov Decision Processes (POMDPs)**. Value-based approaches rely on identifying and deriving the optimal policy based on the identification of the optimal value function. Algorithms such as Q-learning or SARSA(λ) are included within this category, and for tasks involving lookup tables, their implementation leads to convergence on a return that is optimal, globally. As the algorithms rely on a known model of the environment, for partially observable or continuous spaces, there are no guarantees for convergence on a solution that is optimal using these value search methods.

Conversely, policy-based approaches, instead of relying on the value function to maximize the return, use gradient methods (stochastic optimization) to explore the policy space. Gradient-based methods or policy gradient methods map the parametrized space (environment) to the policy space using loss functions, thus enabling the RL agent to explore directly the entirety, or a portion, of the policy space. One of the most widely used methods (which is going to be implemented in this section) is gradient descent.

> **NOTE**
>
> For further reading on gradient descent, please refer to the technical paper by *Marbach, 2001*, at the following link: https://link.springer.com/article/10.1023/A:1022145020786.

The advantages of the gradient approaches (stochastic gradient descent or ascent) are that they are suitable for POMDPs or non-MDPs, especially for solving robotics problems with multiple constraints. However, there are several disadvantages to employing gradient-based methods. The most notable one is that algorithms such as **REINFORCE** and **DPG** determine a local optimum of the expected reward. As the local optimum is found, the RL agent does not expand its search globally. For example, a robot solving a maze problem will get stuck in a corner and will continuously try to move in the same location. Additionally, when dealing with high return variance or noisy input data, algorithm performance is affected as they converge slower. This happens when, for instance, a robotic arm is programmed to pick up and place a blue component in a tray, but the table has blue hues to its color, which interferes with the detection of the component through the sensors (such as a camera).

> **NOTE**
>
> For further reading on the **REINFORCE** algorithm, please refer to the technical paper by *Williams, 1992*, at the following link: https://link.springer.com/article/10.1007/BF00992696.
>
> Similarly, please read about the **DPG** algorithm by *Silvester, 2014*, at the following link: http://proceedings.mlr.press/v32/silver14.pdf.

An alternative to gradient-based methods is the use of gradient-free methods, which rely on evolutionary algorithms to achieve a global optimum for the return.

The following exercise will enable you to understand the potential of gradient methods for converging on optimal solutions and the lengthy process that is undertaken as the method searches step by step for the optimal solution. You will be presented with a mathematical function (loss function) that maps the input values, x, y, to an output value, z. The goal is to identify the optimal values of the inputs that lead to the lowest value of the output; however, this is step-dependent and is at risk of staying at a local optimum. We will be using the **GradientTape()** function to calculate the gradients, which are nothing but differentiation solutions. This will help you understand the limitations of such optimization strategies.

EXERCISE 12.01: OPTIMIZATION USING STOCHASTIC GRADIENT DESCENT

This exercise aims to enable you to apply gradient methods, most notably **Stochastic Gradient Descent** (**SGD**), available in TensorFlow by following the steps required to converge on an optimal solution.

The following loss function has two unknowns, x, y:

$$z(x, y) = x^2 - 8x + y^2 + 3y$$

Figure 12.1: Sample loss function

Find the optimum values for x, y within 100 steps with a learning rate of **0.1**.

Perform the following steps to complete the exercise:

1. Create a new Jupyter Notebook.

2. Import the **tensorflow** package as **tf**:

```
import tensorflow as tf
```

3. Define a function that outputs $z(x, y)$:

```
def funct(x,y):
    return x**2-8*x+y**2+3*y
```

4. Define a function for initializing the **x** and **y** variables and initialize them with the values **5** and **10**:

```
def initialize():
    x = tf.Variable(5.0)
    y = tf.Variable(10.0)
```

```
       return x, y
   x, y= initialize()
```

5. In the preceding code snippet, we have used decimal format for the values assigned to **x** and **y** to start the optimization process, as the **Variable()** constructor needs to have a tensor type of **float32**.

6. Instantiate the optimizer by selecting **SGD** from **keras** in TensorFlow and input the learning rate of 0.1:

```
optimizer = tf.keras.optimizers.SGD(learning_rate = 0.1)
```

7. Set a loop of **100** steps, where you calculate the loss, use the **GradientTape()** function for automatic differentiations, and process the gradients:

```
for i in range(100):
    with tf.GradientTape() as tape:
        # Calculate loss function using x and y values
        loss= funct(x,y)
        # Get gradient values
        gradients = tape.gradient(loss, [x, y])
        # Save gradients in array without altering them
        p_gradients = [grad for grad in gradients]
```

In the preceding code, we have used **GradientTape()** from TensorFlow to calculate the gradients (which essentially are differentiation solutions). We created a loss parameter that stores the $z(x, y)$ value when calling the function. **GradientTape()** is activated when calling the **gradient()** method, which essentially is used to compute multiple gradients in a single computation. The gradients are stored in a **p_gradients** array.

8. Use the **zip()** function to aggregate the gradients to the values:

```
    ag = zip(p_gradients, [x,y])
```

9. Print the step and the values of z, x, y:

```
    print('Step={:.1f} , z ={:.1f},x={:.1f},y={:.1f}'\
        .format(i, loss.numpy(), x.numpy(), y.numpy()))
```

10. Apply the optimizer using the gradients that were processed:

```
    optimizer.apply_gradients(ag)
```

11. Run the application.

   You will get the following output:

```
Step=0.0 , z =115.0,x=5.0,y=10.0
Step=1.0 , z =67.0,x=4.8,y=7.7
Step=2.0 , z =36.3,x=4.6,y=5.9
Step=3.0 , z =16.7,x=4.5,y=4.4
Step=4.0 , z =4.1,x=4.4,y=3.2
Step=5.0 , z =-3.9,x=4.3,y=2.3
Step=6.0 , z =-9.1,x=4.3,y=1.5
Step=7.0 , z =-12.4,x=4.2,y=0.9
Step=8.0 , z =-14.5,x=4.2,y=0.4
Step=9.0 , z =-15.8,x=4.1,y=0.0
Step=10.0 , z =-16.7,x=4.1,y=-0.3
Step=11.0 , z =-17.3,x=4.1,y=-0.5
Step=12.0 , z =-17.6,x=4.1,y=-0.7
Step=13.0 , z =-17.8,x=4.1,y=-0.9
Step=14.0 , z =-18.0,x=4.0,y=-1.0
Step=15.0 , z =-18.1,x=4.0,y=-1.1
Step=16.0 , z =-18.1,x=4.0,y=-1.2
Step=17.0 , z =-18.2,x=4.0,y=-1.2
Step=18.0 , z =-18.2,x=4.0,y=-1.3
Step=19.0 , z =-18.2,x=4.0,y=-1.3
Step=20.0 , z =-18.2,x=4.0,y=-1.4
Step=21.0 , z =-18.2,x=4.0,y=-1.4
Step=22.0 , z =-18.2,x=4.0,y=-1.4
Step=23.0 , z =-18.2,x=4.0,y=-1.4
Step=24.0 , z =-18.2,x=4.0,y=-1.4
Step=25.0 , z =-18.2,x=4.0,y=-1.5
Step=26.0 , z =-18.2,x=4.0,y=-1.5
Step=27.0 , z =-18.2,x=4.0,y=-1.5
Step=28.0 , z =-18.2,x=4.0,y=-1.5
Step=29.0 , z =-18.2,x=4.0,y=-1.5
Step=30.0 , z =-18.2,x=4.0,y=-1.5
```

Figure 12.2: Step-by-step optimization using SGD

You can observe in the output that from **Step=25** onward, the z, x, y values do not change; therefore, they are considered to be the optimum values for the respective loss function.

By printing the steps and values of the inputs and outputs, you can observe that the algorithm converges before the termination of the 100 steps to the optimal values of z, x, y. However, you can observe that the problem is step-dependent: if the optimization is stopped before global optimum convergence, the solution would be sub-optimal.

> **NOTE**
>
> To access the source code for this specific section, please refer to https://packt.live/2C10rXD.
>
> You can also run this example online at https://packt.live/2DIWSqc.

This exercise helped your understanding and application of SGD when solving a loss function, developing your analysis skills as well as your skills in programming using TensorFlow. This will help you in your choice of optimization algorithm, giving you an understanding of the limitations of gradient-based methods.

In this section, we have explored the benefits and disadvantages of gradient methods with respect to RL algorithms, identifying the types of problems that they are suitable for within the context of decision-making processes. The example offered a simple application of gradient descent, where the optimal solution for two unknowns was identified using SGD optimization in TensorFlow. In the next section, we will be exploring an optimization alternative that is gradient-free: genetic algorithms.

INTRODUCTION TO GENETIC ALGORITHMS

As the problem with gradient methods is that the solution can get stuck at a single local optimum, other methods, such as gradient-free algorithms, can be considered as alternatives. In this section, you will learn about gradient-free methods, specifically evolutionary algorithms (for example, genetic algorithms). This section provides an overview of the steps taken for the implementation of genetic algorithms and exercises on how to implement an evolutionary algorithm to solve the loss function given in the previous section.

When multiple local optima exist or function optimization is required, gradient-free methods are recommended. These methods include evolutionary algorithms and particle swarm optimizations. A characteristic of these methods is that they rely on sets of optimization solutions that are commonly referred to as populations. The methods rely on iteratively searching for a good solution or a distribution that can solve a problem or a mathematical function. The search pattern for the optimal solution is modeled based on Darwin's natural selection paradigm and the biological phenomenon of genetic evolution. Evolutionary algorithms draw inspiration from biological evolution patterns such as mutation, reproduction, recombination, and selection. Particle swarm algorithms are inspired by group social behavior, such as a beehive organization or ant farms, where single solutions are termed as particles that can evolve over time.

Natural selection stems from the premise that genetic material (the chromosome) encodes the survival of a species, in a certain way. The evolution of the species relies on how well it adapts to its external environment and the information passed from parents to children. In genetic material, there are variations (mutations) between generations that can lead to successful or unsuccessful adaptation to the environment (especially in dire conditions). Therefore, there are three steps to genetic algorithms: selection, reproduction (crossover), and mutation.

Evolutionary algorithms go about things by creating an original population of solutions, selecting a sub-set, and using recombination or mutation to obtain different solutions. This new set of solutions can replace, partly or fully, the original set. For the replacement to take place, the solutions go through a selection process that relies on analyzing their fitness. This increases the chances of solutions that are more suited to being utilized to develop a new set of solutions.

Other than the development of solutions, evolutionary algorithms can be used for parameter adaptation, using probability distributions. A population is still generated; however, a fitness method is used to select the parameters of the distribution instead of the actual solutions. After the new parameters are identified, the new distribution is used to generate a new set of solutions. Some strategies of parameter selection include the following:

- Using natural gradient ascent after the gradients of the parameters are estimated from the original population, also known as **Natural Evolutionary Strategies (NESes)**

- Selecting solutions with a specific parameter and using the mean of this sub-set to find a new distribution mean, known as **Cross-Entropy Optimization (CEO)**

- Attributing a weight to each solution based on its fitness, using the weighted average as a new distribution mean – **Covariance Matrix Adaptation Evolution Strategies (CMAESes)**

One of the major problems identified with evolutionary strategies is that achieving solution fitness can be computationally expensive and noisy.

Genetic Algorithms (GAs) keep the solution population and conduct searches in multiple directions (through the chromosomes), furthering the exchange of information in these directions. The algorithms are most notably implemented on strings, which are either binary or character-based. The two main operations performed are mutation and crossover. The selection of the progenies is based on how close the solution is to the target (objective function), which denotes their fitness.

As an overview, GAs have the following steps:

1. Population creation.

2. Fitness score creation and assignment to each solution of the population.

3. The selection of two parents to reproduce based on the fitness scores (potentially the solutions with the best performance).

4. The creation of the two child solutions by combining and re-organizing the code of the two parents.

5. The application of a random mutation.

6. Child generation is repeated until the new population size is achieved and weights (fitness scores) for the population are assigned.

7. The process is repeated until the maximum number of generations is reached or the target performance is achieved.

We will be looking at each of these steps in detail further in this chapter.

Among the many differences between gradient-based algorithms and GAs, one difference is the process of development. Gradient-based algorithms rely on differentiation, whereas GAs use the genetic processes of selection, reproduction, and mutation. The following exercise will enable you to implement GAs and evaluate their performance. You will be using a simple genetic algorithm in TensorFlow to identify the GA hyperparameter optimization for finding the optimal solution for **Recurrent Neural Network** (**RNN**) training for the same loss function as for the SGD method. To identify the optimal values, you will need to implement an evolutionary algorithm called **Differential Evolution** (**DE**), available in the `tensorflow_probability` package.

EXERCISE 12.02: IMPLEMENTING FIXED-VALUE AND UNIFORM DISTRIBUTION OPTIMIZATION USING GAS

In this exercise, you will still need to solve the following function, as in the previous exercise:

$$z(x, y) = x^2 - 8x + y^2 + 3y$$

Figure 12.3: Sample loss function

Find the optimum values for x, y for a population size of 100, starting from x, y initialized to 5 and 10, and then extending to random samples from a distribution similar to the gradient-based method.

The goal of this exercise is to enable you to analyze the differences in applying GAs and gradient-descent methods, by starting from a single pair of variables and a variety of potential solutions. The algorithm aids in optimization problems by applying selection, crossover, and mutation to reach an optimal or nearly optimal solution. Additionally, you will sample the values x, y from a uniform distribution for a population of 100. By the end of this exercise, you will have evaluated the differences between starting from a fixed variable and sampling from a distribution:

1. Create a new Jupyter Notebook.

2. Import the **tensorflow** package and download and import **tensorflow_probability**:

```
import tensorflow as tf
import tensorflow_probability as tfp
```

3. Define a function that outputs $z(x, y)$:

```
def funct(x,y):
    return x**2-8*x+y**2+3*y
```

4. Identify the initial step by defining the x, y variables with values of 5 and 10:

```
initial_position = (tf.Variable(5.0), tf.Variable(10.0))
```

5. Instantiate the optimizer by selecting the **tensorflow_probability** optimizer named **differential_evolution_minimize**:

```
optimizer1 = tfp.optimizer.differential_evolution_minimize\
            (funct, initial_position = initial_position, \
            population_size = 100, \
            population_stddev = 1.5, seed = 879879)
```

6. Print the final values of z, x, y, by using the **objective_value** and **position** functions:

```
print('Final solution: z={:.1f}, x={:.1f}, y={:.1f}'\
        .format(optimizer1.objective_value.numpy(),\
        optimizer1.position[0].numpy(), \
        optimizer1.position[1].numpy()))
```

7. Run the application. You will get the following output. You can observe that the final values are identical to the **Step=25.0** value in *Figure 12.2*:

```
Final solution: z=-18.2, x=4.0, y=-1.5
```

In this exercise, the final optimal solution will be displayed. There are no additional optimization steps needed to reach the same solution as the gradient-based method. You can see that you are using fewer lines of code and that the time taken for the algorithm to converge is shorter.

For uniform optimization, the steps to modify the code are as follows:

8. Import the **random** package:

```
import random
```

9. Initialize the population size and create the initial population sampling the *x*, *y* variables from a random uniform distribution of the population size:

```
size = 100
initial_population = (tf.random.uniform([size]), \
                      tf.random.uniform([size]))
```

10. Use the same optimizer, change the **initial_position** parameter to **initial_population**; use the same seed:

```
optimizer2 = tfp.optimizer.differential_evolution_minimize\
             (funct, initial_population= initial_population,\
             seed=879879)
```

11. Print the final values of *Z*, *X*, *Y*, by using the **objective_value** and **position** functions:

```
print('Final solution: z={:.1f}, x={:.1f}, y={:.1f}'\
      .format(optimizer2.objective_value.numpy(),\
      optimizer2.position[0].numpy(),\
      optimizer2.position[1].numpy()))
```

The output will be as follows:

```
Final solution: z=-18.2, x=4.0, y=-1.5
```

You will get the same result despite the variation in values. This means that we can randomly sample or choose a specific set of initial values, and the GA will still converge to the optimal solution faster, meaning we can improve our code by using fewer lines of code than if we'd used a gradient-based method.

> **NOTE**
>
> To access the source code for this specific section, please refer to https://packt.live/2MQmlPr.
>
> You can also run this example online at https://packt.live/2zpH6hJ.

The solution will converge to the optimal values irrespective of the initial starting point, whether using a fixed value for the inputs or a random sampling of the population of chromosomes.

This section offered a general overview of evolutionary algorithms, explaining the differences between evolutionary strategies and GAs. You've had the opportunity to implement differential evolution using the **tensorflow_probabilities** package to optimize the solution of a loss function, analyzing the implementation of two different techniques: starting from fixed input values and using random sampling for the input values. You also had the opportunity to evaluate the implementation of GAs compared to gradient descent methods. GAs can use independent starting values and their convergence to a global optimum is faster and less prone to disturbances that gradient descent methods, whereas gradient descent is step-dependent and has higher sensitivity to the input variable.

In the following section, we will build on the principles of developing GAs, starting with a look at population creation.

COMPONENTS: POPULATION CREATION

In the previous section, you were introduced to evolutionary methods for function optimization. In this section, we will concentrate on population creation, fitness score creation, and the task of creating the genetic algorithm.

The population, $X(i)$, is identified as a group of individuals or chromosomes:

$$X(i) = \left\{ x_1^i, x_2^i, \cdots, x_s^i \right\}$$

Figure 12.4: Expression for the population

Here, **s** represents the total number of chromosomes (population size) and **i** is the iteration. Each chromosome is a possible solution to the presented problem in an abstract form. For a binary problem, the population can be a matrix with randomly generated ones and zeros.

The chromosome is a combination of input variables (genes):

$$x^i_1 = \left[g_{1.1}, g_{1.2}, \cdots, g_{1.m} \right]$$

Figure 12.5: Expression for the chromosome

Here, **m** is the maximum number of genes (or variables).

When translated to code, population creation can be demonstrated as follows:

```
population =  np.zeros((no_chromosomes, no_genes))
for i in range(no_chromosomes):
    ones = random.randint(0, no_genes)
    population[i, 0:ones] = 1
    np.random.shuffle(population[i])
```

Each chromosome is then compared using a fitness function:

$$f\left(x^i_1 \right) = f\left(g_{1.1}, g_{1.2}, \cdots, g_{1.m} \right)$$

Figure 12.6: Fitness function

The fitness function can be translated to code as follows:

```
identical_to_target = population == target
```

The output of the function is a score indicating how close the chromosome is to the target (optimal solution). The target is represented by the maximization of the fitness function. There cases where the optimization problem relies on minimizing a cost function. The function can be a mathematical one, a thermodynamic model, or a computer game. This can be done either by considering the chromosomes with low weightings (scores) or by adapting the cost function into a fitness one.

Once the fitness function is identified and defined, the evolution process can start. The initial population is generated. A characteristic of the initial population is diversity. To offer this diversity, the elements can be randomly generated. To make the population evolve, the iterative process starts by selecting the parents that offer the best fit to start the reproduction process.

EXERCISE 12.03: POPULATION CREATION

In this exercise, we will be creating an original population of binary chromosomes of length 5. Each chromosome should have eight genes. We will define a target solution and output the similarity of each chromosome to it. This exercise aims to allow you to design and establish the first set of steps for a GA and find the binary solution that fits the target. The exercise is similar to matching the output of a control system with a target:

1. Create a new Jupyter Notebook. Import the **random** and **numpy** libraries:

```
import random
import numpy as np
```

2. Create a function for the random population:

```
# create function for random population
def original_population(chromosomes, genes):
    #initialize the population with zeroes
    population =  np.zeros((chromosomes, genes))
    #loop through each chromosome
    for i in range(chromosomes):
        #get random no. of ones to be created
        ones = random.randint(0, genes)
        #change zeroes to ones
        population[i, 0:ones] = 1
        #shuffle rows
        np.random.shuffle(population[i])
    return population
```

3. Define a function for creating the target solution:

```
def create_target_solution(gene):
    #assume that there is an equal number of ones and zeroes
    counting_ones = int(gene/2)

    # build array with equal no. of ones and zeros
    target = np.zeros(gene)
    target[0:counting_ones] = 1
```

```
# shuffle the array to mix zeroes and ones
np.random.shuffle(target)

return target
```

4. Define a function for calculating the fitness weighting for each chromosome:

```
def fitness_function(target,population):
    #create an array of true/false compared to the reference
    identical_to_target = population == target
    #sum no. of genes that are identical
    fitness_weights = identical_to_target.sum(axis = 1)
    return fitness_weights
```

In the preceding code, you are comparing each chromosome of the population with the target and cataloging the similarity as a Boolean – **True** if similar or **False** if different – in the matrix called **identical_to_target**. Count all the elements that are true and output them as the weights.

5. Initialize the population with **5** chromosomes and **8** genes and calculate **weights**:

```
#population of 5 chromosomes, each having 8 genes
population = original_population(5,8)
target = create_target_solution(8)
weights = fitness_function(target,population)
```

In the preceding code, we calculate **population**, **target**, and **weights** based on the three developed functions.

6. Print the target solution, the index of the chromosome, the chromosome, and the weight using a **for** loop:

```
print('\n target:', target)
for i in range(len(population)):
    print('Index:', i, '\n chromosome:', population[i],\
        '\n similarity to target:', weights[i])
```

7. Run the application. You will get a similar output to this, as the population elements are randomized:

```
target: [0. 0. 1. 1. 1. 0. 0. 1.]
Index: 0
 chromosome: [1. 1. 1. 1. 1. 0. 1. 1.]
 similarity to target: 5
```

```
Index: 1
 chromosome: [1. 0. 1. 1. 1. 0. 0. 0.]
  similarity to target: 6
Index: 2
 chromosome: [1. 0. 0. 0. 0. 0. 0. 0.]
  similarity to target: 3
Index: 3
 chromosome: [0. 0. 0. 1. 1. 0. 1. 0.]
  similarity to target: 5
Index: 4
 chromosome: [1. 0. 0. 1. 1. 1. 0. 1.]
  similarity to target: 5
```

You will notice that each chromosome is compared to the target and the similarity (based on the fitness function) is printed out.

> **NOTE**
>
> To access the source code for this specific section, please refer to https://packt.live/2zrjadT.
>
> You can also run this example online at https://packt.live/2BSSeEG.

This section showcased the first steps of genetic algorithm development: the generation of a random population, fitness score assignment for each element of the population (chromosome), and getting the number of elements that are the best fit compared to the target (in this case have the highest similarity with an optimal solution). The following sections will expand on the code generation that occurs until the optimal solution is reached. To do this, in the next section, you will explore the selection of the parents for the reproduction process.

COMPONENTS: PARENT SELECTION

The previous section showcased the concepts of populations; we looked at creating a target solution and comparing that solution with the elements (chromosomes) of the population. These concepts were implemented in an exercise that will be continued in this section. In this section, you will explore the concept of selection and implement two selection strategies.

For the reproduction process (which is the quintessential part of GAs, as they rely on creating future generations of stronger chromosomes), there are three steps:

1. Parent selection

2. Mixing the parents to create new children (crossover)

3. Replacing them with the children in the population

Selection essentially consists of choosing two or more parents for the mixing process. Once a fitness criterion is selected, the way in which the selection of the parents will be performed needs to be chosen, as does how many children will come from the parents. Selection is a vital step in performing genetic evolution, as it involves determining the children with the highest fitness. The most common way to select the best individuals is by the "survival of the fittest." This means the algorithm will improve the population in a step-by-step manner. The convergence of the GA is dependent upon the degree to which chromosomes with higher fitness are chosen. Therefore, the convergence speed is highly dependent on the successful selection of chromosomes. If the chromosomes with the highest fitness are prioritized, there is a chance that a sub-optimal solution will be found; if the candidates have consistently low fitness, then convergence will be extremely slow.

The available selection methods are as follows:

*   **Top-to-bottom pairing**: This refers to creating a list of chromosomes and pairing them two by two. The chromosomes with odd indexes are paired with the even chromosomes, thus generating mother-father couples. The chromosomes at the top of the list are selected.

*   **Random Selection**: This involves using a uniform number generator to select the parents.

*   **Random Weighted Selection or a Roulette Wheel**: This involves calculating the probability of the suitability of a chromosome compared to the entire population. The selection of the parent is done randomly. The probability (weight) can be determined either by rank or fitness. The first approach (see *Figure 12.7*) relies on the rank of the chromosome (r), which can constitute the index of the chromosome in the population list, and k represents the number of required chromosomes (parents):

$$p_r = \frac{k - r + 1}{\sum_{i=1}^{k} r}$$

Figure 12.7: Probability using rank

The second approach (see *Figure 12.8*) relies on the fitness of the chromosome (f_c) compared to the sum of the fitness of the entire population ($\sum_{i=1}^{n} f_{ci}$):

$$Pc = \frac{f_c}{\sum_{i=1}^{n} f_{ci}}$$

Figure 12.8: Probability using chromosome fitness

As an alternative, the probability (see *Figure 12.9*) can also be calculated based on the fitness of the chromosome (f_c) compared with the highest fitness of the population $\left(max\left(f_p\right)\right)$. In all of the cases, the probabilities are compared to the randomly selected numbers to identify the parents with the best weights:

$$Pc = \frac{f_c}{\left(max\left(f_p\right)\right)}$$

Figure 12.9: Probability using the highest fitness of the population

- **Selection by Tournament**: This method is based on the random selection of a subset of chromosomes, out of which the chromosome with the highest fitness is selected as a parent. This repeats until the required number of parents is identified.

The roulette wheel and tournament techniques are among the most popular selection methods implemented in GAs, as they are inspired by biological processes. The problem with the roulette technique is that it can be noisy, and depending on which type of selection is used, the convergence rate can be affected. A benefit of the tournament method is that it can deal with large populations, leading to smoother convergence. The roulette wheel method is used to include random elements in the population, whereas when you are aiming to identify the parents with the highest similarity to the target, you use the tournament method. The following exercise will enable you to implement the tournament and roulette wheel techniques and evaluate your understanding of them.

EXERCISE 12.04: IMPLEMENTING THE TOURNAMENT AND ROULETTE WHEEL TECHNIQUES

In the exercise, you will implement the tournament and roulette wheel methods for the population of binary chromosomes of *Exercise 12.02, Implementing Fixed Value and Uniform Distribution Optimization Using GAs*. Each chromosome should have eight genes. We will define a target solution and print two sets of parents: one based on the tournament method and the other by roulette from the remaining population. Once each parent is chosen, set the fitness rank to the minimum:

1. Create a new Jupyter Notebook. Import the **random** and **numpy** libraries:

```
import random
import numpy as np
```

2. Create a function for the random population:

```
# create  function for random population
def original_population(chromosomes, genes):
    #initialize the population with zeroes
    population =  np.zeros((chromosomes, genes))
    #loop through each chromosome
    for i in range(chromosomes):
        #get random no. of ones to be created
        ones = random.randint(0, genes)
        #change zeroes to ones
        population[i, 0:ones] = 1
        #shuffle rows
        np.random.shuffle(population[i])
    return population
```

3. Define a function for creating the target solution:

```
def create_target_solution(gene):
    #assume that there is an equal number of ones and zeroes
    counting_ones = int(gene/2)

    # build array with equal no. of ones and zeros
    target = np.zeros(gene)
    target[0:counting_ones] = 1
```

```
    # shuffle the array to mix zeroes and ones
    np.random.shuffle(target)


    return target
```

4. Define a function for calculating the fitness weighting for each chromosome:

```
def fitness_function(target,population):
    #create an array of true/false compared to the reference
    identical_to_target = population == target
    #sum no. of genes that are identical
    fitness_weights = identical_to_target.sum(axis = 1)
    return fitness_weights
```

5. Define a function for selecting the pair of parents with the highest weighting (the highest fitness score). Since the population is reduced, the chromosomes are competing more. This method is also known as tournament selection:

```
# select the best parents
def select_parents(population, weights):
    #identify the parent with the highest weight
    parent1 = population[np.argmax(weights)]
    #replace weight with the minimum number
    weights[np.argmax(weights)] = 0
    #identify the parent with the second-highest weight
    parent2 = population[np.argmax(weights)]
    return parent1, parent2
```

6. Create a function for the roulette wheel by selecting a random number from a uniform distribution:

```
def choice_by_roulette(sorted_population, fitness):
    normalised_fitness_sum = 0
    #get a random draw probability
    draw = random.uniform(0,1)
    prob = []
```

7. In the function, calculate the sum of all the fitness scores:

```
    for i in range(len(fitness)):
        normalised_fitness_sum += fitness[i]
```

8. Calculate the probability of the fitness of the chromosome compared to the sum of all fitness scores and compared to the chromosome with the highest fitness score:

```
    ma = 0
    n = 0
# calculate the probability of the fitness selection
    for i in range(len(sorted_population)):
            probability = fitness[i]/normalised_fitness_sum
            #compare fitness to the maximum fitness and track it
            prob_max = fitness[i]/np.argmax(fitness)
            prob.append(probability)
            if ma < prob_max:
                ma = prob_max
                n = i
```

9. Run through all the chromosomes and select the parent that has a higher fitness probability compared to the sum of fitness scores higher than the **draw**, or the parent with the highest probability compared to the maximum fitness score:

```
        for i in range(len(sorted_population)):
            if draw <= prob[i]:
                fitness[i] = 0
                return sorted_population[i], fitness
            else:
                fitness[n] = 0
                return sorted_population[n], fitness
```

10. Initialize **population**, calculate **target** and the fitness scores, and print the scores and **target**:

```
population = original_population(5,8)
target = create_target_solution(8)
weights = fitness_function(target,population)
print(weights)
print('\n target:', target)
```

You will get a similar output to this:

```
[5 1 5 3 4]
```

11. Apply the first selection method and print out the parents and the new scores:

```
print('\n target:', target)
parents = select_parents(population,weights)
print('Parent 1:', parents[0],'\nParent 2:', parents[1])
print(weights)
```

You will get a similar output to this for the tournament selection process:

```
target: [0. 1. 1. 1. 1. 0. 0. 0.]
Parent 1: [1. 1. 1. 1. 1. 0. 1. 1.]
Parent 2: [1. 1. 1. 1. 1. 1. 1. 0.]
[0 1 5 3 4]
```

You can observe that for parent 1, the score has been replaced with **0**. For parent 2, the score stays the same.

12. Use the roulette function to select the next two parents and print out the parents and the weights:

```
parent3, weights = choice_by_roulette(population, weights)
print('Parent 3:', parent3, 'Weights:', weights)
parent4, weights = choice_by_roulette(population, weights)
print('Parent 4:', parent4,'Weights:', weights)
```

You will have a similar output to this:

```
0.8568696148662779
[0.0, 0.07692307692307693, 0.38461538461538464,
 0.23076923076923078, 0.3076923076923077]
Parent 3: [1. 1. 1. 1. 1. 1. 1. 0.] Weights: [0 1 0 3 4]
0.4710306341255527
[0.0, 0.125, 0.0, 0.375, 0.5]
Parent 4: [0. 0. 1. 0. 1. 1. 1. 0.] Weights: [0 1 0 3 0]
```

You can see that parents 2 and 3 are the same. This time, the weight for the respective parent is changed to 0. Additionally, parent 4 is selected and has its weighting changed to 0.

> **NOTE**
>
> To access the source code for this specific section, please refer to https://packt.live/2MTsKJO.
>
> You can also run this example online at https://packt.live/2YrwMhP.

With this exercise, you have implemented a tournament-like method, by selecting the parents with the highest scores, and the roulette wheel selection technique. Also, you have developed a method of avoiding the double-selection of the same chromosome. The first set of parents was chosen using the first method, whereas the second method was used in selecting the second set of parents. We have also identified a need for a method of replacing indexes to avoid double-selection of the same chromosome, which is one of the pitfalls of the selection process. This helped you to understand the differences between the two methods and allowed you to put into practice GA-related methods from population generation to selection.

COMPONENTS: CROSSOVER APPLICATION

This section expands on recombining the genetic code of the parents by means of crossover into children (that is, the creation of the two child solutions by combining and re-organizing the code of the two parents). Various techniques can be used to create new solutions for generating a new population. The binary information of two viable solutions in machine learning can be recombined by a process called crossover, which is similar to biological genetic exchange, where genetic information is transmitted from parents to children. Crossover ensures that the genetic material of a solution is transmitted to the next generation.

Crossover is the most common form of reproduction technique, or mating. Between the first and last bits of the parents (selected chromosomes), the crossover point represents the splitting point of the binary code that will be passed onto the children (offspring): the part to the left of the crossover point of the first parent will be inherited by the first child, and everything to the right side of the crossover point of the second parent will become the part of the first child. The left side of the second parent combined with the right side of the first parent results in the second child:

```
child1 = np.hstack((parent1[0:p],parent2[p:]))
child2 = np.hstack((parent2[0:p], parent1[p:]))
```

There are multiple crossover techniques, as listed follows:

- Single-point crossover (which you can see in the preceding code) involves splitting the genetic code of the parents at one point and passing the first part to the first child, and the second part to the second child. It is used by traditional GAs; the crossover point is identical for both chromosomes and is selected randomly.

- Two-point crossover involves two crossover points impacting the gene exchange between the two parents. The more crossover points are introduced, the more the performance of the GA can be reduced as the genetic makeup is lost. However, introducing two-point crossover can lead to a better exploration of the state or parameter space.

- Multi-point crossover involves a number of splits. If the number of splits is even, then the splits are selected randomly and the sections in the chromosome are exchanged. If the number is odd, then the splits are alternating the exchanges of section.

- Uniform crossover involves the random selection (as in a coin toss) of the parent that will provide an element of the chromosome (gene).

- Three-parent crossover entails the comparison of each gene between two parents. If they have the same value, the child inherits the gene; if not, the child inherits the gene from the third parent.

Consider the following code example:

```
def crossover_reproduction(parents, population):
    #define parents separately
    parent1 = parents[0]
    parent2 = parents[1]

    #randomly assign a point for cross-over
    p = random.randrange(0, len(population))
    print("Crossover point:", p)

    #create children by joining the parents at the cross-over point
    child1 = np.hstack((parent1[0:p],parent2[p:]))
    child2 = np.hstack((parent2[0:p], parent1[p:]))



    return child1, child2
```

In the preceding code, we define the crossover function between two parents. We have defined the parents separately and then randomly assigned a certain point for crossover. Then, we have defined the children to be created by joining the parents at the defined crossover point.

In the following exercise, you will continue the process of implementing the components of GAs to create child chromosomes.

EXERCISE 12.05: CROSSOVER FOR A NEW GENERATION

In this exercise, we will be implementing crossover between two parents, for a new generation. Following the steps from *Exercise 12.04, Implementing Tournament and Roulette Wheel*, and using the chromosomes with the highest weight, we will apply single-point crossover to create the first new set of children:

1. Create a new Jupyter Notebook. Import the **random** and **numpy** libraries:

```
import random
import numpy as np
```

2. Create the function for a random population:

```
def original_population(chromosomes, genes):
    #initialize the population with zeroes
    population =  np.zeros((chromosomes, genes))
    #loop through each chromosome
    for i in range(chromosomes):
        #get random no. of ones to be created
        ones = random.randint(0, genes)
        #change zeroes to ones
        population[i, 0:ones] = 1
        #shuffle rows
        np.random.shuffle(population[i])
    return population
```

As you can see in the previous code, we have created a **population** function.

3. Define a function to create the target solution:

```
def create_target_solution(gene):
    #assume that there is an equal number of ones and zeroes
    counting_ones = int(gene/2)

    # build array with equal no. of ones and zeros
    target = np.zeros(gene)
    target[0:counting_ones] = 1

    # shuffle the array to mix zeroes and ones
    np.random.shuffle(target)

    return target
```

4. Define a function for calculating the fitness weighting for each chromosome:

```
def fitness_function(target,population):
    #create an array of true/false compared to the reference
    identical_to_target = population == target
    #sum no. of genes that are identical
    fitness_weights = identical_to_target.sum(axis = 1)
    return fitness_weights
```

5. Define a function for selecting the pair of parents with the highest weighting (highest fitness score). Since the population is smaller, the chromosomes are competing more. This method is also known as tournament selection:

```
# select the best parents
def select_parents(population, weights):
    #identify the parent with the highest weight
    parent1 = population[np.argmax(weights)]
    #replace weight with the minimum number
    weights[np.argmax(weights)] = 0
    #identify the parent with the second-highest weight
    parent2 = population[np.argmax(weights)]
    return parent1, parent2
```

6. Define a function for crossover by using a randomly selected crossover point:

```
def crossover_reproduction(parents, population):
    #define parents separately
    parent1 = parents[0]
    parent2 = parents[1]

    #randomly assign a point for cross-over
    p = random.randrange(0, len(population))
    print("Crossover point:", p)

    #create children by joining the parents at the cross-over point
    child1 = np.hstack((parent1[0:p],parent2[p:]))
    child2 = np.hstack((parent2[0:p], parent1[p:]))

    return child1, child2
```

7. Initialize the population with **5** chromosomes and **8** genes and calculate `weights`:

```
population = original_population(5,8)
target = create_target_solution(8)
weights = fitness_function(target,population)
```

8. Print the **target** solution:

```
print('\n target:', target)
```

The output will be as follows:

```
target: [1. 0. 0. 1. 1. 0. 1. 0.]
```

9. Select the parents with the highest weight and print the final selection:

```
parents = select_parents(population,weights)
print('Parent 1:', parents[0],'\nParent 2:', parents[1])
```

The output will be as follows:

```
Parent 1: [1. 0. 1. 1. 1. 0. 1. 1.]
Parent 2: [1. 0. 0. 0. 0. 0. 0. 0.]
```

10. Apply the **crossover** function and print the children:

```
children = crossover_reproduction(parents,population)
print('Child 1:', children[0],'\nChild 2:', children[1])
```

The output will be as follows:

```
Crossover point: 4
Child 1: [1. 0. 1. 1. 0. 0. 0. 0.]
Child 2: [1. 0. 0. 0. 1. 0. 1. 1.]
```

11. Run the application.

You will get a similar output to that shown in the following snippet. As you can see, the population elements are randomized. Check that the elements of **Child 1** and **Child 2** are the same as those of **Parent 1** and **Parent 2**:

```
target: [1. 0. 1. 1. 0. 0. 1. 0.]
. . .
Parent 1: [1. 0. 1. 1. 1. 0. 1. 1.]
Parent 2: [0. 0. 1. 1. 0. 1. 0. 0.]
. . .
```

```
Crossover point: 1
Child 1: [1. 0. 1. 1. 0. 1. 0. 0.]
Child 2: [0. 0. 1. 1. 1. 0. 1. 1.]. . .
```

You can check that the starting elements from the crossover point in the array of **Child 1** have the same array elements as **Parent 2**, and that **Child 2** has the same array elements as **Parent 1**.

> **NOTE**
>
> To access the source code for this specific section, please refer to https://packt.live/30zHbup.
>
> You can also run this example online at https://packt.live/3fueZxx.

In this section, we identified the various strategies for the recombination technique known as crossover. A basic implementation of single-point crossover, where the crossover point is randomly generated, was represented. In the following section, we will examine the last element of GA design: population mutation.

COMPONENTS: POPULATION MUTATION

In the previous sections, you have implemented population generation, parent selection, and crossover reproduction. This section will concentrate on the application of random mutation and the repetition of child generations until a new population size is achieved and weights (fitness scores) for the population of the genetic algorithm are assigned. This section will include an explanation of the mutation technique. This will be followed by a presentation of the available mutation techniques as well as a discussion about population replacement. Finally, an exercise implementing mutation techniques will be presented.

A caveat of gradient methods is that the algorithms can stop at a local optimum solution. To prevent this from happening, mutations can be introduced to the population of solutions. Mutation generally occurs after the crossover process. Mutation relies on randomly assigning binary information in either a set of chromosomes or in the entire population. Mutation provides an avenue of problem space exploration by introducing a random change in the population. This technique prevents rapid convergence and encourages the exploration of new solutions. In the final steps (the last generations) or when the optimal solution is reached, mutation ceases to be applied.

There are various mutation techniques, as follows:

- Single-point mutation (flipping) involves randomly selecting genes from different chromosomes and changing their binary values to their opposites (from 0 to 1 and 1 to 0).

- Interchanging involves selecting two sections of the chromosome of one parent and swapping them, thus generating a new child.

- You can also reverse a randomly selected segment within the parent or the population of chromosomes, and all the binary values are changed to their opposites.

The occurrence of a mutation is determined by its probability. The probability defines the frequency at which mutations occur within the population. If the probability is 0%, then after the crossover, the children are unaltered; if a mutation occurs, one or more parts of the chromosome or the population are changed. If the probability is 100%, then the entire chromosome is changed.

After the mutation process occurs, the fitness of the new children is calculated, and the population is altered to include them. This leads to a new generation of the population. Depending on the strategy used, the parents with the lowest fitness scores are discarded to leave room for the newly generated children.

EXERCISE 12.06: NEW GENERATION DEVELOPMENT USING MUTATION

In this exercise, we will be focusing on the development of a new generation. We will again create a new population, select two parent chromosomes, and use crossover to develop two children. We will then add the two new chromosomes to the population and mutate the entire population with a probability of 0.05:

1. Create a new Jupyter Notebook. Import the **random** and **numpy** libraries:

```
import random
import numpy as np
```

2. Create a function for the random population:

```
def original_population(chromosomes, genes):
    #initialize the population with zeroes
    population =  np.zeros((chromosomes, genes))
    #loop through each chromosome
    for i in range(chromosomes):
        #get random no. of ones to be created
        ones = random.randint(0, genes)
```

```
        #change zeroes to ones
        population[i, 0:ones] = 1
        #shuffle rows
        np.random.shuffle(population[i])
    return population
```

3. Define a function to create the target solution:

```
def create_target_solution(gene):
    #assume that there is an equal number of ones and zeroes
    counting_ones = int(gene/2)

    # build array with equal no. of ones and zeros
    target = np.zeros(gene)
    target[0:counting_ones] = 1

    # shuffle the array to mix zeroes and ones
    np.random.shuffle(target)

    return target
```

4. Define a function to calculate the fitness weighting for each chromosome:

```
def fitness_function(target,population):
    #create an array of true/false compared to the reference
    identical_to_target = population == target
    #sum no. of genes that are identical
    fitness_weights = identical_to_target.sum(axis = 1)
    return fitness_weights
```

5. Define a function to select the pair of parents with the highest weighting (highest fitness score). Since the population is small, the chromosomes are competing more. This method is also known as tournament selection:

```
# select the best parents
def select_parents(population, weights):
    #identify the parent with the highest weight
    parent1 = population[np.argmax(weights)]
    #replace weight with the minimum number
    weights[np.argmax(weights)] = 0
    #identify the parent with the second-highest weight
    parent2 = population[np.argmax(weights)]
    return parent1, parent2
```

6. Define a function for crossover by using a randomly selected crossover point:

```
def crossover_reproduction(parents, population):
    #define parents separately
    parent1 = parents[0]
    parent2 = parents[1]

    #randomly assign a point for cross-over
    p = random.randrange(0, len(population))
    print("Crossover point:", p)

    #create children by joining the parents at the cross-over point
    child1 = np.hstack((parent1[0:p],parent2[p:]))
    child2 = np.hstack((parent2[0:p], parent1[p:]))


    return child1, child2
```

7. Define a function for a mutation that uses the probability and the population as inputs:

```
def mutate_population(population, mutation_probability):
    #create array of random mutations that uses the population
    mutation_array = np.random.random(size = (population.shape))
    """
    compare elements of the array with the probability and
    put the results into an array
    """
    mutation_boolean = mutation_array \
                    >= mutation_probability
    """
    convert boolean into binary and store to create a new
    array for the population
    """
    population[mutation_boolean] = np.logical_not\
                                (population[mutation_boolean])
    return population
```

In the preceding code snippet, the condition set for the mutation selection is to check that each element of the array is higher than the mutation probability which acts as a threshold. If the element is higher than the threshold, mutation is applied.

8. Append the array of **children** to the original population, creating a new crossover **population**, and use the **print()** function to display it:

```
population = original_population(5,8)
target = create_target_solution(8)
weights = fitness_function(target,population)
parents = select_parents(population,weights)
children = crossover_reproduction(parents,population)
```

The output will be as follows:

```
Crossover point: 3
```

9. Next, append **population** with **children**:

```
population_crossover = np.append(population, children, axis= 0)
print('\nPopulation after the cross-over:\n', \
      population_crossover)
```

The population will be as follows:

```
Population after the cross-over:
 [[0. 1. 0. 0. 0. 0. 1. 0.]
 [0. 0. 0. 0. 0. 1. 0. 0.]
 [1. 1. 1. 1. 1. 0. 0. 1.]
 [1. 1. 1. 0. 1. 1. 1. 1.]
 [0. 1. 1. 1. 1. 0. 0. 0.]
 [1. 1. 1. 1. 1. 0. 0. 1.]
 [1. 1. 1. 0. 1. 1. 1. 1.]]
```

10. Use the crossover population and the mutation probability of **0.05** to create a new population and display the mutated population:

```
mutation_probability = 0.05
new_population = mutate_population\
              (population_crossover,mutation_probability)
print('\nNext generation of the population:\n',\
      new_population)
```

As you can see, the threshold(mutation_probability) is 0.05. Hence, if the elements are higher than this threshold, they will incur a mutation (so there is a 95% chance of the mutation occurring to the gene).

The output will be as follows:

```
Next generation of the population:
 [[1. 0. 1. 1. 1. 1. 0. 1.]
 [1. 0. 1. 1. 1. 0. 1. 1.]
 [1. 0. 0. 0. 0. 1. 1. 0.]
 [0. 0. 0. 1. 0. 0. 0. 0.]
 [1. 0. 0. 0. 1. 1. 1. 1.]
 [0. 0. 0. 0. 0. 1. 1. 0.]
 [0. 0. 0. 1. 0. 1. 0. 1.]]
```

You will get a similar output as the population elements are randomized. You can see that the chromosomes resulting from crossover are added to the original population and that after mutation, the population has the same number of chromosomes, but the genes are different. The crossover and mutation steps can be repeated until the target solution is reached by looping the functions. These cycles are also known as generations.

> **NOTE**
>
> To access the source code for this specific section, please refer to https://packt.live/3dXaBqi.
>
> You can also run this example online at https://packt.live/2Ysc5Cl.

In this section, mutation was described. The benefit of mutation is that it introduces random variation to chromosomes, encouraging exploration and helping to avoid local optima. Various mutation techniques were presented. The example we used showed the impact of mutation probability by implementing reverse mutation on a population after the crossover process was finalized.

APPLICATION TO HYPERPARAMETER SELECTION

In this section, we will explore the use of GAs for parameter selection, especially when using neural networks. GAs are widely used for optimization problems in scheduling in both production and railway management. The solutions to these types of problems rely on creating a combination of neural networks and GAs as function optimizers.

The exercise in this section provides a platform for tuning hyperparameters for a neural network to predict wind flow patterns. You will apply a simple genetic algorithm to optimize the values of the hyperparameters used to train a neural network.

Artificial Neural Networks (ANNs) model the biological processes and structures of neurons in the brain. The neurons in ANNs rely on a combination of input information (parameters) and weights. The product (which has an added bias) passes through a transfer function, which is a set of neurons arranged in parallel with each other to form a layer.

For weight and bias optimization, ANNs use gradient descent methods for their training processes and backpropagation processes. This impacts the development of the neural network, as before training even commences, the neural network topology needs to be fully designed. Because the design is pre-set, some neurons may not be used in the training process, but they may still be active, therefore making them redundant. Additionally, neural networks using gradient methods can become stuck at a local optimum, and therefore need to rely on alternative methods to help them continue their processes, such as regularization, ridge regression, or lasso regression. ANNs are widely used in speech recognition, feature detection (whether for image, topology, or signal processing), and disease detection.

To prevent these problems and enhance the training of neural networks, GAs can be implemented. GAs are used for function optimization, while crossover and mutation techniques help with problem space exploration. Initially, GAs were used to optimize the weights and number of nodes of neural networks. For this, the chromosomes of the GA are encoded with possible variations of weights and nodes. The fitness function generated by the ANN relies on the mean squared error of the potential values and the exact values of the parameters.

However, research has expanded to implementations of **Recurrent Neural Networks (RNNs)** and combining them with RL, aiming towards multi-processor performance. An RNN is a type of ANN that produces outputs that are not only a result of the weighting process of the input but also of a vector containing previous input and outputs. This enables the neural network to maintain prior knowledge of previous training instances.

GAs serve in expanding the topology of the neural networks beyond weighting adjustments. One example is EDEN, whereby encoding is done within the chromosome and the architecture of the network, and the learning rate achieves high accuracy rates on multiple TensorFlow datasets. One of the most challenging problems in training neural networks is the quality of the features (or input hyperparameters) that are fed to the network. If the parameters are not appropriate, the mapping of inputs and outputs will be erroneous. Therefore, GAs can act as a wrapper alternative to the ANNs by optimizing the selection of features.

The following exercise will teach you how to apply a simple genetic algorithm to identify the optimal parameters (window size and number of units) for an RNN. The genetic algorithm implemented is using the **deap** package, through the **eaSimple()** function, which enables you to create, using toolbox-based code, a simple GA that includes population creation, selection through the **selRandom()** function, reproduction through the **cxTwoPoint()** function, and mutation through the **mutFlipBit()** function. For comparing and hyperparameter selection, the **selBest()** function is used.

EXERCISE 12.07: IMPLEMENTING GA HYPERPARAMETER OPTIMIZATION FOR RNN TRAINING

Our goal in this exercise is to identify the best hyperparameters to use for an RNN using a simple genetic algorithm. In this exercise, we are using a dataset that was part of a weather forecasting challenge in 2012. A single feature, **wp2**, is used in the training and validation of the parameters. The two hyperparameters used are the number of units and the window size. These hyperparameters represent the genetic material for the chromosome:

> **NOTE**
>
> The dataset can be found in the GitHub repository at the following link: https://packt.live/2Ajjz2F.
>
> The original dataset can be found at the following link: https://www.kaggle.com/c/GEF2012-wind-forecasting/data.

1. Create a new Jupyter Notebook. Import the **pandas** and **numpy** libraries and functions:

```
import numpy as np
import pandas as pd


from sklearn.metrics import mean_squared_error
from sklearn.model_selection import train_test_split as split
from tensorflow.keras.layers import SimpleRNN, Input, Dense
from tensorflow.keras.models import Model


from deap import base, creator, tools, algorithms
from scipy.stats import bernoulli
from bitstring import BitArray
```

From the **sklearn** package, import **mean_squared_error** and **train_test_split**. Also, from the **tensorflow** and **keras** packages, import **SimpleRNN**, **Input**, **Dense** (from the **layers** folder), and the model (from the **Model** class). To create the GA, it is necessary to call from the **deap** package **base**, **creator**, **tools**, and **algorithms**. For statistics, we are using the Bernoulli equation; therefore, we will call **bernoulli** from the **scipy.stats** package. From **bitstrings**, we will call **BitArray**.

2. Use a random seed for model development; **998** is an initialization number for the seed:

```
np.random.seed(998)
```

3. Load data from the **train.csv** file, use **np.reshape()** to modify the data into an array that only contains column **wp2**, and select the first 1,501 elements:

```
#read data from csv
data = pd.read_csv('../Dataset/train.csv')
#use column wp2
data = np.reshape(np.array(data['wp2']), (len(data['wp2']), 1))
data = data[0:1500]
```

4. Define a function to split the dataset based on window size:

```
def format_dataset(data, w_size):
    #initialize as empty array
    X, Y = np.empty((0, w_size)), np.empty(0)
    """
```

```
depending on the window size the data is separated in
2 arrays containing each of the sizes
"""
for i in range(len(data)-w_size-1):
    X = np.vstack([X,data[i:(i+w_size),0]])
    Y = np.append(Y, data[i+w_size,0])
X = np.reshape(X, (len(X),w_size,1))
Y = np.reshape(Y,(len(Y), 1))
return X, Y
```

5. Define a function to train the RNN to identify the optimal hyperparameters using a simple genetic algorithm:

```
def training_hyperparameters(ga_optimization):
    """
    decode GA solution to integer window size and number of units
    """
    w_size_bit = BitArray(ga_optimization[0:6])
    n_units_bit = BitArray(ga_optimization[6:])
    w_size = w_size_bit.uint
    n_units = n_units_bit.uint
    print('\nWindow Size: ', w_size, \
            '\nNumber of units: ',n_units)


    """
    return fitness score of 100 if the size or the units are 0
    """
    if w_size == 0 or n_units == 0:
        return 100


    """
    segment train data on the window size splitting it into
    90 train, 10 validation
    """
    X,Y = format_dataset(data, w_size)
    X_train, X_validate, Y_train, Y_validate = \
    split(X, Y, test_size= 0.10, random_state= 998)
```

The first step is identifying the sections of the chromosome pertaining to window size and the number of units. The next step is to return an extremely high fitness score, if there are no window sizes or the number of units. Split the two arrays into training and validation arrays with a 90:10 split.

6. Initialize the input features, and use the **SimpleRNN** model with the training dataset. For optimization, use the Adam algorithm with mean squared error as the loss function. To train the model, use the **fit** function with **5** for **epochs** and a batch size of **4**. To generate the predicted values, use the input values stored in **X_validate** in the **predict** function for the model. Calculate the **Root Mean Squared Error** (**RMSE**) between the validation set and predicted set of output variables. Return **RMSE**:

```python
input_features = Input(shape=(w_size,1))
x = SimpleRNN(n_units,input_shape=(w_size,1))(input_features)
output = Dense(1, activation='linear')(x)
rnnmodel = Model(inputs=input_features, outputs = output)
rnnmodel.compile(optimizer='adam', \
                 loss = 'mean_squared_error')
rnnmodel.fit(X_train, Y_train, epochs=5, \
             batch_size=4, shuffle = True)
Y_predict = rnnmodel.predict(X_validate)

# calculate RMSE score as fitness score for GA
RMSE = np.sqrt(mean_squared_error(Y_validate, Y_predict))
print('Validation RMSE: ', RMSE, '\n')

return RMSE,
```

7. Instantiate the population size, the number of generations used for the genetic algorithm, and the length of the gene with **4**, **5**, and **10**, respectively:

```python
population_size = 4
generations = 5
gene = 10
```

8. Use the toolbox available in the **deap** package to instantiate the genetic algorithm, **eaSimple()**. To do this, use the creator tool to instantiate the fitness function as **RMSE**:

```python
creator.create('FitnessMax', base.Fitness, weights= (-1.0,))
creator.create('Individual', list, fitness = creator.FitnessMax)

toolbox = base.Toolbox()
toolbox.register('bernoulli', bernoulli.rvs, 0.5)
toolbox.register('chromosome', tools.initRepeat, \
            creator.Individual, toolbox.bernoulli, n = gene)
```

```
toolbox.register('population', tools.initRepeat, \
                 list, toolbox.chromosome)

toolbox.register('mate', tools.cxTwoPoint)
toolbox.register('mutate', tools.mutFlipBit, indpb = 0.6)
toolbox.register('select', tools.selRandom)
toolbox.register('evaluate', training_hyperparameters)

population = toolbox.population(n = population_size)
algo = algorithms.eaSimple(population,toolbox,cxpb=0.4, \
                           mutpb=0.1, ngen=generations, \
                           verbose=False)
```

The last few lines of the output will be as follows:

```
Window Size:  48
Number of units:  15
Train on 1305 samples
Epoch 1/5
1305/1305 [==============================] - 3s 2ms/sample
- loss: 0.0106
Epoch 2/5
1305/1305 [==============================] - 3s 2ms/sample
- loss: 0.0066
Epoch 3/5
1305/1305 [==============================] - 3s 2ms/sample
- loss: 0.0057
Epoch 4/5
1305/1305 [==============================] - 3s 2ms/sample
- loss: 0.0051
Epoch 5/5
1305/1305 [==============================] - 3s 2ms/sample
- loss: 0.0049
Validation RMSE:  0.05564985152918074
```

The lower the **RMSE** value, the better the hyperparameters. The Bernoulli distribution serves to randomly initialize the chromosome genes. Based on the chromosome, the population is initialized. Within the toolbox, there are four steps for creating a new population: **mate** (this refers to the crossover process: **cxTwoPoint()** refers to the parents crossing information at two points in a crossover), **mutate** (this refers to the mutation process: **mutFlipBit()** will only mutate one of the elements of the chromosome with a **0.6** probability of occurrence), **select** (selection of the parents happens randomly through the **selRandom()** function), **evaluate** (this uses the RNN training function from *Step 6* and *Step 7*).

9. Use the **selBest()** function for a single optimal solution, **k=1**, compare the solutions to the fitness function, and select the one with the highest similarity. To get the optimal window size and number of units, loop through the chromosome and convert the bit values to unsigned integers and print the optimal hyperparameters:

```
optimal_chromosome = tools.selBest(population, k = 1)
optimal_w_size = None
optimal_n_units = None

for op in optimal_chromosome:
    w_size_bit = BitArray(op[0:6])
    n_units_bit = BitArray(op[6:])
    optimal_w_size = w_size_bit.uint
    optimal_n_units = n_units_bit.uint
    print('\nOptimal window size:', optimal_w_size, \
        '\n Optimal number of units:', optimal_n_units)
```

The output will be as follows:

```
Optimal window size: 48
Optimal number of units: 15
```

10. Run the application. You will get a similar output to what you see here. The initial values for the window size and number of units will be displayed. The GA will run using the RNN for the total number of epochs. At the end of each epoch, the **RMSE** value is displayed. Once all the epochs have executed, the optimal values are displayed:

```
880/1303 [======================>..........] - ETA: 1s - loss: 0.0047
892/1303 [======================>..........] - ETA: 1s - loss: 0.0048
904/1303 [======================>..........] - ETA: 1s - loss: 0.0048
920/1303 [======================>.........] - ETA: 1s - loss: 0.0047
932/1303 [======================>.........] - ETA: 1s - loss: 0.0048
948/1303 [======================>.........] - ETA: 1s - loss: 0.0051
960/1303 [======================>........] - ETA: 1s - loss: 0.0052
972/1303 [======================>........] - ETA: 1s - loss: 0.0051
984/1303 [======================>........] - ETA: 1s - loss: 0.0051
996/1303 [======================>........] - ETA: 1s - loss: 0.0052
1008/1303 [=======================>.......] - ETA: 1s - loss: 0.0052
1020/1303 [=======================>.......] - ETA: 1s - loss: 0.0052
1032/1303 [=======================>.......] - ETA: 1s - loss: 0.0051
1048/1303 [========================>......] - ETA: 1s - loss: 0.0051
1060/1303 [========================>......] - ETA: 0s - loss: 0.0051
1072/1303 [========================>......] - ETA: 0s - loss: 0.0051
1084/1303 [========================>......] - ETA: 0s - loss: 0.0051
1096/1303 [=========================>.....] - ETA: 0s - loss: 0.0051
1108/1303 [=========================>.....] - ETA: 0s - loss: 0.0051
1124/1303 [=========================>.....] - ETA: 0s - loss: 0.0051
1136/1303 [==========================>....] - ETA: 0s - loss: 0.0051
1152/1303 [==========================>....] - ETA: 0s - loss: 0.0050
1168/1303 [==========================>....] - ETA: 0s - loss: 0.0050
1180/1303 [===========================>...] - ETA: 0s - loss: 0.0050
1196/1303 [===========================>...] - ETA: 0s - loss: 0.0049
1208/1303 [===========================>...] - ETA: 0s - loss: 0.0049
1220/1303 [============================>..] - ETA: 0s - loss: 0.0049
1232/1303 [============================>..] - ETA: 0s - loss: 0.0050
1244/1303 [============================>..] - ETA: 0s - loss: 0.0050
1256/1303 [============================>..] - ETA: 0s - loss: 0.0050
1268/1303 [=============================>.] - ETA: 0s - loss: 0.0050
1280/1303 [=============================>.] - ETA: 0s - loss: 0.0050
1292/1303 [=============================>.] - ETA: 0s - loss: 0.0050
1303/1303 [==============================] - 5s 4ms/step - loss: 0.0050
Validation RMSE:   0.05419707364345042

Optimal window size: 28
 Optimal number of units: 4
```

Figure 12.10: Optimization of the window size and number of units using GA

We started with an initial window size of **51** and **15** units; the optimal window size is reduced to **28**, and the number of units to **4**. The difference between the parameters based on **RMSE** is reduced to **0.05**.

NOTE

To access the source code for this specific section, please refer to https://packt.live/37sgQA6.

You can also run this example online at https://packt.live/30AOKRK.

This section has covered combining GAs with neural networks as an alternative to using gradient descent methods. GAs mainly served in optimizing the number of neurons and weights for the neural networks, but their use can be expanded, through hybridization, to optimizing the structure of the network and hyperparameter selection. This exercise tested your ability to apply a genetic algorithm to find the optimal values of two features related to a weather forecasting problem. The features were used to train an RNN to estimate wind flow using RMSE values. In the following section, you will expand your knowledge of hybrid optimization techniques for the entire architecture of a neural network using NEAT.

NEAT AND OTHER FORMULATIONS

Neuroevolution is a term that refers to evolving neural networks using GAs. This branch of machine learning is shown to outperform RL in various problems and can be coupled with RL, as it is a method for unsupervised learning. As mentioned in the previous section, neuroevolution systems concentrate on changing the weights, the number of neurons (in the hidden layers), and the topology of ANNs.

Neuroevolution of Augmented Topologies (**NEAT**) focuses on topology evolution for ANNs. It involves training a simple ANN structure, consisting of input and output neurons and units to represent the bias, but no hidden layers. Each ANN structure is encoded within a chromosome that contains node genes and connection genes (the mapping or link between two node genes). Each connection specifies the input, output, weight node, activation of the connection, and innovation number that serves as a link between genes for the crossover process.

Mutations relate to the weights of the connections or the structure of the full system. Structural mutations can appear either by including a connection between two nodes that are not linked or by including a new node to a pre-existing connection, which causes two new connections to be built (one between the existing pair of nodes and one that includes the newly created node).

The crossover process entails the identification of common genes between different chromosomes within the population. This relies on the historical information about gene derivation, using a global innovation number. The genes resulting from the mutation receive incremented numbers from the gene they mutated, whereas through crossover, the genes keep their original numbers. This technique helps in solving the problems with gene matching that cause issues for neural network topologies. The genes that do not have the same innovation number are selected from the parent with the highest fitness. If both parents have the same fitness, the genes are selected randomly from each of the parents.

Chromosomes that have similar topologies are grouped based on how far apart they are (d); individuals are therefore evaluated based on the genes that are different (D), supplementary genes (S), and the differences in weight (ΔW) for the similar genes compared to the average number of genes (M). Each of the coefficients (C_1, C_2, C_3) acts as a weight that highlights the significance of each parameter:

$$d = \frac{D \times C_1 + S \times C_2}{M} + \Delta W \times C_3$$

Figure 12.11: Topology distance calculation

To categorize the chromosomes into species, the distance (d) is compared with a threshold (t). If $d < t$, then the chromosome belongs to the first species where this condition is fulfilled. To prevent species dominance, all the elements of the species need to have the same fitness level, which is calculated based on the number of members in the species. The evolution of the species (how many new chromosomes are included, (M_{new}) depends on the comparison between the fitness of the species, $\sum_{i=1}^{M_{old}} f_{si}$, and the average fitness of the population, $\bar{f_p}$:

$$M_{new} = \frac{\sum_{i=1}^{M_{old}} f_{si}}{\bar{f_p}}$$

Figure 12.12: Calculation of the number of new chromosomes

The advantage of NEAT is that, unlike neuroevolution algorithms that have a random set of topology parameters, it starts with the simplest topological form of a neural network and progressively evolves it to find the optimal solution, significantly reducing the number of used generations.

Evolving topology algorithms are categorized as **Weight Evolving Artificial Neural Networks (TWEANNs)**, which include EDEN, **Cellular Encoding (CE)**, **Enforced Subpopulations (SE)** – a fixed topology system (out of which NEAT outperforms the latter two on CartPole) – **Parallel Distributed Genetic Programming (PDGP)**, and **Generalized Acquisition of Recurrent Links (GNARL)**.

We will now see an exercise on applying NEAT to solve a simple XNOR gate, a logic gate that has a binary output. The binary inputs and output are quantified using a truth table, which is a representation of the sets of the functional values of Boolean logic expressions showcasing the combination of the logical values.

EXERCISE 12.08: XNOR GATE FUNCTIONALITY USING NEAT

In the exercise, you will see the impact that NEAT has on solving a simple Boolean algebra problem. The problem involves implementing the NEAT algorithm to identify the optimal neural network topology for reproducing the binary output of an exclusive NOR (XNOR) gate. This is a type of logic gate where, when both inputs have the same signal (either 0 or 1 – equivalent to off and on, respectively), the output of the logic gate will be 1 (on), whereas when one of the inputs is high (1) and the other is low (0), the output will be 0 (off).

We have the following truth table for the XNOR logic gate:

X (Input 1)	Y (Input 2)	Z (Output)
0	0	1
0	1	0
1	0	0
1	1	1

Figure 12.13: Truth table for the XNOR gate

Use the NEAT algorithm to create a feedforward neural network that can mimic the output of an XNOR gate.

Perform the following steps to complete the exercise:

1. In your Anaconda environment, execute the following command:

```
conda install neat
```

2. Create a new Jupyter Notebook.

3. Import **print_function** from the **__future__** file, and import the **neat** and **os** packages:

```
from __future__ import print_function
import os
import neat
```

4. Initialize the inputs and the output of the XNOR gate based on the truth table:

```
xnor_inputs = [(0.0, 0.0), (0.0, 1.0), (1.0, 0.0), (1.0, 1.0)]

xnor_output = [(1.0,),(0.0,),(0.0,),(1.0,)]
```

5. Create a fitness function that uses the squared difference between the actual output and the output of a feedforward neural network using NEAT:

```
def fitness_function(chromosomes, configuration):
    for ch_id, chromosome in chromosomes:
        chromosome.fitness = 4.0
        neural_net = neat.nn.FeedForwardNetwork.create\
                    (chromosome, configuration)
        for xnor_i,xnor_o in zip(xnor_inputs, xnor_output):
            output = neural_net.activate(xnor_i)
            squared_diff = (output[0] - xnor_o[0])**2
            chromosome.fitness -= squared_diff
```

6. Create a new text file with the name **config-feedforward-xnor**. Include in the file the following parameters for the NEAT algorithm. For the fitness function, select the maximal value, with a threshold close to **4** and a population size of **200**:

```
[NEAT]
fitness_criterion     = max
fitness_threshold     = 3.9
pop_size              = 200
reset_on_extinction   = False
```

7. In the same **config-feedforward-xnor** file, include the **sigmoid** function for node activation with a mutation rate of **0.01**. The aggregation options are mostly about adding the values, with a mutation rate of 0 for aggregation:

```
[DefaultGenome]
# activation options of the nodes
activation_default       = sigmoid
activation_mutate_rate   = 0.01
activation_options       = sigmoid


# aggregation options for the node
aggregation_default      = sum
aggregation_mutate_rate  = 0.0
aggregation_options      = sum
```

8. Set the **bias** parameters for the algorithm:

```
# bias options for the node
bias_init_mean          = 0.0
bias_init_stdev         = 0.05
bias_max_value          = 30.0
bias_min_value          = -30.0
bias_mutate_power       = 0.5
bias_mutate_rate        = 0.8
bias_replace_rate       = 0.1
```

For the bias, the minimum and maximum values are **−30** and **30**. Set the initial standard deviation at **0.05**, as low as possible, with a power of **0.5**, a mutation rate of **0.8**, and a replacement rate of **0.1**. These values are essential for implementing the genetic algorithm optimization.

9. Define the coefficients c_1, c_3, as we are only considering the difference between the genes (how disjointed they are) and the difference in weights:

```
# compatibility options for the genes in the chromosome
compatibility_disjoint_coefficient = 1.0
compatibility_weight_coefficient   = 0.5
```

10. Include the information about topology, connection, and node inclusion or removal-related parameters:

```
# add/remove rates for connections between nodes
conn_add_prob           = 0.5
conn_delete_prob        = 0.5

# connection enable options
enabled_default         = True
enabled_mutate_rate     = 0.01

feed_forward            = True
initial_connection      = full

# add/remove rates for nodes
node_add_prob           = 0.2
node_delete_prob        = 0.2
```

11. Start with a simple network without any hidden layers and set the response parameters for the nodes and connections:

```
# network parameters
num_hidden              = 0
num_inputs              = 2
num_outputs             = 1

# node response options
response_init_mean      = 1.0
response_init_stdev     = 0.0
response_max_value      = 30.0
response_min_value      = -30.0
response_mutate_power   = 0.0
response_mutate_rate    = 0.0
response_replace_rate   = 0.0

# connection weight options
weight_init_mean        = 0.0
weight_init_stdev       = 1.0
weight_max_value        = 30
```

```
weight_min_value        = -30
weight_mutate_power     = 0.5
weight_mutate_rate      = 0.9
weight_replace_rate     = 0.15
```

12. Select the default parameters for the distance threshold, species fitness function, and parent selection. This is the final set of parameters to be included in the **config-feedforward-xnor** file:

```
[DefaultSpeciesSet]
compatibility_threshold = 3.0


[DefaultStagnation]
species_fitness_func = max
max_stagnation       = 20
species_elitism      = 2


[DefaultReproduction]
Elitism              = 2
survival_threshold = 0.2
```

13. Now, in the main code file, use the **config-feedforward-xnor** file to configure the NEAT formulation of the neural network and output each configuration of the network within **Exercise 12.08**:

```
#load configuration
configuration = neat.Config(neat.DefaultGenome, \
                            neat.DefaultReproduction, \
                            neat.DefaultSpeciesSet, \
                            neat.DefaultStagnation,\
                            "../Dataset/config-feedforward-xnor")
print("Output of file configuration:", configuration)
```

The output will be as follows:

```
Output of file configuration: <neat.config.Config object at
0x0000017618944AC8>
```

14. Get the population based on the configuration of the NEAT algorithm and include the progress to the terminal to monitor the statistical differences:

```
#load the population size
pop = neat.Population(configuration)

#add output for progress in terminal
pop.add_reporter(neat.StdOutReporter(True))
statistics = neat.StatisticsReporter()
pop.add_reporter(statistics)
pop.add_reporter(neat.Checkpointer(5))
```

15. Run the algorithm for **200** generations and select the best solution for the neural network topology:

```
#run for 200 generations using
best = pop.run(fitness_function, 200)
#display the best chromosome
print('\n Best chromosome:\n{!s}'.format(best))
```

The output will be similar to the following:

```
****** Running generation 0 ******

Population's average fitness: 2.45675 stdev: 0.36807
Best fitness: 2.99412 - size: (1, 2) - species 1 - id 28
Average adjusted fitness: 0.585
Mean genetic distance 0.949, standard deviation 0.386
Population of 200 members in 1 species:
    ID   age   size   fitness   adj fit   stag
   ====  ===  ====  =======  =======  ====
     1    0    200      3.0     0.585     0
Total extinctions: 0
Generation time: 0.030 sec


 ****** Running generation 1 ******

Population's average fitness: 2.42136 stdev: 0.28774
Best fitness: 2.99412 - size: (1, 2) - species 1 - id 28
Average adjusted fitness: 0.589
Mean genetic distance 1.074, standard deviation 0.462
Population of 200 members in 1 species:
```

```
    ID    age   size   fitness   adj fit   stag

   ====   ===   ====   =======   =======   ====

     1     1    200       3.0     0.589      1
Total extinctions: 0
Generation time: 0.032 sec (0.031 average)
```

16. Use functions to compare the output of the neural network with the desired output:

```
#show output of the most fit chromosome against the data
print('\n Output:')
best_network = neat.nn.FeedForwardNetwork.create\
                (best, configuration)
for xnor_i, xnor_o in zip(xnor_inputs, xnor_output):
    output = best_network.activate(xnor_i)
    print("input{!r}, expected output {!r}, got: {:.1f}"\
          .format(xnor_i,xnor_o,output[0]))
```

The output will be as follows:

```
Output:
input(0.0, 0.0), expected output (1.0,), got: 0.9
input(0.0, 1.0), expected output (0.0,), got: 0.0
input(1.0, 0.0), expected output (0.0,), got: 0.2
input(1.0, 1.0), expected output (1.0,), got: 0.9
```

17. Run the code and you will get a similar output to what you see here. As the chromosomes are populated randomly, the algorithm will converge to a nearly optimal solution in a different number of generations for you:

```
****** Running generation 41 ******


Population's average fitness: 2.50036 stdev: 0.52561
Best fitness: 3.97351 - size: (8, 16) - species 2 - id 8095


Best individual in generation 41 meets fitness threshold \
- complexity: (8, 16)


Best chromosome:
Key: 8095
Fitness: 3.9735119749933214
```

```
Nodes:
    0 DefaultNodeGene(key=0, bias=-0.02623087593563278, \
                    response=1.0, activation=sigmoid, \
                    aggregation=sum)
    107 DefaultNodeGene(key=107, bias=-1.5209385195946818, \
                    response=1.0, activation=sigmoid, \
                    aggregation=sum) [...]

Connections:
    DefaultConnectionGene(key=(-2, 107), \
                    weight=1.8280370376000628, \
                    enabled=True)
    DefaultConnectionGene(key=(-2, 128), \
                    weight=0.08641968818530771, \
                    enabled=True)
    DefaultConnectionGene(key=(-2, 321), \
                    weight=1.2366021868005421, \
                    enabled=True) [...]
```

By running this experiment, you can see that the conversion to a nearly optimal solution happened in less than the maximum number of generations (**200**). The output of the feedforward neural network is nearly optimal, as the values are integers. Their values are close to 1 and 0. You can also observe that from a neural network with no hidden layers, the ANN has evolved to have **1149** nodes with various connections.

> **NOTE**
>
> To access the source code for this specific section, please refer to https://packt.live/2XTBs0M.
>
> This section does not currently have an online interactive example, and will need to be run locally.

In this section, the NEAT algorithm, a neuroevolution algorithm that varies the topology of neural networks, was presented. What sets the NEAT algorithm apart from alternative TWEANNs is the way in which mutation, crossover, and selection take place to optimize the structure of the neural network, starting from a simple network with no hidden layers and evolving into a more complex one with an increased number of nodes and connections.

This exercise, which involved implementing NEAT to reproduce the output of an XNOR logic gate, enabled you to understand the structure of the NEAT algorithm and analyze the benefits and implications of applying neuroevolutionary techniques as alternatives to simple electronic problems. In the next section, you will test your programming abilities and your knowledge of GAs by solving the cart-pole problem.

ACTIVITY 12.01: CART-POLE ACTIVITY

Automatic control is a challenge, especially when operating specific equipment using robotic arms or carts that are transporting equipment on a shop floor. This problem is often generalized as the cart-pole problem. You are going to program an automated cart to balance a pole. The goal is to maximize the time that the pole is balanced for. To solve this problem, an agent can use a neural network for the state-action mapping. The challenge lies in identifying the structure of the neural network and a solution for determining the optimal values for the weights, bias, and number of neurons for each layer of the neural network. We will use a GA to identify the best values for these parameters.

This activity aims to implement a GA for parameter selection for an ANN that, after 20 generations, can obtain a high average score for 500 trials. You will output the average scores for both the generations and the episodes, and you will monitor the convergence to an optimal policy by tuning the parameters of the neural network using a genetic algorithm in the form of a graph. This activity has the purpose of testing your programming abilities by implementing concepts from previous chapters and the current one. The following are the steps needed to implement this activity:

1. Create a Jupyter Notebook file and import the appropriate packages as follows:

```
import gym
import numpy as np
import math
import tensorflow as tf
from matplotlib import pyplot as plt
from random import randint
from statistics import median, mean
```

2. Initialize the environment and the state and action space shapes.

3. Create a function to generate randomly selected initial network parameters.

4. Create a function to generate the neural network using the set of parameters.

5. Create a function to get the total reward for 300 steps when using the neural network.

6. Create a function to get the fitness scores for each element of the population when running the initial random selection.

7. Create a mutation function.

8. Create a single-point crossover function.

9. Create a function for the next-generation creation by selecting the pair with the highest rewards.

10. Select the parameters within the function to construct the neural network that adds the parameters.

11. Build the neural network using the identified parameters and obtain a new reward based on the constructed neural network.

12. Create a function to output the convergence graph.

13. Create a function for the genetic algorithm that outputs the parameters of the neural network based on the highest average reward.

14. Create a function that decodes the array of parameters to each neural network parameter.

15. Set the generations to 50, the number of trial tests to 15, and the number of steps and trials to 500. You will get a similar output to this (only the first few lines are displayed here):

```
Generation:1, max reward:11.0
Generation:2, max reward:11.0
Generation:3, max reward:10.0
Generation:4, max reward:10.0
Generation:5, max reward:11.0
Generation:6, max reward:10.0
Generation:7, max reward:10.0
Generation:8, max reward:10.0
Generation:9, max reward:11.0
Generation:10, max reward:10.0
Generation:11, max reward:10.0
Generation:12, max reward:10.0
Generation:13, max reward:10.0
Generation:14, max reward:10.0
Generation:15, max reward:10.0
Generation:16, max reward:10.0
Generation:17, max reward:10.0
```

```
Generation:18, max reward:10.0
Generation:19, max reward:11.0
Generation:20, max reward:11.0
```

The plot of rewards against generations will be similar to the following:

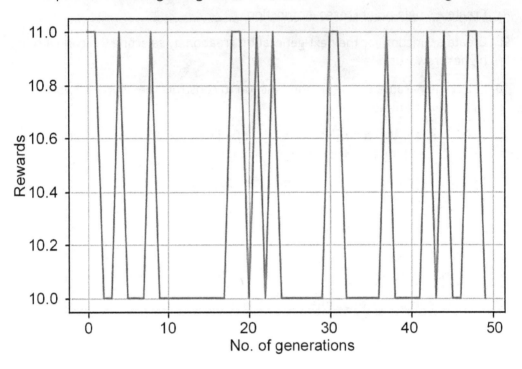

Figure 12.14: Rewards obtained over the generations

The output for the average rewards (just the last few lines are shown here) will be similar to the following:

```
Trial:486, total reward:8.0
Trial:487, total reward:9.0
Trial:488, total reward:10.0
Trial:489, total reward:10.0
Trial:490, total reward:8.0
Trial:491, total reward:9.0
Trial:492, total reward:9.0
Trial:493, total reward:10.0
Trial:494, total reward:10.0
Trial:495, total reward:9.0
Trial:496, total reward:10.0
Trial:497, total reward:9.0
```

```
Trial:498, total reward:10.0
Trial:499, total reward:9.0
Average reward: 9.384
```

> **NOTE**
>
> The solution to this activity can be found on page 774.

SUMMARY

In this chapter, you have explored gradient-based and gradient-free methods of algorithm optimization, with an emphasis on the potential of evolutionary algorithms – in particular, GAs – to solve optimization problems, such as sub-optimal solutions, using a nature-inspired approach. GAs consist of specific elements, such as population generation, parent selection, parent reproduction or crossover, and finally mutation occurrence, which they use to create a binary optimal solution.

Then, the use of GAs for hyperparameter tuning and selection for neural networks was explored, helping us to find the most suitable window size and unit number. We saw implementations of state-of-the-art algorithms that combined deep neural networks and evolutionary strategies, such as NEAT for XNOR output estimation. Finally, you had a chance to implement what was studied in this chapter through an OpenAI Gym cart-pole simulation, where we examined the application of GAs for parameter tuning with action selection using a deep neural network.

The development of hybrid methods in RL systems is one of the most recent optimization developments. You have developed and implemented optimization methods for model-free RL systems. In the bonus chapter (which is available on the interactive version of the workshop at courses.packtpub.com), you will be exploring model-based RL methods and state-of-the-art advances in deep RL for control systems that can be applied in the robotics, manufacturing, and transportation fields.

You are now capable of applying the concepts that you learned about in this book using various coding techniques and various models that can help further enhance your field of expertise and potentially bring new changes and advancements. Your journey has just begun – you have taken the first steps to deciphering the world of RL, and you now have the tools to enhance your Python programming skills for RL, all of which you can independently apply.

APPENDIX

CHAPTER 1: INTRODUCTION TO REINFORCEMENT LEARNING

ACTIVITY 1.01: MEASURING THE PERFORMANCE OF A RANDOM AGENT

1. Import the required libraries – **abc**, **numpy**, and **gym**:

```
import abc
import numpy as np
import gym
```

2. Define the abstract class representing the agent:

```
"""
Abstract class representing the agent
Init with the action space and the function pi returning the action
"""


class Agent:
    def __init__(self, action_space: gym.spaces.Space):
        """
        Constructor of the agent class.

        Args:
            action_space (gym.spaces.Space): environment action space
        """
        raise NotImplementedError("This class cannot be
instantiated.")

    @abc.abstractmethod
    def pi(self, state: np.ndarray) -> np.ndarray:
        """
        Agent's policy.

        Args:
            state (np.ndarray): environment state

        Returns:
            The selected action
        """
        pass
```

An agent is represented by only a constructor and an abstract method, **pi**. This method is the actual policy; it takes as input the environment state and returns the selected action.

3. Define a continuous agent. A continuous agent has to initialize the probability distribution according to the action space passed as an input to the constructor:

```
class ContinuousAgent(Agent):
    def __init__(self, action_space: gym.spaces.Space, seed=46):
        # setup seed
        np.random.seed(seed)
        # check the action space type
        if not isinstance(action_space, gym.spaces.Box):
            raise ValueError\
                ("This is a Continuous Agent pass as "\
                "input a Box Space.")
```

4. If the upper and lower bounds are infinite, the probability distribution is simply a normal distribution centered at 0, with a scale that is equal to 1:

```
        """
        initialize the distribution according to the action space
type
        """

        if (action_space.low == -np.inf) and \
            (action_space.high == np.inf):
            # the distribution is a normal distribution
            self._pi = lambda: np.random.normal\
                            (loc=0, scale=1, \
                            size=action_space.shape)
        return
```

5. If the upper and lower bounds are both finite, the distribution is a uniform distribution defined in that range:

```
if (action_space.low != -np.inf) and \
  (action_space.high != np.inf):
   # the distribution is a uniform distribution
   self._pi = lambda: np.random.uniform\
              (low=action_space.low, \
               high=action_space.high, \
               size=action_space.shape)
   return
```

If the lower bound is $-\infty$, the probability distribution is a shifted negative exponential distribution:

```
if action_space.low == -np.inf:
   # negative exponential distribution
   self._pi = (lambda: -np.random.exponential\
               (size=action_space.shape)
               + action_space.high)
   return
```

If the upper bound is $+\infty$, the probability distribution is a shifted exponential distribution:

```
if action_space.high == np.inf:
   # exponential distribution
   self._pi = (lambda: np.random.exponential\
               (size=action_space.shape)
               + action_space.low)
   return
```

6. Define the `pi` method, which is simply a call to the distribution defined in the constructor:

```
def pi(self, observation: np.ndarray) -> np.ndarray:
   """
   Policy: simply call the internal _pi().

   This is a random agent, so the action is independent
   from the observation.
```

```
        For real agents the action depends on the observation.
        """
        return self._pi()
```

7. We are ready to define the discrete agent. As before, the agent has to correctly initialize the action distribution according to the action space that is passed as a parameter:

```
class DiscreteAgent(Agent):
    def __init__(self, action_space: gym.spaces.Space, seed=46):
        # setup seed
        np.random.seed(seed)
        # check the action space type
        if not isinstance(action_space, gym.spaces.Discrete):
            raise ValueError("This is a Discrete Agent pass "\
                             "as input a Discrete Space.")

        """
        initialize the distribution according to the action
        space n attribute
        """
        # the distribution is a uniform distribution
        self._pi = lambda: np.random.randint\
                (low=0, high=action_space.n)

    def pi(self, observation: np.ndarray) -> np.ndarray:
        """
        Policy: simply call the internal _pi().

        This is a random agent, so the action is independent
        from the observation.

        For real agents the action depends on the observation.
        """
        return self._pi()
```

8. Now it is useful to define a utility function to create the correct agent type based on the action space:

```
def make_agent(action_space: gym.spaces.Space, seed=46):
    """

    Returns the correct agent based on the action space type
    """
    if isinstance(action_space, gym.spaces.Discrete):
        return DiscreteAgent(action_space, seed)
    if isinstance(action_space, gym.spaces.Box):
        return ContinuousAgent(action_space, seed)
    raise ValueError("Only Box spaces or Discrete Spaces "\
                     "are allowed, check the action space of "\
                     "the environment")
```

9. The last step is to define the RL loop in which the agent interacts with the environment and collects rewards.

   Define the parameters, and then create the environment and the agent:

```
# Environment Name
env_name = "CartPole-v0"
# Number of episodes
episodes = 10
# Number of Timesteps of each episode
timesteps = 100
# Discount factor
gamma = 1.0
# seed environment
seed = 46


# Needed to show the environment in a notebook
from gym import wrappers
env = gym.make(env_name)
env.seed(seed)
# the last argument is needed to record all episodes
# otherwise gym would record only some of them
# The monitor saves the episodes inside the folder ./gym-results
env = wrappers.Monitor(env, "./gym-results", force=True, \
                       video_callable=lambda episode_id: True)

agent = make_agent(env.action_space, seed)
```

10. We have to track the returns for each episode; to do this, we can use a simple list:

```
# list of returns
episode_returns = []
```

11. Start a loop for each episode:

```
# loop for the episodes
for episode_number in range(episodes):
    # here we are inside an episode
```

12. Initialize the variables for the calculation of the cumulated discount factor and the current episode return:

```
# reset cumulated gamma
gamma_cum = 1


# return of the current episode
episode_return = 0
```

13. Reset the environment and get the first observation:

```
# the reset function resets the environment and returns
# the first environment observation
observation = env.reset()
```

14. Loop for the number of timesteps:

```
# loop for the given number of timesteps or
# until the episode is terminated
for timestep_number in range(timesteps):
```

15. Render the environment, select the action, and then apply it:

```
# if you want to render the environment
# uncomment the following line
# env.render()


# select the action
action = agent.pi(observation)


# apply the selected action by calling env.step
observation, reward, done, info = env.step(action)
```

16. Increment the return, and calculate the cumulated discount factor:

```
# increment the return
episode_return += reward * gamma_cum


# update the value of cumulated discount factor
gamma_cum = gamma_cum * gamma
```

17. If the episode is terminated, break from the timestep's loop:

```
"""
if done the episode is terminated, we have to reset
the environment
"""
if done:
    print(f"Episode Number: {episode_number}, \
Timesteps: {timestep_number}, Return: {episode_return}")
    # break from the timestep loop
    break
```

18. After the timestep loop, we have to record the current return by appending it to the list of returns for each episode:

```
episode_returns.append(episode_return)
```

19. After the episode loop, close the environment and calculate **statistics**:

```
# close the environment
env.close()


# Calculate return statistics
avg_return = np.mean(episode_returns)
std_return = np.std(episode_returns)
var_return = std_return ** 2  # variance is std^2


print(f"Statistics on Return: Average: {avg_return}, \
Variance: {var_return}")
```

You will get the following results:

```
Episode Number: 0, Timesteps: 27, Return: 28.0
Episode Number: 1, Timesteps: 9, Return: 10.0
Episode Number: 2, Timesteps: 13, Return: 14.0
Episode Number: 3, Timesteps: 16, Return: 17.0
Episode Number: 4, Timesteps: 31, Return: 32.0
Episode Number: 5, Timesteps: 10, Return: 11.0
Episode Number: 6, Timesteps: 14, Return: 15.0
Episode Number: 7, Timesteps: 11, Return: 12.0
Episode Number: 8, Timesteps: 10, Return: 11.0
Episode Number: 9, Timesteps: 30, Return: 31.0
Statistics on Return: Average: 18.1, Variance: 68.89000000000001
```

In this activity, we implemented two different types of agents: a discrete agent, working with discrete environments, and a continuous agent, working with continuous environments.

Additionally, you can render the episodes inside a notebook using the following code:

```
# Render the episodes
import io
import base64
from IPython.display import HTML, display

episodes_to_watch = 1
for episode in range(episodes_to_watch):
    video = io.open(f"./gym-results/openaigym.video\
.{env.file_infix}.video{episode:06d}.mp4", "r+b").read()
    encoded = base64.b64encode(video)
    display(
        HTML(
            data="""
        <video width="360" height="auto" alt="test" controls>
        <source src="data:video/mp4;base64,{0}" type="video/mp4" />
        </video>""".format(
                encoded.decode("ascii")
            )
        )
    )
```

You can see the episode duration is not too long. This is because the actions are taken at random, so the pole falls after some timesteps.

> **NOTE**
>
> To access the source code for this specific section, please refer to https://packt.live/3fbxR3Y.
>
> This section does not currently have an online interactive example and will need to be run locally.

Discrete and continuous agents are two different possibilities when facing a new RL problem.

We have designed our agents in a very flexible way so that they can be applied to almost all environments without having to change the code.

We also implemented a simple RL loop and measured the performance of our agent on a classical RL problem.

CHAPTER 2: MARKOV DECISION PROCESSES AND BELLMAN EQUATIONS

ACTIVITY 2.01: SOLVING GRIDWORLD

1. Import the required libraries:

```
from enum import Enum, auto
import matplotlib.pyplot as plt
import numpy as np
from scipy import linalg
from typing import Tuple
```

2. Define the **visualization** function:

```
# helper function
def vis_matrix(M, cmap=plt.cm.Blues):
    fig, ax = plt.subplots()
    ax.matshow(M, cmap=cmap)
    for i in range(M.shape[0]):
        for j in range(M.shape[1]):
            c = M[j, i]
            ax.text(i, j, "%.2f" % c, va="center", ha="center")
```

3. Define the possible actions:

```
# Define the actions
class Action(Enum):
    UP = auto()
    DOWN = auto()
    LEFT = auto()
    RIGHT = auto()
```

4. Define the **Policy** class, representing the random policy:

```
# Agent Policy, random
class Policy:
    def __init__(self):
        self._possible_actions = [action for action in Action]
        self._action_probs = {a: 1 / len(self._possible_actions) \
                              for a in self._possible_actions}

    def __call__(self, state: Tuple[int, int], \
                action: Action) -> float:
        """
```

```
        Returns the action probability
        """
        assert action in self._possible_actions
        # state is unused for this policy
        return self._action_probs[action]
```

5. Define the **Environment** class and the **step** function:

```
class Environment:
    def __init__(self):
        self.grid_width = 5
        self.grid_height = 5
        self._good_state1 = (0, 1)
        self._good_state2 = (0, 3)
        self._to_state1 = (4, 2)
        self._to_state2 = (2, 3)
        self._bad_state1 = (1, 1)
        self._bad_state2 = (4, 4)
        self._bad_states = [self._bad_state1, self._bad_state2]
        self._good_states = [self._good_state1, self._good_state2]
        self._to_states = [self._to_state1, self._to_state2]
        self._good_rewards = [10, 5]

    def step(self, state, action):
        i, j = state

        # search among good states
        for good_state, reward, \
            to_state in zip(self._good_states, \
                            self._good_rewards, \
                            self._to_states):
            if (i, j) == good_state:
                return (to_state, reward)
        reward = 0

        # if the state is a bad state, the reward is -1
        if state in self._bad_states:
            reward = -1
```

```
        # calculate next state based on the action
        if action == Action.LEFT:
            j_next = max(j - 1, 0)
            i_next = i
            if j - 1 < 0:
                reward = -1
        elif action == Action.RIGHT:
            j_next = min(j + 1, self.grid_width - 1)
            i_next = i
            if j + 1 > self.grid_width - 1:
                reward = -1
        elif action == Action.UP:
            j_next = j
            i_next = max(i - 1, 0)
            if i - 1 < 0:
                reward = -1
        elif action == Action.DOWN:
            j_next = j
            i_next = min(i + 1, self.grid_height - 1)
            if i + 1 > self.grid_height - 1:
                reward = -1
        else:
            raise ValueError("Invalid action")
        return ((i_next, j_next), reward)
```

6. Loop for all states and actions and build the transition and reward matrices:

```
pi = Policy()
env = Environment()

# setup probability matrix and reward matrix
P = np.zeros((env.grid_width*env.grid_height, \
            env.grid_width*env.grid_height))
R = np.zeros_like(P)
possible_actions = [action for action in Action]
```

```
# Loop for all states and fill up P and R
for i in range(env.grid_height):
    for j in range(env.grid_width):
        state = (i, j)
        # loop for all action and setup P and R
        for action in possible_actions:
            next_state, reward = env.step(state, action)
            (i_next, j_next) = next_state
            P[i*env.grid_width+j, \
              i_next*env.grid_width \
              + j_next] += pi(state, action)
            """
            the reward depends only on the starting state and
            the final state
            """
            R[i*env.grid_width+j, \
              i_next*env.grid_width + j_next] = reward
```

7. Check the correctness of the matrix:

```
# check the correctness
assert((np.sum(P, axis=1) == 1).all())
```

8. Calculate the expected reward for each state:

```
# expected reward for each state
R_expected = np.sum(P * R, axis=1, keepdims=True)
```

9. Use the function to visualize the expected reward:

```
# reshape the state values in a matrix
R_square = R_expected.reshape((env.grid_height,env.grid_width))
# Visualize
vis_matrix(R_square, cmap=plt.cm.Reds)
```

The function visualizes the matrix using Matplotlib. You should see something similar to this:

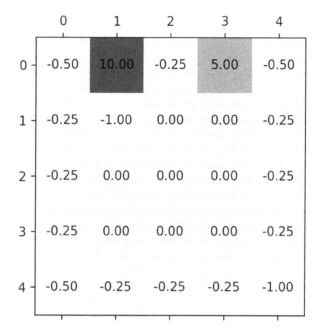

Figure 2.62: The expected reward for each state

The previous figure is a color representation of the expected reward associated with each state considering the current policy. Notice that the expected reward of bad states is exactly equal to **-1**. The expected reward of good states is exactly equal to **10** and **5**, respectively.

10. Now set up the matrix form of the Bellman expectation equation:

```
# define the discount factor
gamma = 0.9
# Now it is possible to solve the Bellman Equation
A = np.eye(env.grid_width*env.grid_height) - gamma * P
B = R_expected
```

11. Solve the Bellman equation:

```
# solve using scipy linalg
V = linalg.solve(A, B)
```

12. Visualize the result:

```
# reshape the state values in a matrix
V_square = V.reshape((env.grid_height,env.grid_width))
# visualize results
vis_matrix(V_square, cmap=plt.cm.Reds)
```

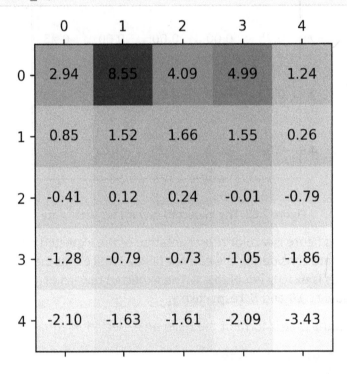

Figure 2.63: State values of Gridworld

Note that the value of good states is less than the expected reward from those states. This is because landing states have an expected reward that is negative or because landing states are close to states for which the reward is negative. You can see that the state with the higher value is state G_1, followed by state G_2. It is also interesting to note the high value of the state in position (**0, 2**), which is close to the good states.

> **NOTE**
>
> To access the source code for this specific section, please refer to https://packt.live/2Al9xOB.
>
> You can also run this example online at https://packt.live/2UChxBy.

In this activity, we experimented with the Gridworld environment, one of the most common toy RL environments. We defined a random policy, and we solved the Bellman expectation equation using `scipy.linalg.solve` to find the state values of the policy.

It is important to visualize the results, when possible, to get a better understanding and to spot any errors.

CHAPTER 3: DEEP LEARNING IN PRACTICE WITH TENSORFLOW 2

ACTIVITY 3.01: CLASSIFYING FASHION CLOTHES USING A TENSORFLOW DATASET AND TENSORFLOW 2

1. Import all the required modules:

```
from __future__ import absolute_import, division, \
print_function, unicode_literals

import numpy as np
import matplotlib.pyplot as plt


# TensorFlow
import tensorflow as tf
import tensorflow_datasets as tfds
```

2. Import the Fashion MNIST dataset using TensorFlow datasets and split it into train and test splits. Then, create a list of classes:

```
# Construct a tf.data.Dataset
(train_images, train_labels), (test_images, test_labels) = \
tfds.as_numpy(tfds.load('fashion_mnist', \
                        split=['train', 'test'],\
                        batch_size=-1, as_supervised=True,))

train_images = np.squeeze(train_images)
test_images = np.squeeze(test_images)

classes = ['T-shirt/top', 'Trouser', 'Pullover', 'Dress', \
           'Coat','Sandal', 'Shirt', 'Sneaker', 'Bag', \
           'Ankle boot']
```

3. Explore the dataset to get familiar with the input features, that is, shapes, labels, and classes:

```
print("Training dataset shape =", train_images.shape)
print("Training labels length =", len(train_labels))
print("Some training labels =", train_labels[:5])
print("Test dataset shape =", test_images.shape)
print("Test labels length =", len(test_labels))
```

The output will be as follows:

```
Training dataset shape = (60000, 28, 28)
Training labels length = 60000
Some training labels = [2 1 8 4 1]
Test dataset shape = (10000, 28, 28)
Test labels length = 10000
```

4. Visualize some instances of the training set.

   It is also useful to take a look at how the images will appear. The following code snippet shows the first training set instance:

```
plt.figure()
plt.imshow(train_images[0])
plt.colorbar()
plt.grid(False)
plt.show()
```

The output image will be as follows:

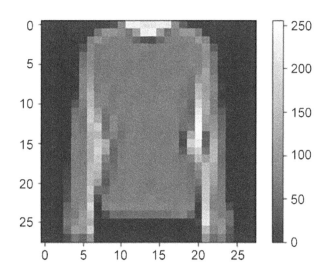

Figure 3.30: First training image plot

5. Perform feature normalization:

```
train_images = train_images / 255.0
test_images = test_images / 255.0
```

6. Now, let's take a look at some instances of our training set by plotting **25** of them with their corresponding labels:

```
plt.figure(figsize=(10,10))
for i in range(25):
    plt.subplot(5,5,i+1)
    plt.xticks([])
    plt.yticks([])
    plt.grid(False)
    plt.imshow(train_images[i], cmap=plt.cm.binary)
    plt.xlabel(classes[train_labels[i]])
plt.show()
```

The output image will be as follows:

Figure 3.31: A set of 25 training samples and their corresponding labels

7. Build the classification model. First, create a model using a layers' sequence:

```
model = tf.keras.Sequential\
        ([tf.keras.layers.Flatten(input_shape=(28, 28)),\
          tf.keras.layers.Dense(128, activation='relu'),\
          tf.keras.layers.Dense(10)])
```

8. Then, associate the model with an **optimizer**, a **loss** function, and a **metrics**:

```
model.compile(optimizer='adam',\
              loss=tf.keras.losses.SparseCategoricalCrossentropy\
              (from_logits=True), metrics=['accuracy'])
```

9. Train the deep neural network:

```
model.fit(train_images, train_labels, epochs=10)
```

The last output lines will be as follows:

```
Epoch 9/1060000/60000 [===============================] \
- 2s 40us/sample - loss: 0.2467 - accuracy: 0.9076
Epoch 10/1060000/60000 [===============================] \
- 2s 40us/sample - loss: 0.2389 - accuracy: 0.9103
```

10. Test the model's accuracy. The accuracy should be in excess of 88%.

11. Evaluate the model on the test set and print the accuracy score:

```
test_loss, test_accuracy = model.evaluate\
                          (test_images, test_labels, verbose=2)
print('\nTest accuracy:', test_accuracy)
```

The output will be as follows:

```
10000/10000 - 0s - loss: 0.3221 - accuracy: 0.8878
Test accuracy: 0.8878
```

> **NOTE**
>
> The accuracy may show slightly different values due to random sampling with a variable random seed.

12. Perform inference and check the predictions against the ground truth.

    As a first step, add a **softmax** layer to the model so that it outputs probabilities instead of logits. Then, print out the probabilities of the first test instance with the following code:

```
probability_model = tf.keras.Sequential\
                    ([model,tf.keras.layers.Softmax()])
predictions = probability_model.predict(test_images)
print(predictions[0:3])
```

    The output will be as follows:

```
[[3.85897374e-06 2.33953915e-06 2.30801385e-02 4.74092474e-07
  9.55752671e-01 1.56392260e-10 2.11589299e-02 8.57651870e-08
  1.49855202e-06 1.05843508e-10]
```

13. Next, compare one model prediction (that is, the class with the highest predicted probability), the one on the first test instance, with its ground truth:

```
print("Class ID, predicted | real =", \
      np.argmax(predictions[0]), "|", test_labels[0])
```

    The output will be as follows:

```
Class ID, predicted | real = 4 | 4
```

14. In order to perform a comparison that's even clearer, create the following two functions. The first one plots the **i**-th test set instance image with a caption showing the predicted class with the highest probability, its probability in percent, and the ground truth between round brackets. This caption will be **blue** for correct predictions, and **red** for incorrect ones:

```
def plot_image(i, predictions_array, true_label, img):
    predictions_array, true_label, img = predictions_array,\
                                         true_label[i], img[i]
    plt.grid(False)
    plt.xticks([])
    plt.yticks([])

    plt.imshow(img, cmap=plt.cm.binary)

    predicted_label = np.argmax(predictions_array)
    if predicted_label == true_label:
        color = 'blue'
```

```
    else:
        color = 'red'
    plt.xlabel("{} {:2.0f}% ({})".format\
                (classes[predicted_label], \
                 100*np.max(predictions_array),\
                 classes[true_label]),\
                 color=color)
```

15. The second function creates a second image showing a bar plot of all classes' predicted probabilities. It will color the highest probable one in **blue** if the prediction is correct, or in **red** if it is incorrect. In this second case, the bar corresponding to the correct label is colored in **blue**:

```
def plot_value_array(i, predictions_array, true_label):
    predictions_array, true_label = predictions_array,\
                                       true_label[i]
    plt.grid(False)
    plt.xticks(range(10))
    plt.yticks([])
    thisplot = plt.bar(range(10), predictions_array,\
              color="#777777")
    plt.ylim([0, 1])
    predicted_label = np.argmax(predictions_array)

    thisplot[predicted_label].set_color('red')
    thisplot[true_label].set_color('blue')
```

16. Using these two functions, we can examine every instance of the test set. In the following snippet, the first test instance is being plotted:

```
i = 0
plt.figure(figsize=(6,3))
plt.subplot(1,2,1)
plot_image(i, predictions[i], test_labels, test_images)
plt.subplot(1,2,2)
plot_value_array(i, predictions[i],  test_labels)
plt.show()
```

The output will be as follows:

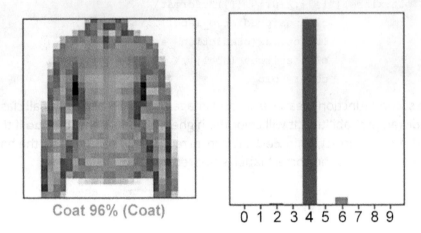

Figure 3.32: First test instance, correctly predicted

17. The very same approach can be used to plot a user-defined number of test instances, arranging the output in subplots, as follows:

```
"""
Plot the first X test images, their predicted labels, and the true
labels.
Color correct predictions in blue and incorrect predictions in red.
"""

num_rows = 5
num_cols = 3
num_images = num_rows*num_cols
plt.figure(figsize=(2*2*num_cols, 2*num_rows))
for i in range(num_images):
    plt.subplot(num_rows, 2*num_cols, 2*i+1)
    plot_image(i, predictions[i], test_labels, test_images)
    plt.subplot(num_rows, 2*num_cols, 2*i+2)
    plot_value_array(i, predictions[i], test_labels)
plt.tight_layout()
plt.show()
```

The output will be as follows:

Figure 3.33: First 25 test instances with their predicted classes and ground truth comparison

NOTE

To access the source code for this specific section, please refer to https://packt.live/3dXv3am.

You can also run this example online at https://packt.live/2Ux5JR5.

In this activity, we faced a problem that is quite similar to a real-world one. We had to deal with complex high dimensional inputs – in our case, grayscale images – and we wanted to build a model capable of autonomously grouping them into 10 different categories. Thanks to the power of deep learning and state-of-the-art machine learning frameworks, we were able to build a fully connected neural network that achieves a classification accuracy in excess of 88%.

CHAPTER 4: GETTING STARTED WITH OPENAI AND TENSORFLOW FOR REINFORCEMENT LEARNING

ACTIVITY 4.01: TRAINING A REINFORCEMENT LEARNING AGENT TO PLAY A CLASSIC VIDEO GAME

1. Import all the required modules from OpenAI Baselines and TensorFlow in order to use the **PPO** algorithm:

```
from baselines.ppo2.ppo2 import learn
from baselines.ppo2 import defaults
from baselines.common.vec_env import VecEnv, VecFrameStack
from baselines.common.cmd_util import make_vec_env, make_env
from baselines.common.models import register
import tensorflow as tf
```

2. Define and register a custom convolutional neural network for the policy network:

```
@register("custom_cnn")
def custom_cnn():
    def network_fn(input_shape, **conv_kwargs):
        """
        Custom CNN
        """
        print('input shape is {}'.format(input_shape))
        x_input = tf.keras.Input\
                (shape=input_shape, dtype=tf.uint8)
        h = x_input
        h = tf.cast(h, tf.float32) / 255.

        h = tf.keras.layers.Conv2D\
            (filters=32,kernel_size=8,strides=4, \
            padding='valid', data_format='channels_last',\
            activation='relu')(h)
        h2 = tf.keras.layers.Conv2D\
            (filters=64, kernel_size=4,strides=2,\
            padding='valid', data_format='channels_last',\
            activation='relu')(h)
        h3 = tf.keras.layers.Conv2D\
            (filters=64, kernel_size=3,strides=1,\
```

```
                padding='valid', data_format='channels_last',\
                activation='relu')(h2)
        h3 = tf.keras.layers.Flatten()(h3)
        h3 = tf.keras.layers.Dense\
            (units=512, name='fc1', activation='relu')(h3)

        network = tf.keras.Model(inputs=[x_input], outputs=[h3])
        network.summary()
        return network

    return network_fn
```

3. Create a function to build the environment in the format required by OpenAI Baselines:

```
def build_env(env_id, env_type):

    if env_type in {'atari', 'retro'}:
        env = make_vec_env(env_id, env_type, 1, None, \
                        gamestate=None, reward_scale=1.0)
        env = VecFrameStack(env, 4)

    else:
        env = make_vec_env(env_id, env_type, 1, None,\
                        reward_scale=1.0,\
                        flatten_dict_observations=True)

    return env
```

4. Build the **PongNoFrameskip-v4** environment, choose the required policy network parameters, and train it:

```
env_id = 'PongNoFrameskip-v0'
env_type = 'atari'
print("Env type = ", env_type)

env = build_env(env_id, env_type)

model = learn(network="custom_cnn", env=env, total_timesteps=1e4)
```

While training, the model produces an output similar to the following (only a few lines have been reported here):

```
Env type =  atari
Logging to /tmp/openai-2020-05-11-16-19-42-770612
input shape is (84, 84, 4)
Model: "model"
_____
Layer (type)                 Output Shape              Param #
=================================================================
input_1 (InputLayer)         [(None, 84, 84, 4)]       0
_____
tf_op_layer_Cast (TensorFlow [(None, 84, 84, 4)]       0
_____
tf_op_layer_truediv (TensorF [(None, 84, 84, 4)]       0
_____
conv2d (Conv2D)              (None, 20, 20, 32)        8224
_____
conv2d_1 (Conv2D)            (None, 9, 9, 64)          32832
_____
conv2d_2 (Conv2D)            (None, 7, 7, 64)          36928
_____
flatten (Flatten)            (None, 3136)              0
_____
fc1 (Dense)                  (None, 512)               1606144
=================================================================
Total params: 1,684,128
Trainable params: 1,684,128
Non-trainable params: 0
_____

-------------------------------------------------
| eplenmean               | 1e+03         |
| eprewmean               | -20           |
| fps                     | 213           |
| loss/approxkl           | 0.00012817292 |
| loss/clipfrac           | 0.0           |
| loss/policy_entropy     | 1.7916294     |
| loss/policy_loss        | -0.00050599687 |
| loss/value_loss         | 0.06880974    |
| misc/explained_variance | 0.000675      |
```

```
| misc/nupdates          | 1                |
| misc/serial_timesteps  | 2048             |
| misc/time_elapsed      | 9.6              |
| misc/total_timesteps   | 2048             |
-------------------------------------------
```

5. Run the trained agent in the environment and print the cumulative reward:

```
obs = env.reset()
if not isinstance(env, VecEnv):
    obs = np.expand_dims(np.array(obs), axis=0)


episode_rew = 0

while True:
    actions, _, state, _ = model.step(obs)
    obs, reward, done, info = env.step(actions.numpy())
    if not isinstance(env, VecEnv):
        obs = np.expand_dims(np.array(obs), axis=0)
    env.render()
    print("Reward = ", reward)
    episode_rew += reward

    if done:
        print('Episode Reward = {}'.format(episode_rew))
        break

env.close()
```

The following lines show the last part of the output:

```
[...]
Reward =   [0.]
Reward =   [0.]
Reward =   [0.]
Reward =   [0.]
Reward =   [0.]
Reward =   [0.]
Reward =   [0.]
Reward =   [0.]
Reward =   [0.]
Reward =   [-1.]
Episode Reward = [-17.]
```

It also renders the environment, showing what happens in the environment in real time:

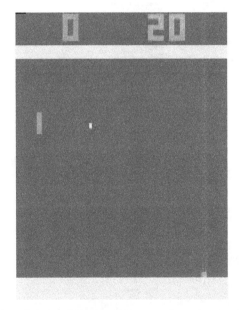

Figure 4.14: One frame of the real-time environment, after rendering

6. Use the built-in OpenAI Baselines run script to train PPO on the **PongNoFrameskip-v0** environment:

```
!python -m baselines.run --alg=ppo2 --env=PongNoFrameskip-v0
--num_timesteps=2e7 --save_path=./models/Pong_20M_ppo2
--log_path=./logs/Pong/
```

The last few lines of the output will be similar to the following:

```
Stepping environment...
-------------------------------------------
| eplenmean               | 867           |
| eprewmean               | -20.8         |
| fps                     | 500           |
| loss/approxkl           | 4.795634e-05  |
| loss/clipfrac           | 0.0           |
| loss/policy_entropy     | 1.7456135     |
| loss/policy_loss        | -0.0005875508 |
| loss/value_loss         | 0.050125826   |
| misc/explained_variance | 0.145         |
| misc/nupdates           | 19            |
| misc/serial_timesteps   | 2432          |
| misc/time_elapsed       | 22            |
| misc/total_timesteps    | 9728          |
-------------------------------------------
```

7. Use the built-in OpenAI Baselines run script to run the trained model on the **PongNoFrameskip-v0** environment:

    ```
    !python -m baselines.run --alg=ppo2 --env=PongNoFrameskip-v0
        --num_timesteps=0 --load_path=./models/Pong_20M_ppo2 --play
    ```

    The output will be similar to the following:

    ```
    episode_rew=-21.0
    episode_rew=-20.0
    episode_rew=-20.0
    episode_rew=-19.0
    ```

8. Use the pretrained weights provided to see the trained agent in action:

    ```
    !wget -O pong_20M_ppo2.tar.gz \
    https://github.com/PacktWorkshops\
    /The-Reinforcement-Learning-Workshop/blob/master\
    /Chapter04/pong_20M_ppo2.tar.gz?raw=true
    ```

The output will be as follows:

```
Saving to: 'pong_20M_ppo2.tar.gz'

pong_20M_ppo2.tar.g 100%[====================>]  17,44M  15,
1MB/s     in 1,2s

2020-05-11 16:19:11 (15,1 MB/s) - 'pong_20M_ppo2.tar.gz' saved
[18284569/18284569]
```

You can read the **.tar** file by using the following command:

```
!tar xvzf pong_20M_ppo2.tar.gz
```

The output will be as follows:

```
pong_20M_ppo2/ckpt-1.data-00000-of-00001
pong_20M_ppo2/ckpt-1.index
pong_20M_ppo2/
pong_20M_ppo2/checkpoint
```

9. Use the built-in OpenAI Baselines run script to train PPO on **PongNoFrameskip-v0**:

```
!python -m baselines.run --alg=ppo2 --env=PongNoFrameskip-v0 --num_
timesteps=0 --load_path=./pong_20M_ppo2 -play
```

> **NOTE**
>
> To access the source code for this specific section, please refer
> to https://packt.live/30yFmOi.
>
> This section does not currently have an online interactive example,
> and will need to be run locally.

In this activity, we learned how to train a state-of-the-art reinforcement learning agent that, by only looking at screen pixels, is able to achieve better-than-human performance when playing a classic Atari video game. We made use of a convolutional neural network to encode environment observations and leveraged the state-of-the-art OpenAI tool to successfully train a PPO algorithm.

CHAPTER 5: DYNAMIC PROGRAMMING

ACTIVITY 5.01: IMPLEMENTING POLICY AND VALUE ITERATION ON THE FROZENLAKE-V0 ENVIRONMENT

1. Import the required libraries:

```
import numpy as np
import gym
```

2. Initialize the environment and reset the current one. Set **is_slippery=False** in the initializer. Show the size of the action space and the number of possible states:

```
def initialize_environment():
    """initialize the OpenAI Gym environment"""
    env = gym.make("FrozenLake-v0", is_slippery=False)
    print("Initializing environment")
    # reset the current environment
    env.reset()
    # show the size of the action space
    action_size = env.action_space.n
    print(f"Action space: {action_size}")
    # Number of possible states
    state_size = env.observation_space.n
    print(f"State space: {state_size}")
    return env
```

3. Perform policy evaluation iterations until the smallest change is less than **smallest_change**:

```
def policy_evaluation(V, current_policy, env, \
                      gamma, small_change):
    """
    Perform policy evaluation iterations until the smallest
    change is less than
    'smallest_change'
    Args:
        V: the value function table
        current_policy: current policy
        env: the OpenAI FrozenLake-v0 environment
        gamma: future reward coefficient
```

```
            small_change: how small should the change be for the
                iterations to stop
        Returns:
            V: the value function after convergence of the evaluation
        """

        state_size = env.observation_space.n

        while True:
            biggest_change = 0
            # loop through every state present
            for state in range(state_size):
                old_V = V[state]
```

4. Take the action according to the current policy:

```
                action = current_policy[state]
                prob, new_state, reward, done = env.env.P[state]\
                                                        [action][0]
```

5. Use the Bellman optimality equation to update $v_{(s)}$:

```
                V[state] = reward + gamma * V[new_state]
                # if the biggest change is small enough then it means
                # the policy has converged, so stop.
                biggest_change = max(biggest_change, abs(V[state] \
                                     - old_V))
            if biggest_change < small_change:
                break
        return V
```

6. Perform policy improvement using the Bellman optimality equation:

```
def policy_improvement(V, current_policy, env, gamma):
    """
    Perform policy improvement using the Bellman Optimality Equation.
    Args:
        V: the value function table
        current_policy: current policy
        env: the OpenAI FrozenLake-v0 environment
        gamma: future reward coefficient
    Returns:
        current_policy: the updated policy
        policy_changed: True, if the policy was changed, else,
```

```
            False
    """
    state_size = env.observation_space.n
    action_size = env.action_space.n
    policy_changed = False
    for state in range(state_size):
        best_val = -np.inf
        best_action = -1
        # loop over all actions and select the best one
        for action in range(action_size):
            prob, new_state, reward, done = env.env.\
                                        P[state][action][0]
```

7. Calculate the future reward by taking this action. Note that we are using the simplified equation because we don't have non-one transition probabilities:

```
        future_reward = reward + gamma * V[new_state]
        if future_reward > best_val:
            best_val = future_reward
            best_action = action
```

8. Using **assert** statements, we can avoid getting into unwanted situations:

```
        assert best_action != -1
        if current_policy[state] != best_action:
            policy_changed = True
```

9. Update the best action for this current state:

```
        current_policy[state] = best_action
    # if the policy didn't change, it means we have converged
    return current_policy, policy_changed
```

10. Find the most optimal policy for the FrozenLake-v0 environment using policy iteration:

```
def policy_iteration(env):
    """
    Find the most optimal policy for the FrozenLake-v0
    environment using Policy
    Iteration
    Args:
        env: FrozenLake-v0 environment
    Returns:
```

```python
        policy: the most optimal policy
    """
    V = dict()
    """
    initially the value function for all states
    will be random values close to zero
    """
    state_size = env.observation_space.n
    for i in range(state_size):
        V[i] = np.random.random()

    # when the change is smaller than this, stop
    small_change = 1e-20
    # future reward coefficient
    gamma = 0.9
    episodes = 0
    # train for these many episodes
    max_episodes = 50000

    # initially we will start with a random policy
    current_policy = dict()
    for s in range(state_size):
        current_policy[s] = env.action_space.sample()

    while episodes < max_episodes:
        episodes += 1
        # policy evaluation
        V = policy_evaluation(V, current_policy,\
                              env, gamma, small_change)
        # policy improvement
        current_policy, policy_changed = policy_improvement\
                                        (V, current_policy, \
                                         env, gamma)
        # if the policy didn't change, it means we have converged
        if not policy_changed:
            break
    print(f"Number of episodes trained: {episodes}")
    return current_policy
```

11. Perform a test pass on the FrozenLake-v0 environment:

```
def play(policy, render=False):
    """

    Perform a test pass on the FrozenLake-v0 environment
    Args:
        policy: the policy to use
        render: if the result should be rendered at every step.
          False by default
    """
    env = initialize_environment()
    rewards = []
```

12. Define the maximum number of steps the agent is allowed to take. If it doesn't reach a solution in this time, then we call it an episode and proceed ahead:

```
    max_steps = 25
    test_episodes = 50
    for episode in range(test_episodes):
        # reset the environment every new episode
        state = env.reset()
        total_rewards = 0
        print("*" * 100)
        print("Episode {}".format(episode))

        for step in range(max_steps):
```

13. Take the action that has the highest Q value in the current state:

```
            action = policy[state]
            new_state, reward, done, info = env.step(action)
            if render:
                env.render()
            total_rewards += reward
            if done:
                rewards.append(total_rewards)
                print("Score", total_rewards)
                break
            state = new_state
    env.close()
    print("Average Score", sum(rewards) / test_episodes)
```

14. Step through the **FrozenLake-v0** environment randomly:

```python
def random_step(n_steps=5):
    """

    Steps through the FrozenLake-v0 environment randomly
    Args:
        n_steps: Number of steps to step through
    """
    # reset the environment
    env = initialize_environment()
    state = env.reset()
    for i in range(n_steps):
        # choose an action at random
        action = env.action_space.sample()
        env.render()
        new_state, reward, done, info = env.step(action)
        print(f"New State: {new_state}\n"\
              f"reward: {reward}\n"\
              f"done: {done}\n"\
              f"info: {info}\n")
        print("*" * 20)
```

15. Perform value iteration to find the most optimal policy for the
    FrozenLake-v0 environment:

```python
def value_iteration(env):
    """

    Performs Value Iteration to find the most optimal policy for the
    FrozenLake-v0 environment
    Args:
        env: FrozenLake-v0 Gym environment
    Returns:
        policy: the most optimum policy
    """
    V = dict()
    gamma = 0.9
    state_size = env.observation_space.n
    action_size = env.action_space.n
    policy = dict()
```

16. Initialize the value table randomly and initialize the policy randomly:

```
for x in range(state_size):
    V[x] = -1
    policy[x] = env.action_space.sample()
"""
this loop repeats until the change in value function
is less than delta
"""
while True:
    delta = 0
    for state in reversed(range(state_size)):
        old_v_s = V[state]
        best_rewards = -np.inf
        best_action = None
        # for all the actions in current state
        for action in range(action_size):
```

17. Check the reward obtained if we were to perform this action:

```
            prob, new_state, reward, done = env.env.P[state]\
                                        [action][0]
            potential_reward = reward + gamma * V[new_state]
            """
            select the one that has the best reward
            and also save the action to the policy
            """
            if potential_reward > best_rewards:
                best_rewards = potential_reward
                best_action = action
        policy[state] = best_action
        V[state] = best_rewards
        # terminate if the change is not high
        delta = max(delta, abs(V[state] - old_v_s))
    if delta < 1e-30:
        break
print(policy)
print(V)
return policy
```

18. Run the code and make sure the output matches the expectation by running it in the **main** block:

```
if __name__ == '__main__':
    env = initialize_environment()
    # policy = policy_iteration(env)
    policy = value_iteration(env)
    play(policy, render=True)
```

After running this, you should be able to see the following output:

Figure 5.27: FrozenLake-v0 environment output

As can be seen from the output, we have successfully achieved the goal of retrieving the frisbee.

> **NOTE**
>
> To access the source code for this specific section, please refer to https://packt.live/3fxtZuq.
>
> You can also run this example online at https://packt.live/2Chl1Ss.

CHAPTER 6: MONTE CARLO METHODS

ACTIVITY 6.01: EXPLORING THE FROZEN LAKE PROBLEM – THE REWARD FUNCTION

1. Import the necessary libraries:

```
import gym
import numpy as np
from collections import defaultdict
```

2. Select the environment as **FrozenLake**. **is_slippery** is set to **False**. The environment is reset with the line **env.reset()** and rendered with the line **env.render()**:

```
env = gym.make("FrozenLake-v0", is_slippery=False)
env.reset()
env.render()
```

You will get the following output:

Figure 6.15: Frozen Lake state rendered

This is a text grid with the letters **S**, **F**, **G**, and **H** used to represent the current environment of **FrozenLake**. The highlighted cell **S** is the current state of the agent.

3. Print the possible values in the observation space and the number of action values using the **print(env.observation_space)** and **print(env.action_space)** functions respectively:

```
print(env.observation_space)
print(env.action_space)
name_action = {0:'Left',1:'Down',2:'Right',3:'Up'}
```

You will get the following output:

```
Discrete(16)
Discrete(4)
```

16 is the number of cells in the grid, so **print(env.observation_space)** prints **16**. **4** is the number of possible actions, so **print(env.action_space)** prints **4**. **Discrete** shows the observation space and action space take only discrete values and do not take continuous values.

4. The next step is to define a function to generate a frozen lake episode. We initialize **episodes** and the environment:

```
def generate_frozenlake_episode():
    episode = []
    state = env.reset()
    step = 0;
```

5. Navigate step by step and store **episode** and return **reward**:

```
    while (True):
        action = env.action_space.sample()
        next_state, reward, done, info = env.step(action)
        episode.append((next_state, action, reward))
        if done:
            break
        state = next_state
        step += 1
    return episode, reward
```

The action is obtained with **env.action_space.sample()**. **next_state**, **action**, and **reward** are obtained by calling the **env_step(action)** function. They are then appended to an episode. The **episode** is now a list of states, actions, and rewards.

The key is now to calculate the success rate, which is the likelihood of success for a batch of episodes. The way we do this is by calculating the total number of attempts in a batch of episodes. We calculate how many of them successfully reached the goal. The ratio of the agent successfully reaching the goal to the number of attempts made by the agent is the success ratio.

6. First, we initialize the total reward:

```
def frozen_lake_prediction(batch):
    for batch_number in range(batch+1):
        total_reward = 0
```

7. Generate the episode and reward for every iteration and calculate the total reward:

```
for i_episode in range(100):
    episode, reward = generate_frozenlake_episode()
    total_reward += reward
```

8. The success ratio is calculated by dividing **total_reward** by **100** and is printed:

```
success_percent = total_reward/100
print("Episode", batch_number*100, \
    "Policy Win Rate=>", float(success_percent*100), \
    "%")
```

9. The frozen lake prediction is calculated using the **frozen_lake_prediction** function:

```
frozen_lake_prediction(100)
```

You will get the following output:

```
Episode 0 Policy Win Rate=> 1.0 %
Episode 100 Policy Win Rate=> 1.0 %
Episode 200 Policy Win Rate=> 1.0 %
Episode 300 Policy Win Rate=> 2.0 %
Episode 400 Policy Win Rate=> 1.0 %
Episode 500 Policy Win Rate=> 0.0 %
Episode 600 Policy Win Rate=> 1.0 %
Episode 700 Policy Win Rate=> 1.0 %
Episode 800 Policy Win Rate=> 2.0 %
Episode 900 Policy Win Rate=> 0.0 %
Episode 1000 Policy Win Rate=> 0.0 %
Episode 1100 Policy Win Rate=> 2.0 %
Episode 1200 Policy Win Rate=> 0.0 %
Episode 1300 Policy Win Rate=> 0.0 %
```

Figure 6.16: Output of Frozen Lake without learning

The output prints the policy win ratio for the various episodes in batches of 100. The ratios are quite low as this is the simulation of an agent following a random policy. We will see in the next exercise how this can be improved by learning to a higher level by using a combination of a greedy policy and an epsilon soft policy.

> **NOTE**
>
> To access the source code for this specific section, please refer to https://packt.live/2Akh8Nm.
>
> You can also run this example online at https://packt.live/2zruU07.

ACTIVITY 6.02 SOLVING FROZEN LAKE USING MONTE CARLO CONTROL EVERY VISIT EPSILON SOFT

1. Import the necessary libraries:

```
import gym
import numpy as np
```

2. Select the environment as **FrozenLake**. **is_slippery** is set to **False**:

```
#Setting up the Frozen Lake environment
env = gym.make("FrozenLake-v0", is_slippery=False)
```

3. Initialize the **Q** value and **num_state_action** to zeros:

```
#Initializing the Q and num_state_action
Q = np.zeros([env.observation_space.n, env.action_space.n])
num_state_action = np.zeros([env.observation_space.n, \
                             env.action_space.n])
```

4. Set the value of **num_episodes** to **100000** and create **rewardsList**. We set **epsilon** to **0.30**:

```
num_episodes = 100000
epsilon = 0.30
rewardsList = []
```

Setting epsilon to **0.30** means we will explore with a likelihood of 0.30 and be greedy with a likelihood of 1-0.30 or 0.70.

5. Run the loop till **num_episodes**. We initialize the environment, **results_List**, and **result_sum** to zero. Also, reset the environment:

```
for x in range(num_episodes):
    state = env.reset()

    done = False
    results_list = []
    result_sum = 0.0
```

6. Start a **while** loop, and check whether you need to pick a random action with a probability epsilon or greedy policy with a probability of 1-epsilon:

```
while not done:

    #random action less than epsilon
    if np.random.rand() < epsilon:
        #we go with the random action
        action = env.action_space.sample()
    else:
        """
        1 - epsilon probability, we go with the greedy algorithm
        """
        action = np.argmax(Q[state, :])
```

7. Now step through the **action** and get **new_state** and **reward**:

```
    #action is performed and assigned to new_state, reward
    new_state, reward, done, info = env.step(action)
```

8. The result list is appended with the **state** and **action** pair. **result_sum** is incremented by the value of the result:

```
    results_list.append((state, action))
    result_sum += reward
```

9. **new_state** is assigned to **state** and **result_sum** is appended to **rewardsList**:

```
    #new state is assigned as state
    state = new_state

    #appending the results sum to the rewards list
    rewardsList.append(result_sum)
```

10. Calculate `Q[s,a]` using the incremental method, as `Q[s,a] + (result_sum - Q[s,a]) / N(s,a)`:

```
for (state, action) in results_list:
    num_state_action[state, action] += 1.0
    sa_factor = 1.0 / num_state_action[state, action]
    Q[state, action] += sa_factor * \
                        (result_sum - Q[state, action])
```

11. Print the value of the success rates in batches of **1000**:

```
if x % 1000 == 0 and x is not 0:
    print('Frozen Lake Success rate=>', \
        str(sum(rewardsList) * 100 / x ), '%')
```

12. Print the final success rate:

```
print("Frozen Lake Success rate=>", \
    str(sum(rewardsList)/num_episodes * 100), "%")
```

You will get the following output initially:

```
Frozen Lake Success rate=> 0.0 %
Frozen Lake Success rate=> 0.0 %
Frozen Lake Success rate=> 0.0 %
Frozen Lake Success rate=> 0.0 %
Frozen Lake Success rate=> 0.0 %
Frozen Lake Success rate=> 0.0 %
Frozen Lake Success rate=> 8.4 %
Frozen Lake Success rate=> 15.0625 %
Frozen Lake Success rate=> 20.27777777777778 %
Frozen Lake Success rate=> 24.53 %
Frozen Lake Success rate=> 28.318181818181817 %
Frozen Lake Success rate=> 31.525 %
Frozen Lake Success rate=> 34.06153846153846 %
Frozen Lake Success rate=> 36.121428571428574 %
Frozen Lake Success rate=> 38.02 %
Frozen Lake Success rate=> 39.5375 %
Frozen Lake Success rate=> 40.81764705882353 %
Frozen Lake Success rate=> 41.87777777777778 %
```

Figure 6.17: Initial output of the Frozen Lake success rate

You will get the following output finally:

```
Frozen Lake Success rate=> 60.38953488372093 %
Frozen Lake Success rate=> 60.44712643678161 %
Frozen Lake Success rate=> 60.492045454545455 %
Frozen Lake Success rate=> 60.53370786516854 %
Frozen Lake Success rate=> 60.59222222222222 %
Frozen Lake Success rate=> 60.62967032967033 %
Frozen Lake Success rate=> 60.69565217391305 %
Frozen Lake Success rate=> 60.75268817204301 %
Frozen Lake Success rate=> 60.82127659574468 %
Frozen Lake Success rate=> 60.87684210526316 %
Frozen Lake Success rate=> 60.95104166666667 %
Frozen Lake Success rate=> 61.00309278350515 %
Frozen Lake Success rate=> 61.039795918367346 %
Frozen Lake Success rate=> 61.07575757575758 %
Frozen Lake Success rate=> 61.117999999999995 %
```

Figure 6.18: Final output of the Frozen Lake success rate

The success rate starts with a very low value close to 0% but with reinforcement learning, it learns, and the success rate increases incrementally going up to 60%.

> **NOTE**
>
> To access the source code for this specific section, please refer to https://packt.live/2Ync9Dq.
>
> You can also run this example online at https://packt.live/3cUJLxQ.

CHAPTER 7: TEMPORAL DIFFERENCE LEARNING

ACTIVITY 7.01: USING TD(0) Q-LEARNING TO SOLVE FROZENLAKE-V0 STOCHASTIC TRANSITIONS

1. Import the required modules:

```
import numpy as np
import matplotlib.pyplot as plt
%matplotlib inline

import gym
```

2. Instantiate the **gym** environment called **FrozenLake-v0** using the **is_slippery** flag set to **True** in order to enable stochasticity:

```
env = gym.make('FrozenLake-v0', is_slippery=True)
```

3. Take a look at the action and observation spaces:

```
print("Action space = ", env.action_space)
print("Observation space = ", env.observation_space)
```

This will print out the following:

```
Action space =  Discrete(4)
Observation space =  Discrete(16)
```

4. Create two dictionaries to easily translate the **actions** numbers into moves:

```
actionsDict = {}
actionsDict[0] = " L "
actionsDict[1] = " D "
actionsDict[2] = " R "
actionsDict[3] = " U "

actionsDictInv = {}
actionsDictInv["L"] = 0
actionsDictInv["D"] = 1
actionsDictInv["R"] = 2
actionsDictInv["U"] = 3
```

5. Reset the environment and render it to take a look at the grid problem:

```
env.reset()
env.render()
```

Its initial state is as follows:

Figure 7.39: Environment's initial state

6. Visualize the optimal policy for this environment:

```
optimalPolicy = ["  *  "," U ","L/R/D"," U ",\
                 " L "," - "," L/R "," - ",\
                 " U "," D "," L "," - ",\
                 " - "," R ","R/D/U"," ! ",]

print("Optimal policy:")
idxs = [0,4,8,12]
for idx in idxs:
    print(optimalPolicy[idx+0], optimalPolicy[idx+1], \
          optimalPolicy[idx+2], optimalPolicy[idx+3])
```

This prints out the following output:

```
Optimal policy:
  L/R/D  U    U    U
    L    -   L/R   -
    U    D    L    -
    -    R    D    !
```

7. Define the functions that will take ε-greedy actions:

```
def action_epsilon_greedy(q, s, epsilon=0.05):
    if np.random.rand() > epsilon:
        return np.argmax(q[s])
    return np.random.randint(4)
```

8. Define a function that will take greedy actions:

```
def greedy_policy(q, s):
    return np.argmax(q[s])
```

9. Define a function that will calculate the agent's average performance:

```
def average_performance(policy_fct, q):
    acc_returns = 0.
    n = 500
    for i in range(n):
        done = False
        s = env.reset()
        while not done:
            a = policy_fct(q, s)
            s, reward, done, info = env.step(a)
            acc_returns += reward
    return acc_returns/n
```

10. Initialize the Q-table so that all the values are equal to **1**, except for the values at the terminal states:

```
q = np.ones((16, 4))
# Set q(terminal,*) equal to 0
q[5,:] = 0.0
q[7,:] = 0.0
q[11,:] = 0.0
q[12,:] = 0.0
q[15,:] = 0.0
```

11. Set the number of total episodes, the number of steps representing the interval by which we're evaluating the agent's average performance, the learning rate, the discounting factor, the **ε** value for the exploration policy, and an array to collect all the agent's performance evaluations during training:

```
nb_episodes = 80000
STEPS = 2000
alpha = 0.01
gamma = 0.99
epsilon_expl = 0.2


q_performance = np.ndarray(nb_episodes//STEPS)
```

12. Train the Q-learning algorithm. Loop among all episodes:

```
for i in range(nb_episodes):
```

13. Reset the environment and start the in-episode loop:

```
done = False
s = env.reset()
while not done:
```

14. Select the exploration action with an ε-greedy policy:

```
# behavior policy
a = action_epsilon_greedy(q, s, epsilon=epsilon_expl)
```

15. Step the environment with the selected exploration action and retrieval of the new state, reward, and done conditions:

```
new_s, reward, done, info = env.step(a)
```

16. Select a new action with the greedy policy:

```
a_max = np.argmax(q[new_s]) # estimation policy
```

17. Update the Q-table with the Q-learning TD(0) rule:

```
q[s, a] = q[s, a] + alpha * \
            (reward + gamma * q[new_s, a_max] - q[s, a])
```

18. Update the state with a new value:

```
s = new_s
```

19. Evaluate the agent's average performance for every step:

```
if i%STEPS == 0:
    q_performance[i//STEPS] = average_performance\
                             (greedy_policy, q)
```

20. Plot the Q-learning agent's mean reward history during training:

```
plt.plot(STEPS * np.arange(nb_episodes//STEPS), q_performance)
plt.xlabel("Epochs")
plt.ylabel("Average reward of an epoch")
plt.title("Learning progress for Q-Learning")
```

This generates the following output, showing the learning progress for the Q-learning algorithm:

```
Text(0.5, 1.0, 'Learning progress for Q-Learning')
```

The plot for this can be visualized as follows:

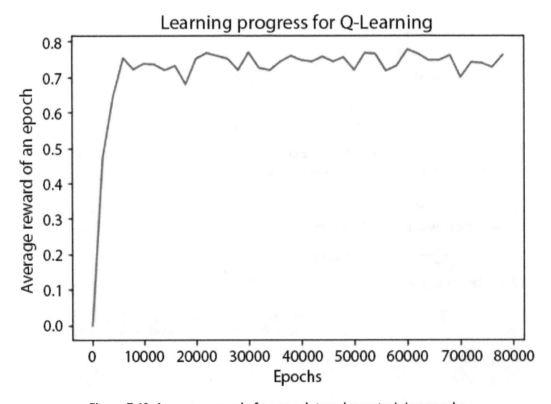

Figure 7.40: Average reward of an epoch trend over training epochs

In this case, as in the case of Q-learning applied to the deterministic environment, the plot shows how quickly Q-learning performance grows over epochs as the agent collects more and more experience. It also demonstrates that the algorithm is not capable of reaching 100% success after learning due to the limitations of stochasticity. When compared with using the SARSA method on a stochastic environment, as seen in *Figure 7.15*, the algorithm's performance grows faster and more steadily.

21. Evaluate the greedy policy's performance for the trained agent (Q-table):

```
greedyPolicyAvgPerf = average_performance(greedy_policy, q=q)
print("Greedy policy Q-learning performance =", \
        greedyPolicyAvgPerf)
```

This prints out the following:

```
Greedy policy Q-learning performance = 0.708
```

22. Display the Q-table values:

```
q = np.round(q,3)
print("(A,S) Value function =", q.shape)
print("First row")
print(q[0:4,:])
print("Second row")
print(q[4:8,:])
print("Third row")
print(q[8:12,:])
print("Fourth row")
print(q[12:16,:])
```

This generates the following output:

```
(A,S) Value function = (16, 4)
First row
[[0.543 0.521 0.516 0.515]
 [0.319 0.355 0.322 0.493]
 [0.432 0.431 0.425 0.461]
 [0.32  0.298 0.296 0.447]]

Second row
[[0.559 0.392 0.396 0.393]
 [0.    0.    0.    0.    ]
 [0.296 0.224 0.327 0.145]
```

```
 [0.    0.    0.    0.   ]]

Third row
[[0.337 0.366 0.42  0.595]
 [0.484 0.639 0.433 0.415]
 [0.599 0.511 0.342 0.336]
 [0.    0.    0.    0.   ]]

Fourth row
[[0.    0.    0.    0.   ]
 [0.46  0.53  0.749 0.525]
 [0.711 0.865 0.802 0.799]
 [0.    0.    0.    0.   ]]
```

23. Print out the greedy policy that was found and compare it with the optimal policy:

```
policyFound = [actionsDict[np.argmax(q[0,:])],\
               actionsDict[np.argmax(q[1,:])],\
               actionsDict[np.argmax(q[2,:])],\
               actionsDict[np.argmax(q[3,:])],\
               actionsDict[np.argmax(q[4,:])],\
               " - ",\
               actionsDict[np.argmax(q[6,:])],\
               " - ",\
               actionsDict[np.argmax(q[8,:])],\
               actionsDict[np.argmax(q[9,:])],\
               actionsDict[np.argmax(q[10,:])],\
               " - ",\
               " - ",\
               actionsDict[np.argmax(q[13,:])],\
               actionsDict[np.argmax(q[14,:])],\
               " ! "]

print("Greedy policy found:")
idxs = [0,4,8,12]
for idx in idxs:
    print(policyFound[idx+0], policyFound[idx+1], \
          policyFound[idx+2], policyFound[idx+3])
```

```
print(" ")

print("Optimal policy:")
idxs = [0,4,8,12]
for idx in idxs:
    print(optimalPolicy[idx+0], optimalPolicy[idx+1], \
            optimalPolicy[idx+2], optimalPolicy[idx+3])
```

This generates the following output:

```
Greedy policy found:
    L    U    U    U
    L    -    R    -
    U    D    L    -
    -    R    D    !


Optimal policy:
  L/R/D  U    U    U
    L    -   L/R   -
    U    D    L    -
    -    R    D    !
```

This output shows that, as for all the exercises in this chapter, the off-policy, one-step Q-learning algorithm is able to find the optimal policy by simply exploring the environment, even in the context of stochastic environment transitions. As anticipated, for this setting, it is not possible to achieve the maximum reward 100% of the time.

As we can see, for every state of the grid world that the greedy policy obtained with the Q-table that was calculated by our algorithm, it prescribes an action that is in accordance with the optimal policy that was defined by analyzing the environment problem. As we already saw, there are two states in which many different actions are equally optimal, and the agent correctly implements one of them.

> **NOTE**
>
> To access the source code for this specific section, please refer to https://packt.live/3elMxxu.
>
> You can also run this example online at https://packt.live/37HSDWx.

CHAPTER 8: THE MULTI-ARMED BANDIT PROBLEM

ACTIVITY 8.01: QUEUEING BANDITS

1. Import the necessary libraries and tools, as follows:

```
import numpy as np

from utils import QueueBandit
```

2. Declare the bandit object, as follows:

```
N_CLASSES = 3

queue_bandit = QueueBandit(filename='data.csv')
```

The **N_CLASSES** variable will be used by our subsequent code.

3. Implement the Greedy algorithm, as follows:

```
class GreedyQueue:
    def __init__(self, n_classes=3):
        self.n_classes = n_classes
        self.time_history = [[] for _ in range(n_classes)]

    def decide(self, queue_lengths):
        for class_ in range(self.n_classes):
            if queue_lengths[class_] > 0 and \
                len(self.time_history[class_]) == 0:
                return class_

        mean_times = [np.mean(self.time_history[class_])\
                        if queue_lengths[class_] > 0 else np.inf\
                        for class_ in range(self.n_classes)]

        return int(np.random.choice\
                    (np.argwhere\
                    (mean_times == np.min(mean_times)).flatten()))

    def update(self, class_, time):
        self.time_history[class_].append(time)
```

Notice that we are taking care to avoid choosing a class that does not have any customers left in it by checking if **queue_lengths[class_]** is greater than 0 or not. The remaining code is analogous to what we had in our earlier discussion of Greedy.

Subsequently, apply the algorithm to the bandit object, as follows:

```
cumulative_times = queue_bandit.repeat\
                   (GreedyQueue, [N_CLASSES], \
                    visualize_cumulative_times=True)

np.max(cumulative_times), np.mean(cumulative_times)
```

This will generate the following graph:

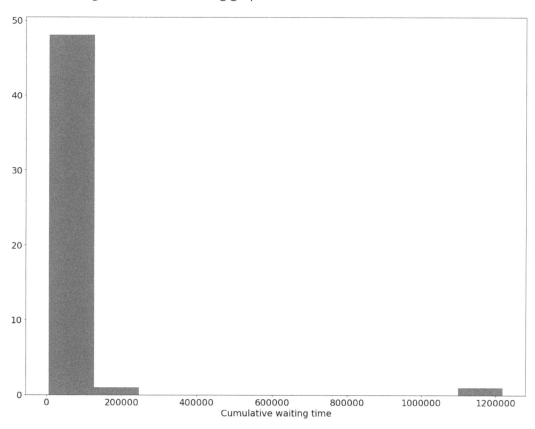

Figure 8.24: Distribution of cumulative waiting time from Greedy

Additionally, the following will be printed out as the max and mean cumulative waiting times:

```
(1218887.7924350922, 45155.236786598274)
```

While these values might appear large compared to our earlier discussions, this is because the reward/cost distributions we are working with here take on higher values. We will use these values from Greedy as a frame of reference to analyze the performance of later algorithms.

4. Implement the Explore-then-commit algorithm using the following code:

```
class ETCQueue:
    def __init__(self, n_classes=3, T=3):
        self.n_classes = n_classes
        self.T = T
        self.time_history = [[] for _ in range(n_classes)]

    def decide(self, queue_lengths):
        for class_ in range(self.n_classes):
            if queue_lengths[class_] > 0 and \
            len(self.time_history[class_]) < self.T:
                return class_

        mean_times = [np.mean(self.time_history[class_])\
                        if queue_lengths[class_] > 0 else np.inf\
                        for class_ in range(self.n_classes)]

        return int(np.random.choice\
                    (np.argwhere(mean_times == np.min(mean_times))\
                    .flatten()))

    def update(self, class_, time):
        self.time_history[class_].append(time)
```

5. Apply the algorithm to the bandit object, as follows:

```
cumulative_times = queue_bandit.repeat\
                    (ETCQueue, [N_CLASSES, 2],\
                    visualize_cumulative_times=True)

np.max(cumulative_times), np.mean(cumulative_times)
```

This will produce the following graph:

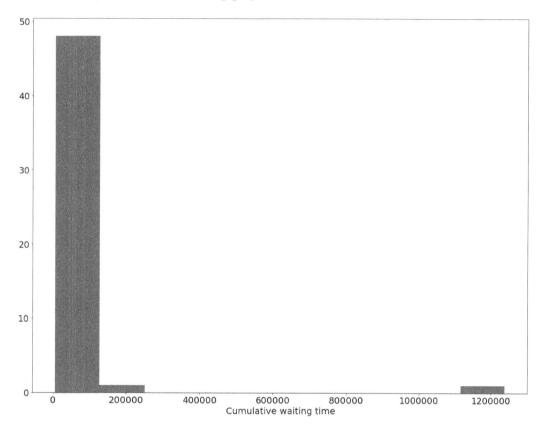

Figure 8.25: Distribution of cumulative waiting time from Explore-then-commit

This will also produce the max and average cumulative waiting times:
`(1238591.3208636027, 45909.77140562623)`. Compared to Greedy
`(1218887.7924350922, 45155.236786598274)`, Explore-then-commit
did relatively worse on this queueing bandit problem.

6. Implement Thompson Sampling, as follows:

```
class ExpThSQueue:
    def __init__(self, n_classes=3):
        self.n_classes = n_classes
        self.time_history = [[] for _ in range(n_classes)]
        self.temp_beliefs = [(0, 0) for _ in range(n_classes)]

    def decide(self, queue_lengths):
        for class_ in range(self.n_classes):
```

```
            if queue_lengths[class_] > 0 and \
        len(self.time_history[class_]) == 0:
            return class_

    rate_draws = [np.random.gamma\
                (self.temp_beliefs[class_][0],1 \
                / self.temp_beliefs[class_][1])\
                if queue_lengths[class_] > 0 else -np.inf\
                for class_ in range(self.n_classes)]

    return int(np.random.choice\
                (np.argwhere(rate_draws == np.max(rate_draws))\
                .flatten()))

def update(self, class_, time):
    self.time_history[class_].append(time)

    # Update parameters according to Bayes rule
    alpha, beta = self.temp_beliefs[class_]
    alpha += 1
    beta += time
    self.temp_beliefs[class_] = alpha, beta
```

Recall that in our initial discussion of Thompson Sampling, we draw random samples to estimate the reward expectation for each arm. Here, we drew random samples from the corresponding Gamma distributions (which are being used to model service rates) to estimate the rates (or the inverse job lengths) and choose the largest drawn sample.

7. This can be applied to solve the bandit problem using the following code:

```
cumulative_times = queue_bandit.repeat\
                (ExpThSQueue, [N_CLASSES], \
                visualize_cumulative_times=True)

np.max(cumulative_times), np.mean(cumulative_times)
```

The following plot will be produced:

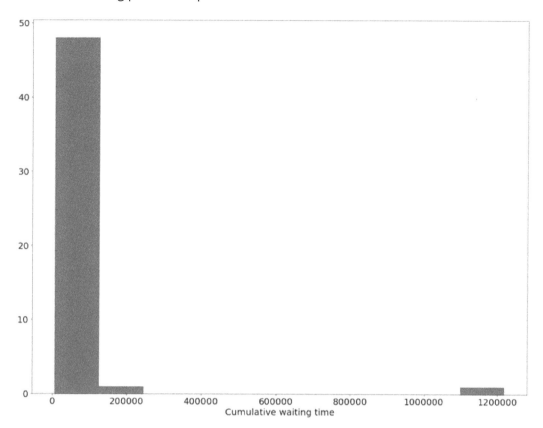

Figure 8.26: Distribution of cumulative waiting time from Thompson Sampling

From the max and mean waiting time **(1218887.7924350922, 45129.343871806814)**, we can see that Thompson Sampling is able to improve on Greedy.

8. The modified version of Thompson Sampling can be implemented as follows:

```
class ExploitingThSQueue:
    def __init__(self, n_classes=3, r=1):
        self.n_classes = n_classes
        self.time_history = [[] for _ in range(n_classes)]
        self.temp_beliefs = [(0, 0) for _ in range(n_classes)]
        self.t = 0
        self.r = r

    def decide(self, queue_lengths):
```

```
        for class_ in range(self.n_classes):
            if queue_lengths[class_] > 0 and \
            len(self.time_history[class_]) == 0:
                return class_

        if self.t > self.r * np.sum(queue_lengths):
            mean_times = [np.mean(self.time_history[class_])\
                            if queue_lengths[class_] > 0 \
                            else np.inf\
                            for class_ in range(self.n_classes)]

            return int(np.random.choice\
                    (np.argwhere\
                    (mean_times == np.min(mean_times))\
                    .flatten()))

        rate_draws = [np.random.gamma\
                    (self.temp_beliefs[class_][0],\
                    1 / self.temp_beliefs[class_][1])\
                    if queue_lengths[class_] > 0 else -np.inf\
                    for class_ in range(self.n_classes)]

        return int(np.random.choice\
                    (np.argwhere\
                    (rate_draws == np.max(rate_draws)).flatten()))
```

The initialization method of this class implementation has an additional attribute, **r**, which we will use to implement the exploitation logic.

In the **decide()** method, right before we draw samples to estimate the rates, we check to see if the current time (**t**) is greater than the current queue length (the sum of **queue_lengths**). This Boolean indicates whether we have processed more than half of the customers or not. If so, we simply implement the logic of the Greedy algorithm and return the arm with the optimal average rate. Otherwise, we have our actual Thompson Sampling logic.

The **update()** method should be the same as the actual Thompson Sampling algorithm from the previous step, as follows:

```
def update(self, class_, time):
    self.time_history[class_].append(time)
    self.t += 1
```

```
# Update parameters according to Bayes rule
alpha, beta = self.temp_beliefs[class_]
alpha += 1
beta += time
self.temp_beliefs[class_] = alpha, beta
```

9. Finally, apply the algorithm to the bandit problem:

```
cumulative_times = queue_bandit.repeat\
                (ExploitingThSQueue, [N_CLASSES, 1], \
                visualize_cumulative_times=True)


np.max(cumulative_times), np.mean(cumulative_times)
```

We will obtain the following graph:

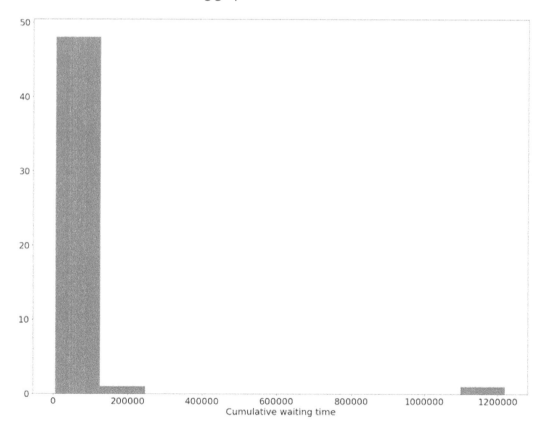

Figure 8.27: Distribution of cumulative waiting time from modified Thompson Sampling

Together with the max and mean waiting time `(1218887.7924350922,`
`45093.244027644556)`, we can see that this modified version of Thompson
Sampling is more effective than the original at minimizing the cumulative waiting
time across the experiments.

This speaks to the potential benefit of designing algorithms that are tailored to
the contextual bandit problem that they are trying to solve.

> **NOTE**
>
> To access the source code for this specific section, please refer
> to https://packt.live/2Yuw2IQ.
>
> You can also run this example online at https://packt.live/3hnK5Z5.

Throughout this activity, we have learned how to apply the approaches discussed
in this chapter to a queueing bandit problem, that is, exploring an example of a
potential contextual bandit process. Most notably, we have considered a variant
of Thompson Sampling that has been modified to fit the context of the queueing
problem, thus successfully lowering our cumulative regret compared to other
algorithms. This activity also marks the end of this chapter.

CHAPTER 9: WHAT IS DEEP Q-LEARNING?

ACTIVITY 9.01: IMPLEMENTING A DOUBLE DEEP Q NETWORK IN PYTORCH FOR THE CARTPOLE ENVIRONMENT

1. Open a new Jupyter notebook and import all of the required libraries:

```
import gym
import matplotlib.pyplot as plt
import torch
import torch.nn as nn
from torch import optim
import numpy as np
import random
import math
```

2. Write code that will create a device based on the availability of a GPU environment:

```
use_cuda = torch.cuda.is_available()

device = torch.device("cuda:0" if use_cuda else "cpu")

print(device)
```

3. Create a **gym** environment using the **'CartPole-v0'** environment:

```
env = gym.make('CartPole-v0')
```

4. Set the **seed** for torch and the environment for reproducibility:

```
seed = 100
env.seed(seed)
torch.manual_seed(seed)
random.seed(seed)
```

5. Fetch the number of states and actions from the environment:

```
number_of_states = env.observation_space.shape[0]
number_of_actions = env.action_space.n

print('Total number of States : {}'.format(number_of_states))
print('Total number of Actions : {}'.format(number_of_actions))
```

The output is as follows:

```
Total number of States : 4
Total number of Actions : 2
```

6. Set all of the hyperparameter values required for the DDQN process:

```
NUMBER_OF_EPISODES = 500
MAX_STEPS = 1000
LEARNING_RATE = 0.01
DISCOUNT_FACTOR = 0.99
HIDDEN_LAYER_SIZE = 64


EGREEDY = 0.9
EGREEDY_FINAL = 0.02
EGREEDY_DECAY = 500


REPLAY_BUFFER_SIZE = 6000
BATCH_SIZE = 32


UPDATE_TARGET_FREQUENCY = 200
```

7. Implement the **calculate_epsilon** function, as described in the previous exercises:

```
def calculate_epsilon(steps_done):
    """
    Decays epsilon with increasing steps
    Parameter:
    steps_done (int) : number of steps completed
    Returns:
    int - decayed epsilon
    """
    epsilon = EGREEDY_FINAL + (EGREEDY - EGREEDY_FINAL) \
              * math.exp(-1. * steps_done / EGREEDY_DECAY )
    return epsilon
```

8. Create a class, called **DQN**, that accepts the number of states as inputs and outputs Q values for the number of actions present in the environment, with the network that has a hidden layer of size **64**:

```
class DQN(nn.Module):
    def __init__(self , hidden_layer_size):
        super().__init__()
        self.hidden_layer_size = hidden_layer_size
        self.fc1 = nn.Linear(number_of_states,\
                             self.hidden_layer_size)
        self.fc2 = nn.Linear(self.hidden_layer_size,\
                             number_of_actions)

    def forward(self, x):
        output = torch.tanh(self.fc1(x))
        output = self.fc2(output)
        return output
```

9. Implement the **ExperienceReplay** class, as described in the previous exercises:

```
class ExperienceReplay(object):
    def __init__(self , capacity):

        self.capacity = capacity
        self.buffer = []
        self.pointer = 0

    def push(self , state, action, new_state, reward, done):
        experience = (state, action, new_state, reward, done)

        if self.pointer >= len(self.buffer):
            self.buffer.append(experience)
        else:
            self.buffer[self.pointer] = experience

        self.pointer = (self.pointer + 1) % self.capacity
```

```
    def sample(self , batch_size):
        return zip(*random.sample(self.buffer , batch_size))

    def __len__(self):
        return len(self.buffer)
```

10. Instantiate the **ExperienceReplay** class by passing the buffer size as input:

```
memory = ExperienceReplay(REPLAY_BUFFER_SIZE)
```

11. Implement the DQN agent class with the changes discussed for the
    optimize function (from the code example given in the *Double Deep
    Q Network (DDQN)* section):

```
class DQN_Agent(object):
    def __init__(self):

        self.dqn = DQN(HIDDEN_LAYER_SIZE).to(device)
        self.target_dqn = DQN(HIDDEN_LAYER_SIZE).to(device)

        self.criterion = torch.nn.MSELoss()

        self.optimizer = optim.Adam\
                        (params=self.dqn.parameters(), \
                         lr=LEARNING_RATE)

        self.target_dqn_update_counter = 0

    def select_action(self,state,EGREEDY):

        random_for_egreedy = torch.rand(1)[0]

        if random_for_egreedy > EGREEDY:

            with torch.no_grad():

                state = torch.Tensor(state).to(device)
                q_values = self.dqn(state)
                action = torch.max(q_values,0)[1]
                action = action.item()
        else:
```

```
            action = env.action_space.sample()

    return action

def optimize(self):

    if (BATCH_SIZE > len(memory)):
        return

    state, action, new_state, reward, done = memory.sample\
                                    (BATCH_SIZE)

    state = torch.Tensor(state).to(device)
    new_state = torch.Tensor(new_state).to(device)
    reward = torch.Tensor(reward).to(device)
    action = torch.LongTensor(action).to(device)
    done = torch.Tensor(done).to(device)

    """
    select action : get the index associated with max q
    value from prediction network
    """
    new_state_indxs = self.dqn(new_state).detach()
    # to get the max new state indexes
    max_new_state_indxs = torch.max(new_state_indxs, 1)[1]
    """
    Using the best action from the prediction nn get
    the max new state value in target dqn
    """
    new_state_values = self.target_dqn(new_state).detach()
    max_new_state_values = new_state_values.gather\
                            (1, max_new_state_indxs\
                                .unsqueeze(1))\
                            .squeeze(1)
    #when done = 1 then target = reward
    target_value = reward + (1 - done) * DISCOUNT_FACTOR \
                    * max_new_state_values

    predicted_value = self.dqn(state).gather\
                        (1, action.unsqueeze(1))\
```

```
                          .squeeze(1)
        loss = self.criterion(predicted_value, target_value)


        self.optimizer.zero_grad()
        loss.backward()
        self.optimizer.step()


        if self.target_dqn_update_counter \
        % UPDATE_TARGET_FREQUENCY == 0:
            self.target_dqn.load_state_dict(self.dqn.state_dict())


        self.target_dqn_update_counter += 1
```

12. Write the training process loop with the help of the following steps. First, instantiate the DQN agent using the class created earlier. Create a **steps_total** empty list to collect the total number of steps for each episode. Initialize **steps_counter** with zero and use it to calculate the decayed epsilon value for each step:

```
dqn_agent = DQN_Agent()


steps_total = []
steps_counter = 0
```

Use two loops during the training process; the first one is to play the game for a certain number of steps. The second loop ensures that each episode goes on for a fixed number of steps. Inside the second **for** loop, the first step is to calculate the epsilon value for the current step.

Using the present state and epsilon value, you can select the action to perform. The next step is to take the action. Once you take the action, the environment returns the **new_state**, **reward**, and **done** flags.

Using the **optimize** function, perform one step of gradient descent to optimize the DQN. Now make the new state the present state for the next iteration. Finally, check whether the episode is over. If the episode is over, then you can collect and record the reward for the current episode:

```
for episode in range(NUMBER_OF_EPISODES):


    state = env.reset()
    done = False
```

```
step = 0

for i in range(MAX_STEPS):
    step += 1
    steps_counter += 1

    EGREEDY = calculate_epsilon(steps_counter)

    action = dqn_agent.select_action(state, EGREEDY)

    new_state, reward, done, info = env.step(action)

    memory.push(state, action, new_state, reward, done)

    dqn_agent.optimize()

    state = new_state

    if done:
        steps_total.append(step)
        break
```

13. Now observe the reward. As the reward is scalar feedback and gives an indication of how well the agent is performing, you should look at the average reward and the average reward for the last 100 episodes. Also, perform the graphical representation of rewards. Check how the agent is performing while playing more episodes and what the reward average is for the last 100 episodes:

```
print("Average reward: %.2f" \
    % (sum(steps_total)/NUMBER_OF_EPISODES))
print("Average reward (last 100 episodes): %.2f" \
    % (sum(steps_total[-100:])/100))
```

The output will be as follows:

```
Average reward: 174.09
Average reward (last 100 episodes): 186.06
```

14. Plot the rewards collected in the y axis and the number of episodes in the x axis to visualize how the rewards have been collected with the increasing number of episodes:

```
Plt.figure(figsize=(12,5))
plt.title("Rewards Collected")
plt.xlabel('Steps')
plt.ylabel('Reward')
plt.bar(np.arange(len(steps_total)), steps_total, \
        alpha=0.5, color='green', width=6)
plt.show()
```

The output will be as follows:

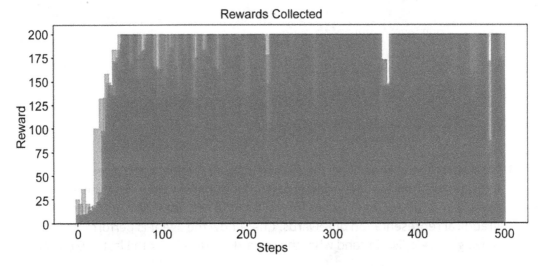

Figure 9.37: Plot for the rewards collected by the agent

> **NOTE**
>
> To access the source code for this specific section, please refer to https://packt.live/3hnLDTd.
>
> You can also run this example online at https://packt.live/37ol5MK.

The following is a comparison between different DQN techniques and DDQN:

Vanilla DQN Outputs:

```
Average reward: 158.83
Average reward (last 100 episodes): 176.28
```

DQN with Experience Replay and Target Network Outputs:

```
Average reward: 154.41
Average reward (last 100 episodes): 183.28
```

DDQN Outputs:

```
Average reward: 174.09
Average reward (last 100 episodes): 186.06
```

As you can see from the preceding figure, along with the comparison of the results shown earlier, DDQN has the highest average reward, compared to other DQN implementations, and the average reward for the last 100 episodes is also higher. We can say that DDQN improves performance significantly in comparison to the other two DQN techniques. After completing this whole activity, we have learned how to combine a DDQN network with experience replay to overcome the issues of a vanilla DQN and achieve more stable rewards.

CHAPTER 10: PLAYING AN ATARI GAME WITH DEEP RECURRENT Q-NETWORKS

ACTIVITY 10.01: TRAINING A DQN WITH CNNS TO PLAY BREAKOUT

Solution

1. Open a new Jupyter Notebook and import the relevant packages: **gym**, **random**, **tensorflow**, **numpy**, and **collections**:

```
import gym
import random
import numpy as np
from collections import deque
import tensorflow as tf
from tensorflow.keras.models import Sequential
from tensorflow.keras.layers import Dense, Conv2D, \
MaxPooling2D, Flatten
from tensorflow.keras.optimizers import RMSprop
import datetime
```

2. Set the seed for NumPy and TensorFlow to **168**:

```
np.random.seed(168)
tf.random.set_seed(168)
```

3. Create the **DQN** class with the following methods: the **build_model()** method to instantiate a CNN, the **get_action()** method to apply the epsilon-greedy algorithm to choose the action to be played, the **add_experience()** method to store in memory the experience acquired by playing the game, the **replay()** method, which will perform experience replay by sampling experiences from the memory and train the DQN model with a callback to save the model every two episodes, and the **update_epsilon()** method to gradually decrease the epsilon value for epsilon-greedy:

`Activity10_01.ipynb`

```
class DQN():
    def __init__(self, env, batch_size=64, max_experiences=5000):
        self.env = env
        self.input_size = self.env.observation_space.shape[0]
        self.action_size = self.env.action_space.n
        self.max_experiences = max_experiences
        self.memory = deque(maxlen=self.max_experiences)
        self.batch_size = batch_size
        self.gamma = 1.0
        self.epsilon = 1.0
        self.epsilon_min = 0.01
        self.epsilon_decay = 0.995

        self.model = self.build_model()
        self.target_model = self.build_model()

    def build_model(self):
        model = Sequential()
        model.add(Conv2D(32, 8, (4,4), activation='relu', \
                         padding='valid',\
                         input_shape=(IMG_SIZE, IMG_SIZE, 1)))
        model.add(Conv2D(64, 4, (2,2), activation='relu', \
                         padding='valid'))
        model.add(Conv2D(64, 3, (1,1), activation='relu', \
                         padding='valid'))
        model.add(Flatten())
        model.add(Dense(256, activation='relu'))
        model.add(Dense(self.action_size))
        model.compile(loss='mse', \
                      optimizer=RMSprop(lr=0.00025, \
                      epsilon=self.epsilon_min), \
                      metrics=['accuracy'])

        return model
```

The complete code for this step can be found at https://packt.live/3hoZXdV.

4. Create the **initialize_env()** function, which will initialize the Breakout environment:

```
def initialize_env(env):
    initial_state = env.reset()
    initial_done_flag = False
    initial_rewards = 0
    return initial_state, initial_done_flag, initial_rewards
```

5. Create the **preprocess_state()** function to preprocess the input images:

```
def preprocess_state(image, img_size):
    img_temp = image[31:195]
    img_temp = tf.image.rgb_to_grayscale(img_temp)
    img_temp = tf.image.resize\
                (img_temp, [img_size, img_size],\
                 method=tf.image.ResizeMethod.NEAREST_NEIGHBOR)
    img_temp = tf.cast(img_temp, tf.float32)
    return img_temp
```

6. Create the **play_game()** function, which will play an entire game of Breakout:

```
def play_game(agent, state, done, rewards):
    while not done:
        action = agent.get_action(state)
        next_state, reward, done, _ = env.step(action)

        next_state = preprocess_state(next_state, IMG_SIZE)

        agent.add_experience(state, action, reward, \
                             next_state, done)

        state = next_state
        rewards += reward
    return rewards
```

7. Create the **train_agent()** function, which will iterate through a number of episodes where the agent will play a game and perform experience replay:

```
def train_agent(env, episodes, agent):
  from collections import deque
  import numpy as np

  scores = deque(maxlen=100)

  for episode in range(episodes):
    state, done, rewards = initialize_env(env)
    state = preprocess_state(state, IMG_SIZE)

    rewards = play_game(agent, state, done, rewards)
    scores.append(rewards)
    mean_score = np.mean(scores)

    if episode % 50 == 0:
        print(f'[Episode {episode}] \
- Average Score: {mean_score}')
        agent.target_model.set_weights(agent.model.get_weights())
        agent.target_model.save_weights\
        (f'dqn/dqn_model_weights_{episode}')

    agent.replay(episode)

  print(f"Average Score: {np.mean(scores)}")
```

8. Instantiate a Breakout environment called **env** with the **gym.make()** function:

```
env = gym.make('BreakoutDeterministic-v4')
```

9. Create two variables, **IMG_SIZE** and **SEQUENCE**, that will take the values **84** and **4**, respectively:

```
IMG_SIZE = 84
SEQUENCE = 4
```

10. Instantiate a **DQN** object called **agent**:

```
agent = DQN(env)
```

11. Create a variable called **episodes** that will take the value **50**:

```
episodes = 50
```

12. Call the **train_agent** function by providing **env**, **episodes**, and **agent**:

```
train_agent(env, episodes, agent)
```

The following is the output of the code:

```
[Episode 0] - Average Score: 3.0
Average Score: 0.59
```

> **NOTE**
>
> To access the source code for this specific section, please refer to https://packt.live/3hoZXdV.
>
> You can also run this example online at https://packt.live/3dWLwfa.

You just completed the first activity of this chapter. You successfully built and trained a DQN agent combined with CNNs to play the game Breakout. The performance of this model is very similar to the random agent (average score of 0.6). However, if you train it for longer (by increasing the number of episodes), it may achieve a better score.

ACTIVITY 10.02: TRAINING A DRQN TO PLAY BREAKOUT

Solution

1. Open a new Jupyter Notebook and import the relevant packages: **gym**, **random**, **tensorflow**, **numpy**, and **collections**:

```
import gym
import random
import numpy as np
from collections import deque
import tensorflow as tf
from tensorflow.keras.models import Sequential
from tensorflow.keras.layers import Dense, Conv2D, \
MaxPooling2D, TimeDistributed, Flatten, LSTM
from tensorflow.keras.optimizers import RMSprop
import datetime
```

2. Set the seed for NumPy and TensorFlow to **168**:

```
np.random.seed(168)
tf.random.set_seed(168)
```

3. Create the **DRQN** class with the following methods: the **build_model()** method to instantiate a CNN combined with a RNN model, the **get_action()** method to apply the epsilon-greedy algorithm to choose the action to be played, the **add_experience()** method to store in memory the experience acquired by playing the game, the **replay()** method, which will perform experience replay by sampling experiences from the memory and train the DRQN model with a callback to save the model every two episodes, and the **update_epsilon()** method to gradually decrease the epsilon value for epsilon-greedy:

Activity10_02.ipynb

```
class DRQN():
    def __init__(self, env, batch_size=64, max_experiences=5000):
        self.env = env
        self.input_size = self.env.observation_space.shape[0]
        self.action_size = self.env.action_space.n
        self.max_experiences = max_experiences
        self.memory = deque(maxlen=self.max_experiences)
        self.batch_size = batch_size
        self.gamma = 1.0
        self.epsilon = 1.0
        self.epsilon_min = 0.01
        self.epsilon_decay = 0.995

        self.model = self.build_model()
        self.target_model = self.build_model()

    def build_model(self):
        model = Sequential()
        model.add(TimeDistributed(Conv2D(32, 8, (4,4), \
                                  activation='relu', \
                                  padding='valid'), \
                input_shape=(SEQUENCE, IMG_SIZE, IMG_SIZE, 1)))
        model.add(TimeDistributed(Conv2D(64, 4, (2,2), \
                                  activation='relu', \
                                  padding='valid')))
        model.add(TimeDistributed(Conv2D(64, 3, (1,1), \
                                  activation='relu', \
                                  padding='valid')))
        model.add(TimeDistributed(Flatten()))
        model.add(LSTM(512))
        model.add(Dense(128, activation='relu'))
        model.add(Dense(self.action_size))
        model.compile(loss='mse', \
                      optimizer=RMSprop(lr=0.00025, \
                                        epsilon=self.epsilon_min), \
                      metrics=['accuracy'])
        return model
```

The complete code for this step can be found at https://packt.live/2AjdgMx .

4. Create the **initialize_env()** function, which will initialize the Breakout environment:

```
def initialize_env(env):
  initial_state = env.reset()
  initial_done_flag = False
  initial_rewards = 0
  return initial_state, initial_done_flag, initial_rewards
```

5. Create the **preprocess_state()** function to preprocess the input images:

```
def preprocess_state(image, img_size):
  img_temp = image[31:195]
  img_temp = tf.image.rgb_to_grayscale(img_temp)
  img_temp = tf.image.resize\
             (img_temp, [img_size, img_size], \
              method=tf.image.ResizeMethod.NEAREST_NEIGHBOR)
  img_temp = tf.cast(img_temp, tf.float32)
  return img_temp
```

6. Create the **combine_images()** function to stack the previous four screenshots:

```
def combine_images(new_img, prev_img, img_size, seq=4):
    if len(prev_img.shape) == 4 and prev_img.shape[0] == seq:
        im = np.concatenate\
             ((prev_img[1:, :, :], \
               tf.reshape(new_img, [1, img_size, img_size, 1])), \
              axis=0)
    else:
        im = np.stack([new_img] * seq, axis=0)
    return im
```

7. Create the **play_game()** function, which will play an entire game of Breakout:

```python
def play_game(agent, state, done, rewards):
    while not done:
        action = agent.get_action(state)
        next_state, reward, done, _ = env.step(action)

        next_state = preprocess_state(next_state, IMG_SIZE)
        next_state = combine_images\
                    (new_img=next_state, prev_img=state, \
                     img_size=IMG_SIZE, seq=SEQUENCE)

        agent.add_experience(state, action, \
                             reward, next_state, done)

        state = next_state
        rewards += reward
    return rewards
```

8. Create the **train_agent()** function, which will iterate through a number of episodes where the agent will play a game and perform experience replay:

```python
def train_agent(env, episodes, agent):
    from collections import deque
    import numpy as np

    scores = deque(maxlen=100)

    for episode in range(episodes):
        state, done, rewards = initialize_env(env)
        state = preprocess_state(state, IMG_SIZE)
        state = combine_images(new_img=state, prev_img=state, \
                               img_size=IMG_SIZE, seq=SEQUENCE)

        rewards = play_game(agent, state, done, rewards)
        scores.append(rewards)
        mean_score = np.mean(scores)

        if episode % 50 == 0:
            print(f'[Episode {episode}] - Average Score: {mean_score}')
            agent.target_model.set_weights\
```

```
            (agent.model.get_weights())
            agent.target_model.save_weights\
            (f'drqn_model_weights_{episode}')


    agent.replay(episode)


  print(f"Average Score: {np.mean(scores)}")
```

9. Instantiate a Breakout environment called **env** with **gym.make()**:

```
env = gym.make('BreakoutDeterministic-v4')
```

10. Create two variables, **IMG_SIZE** and **SEQUENCE**, that will take the values **84** and **4**, respectively:

```
IMG_SIZE = 84
SEQUENCE = 4
```

11. Instantiate a **DRQN** object called **agent**:

```
agent = DRQN(env)
```

12. Create a variable called **episodes** that will take the value **200**:

```
episodes = 200
```

13. Call the **train_agent** function by providing **env**, **episodes**, and **agent**:

```
train_agent(env, episodes, agent)
```

The following is the output of the code:

```
[Episode 0] - Average Score: 0.0
[Episode 50] - Average Score: 0.43137254901960786
[Episode 100] - Average Score: 0.4
[Episode 150] - Average Score: 0.54
Average Score: 0.53
```

> **NOTE**
>
> To access the source code for this specific section, please refer to https://packt.live/2AjdgMx.
>
> You can also run this example online at https://packt.live/37mhlLM.

In this activity, we added an LSTM layer and built a DRQN agent. It learned how to play the Breakout game, but didn't achieve satisfactory results even after 200 episodes. It seems this is still at the exploratory stage. You may try to train it for more episodes.

ACTIVITY 10.03: TRAINING A DARQN TO PLAY BREAKOUT

Solution

1. Open a new Jupyter Notebook and import the relevant packages: **gym**, **random**, **tensorflow**, **numpy**, and **collections**:

```
import gym
import random
import numpy as np
from collections import deque
import tensorflow as tf
from tensorflow.keras.models import Sequential
from tensorflow.keras.layers import Dense, Conv2D, \
MaxPooling2D, TimeDistributed, Flatten, GRU, Attention
from tensorflow.keras.optimizers import RMSprop
import datetime
```

2. Set the seed for NumPy and TensorFlow to **168**:

```
np.random.seed(168)
tf.random.set_seed(168)
```

3. Create the **DARQN** class and create the following methods: the **build_model()** method to instantiate a CNN combined with an RNN model, the **get_action()** method to apply the epsilon-greedy algorithm to choose the action to be played, the **add_experience()** method to store in memory the experience acquired by playing the game, the **replay()** method, which will perform experience replay by sampling experiences from the memory and train the DARQN model with a callback to save the model every two episodes, and the **update_epsilon()** method to gradually decrease the epsilon value for epsilon-greedy:

Activity10_03.ipynb

```python
class DARQN():
    def __init__(self, env, batch_size=64, max_experiences=5000):
        self.env = env
        self.input_size = self.env.observation_space.shape[0]
        self.action_size = self.env.action_space.n
        self.max_experiences = max_experiences
        self.memory = deque(maxlen=self.max_experiences)
        self.batch_size = batch_size
        self.gamma = 1.0
        self.epsilon = 1.0
        self.epsilon_min = 0.01
        self.epsilon_decay = 0.995

        self.model = self.build_model()
        self.target_model = self.build_model()

    def build_model(self):
        inputs = Input(shape=(SEQUENCE, IMG_SIZE, IMG_SIZE, 1))
        conv1 = TimeDistributed(Conv2D(32, 8, (4,4), \
                                activation='relu', \
                                padding='valid'))(inputs)
        conv2 = TimeDistributed(Conv2D(64, 4, (2,2), \
                                activation='relu', \
                                padding='valid'))(conv1)
        conv3 = TimeDistributed(Conv2D(64, 3, (1,1), \
                                activation='relu', \
                                padding='valid'))(conv2)
        flatten = TimeDistributed(Flatten())(conv3)
        out, states = GRU(512, return_sequences=True, \
                          return_state=True)(flatten)
        att = Attention()([out, states])
        output_1 = Dense(256, activation='relu')(att)
        predictions = Dense(self.action_size)(output_1)

        model = Model(inputs=inputs, outputs=predictions)

        model.compile(loss='mse', \
                      optimizer=RMSprop(lr=0.00025, \
                                        epsilon=self.epsilon_min), \
                      metrics=['accuracy'])
        return model
```

The complete code for this step can be found at https://packt.live/2XUDZrH.

4. Create the **initialize_env()** function, which will initialize the Breakout environment:

```python
def initialize_env(env):
    initial_state = env.reset()
    initial_done_flag = False
    initial_rewards = 0
    return initial_state, initial_done_flag, initial_rewards
```

5. Create the **preprocess_state()** function to preprocess the input images:

```
def preprocess_state(image, img_size):
    img_temp = image[31:195]
    img_temp = tf.image.rgb_to_grayscale(img_temp)
    img_temp = tf.image.resize\
                (img_temp, [img_size, img_size],\
                method=tf.image.ResizeMethod.NEAREST_NEIGHBOR)
    img_temp = tf.cast(img_temp, tf.float32)
    return img_temp
```

6. Create the **combine_images()** function to stack the previous four screenshots:

```
def combine_images(new_img, prev_img, img_size, seq=4):
    if len(prev_img.shape) == 4 and prev_img.shape[0] == seq:
        im = np.concatenate((prev_img[1:, :, :], \
                            tf.reshape\
                            (new_img, [1, img_size, \
                                        img_size, 1])), axis=0)
    else:
        im = np.stack([new_img] * seq, axis=0)
    return im
```

7. Create the **preprocess_state()** function to preprocess the input images:

```
def play_game(agent, state, done, rewards):
    while not done:
        action = agent.get_action(state)
        next_state, reward, done, _ = env.step(action)

        next_state = preprocess_state(next_state, IMG_SIZE)
        next_state = combine_images\
                    (new_img=next_state, prev_img=state, \
                    img_size=IMG_SIZE, seq=SEQUENCE)

        agent.add_experience(state, action, reward, \
                            next_state, done)

        state = next_state
        rewards += reward
    return rewards
```

8. Create the **train_agent()** function, which will iterate through a number of episodes where the agent will play a game and perform experience replay:

```
def train_agent(env, episodes, agent):
  from collections import deque
  import numpy as np

  scores = deque(maxlen=100)

  for episode in range(episodes):
    state, done, rewards = initialize_env(env)
    state = preprocess_state(state, IMG_SIZE)
    state = combine_images\
            (new_img=state, prev_img=state, \
             img_size=IMG_SIZE, seq=SEQUENCE)

    rewards = play_game(agent, state, done, rewards)
    scores.append(rewards)
    mean_score = np.mean(scores)

    if episode % 50 == 0:
        print(f'[Episode {episode}] - Average Score: {mean_score}')
        agent.target_model.set_weights\
        (agent.model.get_weights())
        agent.target_model.save_weights\
        (f'drqn_model_weights_{episode}')

    agent.replay(episode)

  print(f"Average Score: {np.mean(scores)}")
```

9. Instantiate a Breakout environment called **env** with **gym.make()**:

```
env = gym.make('BreakoutDeterministic-v4')
```

10. Create two variables, **IMG_SIZE** and **SEQUENCE**, that will take the values **84** and **4**, respectively:

```
IMG_SIZE = 84
SEQUENCE = 4
```

11. Instantiate a **DRQN** object called **agent**:

```
agent = DRQN(env)
```

12. Create a variable called **episodes** that will take the value **400**:

```
episodes = 400
```

13. Call the **train_agent** function by providing **env**, **episodes**, and **agent**:

```
train_agent(env, episodes, agent)
```

The following is the output of the code:

```
[Episode 0] - Average Score: 1.0
[Episode 50] - Average Score: 2.4901960784313726
[Episode 100] - Average Score: 3.92
[Episode 150] - Average Score: 7.37
[Episode 200] - Average Score: 7.76
[Episode 250] - Average Score: 7.91
[Episode 300] - Average Score: 10.33
[Episode 350] - Average Score: 10.94
Average Score: 10.83
```

In this activity, we built and trained a **DARQN** agent. It successfully learned how to play the Breakout game. It started with a score of **1.0** and achieved a final score of over **10** after **400** episodes, as shown in the preceding results. This is quite remarkable performance.

> **NOTE**
>
> To access the source code for this specific section, please refer to https://packt.live/2XUDZrH.
>
> You can also run this example online at https://packt.live/2UDCsUP.

CHAPTER 11: POLICY-BASED METHODS FOR REINFORCEMENT LEARNING

ACTIVITY 11.01: CREATING AN AGENT THAT LEARNS A MODEL USING DDPG

1. Import the necessary libraries (**os**, **gym**, and **ddpg**):

```
import os
import gym

from ddpg import *
```

2. First, we create our Gym environment (**LunarLanderContinuous-v2**), as we did previously:

```
env = gym.make("LunarLanderContinuous-v2")
```

3. Initialize the agent with some sensible hyperparameters, as in *Exercise 11.02, Creating a Learning Agent*:

```
agent = Agent(alpha=0.000025, beta=0.00025, \
              inp_dimensions=[8], tau=0.001,\
              env=env, bs=64, l1_size=400, l2_size=300, \
              nb_actions=2)
```

4. Set up a random seed so that our experiments are reproducible.

```
np.random.seed(0)
```

5. Create a blank array to story the scores; you can name it **history**. Iterate for at least **1000** episodes and in each episode, set a running score variable to **0** and the **done** flag to **False**, then reset the environment. Then, when the **done** flag is not **True**, carry out the following step:

```
history = []
for i in np.arange(1000):
    observation = env.reset()
    score = 0
    done = False
    while not done:
```

6. Select the observations and get the new **state**, **reward**, and **done** flags. Save the **observation**, **action**, **reward**, **state_new**, and **done** flags. Call the **learn** function of the agent and add the current reward to the running score. Set the new state as the observation and finally, when the done flag is **True**, append **score** to **history**:

```
history = []
for i in np.arange(1000):
    observation = env.reset()
    score = 0
    done = False
    while not done:
        action = agent.select_action(observation)
        state_new, reward, done, info = env.step(action)
        agent.remember(observation, action, reward, \
                       state_new, int(done))
        agent.learn()
        score += reward
        observation = state_new
        # env.render() # Uncomment to see the game window
    history.append(score)
```

You can print out **score** and mean **score_history** results to see how the agent is learning over time.

> **NOTE**
>
> To observe the rewards, we can simply add the **print** statement. The rewards will be similar to those in the previous exercise.

Run the code for at least 1,000 iterations and watch your lander attempt to land on the lunar surface.

> **NOTE**
>
> To see the Lunar Lander simulation once the policy is learned, we just need to uncomment the **env.render()** code from the preceding code block. As seen in the previous exercise, this will open another window, where we will be able to see the game simulation.

Here's a glimpse of how your lunar lander might behave once it has learned the policy:

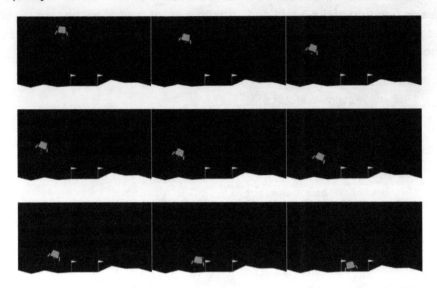

Figure 11.16: Screenshots from the environment after 1,000 rounds of training

ACTIVITY 11.02: LOADING THE SAVED POLICY TO RUN THE LUNAR LANDER SIMULATION

1. Import the essential Python libraries:

```
import os

import gym
import torch as T
import numpy as np

from PIL import Image
```

2. Set your device using the **device** parameter:

```
device = T.device("cuda:0" if T.cuda.is_available() else "cpu")
```

3. Define the **ReplayBuffer** class, as we did in the previous exercise:

```
class ReplayBuffer:
    def __init__(self):
        self.memory_actions = []
        self.memory_states = []
        self.memory_log_probs = []
        self.memory_rewards = []
        self.is_terminals = []

    def clear_memory(self):
        del self.memory_actions[:]
        del self.memory_states[:]
        del self.memory_log_probs[:]
        del self.memory_rewards[:]
        del self.is_terminals[:]
```

4. Define the **ActorCritic** class, as we did in the previous exercise:

Activity11_02.ipynb

```
class ActorCritic(T.nn.Module):
    def __init__(self, state_dimension, action_dimension, \
                 nb_latent_variables):
        super(ActorCritic, self).__init__()

        self.action_layer = T.nn.Sequential\
                            (T.nn.Linear(state_dimension, \
                                         nb_latent_variables),\
                             T.nn.Tanh(),\
                             T.nn.Linear(nb_latent_variables, \
                                         nb_latent_variables),\
                             T.nn.Tanh(),\
                             T.nn.Linear(nb_latent_variables, \
                                         action_dimension),\
                             T.nn.Softmax(dim=-1))
```

The complete code for this step can be found at https://packt.live/2YhzrvD.

5. Define the **Agent** class, as we did in the previous exercise:

Activity11_02.ipynb

```
class Agent:
    def __init__(self, state_dimension, action_dimension, \
    nb_latent_variables, lr, betas, gamma, K_epochs, eps_clip):\
        self.lr = lr
        self.betas = betas
        self.gamma = gamma
        self.eps_clip = eps_clip
        self.K_epochs = K_epochs

        self.policy = ActorCritic(state_dimension,\
                                  action_dimension,\
                                  nb_latent_variables).to(device)
        self.optimizer = T.optim.Adam\
                        (self.policy.parameters(), \
                         lr=lr, betas=betas)
        self.policy_old = ActorCritic(state_dimension,\
                                      action_dimension,\
                                      nb_latent_variables)\
                                      .to(device)
        self.policy_old.load_state_dict(self.policy.state_dict())
```

The complete code for this step can be found at https://packt.live/2YhzrvD.

6. Create the Lunar Lander environment. Initialize the random seed:

```
env = gym.make("LunarLander-v2")
np.random.seed(0)
render = True
```

7. Create the memory buffer and initialize the agent with hyperparameters, as in the previous exercise:

```
memory = ReplayBuffer()
agent = Agent(state_dimension=env.observation_space.shape[0],\
              action_dimension=4, nb_latent_variables=64,\
              lr=0.002, betas=(0.9, 0.999), gamma=0.99,\
              K_epochs=4, eps_clip=0.2)
```

8. Load the saved policy as an old policy from the **Exercise11.03** folder:

```
agent.policy_old.load_state_dict\
(T.load("../Exercise11.03/PPO_LunarLander-v2.pth"))
```

9. Finally, loop through your desired number of episodes. In every iteration, start by initializing the episode reward as **0**. Do not forget to reset the state. Run another loop, specifying the **max** timestamp. Get the **state**, **reward**, and **done** flags for each action taken and add the reward to the episode reward. Render the environment to see how your Lunar Lander is doing:

```
for ep in range(5):
    ep_reward = 0
    state = env.reset()

    for t in range(300):
        action = agent.policy_old.act(state, memory)
        state, reward, done, _ = env.step(action)

        ep_reward += reward

        if render:
            env.render()

            img = env.render(mode = "rgb_array")
            img = Image.fromarray(img)
            image_dir = "./gif"
            if not os.path.exists(image_dir):
```

```
            os.makedirs(image_dir)
          img.save(os.path.join(image_dir, "{}.jpg".format(t)))
      if done:
          break


    print("Episode: {}, Reward: {}".format(ep, int(ep_reward)))
    ep_reward = 0
env.close()
```

The following is the output of the code:

```
Episode: 0, Reward: 272
Episode: 1, Reward: 148
Episode: 2, Reward: 249
Episode: 3, Reward: 169
Episode: 4, Reward: 35
```

You'll see the reward oscillate in the positive zone as our Lunar Lander now has some idea of what a good policy can be. The reward may oscillate as there is more scope for learning. You might iterate over a few thousand more iterations to make your agent learn a better policy. Do not hesitate to tinker with the parameters specified in the code. The following screenshot shows the simulation output of some of the stages:

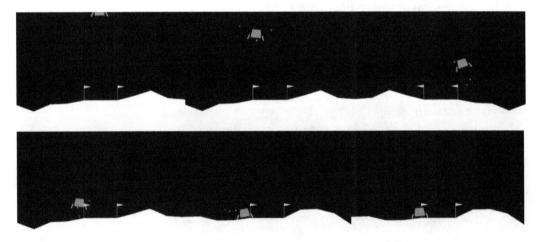

Figure 11.17: The environment showing the simulation of the Lunar Lander

Before this activity, we explained some necessary concepts, such as creating a learning agent, training a policy, saving and loading the learned policies, and so on, in isolation. Through carrying out this activity, you learned how to build a complete RL project or a working prototype on your own by combining all that you have learned in this chapter.

> **NOTE**
>
> The complete simulation output can be found in the form of images at https://packt.live/3ehPaAj.
>
> To access the source code for this specific section, please refer to https://packt.live/2YhzrvD.
>
> This section does not currently have an online interactive example and will need to be run locally.

CHAPTER 12: EVOLUTIONARY STRATEGIES FOR RL

ACTIVITY 12.01: CART-POLE ACTIVITY

1. Import the required packages as follows:

```
import gym
import numpy as np
import math
import tensorflow as tf
from matplotlib import pyplot as plt
from random import randint
from statistics import median, mean
```

2. Initialize the environment and the state and action space shapes:

```
env = gym.make('CartPole-v0')
no_states = env.observation_space.shape[0]
no_actions = env.action_space.n
```

3. Create a function to generate randomly selected initial network parameters:

```
def initial(run_test):
    #initialize arrays
    i_w = []
    i_b = []
    h_w = []
    o_w = []
    no_input_nodes = 8
    no_hidden_nodes = 4

    for r in range(run_test):
        input_weight = np.random.rand(no_states, no_input_nodes)
        input_bias = np.random.rand((no_input_nodes))
        hidden_weight = np.random.rand(no_input_nodes,\
                                    no_hidden_nodes)
        output_weight = np.random.rand(no_hidden_nodes, \
                                    no_actions)
```

```
        i_w.append(input_weight)
        i_b.append(input_bias)
        h_w.append(hidden_weight)
        o_w.append(output_weight)


    chromosome =[i_w, i_b, h_w, o_w]
    return chromosome
```

4. Create a function to generate the neural network using the set of parameters:

```
def nnmodel(observations, i_w, i_b, h_w, o_w):
    alpha = 0.199
    observations = observations/max\
                   (np.max(np.linalg.norm(observations)),1)
    #apply relu on layers
    funct1 = np.dot(observations, i_w)+ i_b.T
    layer1= tf.nn.relu(funct1)-alpha*tf.nn.relu(-funct1)
    funct2 = np.dot(layer1,h_w)
    layer2 = tf.nn.relu(funct2) - alpha*tf.nn.relu(-funct2)
    funct3 = np.dot(layer2, o_w)
    layer3 = tf.nn.relu(funct3)-alpha*tf.nn.relu(-funct3)
    #apply softmax
    layer3 = np.exp(layer3)/np.sum(np.exp(layer3))
    output = layer3.argsort().reshape(1,no_actions)
    action = output[0][0]

    return action
```

5. Create a function to get the total reward for **300** steps when using the neural network:

```
def get_reward(env, i_w, i_b, h_w, o_w):
    current_state = env.reset()
    total_reward = 0
    for step in range(300):
        action = nnmodel(current_state, i_w, i_b, h_w, o_w)
        next_state, reward, done, info = env.step(action)
        total_reward += reward
        current_state = next_state
        if done:
            break
    return total_reward
```

6. Create a function to get the fitness scores for each element of the population when running the initial random selection:

```
def get_weights(env, run_test):
    rewards = []
    chromosomes = initial(run_test)
    for trial in range(run_test):
        i_w = chromosomes[0][trial]
        i_b = chromosomes[1][trial]
        h_w = chromosomes[2][trial]
        o_w = chromosomes[3][trial]
        total_reward = get_reward(env, i_w, i_b, h_w, o_w)
        rewards = np.append(rewards, total_reward)
    chromosome_weight = [chromosomes, rewards]
    return chromosome_weight
```

7. Create a mutation function:

```
def mutate(parent):
    index = np.random.randint(0, len(parent))
    if(0 < index < 10):
        for idx in range(index):
            n = np.random.randint(0, len(parent))
            parent[n] = parent[n] + np.random.rand()
    mutation = parent
    return mutation
```

8. Create a single-point crossover function:

```
def crossover(list_chr):
    gen_list = []
    gen_list.append(list_chr[0])
    gen_list.append(list_chr[1])

    for i in range(10):
        m = np.random.randint(0, len(list_chr[0]))
        parent = np.append(list_chr[0][:m], list_chr[1][m:])
        child = mutate(parent)
        gen_list.append(child)
    return gen_list
```

9. Create a function for creating the next generation by selecting the pair with the highest rewards:

```
def generate_new_population(rewards, chromosomes):
    #2 best reward indexes selected
    best_reward_idx = rewards.argsort()[-2:][::-1]
    list_chr = []
    new_i_w =[]
    new_i_b = []
    new_h_w = []
    new_o_w = []
    new_rewards = []
```

10. Get the current parameters for the weights and bias using a **for** loop to go through the indices:

```
for ind in best_reward_idx:
    weight1 = chromosomes[0][ind]
    w1 = weight1.reshape(weight1.shape[1], -1)

    bias1 = chromosomes[1][ind]
    b1 = np.append(w1, bias1)

    weight2 = chromosomes[2][ind]
    w2 = np.append\
        (b1, weight2.reshape(weight2.shape[1], -1))

    weight3 = chromosomes[3][ind]
    chr = np.append(w2, weight3)
    #the 2 best parents are selected
    list_chr.append(chr)

gen_list = crossover(list_chr)
```

11. Build the neural network using the identified parameters and obtain a new reward based on the constructed neural network:

```
for l in gen_list:
    chromosome_w1 = np.array(l[:chromosomes[0][0].size])
    new_input_weight = np.reshape(chromosome_w1,(-
1,chromosomes[0][0].shape[1]))
    new_input_bias = np.array\
                      ([l[chromosome_w1.size:chromosome_w1\
                        .size+chromosomes[1][0].size]]).T
    hidden = chromosome_w1.size + new_input_bias.size
    chromosome_w2 = np.array\
                      ([l[hidden:hidden \
                       + chromosomes[2][0].size]])
    new_hidden_weight = np.reshape\
                        (chromosome_w2, \
                        (-1, chromosomes[2][0].shape[1]))
    final = chromosome_w1.size+new_input_bias.size\
            +chromosome_w2.size
    new_output_weight = np.array([l[final:]]).T
    new_output_weight = np.reshape\
                        (new_output_weight,\
                        (-1, chromosomes[3][0].shape[1]))

    new_i_w.append(new_input_weight)
    new_i_b.append(new_input_bias)
    new_h_w.append(new_hidden_weight)
    new_o_w.append(new_output_weight)

    new_reward = get_reward(env, new_input_weight, \
                            new_input_bias, new_hidden_weight, \
                            new_output_weight)
    new_rewards = np.append(new_rewards, new_reward)

generation = [new_i_w, new_i_b, new_h_w, new_o_w]

return generation, new_rewards
```

12. Create a function to output the convergence graph:

```
def graphics(act):
    plt.plot(act)
    plt.xlabel('No. of generations')
    plt.ylabel('Rewards')
    plt.grid()

    print('Mean rewards:', mean(act))
    return plt.show()
```

13. Create a function for the genetic algorithm that outputs the parameters of the neural network based on the highest average reward:

```
def ga_algo(env, run_test, no_gen):
    weights = get_weights(env, run_test)

    chrom = weights[0]
    current_rewards = weights[1]
    act = []

    for n in range(no_gen):
        gen, new_rewards = generate_new_population\
                              (current_rewards, chrom)
        average = np.average(current_rewards)
        new_average = np.average(new_rewards)
        if average > new_average:
            parameters = [chrom[0][0], chrom[1][0], \
                            chrom[2][0], chrom[3][0]]
        else:
            parameters = [gen[0][0], gen[1][0], \
                            gen[2][0], gen[3][0]]
        chrom = gen
        current_rewards = new_rewards
        max_arg = np.amax(current_rewards)
```

```
        print('Generation:{}, max reward:{}'.format(n+1, max_arg))
        act = np.append(act, max_arg)



    graphics(act)

    return parameters
```

14. Create a function that decodes the array of parameters to each neural network parameter:

```
def params(parameters):
    i_w = parameters[0]
    i_b = parameters[1]
    h_w = parameters[2]
    o_w = parameters[3]

    return i_w,i_b,h_w,o_w
```

15. Set the generations to **50**, the number of trial tests to **15**, and the number of steps and trials to **500**:

```
generations = []
no_gen = 50
run_test = 15
trial_length = 500
no_trials = 500
rewards = []
final_reward = 0

parameters = ga_algo(env, run_test, no_gen)
i_w, i_b, h_w, o_w = params(parameters)

for trial in range(no_trials):
    current_state = env.reset()
    total_reward = 0
```

```
    for step in range(trial_length):
        env.render()
        action = nnmodel(current_state, i_w,i_b, h_w, o_w)
        next_state,reward, done, info = env.step(action)
        total_reward += reward
        current_state = next_state
        if done:
            break
    print('Trial:{}, total reward:{}'.format(trial, total_reward))
    final_reward +=total_reward
print('Average reward:', final_reward/no_trials)
env.close()
```

The output (just the first few lines are shown here) will be similar to
the following:

```
Generation:1, max reward:11.0
Generation:2, max reward:11.0
Generation:3, max reward:10.0
Generation:4, max reward:10.0
Generation:5, max reward:11.0
Generation:6, max reward:10.0
Generation:7, max reward:10.0
Generation:8, max reward:10.0
Generation:9, max reward:11.0
Generation:10, max reward:10.0
Generation:11, max reward:10.0
Generation:12, max reward:10.0
Generation:13, max reward:10.0
Generation:14, max reward:10.0
Generation:15, max reward:10.0
Generation:16, max reward:10.0
Generation:17, max reward:10.0
Generation:18, max reward:10.0
Generation:19, max reward:11.0
Generation:20, max reward:11.0
```

The output can be visualized in a plot as follows:

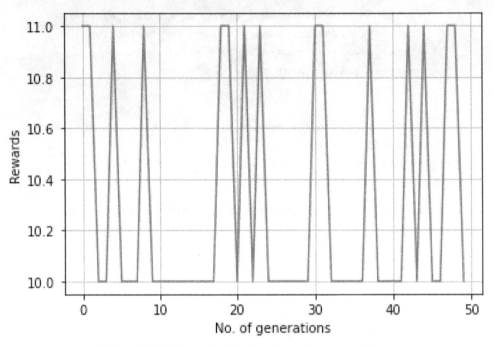

Figure 12.15: Rewards obtained over the generations

The average of the rewards output (just the last few lines are shown here) will be similar to the following:

```
Trial:486, total reward:8.0
Trial:487, total reward:9.0
Trial:488, total reward:10.0
Trial:489, total reward:10.0
Trial:490, total reward:8.0
Trial:491, total reward:9.0
Trial:492, total reward:9.0
Trial:493, total reward:10.0
Trial:494, total reward:10.0
Trial:495, total reward:9.0
Trial:496, total reward:10.0
Trial:497, total reward:9.0
Trial:498, total reward:10.0
Trial:499, total reward:9.0
Average reward: 9.384
```

You will notice that depending on the start state, the convergence of the GA algorithm to the highest score will vary; also, the neural network model will not always achieve the optimal solution. The purpose of this activity was for you to implement the genetic algorithm techniques studied in this chapter and to see how you can combine evolutionary methods of neural network parameter tuning for action selection.

> **NOTE**
>
> To access the source code for this specific section, please refer to https://packt.live/2AmKR8m.
>
> This section does not currently have an online interactive example and will need to be run locally.

INDEX

A

adadelta:148
adagrad: 148, 493
adamax:148
aggregate:627
alphago: 8, 65-66, 308
alphastar:68
arange: 332, 358,
    372, 384, 405, 414,
    495, 512-513, 524,
    540, 589, 604
arbitrary: 95, 144,
    394, 582
argmax: 109, 111,
    125-126, 139, 240,
    329-331, 333, 340,
    355, 361, 369, 375,
    379, 382-383, 386,
    402, 407, 411,
    416, 511, 516,
    642-643, 648, 652
auto-mpg:171

B

baseline: 586, 608
bayesian: 451,
    459-467, 481
bellman: 73-74,
    84-88, 90-91, 95,
    101, 113-116,
    118, 127-129, 132,
    282-283, 290, 296,
    300-301, 303, 305,
    500, 503, 515, 517,
    522, 529-531
bernoulli: 428-431,
    436, 439, 444,
    454, 456, 460-462,
464-468, 658,
660, 662

bertrand:583
binary: 154, 188-189,
    191, 193-195,
    198, 203-205,
    269, 485-486, 630,
    635-636, 641,
    645, 650-651,
    653, 666, 677
binomial: 436,
    444, 456, 468
bitarray: 658-659, 662
bitstring:658
boltzmann:36
boolean: 432, 637,
    653, 666

C

cartesian:45
cartpole: 44-45,
    52-54, 59-61, 63,
    214, 221-224, 226,
    236-238, 246,
    251-255, 312,
    507, 518, 525,
    533, 546, 666
coherence:166
conjugate: 464, 479
convergent:396
coordinate: 10, 47, 514
corpora:566
covariance:630
criterion: 308,
    492-494, 517,
    521-522, 529,
    537-538, 546,
    639, 667
crossover: 630, 632,
639, 645-651,
653-656, 662,
664-665, 673,
675, 677

cumulative: 21, 40,
    42, 82, 236, 250,
    255, 329, 352,
    425, 427-428,
    430-431, 434,
    437-440, 444-445,
    447, 449, 456-457,
    459, 468-469, 471,
    474-481, 582
customenv:228

D

dataframe: 177,
    179, 194, 281
dataset: 6-7, 9, 11,
    133, 135, 152-155,
    157, 160, 170-175,
    182-185, 188-189,
    197, 201-204, 207,
    233, 488, 491, 558,
    657-660, 670
debugger:270
decoder:574
divergence:601
docker: 232-235
dqn-output:515

E

egreedy: 443-444, 446,
    448, 510-511, 513,
    516-517, 519-521,
    523, 532, 534-535,
    537, 539, 547
emulators:232

www.ingramcontent.com/pod-product-compliance
Lightning Source LLC
Chambersburg PA
CBHW081447050326
40690CB00015B/2707